Applied Mathematical Sciences
Volume 151

Springer
Berlin
Heidelberg
New York
Barcelona
Hong Kong
London
Milan
Paris
Tokyo

Applied Mathematical Sciences

(continued following index)

Alain Bensoussan   Jens Frehse

Regularity Results for Nonlinear Elliptic Systems and Applications

 Springer

Alain Bensoussan
CNES
2 Place Maurice Quentin
75039 Paris, France
alain.bensoussan@cnes.fr

Jens Frehse
Institut für Angewandte Mathematik
Universität Bonn
Wegelerstrasse 10
53115 Bonn, Germany

Editors

Stuart S. Antman
Department of Mathematics
and
Institute for Physical Science
and Technology
University of Maryland
College Park, MD 20742-4015
USA

Jerrold E. Marsden
Control and Dynamical
Systems 107-81
California Institute of
Technology
Pasadena, CA 91125
USA

Lawrence Sirovich
Division of Applied
Mathematics
Brown University
Providence, RI 02912
USA

Mathematics Subject Classification (2000):
35-XX, 49-XX, 74-XX, 91A15, 76-XX

Cataloging-in-Publication data applied for

Die Deutsche Bibliothek – CIP-Einheitsaufnahme

Bensoussan, Alain: Regularity results for nonlinear elliptic systems and applications /
Alain Bensoussan; Jens Frehse. – Berlin; Heidelberg; NewYork; Barcelona; Hong Kong; London;
Milan; Paris; Tokyo: Springer, 2002
(Applied mathematical sciences; Vol. 151)
ISBN 3-540-67756-9

0101 deutsche buecherei

ISBN 3-540-67756-9 Springer-Verlag Berlin Heidelberg New York

Springer-Verlag Berlin Heidelberg New York
a member of BertelsmannSpringer Science+Business Media GmbH

http://www.springer.de

© Springer-Verlag Berlin Heidelberg 2002
Printed in Germany

TeX-editing by H.-J. Wolf and Educational Media Centre, New Delhi, India
SPIN 10701624   41/3142 - 5 4 3 2 1 0 – Printed on acid-free paper

Preface

Nonlinear elliptic equations and systems are a classical field of analysis, with many applications in differential geometry, continuum mechanics, and probability theory; an important future branch will be their applications to microelectronics.

The most important analytical tools in the field of nonlinear partial differential equations and systems up to, say, 1955 are presented in the books of C. B. Morrey [79] and O.A. Ladyzhenskaya, N.N. Ural'tseva [66]. The bulk of the development for general nonlinear elliptic systems is presented in M. Giaquinta, E. Giusti [41], D. Gilbarg, N.S. Trudinger [46], later in M. Giaquinta [40]. Concerning applications to differential geometry, we mention the books of M. Giaquinta, S. Hildebrandt [42].

The purpose of this book is to present some of the developments that are not covered in the above books and are promising fields for applications and research.

The book is to a large extent self-contained, with the restriction that the linear theory—Schauder estimates and Campanato theory—is not presented. The reader is expected to be familiar with functional-analytic tools, like the theory of monotone operators. References are given in the text to any techniques that are used. The first two chapters contain general methods and auxiliary lemmas. The expert might like our approach to the theorem of De Giorgi–Nash concerning C^α-regularity of solutions to nonlinear scalar equations via the hole-filling method, and our proof of Harnack's inequality without using the John–Nirenberg theorem on functions with bounded mean oscillation.

Chapters 4 and 5 deal with diagonal elliptic systems, which have important applications to differential geometry; however, in order to be complementary to the books of Giaquinta–Hildebrandt, we present only the applications to stochastic problems, where the researcher finds challenging open problems with a broad range of degree of difficulty. In fact, the treatment here is more complete than what is available in the literature.

Chapter 6 deals with Helein's proof of the regularity of harmonic mappings on two-dimensional manifolds. We avoid a more extensive study of harmonic mappings, for which we refer to the books of J. Jost [60], M. Giaquinta, S. Hildebrandt [42] (see also J. Eells, J.H. Sampson [22]).

Chapter 7 presents the standard Van Roosbroek equations in semi-conductor theory and a special model that is related to the avalanche effect. We admit that this choice represents a limited sample compared with the range of interesting new open problems waiting to be solved, but in the interest of brevity we have cut the exposition short. In chapter 8 we present recent results for the regularity problem of the Navier–Stokes equation. Clearly, this chapter is not an introduction to mathematical fluid dynamics, for which the reader should refer to the standard book of O.A. Ladyzhenskaya, V.A. Solonnikov, N.N. Ural'tseva [67] or, recently, P.L. Lions [72] and G.P. Galdi [36]. We have included this chapter in the book because of surprising similarities of the analytical tools to those in the chapter on diagonal systems. In Chapter 9 we collect results concerning strongly coupled elliptic systems, in particular the theory of A. Koshelev [63] concerning sufficient conditions for regularity involving eigenvalues. Chapter 10 presents elements of a dual theory of elliptic systems, the motivation coming from simple models in elasto plasticity. It seems that many techniques in elliptic analysis have a dual analogue. For example, we present a dual proof and formulation of the almost everywhere regularity of solutions of elliptic systems. Chapter 11 contains a short approach to plasticity theory; for the physical background we refer to the books of G. Duvaut, J.L. Lions [15], R. Temam [101] and P. Le Tallec [69]. We believe that the approach via the Norton–Hoff approximation is a recommendable introduction for new-comers who have knowledge of Sobolev spaces. We would like to emphasize that much of the progress concerning the time-dependent Prandtl–Reuss law and regularity properties of its solution has been made by using the dual theory of elliptic equations. This is why it is presented here, although it is a "time"-dependent model, which is in principle outside the scope of this book.

We would like to thank Zamin Iqbal, who carefully read the draft of the book and improved the English to a great extent, and also Josef Malek, who read various parts.

A warm thank you to Chantal Delabarre, who improved the limited LaTeX of the authors, and to Springer-Verlag for publishing this book.

Alain Bensoussan
Jens Frehse

Contents

1. General Technical Results

1.1 Introduction

In this chapter we assemble many technical results that will be used throughout the book. Function spaces play a central role, and we recall first some notation, fundamental definitions, and properties.

1.1.1 Function Spaces

• Hölder Spaces

Let Ω be an open set in R^n, k an integer, and $0 \leq \alpha < 1$. The Hölder Spaces $C^{k,\alpha}(\overline{\Omega})$ are the spaces of functions whose kth-order partial derivatives are uniformly Hölder continuous on $\overline{\Omega}$. We abbreviate

$$C^k(\overline{\Omega}) = C^{k,0}(\overline{\Omega}).$$

By using Ω rather than $\overline{\Omega}$ we enable ourselves to replace "locally" with "uniformly."

Using the notation

$$[f]_{\alpha;\overline{\Omega}} = \sup_{\substack{x,y \in \overline{\Omega} \\ x \neq y}} \frac{|f(x) - f(y)|}{|x-y|^\alpha}, \quad 0 < \alpha < 1,$$

the norm on $C^{k,\alpha}(\overline{\Omega})$ is defined by

$$\|u\|_{C^{k,\alpha}(\overline{\Omega})} = \sup_{x \in \overline{\Omega}} |u(x)| + \sum_{j=0}^{k} \sum_{\beta_1 + \cdots + \beta_n = j} \left[\frac{\partial^j u}{\partial^{\beta_i} x_i} \right]_{\alpha;\overline{\Omega}}.$$

• Sobolev Spaces

We assume that Ω is a bounded domain of R^n. For $1 \leq p < \infty$, let $L^p(\Omega)$ be the space of p-integrable functions on Ω, with the norm

$$\|u\|_{L^p} = \left(\int_\Omega |u|^p \, dx \right)^{1/p}, \quad \forall 1 \leq p < \infty.$$

If we consider weak derivatives (in the sense of distributions), then the Sobolev space $W^{k,p}(\Omega)$ is defined by

$$W^{k,p}(\Omega) = \{u \in L^p \mid D^\alpha u \in L^p \text{ for all } |\alpha| \leq k\},$$

where we use the multi-index notation

$$D^\alpha u = \frac{\partial^{|\alpha|}}{\partial^{\alpha_i} x_i}, \quad \alpha_1 + \cdots + \alpha_n = |\alpha|,$$

and we equip this space with the norm

$$\|u\|_{W^{k,p}} = \left(\sum_{|\alpha| \leq k} |D^\alpha u|^p \, dx \right)^{1/p}.$$

The closure in $W^{k,p}(\Omega)$ of smooth functions with compact support in Ω is denoted by $W_0^{k,p}(\Omega)$. When $p = 2$, we write $H^k(\Omega)$ and $H_0^k(\Omega)$ for these two spaces. They are Hilbert spaces.

When the domain Ω is Lipschitz (its boundary can be represented locally by a Lipschitz function; see Section 1.1.2), then important embedding properties hold. They are:

If $kp < n$, the space $W^{k,p}(\Omega)$ is continuously embedded in $L^{p^*}(\Omega)$, with $p* = \frac{np}{n-kp}$, and compactly embedded in $L^q(\Omega)$ for any $q < p^*$,

If $0 \leq m < k - \frac{n}{p} < m+1$, then $W^{k,p}(\Omega)$ is continuously embedded (1.1) in $C^{m,\alpha}(\overline{\Omega}), \alpha = k - \frac{n}{p} - m$, and compactly embedded in $C^{m,\beta}(\overline{\Omega})$, $\forall \beta < \alpha$.

We can say little about the limiting case

$$n = kp,$$

except, of course, that $W^{k,p}(\Omega)$ is continuously embedded in $L^q(\Omega)$ for every $q < \infty$. Also, embedding into Orlicz spaces can be stated (see [46]). Nevertheless, we can state the following useful result:

If $u \in W^{n,1}(R^n)$ with compact support, then $u \in C^0(R^n)$. (1.2)

It is an easy consequence of the formula

$$u(x_1,\ldots,x_n) = \int_{-\infty}^{x_1} \cdots \int_{-\infty}^{x_n} D_1 \cdots D_n u(\xi_1,\ldots,\xi_n) d\xi_1 \cdots d\xi_n,$$

which holds almost everywhere. The dual of $W^{1,p}(\Omega)$ is isomorphic to $(L^{p'}(\Omega))^{n+1}$. So, if ξ belongs to $(W^{1,p}(\Omega))'$, there exist $h_0, h_1, \ldots, h_n \in L^{p'}$ such that

$$\langle \xi, u \rangle = \int_\Omega \left(h_0 u + \sum_{i=1}^n h_i D_i u \right)(x)\, dx,$$

$$\left(\int_\Omega \left(|h_0|^{p'} + \sum_{i=1}^n |h_i|^{p'} \right)(x)\, dx \right)^{1/p'} \leq C \|\xi\|_{(W^{1,p}(\Omega))'}$$

Combining Sobolev embedding results and the representation of $(W^{1,p}(\Omega))'$, we obtain a useful representation of a function $f \in L^\sigma$. Namely, to f are associated $h_0, h_1, \ldots, h_n \in L^{\sigma^*}$ such that

$$\int_\Omega f u\, dx = \int_\Omega \left(h_0 u + \sum_{i=1}^n h_i D_i u \right)(x)\, dx \qquad (1.3)$$

for every

$$u \in H^{1,(\sigma^*)'},$$

where

$$\sigma^* = \frac{\sigma n}{n - \sigma},$$

$$(\sigma^*)' = \frac{\sigma^*}{\sigma^* - 1} = \frac{\sigma n}{\sigma n - n + \sigma},$$

and

$$\left(\int_\Omega \left(|h_0|^{\sigma^*} + \sum_{i=1}^n |h_i|^{\sigma^*} \right)(x)\, dx \right)^{1/\sigma^*} \leq C \|f\|_{L^\sigma}.$$

Note that

$$\sigma^* > \sigma, \quad (\sigma^*)' < (\sigma)'.$$

The property (1.3) follows from

$$\left| \int_\Omega f u\, dx \right| \leq \|f\|_{L^\sigma} \|u\|_{L^{\sigma'}},$$

and since (as is easily checked)

$$\sigma' = ((\sigma^*)')^*,$$

we get from the embedding Sobolev theorem

$$\left| \int_\Omega f u\, dx \right| \leq \|f\|_{L^\sigma} \|u\|_{W^{1,(\sigma^*)'}},$$

and the representation of $(W^{1,(\sigma^*)'})'$ yields (1.3). In particular, whenever $\sigma > \frac{2n}{n+2}$, then $(\sigma^*)' \leq 2$, and (1.3) holds with $u \in H^1(\Omega)$.

• Morrey and Campanato Spaces

Let us denote by $B_R(x_0)$ the open ball of center x_0 and radius R. The Morrey spaces are defined as follows:

$$L_M^{p,\lambda}(\Omega) = \left\{ u \in L^p(\Omega) | \sup_{x_0,R} \frac{\int_{\Omega \cap B_R(x_0)} |u|^p \, dx}{R^\lambda} < \infty \right\}. \tag{1.4}$$

The Campanato spaces are the following spaces:

$$L_C^{p,\lambda}(\Omega) = \left\{ u \in L^p(\Omega) | \sup_{x_0,R} \frac{\int_{\Omega \cap B_R(x_0)} |u - u_{x_0,R}|^p \, dx}{R^\lambda} < \infty \right\}, \tag{1.5}$$

where

$$u_{x_0,R} = \frac{\int_{\Omega \cap B_R(x_0)} u(x) \, dx}{|\Omega \cap B_R(x_0)|}.$$

We define a norm on $L_M^{p,\lambda}(\Omega)$ by

$$\|u\|_{p,\lambda;M} = \sup_{x_0,R} \left(\frac{\int_{\Omega \cap B_R(x_0)} |u|^p \, dx}{R^\lambda} \right)^{1/p}$$

and a seminorm on $L_C^{p,\lambda}(\Omega)$ by setting

$$[u]_{p,\lambda} = \left(\sup_{x_0,R} \frac{\int_{\Omega \cap B_R(x_0)} |u - u_{x_0,R}|^p \, dx}{R^\lambda} \right)^{1/p}.$$

A norm in $L_C^{p,\lambda}(\Omega)$ is then defined by

$$\|u\|_{p,\lambda;C} = \|u\|_{L^p} + [u]_{p,\lambda}.$$

With these norms, the Morrey and Campanato spaces are Banach spaces, with the following properties

$$L_M^{p,0}(\Omega) = L^p(\Omega),$$
$$L_M^{p,n}(\Omega) = L^\infty(\Omega).$$

The Morrey spaces reduce to $\{0\}$ when $\lambda > n$, and the Campanato spaces reduce to constants when $\lambda > n + p$. Moreover, we have

$$L_M^{p,\lambda}(\Omega) = L_C^{p,\lambda}(\Omega), \quad 0 \leq \lambda < n,$$

and the important property

$$L_C^{p,\lambda}(\Omega) = C^{0,\alpha}(\overline{\Omega}), \quad \alpha = \frac{\lambda - n}{p}, \quad n < \lambda \leq n + p .$$

This last property implies the following theorem of C.B. Morrey [79], [46].

Let $u \in W^{1,1}(\Omega)$. Suppose there exist constants $K, \alpha \leq 1$ such that

$$\int_{B_R} |Du|\, dx \leq KR^{n-1+\alpha} \quad \forall B_R \subset \Omega.$$

Then

$$u \in C^{0,\alpha}(\Omega)$$

and

$$\sup_{x_1, x_2 \in B_R} |u(x_1) - u(x_2)| \leq C(n, \alpha)KR^\alpha.$$

We shall also use the following version of Morrey's result. Setting

$$\|u\|_{\alpha,\Omega}^2 = \sup_{\substack{x_0, R \\ B_R(x_0) \in \Omega}} \frac{1}{R^{n-2+2\alpha}} \int_{B_R(x_0)} |Du|^2\, dx,$$

one then has

$$[u]_{\alpha,\Omega} \leq C\|u\|_{\alpha,\Omega}. \tag{1.6}$$

A related useful norm is the following:

$$\|u\|_{\alpha,\Omega} = \sup_{x_0 \in \Omega} \left(\int_\Omega \frac{|Du|^2}{|x - x_0|^{n-2+2\alpha}}\, dx \right)^{1/2} .$$

It is clear that

$$\|u\|_{\alpha,\Omega} \leq \|u\|_{\alpha,\Omega}. \tag{1.7}$$

We shall prove directly the estimate

$$[u]_{\alpha,\Omega} \leq C\|u\|_{\alpha,\Omega}, \tag{1.8}$$

or, more precisely, a simpler weaker version of it, sufficient to prove local Hölderianity, and avoiding assumptions on the boundary. For every $\tilde{\Omega}$ with

$$\overline{\tilde{\Omega}} \subset \Omega$$

one has the estimate

$$[u]_{\alpha,\overline{\tilde{\Omega}}} \leq C_{\tilde{\Omega},\alpha} \sup_{x_0 \in \overline{\tilde{\Omega}}} \left(\int_\Omega \frac{|Du|^2}{|x - x_0|^{n-2+2\alpha}}\, dx \right)^{1/2} . \tag{1.9}$$

One relies on the important Hardy's inequality; see G. Hardy, J. Littlewood, G. Pólya [55], or E.M. Stein, G. Weiss [97]. Let $p > 1, s \neq 1$, and let

$$
F(r) = \begin{cases} \int_0^r f(\rho)\, d\rho & \text{if } s > 1 \\ \int_r^{+\infty} f(\rho)\, d\rho & \text{if } s < 1. \end{cases}
$$

Then the following inequality holds:

$$
\int_0^{+\infty} r^{-s}|F(r)|^p\, dr \leq \left(\frac{p}{|s-1|}\right)^p \int_0^{+\infty} r^{-s+p}|f(r)|^p\, dr, \tag{1.10}
$$

which can be obtained as a consequence of Jensen's inequality. Indeed, first notice that the case $s < 1$ can be reduced to the case $s > 1$ by applying the inequality for $s > 1$ to the function

$$
f_1(r) = \frac{1}{r^2} f\left(\frac{1}{r}\right).
$$

Thus, we may assume $s > 1$. We write

$$
\int_0^r f(\rho)\, d\rho = \frac{p}{s-1} r^{\frac{s-1}{p}} \frac{\int_0^r f(\rho)\rho^{1-\frac{s-1}{p}} \rho^{\frac{s-1}{p}-1}\, d\rho}{\int_0^r \rho^{\frac{s-1}{p}-1}\, d\rho}
$$

and we apply Jensen's inequality to the convex function $|u|^p$ to derive

$$
\left|\int_0^r f(\rho)\, d\rho\right|^p \leq \left(\frac{p}{s-1}\right)^{p-1} r^{(s-1)(1-\frac{1}{p})} \int_0^r |f(\rho)|^p \rho^{p-s+\frac{s-1}{p}}\, d\rho.
$$

From this estimate the result follows easily.

Let us turn to (1.9). We first notice that there exists δ_0 such that

$$
\forall x_0 \in \overline{\overline{\Omega}}, \quad B_{\delta_0}(x_0) \subset \Omega.
$$

Let $x_0 \in \overline{\overline{\Omega}}$. We represent a point $x \in B_{\delta_0}(x_0)$ by its spherical coordinates $(r, \theta), 0 \leq \delta_0, \theta \in S$, where S is the unit sphere. So we write

$$
u(x_0) - \fint_\Omega u(y)\, dy = u(x) - \fint_\Omega u(y)\, dy - \int_0^r \frac{\partial u}{\partial \rho}(\rho, \theta)\, d\rho.
$$

where $\fint_\Omega u(y)dy$ represents the average of the function u on the set Ω.

We take the square and majorize by Schwarz inequality. Then we integrate over $B_{\delta_0}(x_0)$. Collecting results, we get

$$
|S|\delta_0^n \left|u(x_0) - \fint_\Omega u(y)\, dy\right|^2 \leq 2 \int_{B_{\delta_0}(x_0)} \left|u(x) - \fint_\Omega u(y)\, dy\right|^2 dx
$$

$$
+ 2\delta_0^{n+2\alpha} \int_S dS \int_0^{\delta_0} r^{-1-2\alpha} \left|\int_0^r \frac{\partial u}{\partial \rho}(\rho, \theta)\, d\rho\right|^2 dr.
$$

Using Hardy's inequality we assert that

$$\int_0^{\delta_0} r^{-1-2\alpha} \left| \int_0^r \frac{\partial u}{\partial \rho}(\rho, \theta) \, d\rho \right|^2 dr \leq \frac{1}{\alpha^2} \int_0^{\delta_0} r^{1-2\alpha} \left| \frac{\partial u}{\partial r}(r, \theta) \right|^2 dr.$$

Recalling that

$$\frac{\partial u}{\partial r} = Du \frac{x - x_0}{r}$$

and inserting in previous estimates yields

$$|S| \delta_0^n \left| u(x_0) - \fint_\Omega u(y) \, dy \right|^2 \leq 2 \int_{B_{\delta_0}(x_0)} \left| u(x) - \fint_\Omega u(y) \, dy \right|^2 dx$$

$$\tag{1.11}$$

$$+ \frac{2}{\alpha^2} \delta_0^{n+2\alpha} \int_{B_{\delta_0}(x_0)} \frac{|Du|^2}{r^{n-2+2\alpha}} \, dx.$$

Extending the first integral to Ω and using the Poincaré inequality (see Section 1.1.3) we get

$$\left| u(x_0) - \fint_\Omega u(y) \, dy \right|^2 \leq \frac{C}{\delta_0^n} \int_\Omega |Du(x)|^2 \, dx + C\delta_0^{2\alpha} \int_{B_{\delta_0}(x_0)} \frac{|Du|^2}{r^{n-2+2\alpha}} \, dx.$$

$$\tag{1.12}$$

Now pick $x_1, x_2 \in \overline{\Omega}$. We consider two cases:

Case 1:

$$|x_1 - x_2| \geq \delta_0.$$

We have

$$\frac{|u(x_1) - u(x_2)|^2}{|x_1 - x_2|^{2\alpha}} \leq \frac{|u(x_1) - u(x_2)|^2}{\delta_0^{2\alpha}},$$

and using (1.12) applied successively to $x_0 = x_1$ and $x_0 = x_2$, we obtain

$$\frac{|u(x_1) - u(x_2)|^2}{|x_1 - x_2|^{2\alpha}} \leq \frac{C}{\delta_0^{n+2\alpha}} \int_\Omega |Du(x)|^2 \, dx$$

$$+ \int_{B_{\delta_0}(x_1)} \frac{|Du|^2}{|x - x_1|^{n-2+2\alpha}} \, dx + \int_{B_{\delta_0}(x_2)} \frac{|Du|^2}{|x - x_2|^{n-2+2\alpha}} \, dx$$

and thus finally

$$\frac{|u(x_1) - u(x_2)|^2}{|x_1 - x_2|^{2\alpha}} \leq C(\delta_0) \sup_{x_0 \in \overline{\Omega}} \int_\Omega \frac{|Du|^2}{|x - x_0|^{n-2+2\alpha}} \, dx. \tag{1.13}$$

Case 2:

$$|x_1 - x_2| < \delta_0.$$

We consider

$$\overline{x} = \frac{x_1 + x_2}{2}.$$

We note the inclusions

$$x_1, x_2 \in B_{\delta_0/2}(\overline{x}) \subset B_{\delta_0}(x_1), \quad B_{\delta_0}(x_2) \subset \Omega, \quad B_{|x_1-x_2|/2}(\overline{x}) \subset B_{\delta_0/2}(\overline{x}).$$

Pick $y \in B_{|x_1-x_2|/2}(\overline{x})$, and using

$$|x_1 - x_2| > |x_1 - y|, |x_2 - y|$$

Write

$$\frac{|u(x_1) - u(x_2)|^2}{|x_1 - x_2|^{n+2\alpha}} \leq 2 \left[\frac{|u(x_1) - u(y)|^2}{|x_1 - y|^{n+2\alpha}} + \frac{|u(x_2) - u(y)|^2}{|x_2 - y|^{n+2\alpha}} \right].$$

Thus integrating over $B_{|x_1-x_2|/2}(\overline{x})$ and majorizing we obtain

$$\frac{|S|}{2^n} \frac{|u(x_1) - u(x_2)|^2}{|x_1 - x_2|^{2\alpha}} \tag{1.14}$$

$$\leq 2 \left[\int_{B_{\delta_0}(x_1)} \frac{|u(x_1) - u(y)|^2}{|x_1 - y|^{n+2\alpha}} \, dy + \int_{B_{\delta_0}(x_2)} \frac{|u(x_2) - u(y)|^2}{|x_2 - y|^{n+2\alpha}} \, dy \right].$$

It remains to use the estimate

$$\int_{B_{\delta_0}(x_0)} \frac{|u(x_0) - u(x)|^2}{|x_0 - x|^{n+2\alpha}} \, dx \leq \frac{1}{\alpha^2} \int_{B_{\delta_0}(x_0)} \frac{|Du|^2}{|x_0 - x|^{n+2\alpha-2}} \, dx,$$

which can be obtained, thanks to Hardy's inequality, as in (1.11). Using it in (1.14) yields

$$\frac{|S|}{2^n} \frac{|u(x_1) - u(x_2)|^2}{|x_1 - x_2|^{2\alpha}} \leq \frac{2}{\alpha^2} \left[\int_{B_{\delta_0}(x_1)} \frac{|Du|^2}{|x_1 - x|^{n-2+2\alpha}} \, dx \right.$$

$$\left. + \int_{B_{\delta_0}(x_2)} \frac{|Du|^2}{|x_2 - x|^{n-2+2\alpha}} \, dx \right], \tag{1.15}$$

and thus (1.13) is also satisfied. Hence the proof of (1.9) is complete. $\diamond$

The space $L_C^{p,n}(\Omega)$ does not coincide with L^∞. We have, however,

$$L^\infty(\Omega) \subset L_C^{p,n}(\Omega).$$

This last space is known as B.M.O., the space of functions with bounded mean oscillations, and also as the John–Nirenberg space.

The Morrey and Campanato spaces coincide for $\lambda < n$, and consequently, in this case we omit the notation M or C. We state an embedding theorem, which generalizes the Sobolev embedding theorems as follows. First, for $\lambda < n$ define

$$W^{k,p,\lambda}(\Omega) = \{u \in W^{k,p}(\Omega)|D^\alpha u \in L^{p,\lambda}(\Omega), |\alpha| = k\}$$

Then we have that if $kp < n - \lambda$, the space $W^{k,p,\lambda}(\Omega)$ is continuously embedded in $L^{p^{*\lambda}}(\Omega)$,

$$p^* = \frac{p(n - \lambda)}{n - \lambda - kp} \tag{1.16}$$

This embedding property can be proved by estimating Riesz potentials in the Morrey spaces; see [2], [10]. If $kp \geq n - \lambda$, then one can take p^* arbitrary (and finite).

• Interpolation Between Sobolev spaces and Hölder Spaces
We state here a very useful result, due to C. Miranda and L. Nirenberg [78], [87]. Any function $u \in W^{2,r}(\Omega) \cap C^{0,\alpha}(\overline{\Omega})$ also belongs to $W^{1,q}(\Omega)$, with

$$q = \frac{rn}{n\theta + r(1 - 2\theta - \alpha(1 - \theta))}, \quad \forall \frac{1 - \alpha}{2 - \alpha} \leq \theta \leq 1, \tag{1.17}$$

and we have the estimate

$$\|Du\|_{L^q} \leq C\|D^2u\|_r^\theta [u]_\alpha^{1-\theta} + C[u]_\alpha. \tag{1.18}$$

Note that the embedding properties of Sobolev spaces (see (1.1)) will imply $u \in W^{1,r^*}(\Omega)$, with $r^* = rn/(n - r)$. So there is a gain for $n > r + r(1 - \alpha)$ and not for $r < n \leq r + r(1 - \alpha)$.

• Spaces of Fractional Derivatives
There are several approaches to fractional derivatives, and they are not completely equivalent (see [1], [65], for details). We shall stick to some essential definitions and properties. Let us introduce the notation

$$\Omega_\rho = \{x \in \Omega|d(x, \partial\Omega) > \rho\}$$

and consider the finite difference operator

$$\Delta_h v(x) = v(x + h) - v(x), x \in \Omega_{|h|}.$$

Note that

$$\Delta_h \in \mathcal{L}(L^p(\Omega); L^p(\Omega_{|h|})).$$

Letting $0 < \lambda < 1$ and with p an integer, we define

$$W^{\lambda,p}(\Omega) = \left\{ v \in L^p(\Omega) \,\bigg|\, \sup_{h \neq 0} \frac{\|\Delta_h v\|_{L^p(\Omega_{|h|})}}{|h|^\lambda} < \infty \right\}.$$

with the seminorm

$$[v]_{W^{\lambda,p}} = \sup_{h \neq 0} \frac{\|\Delta_h v\|_{L^p(\Omega_{|h|})}}{|h|^\lambda}.$$

Then $W^{\lambda,p}(\Omega)$ is a Banach space for the norm

$$\|v\|_{W^{\lambda,p}} = \|v\|_{L^p} + [v]_{W^{\lambda,p}}.$$

Similarly, we define

$$W^{k+\lambda,p}(\Omega) = \{v \in W^{k,p}(\Omega) | D_{k_1,\dots,k_n} v \in W^{\lambda,p}, k_1 + \cdots + k_n = k\}$$

and the norm

$$\|v\|_{W^{k+\lambda,p}} = \|v\|_{L^p} + \sum_{\substack{k_1,\dots,k_n \\ k_1 + \cdots + k_n = k}} [D_{k_1,\dots,k_n} v]_{W^{\lambda,p}}.$$

The spaces $W^{k+\lambda,p}(\Omega)$, k integer, $\lambda \in (0,1)$, will be called, Besov–Nikol'skiĭ spaces. When s is not an integer, the notation is consistent with that of the Sobolev spaces $W^{s,p}$. We have the continuous embedding

$$W^{k+\lambda,p}(\Omega) \subset W^{k,p}(\Omega) \subset W^{k-1+\mu,p}(\Omega),$$

the second inclusion making sense for $k \geq 1$. We have also embedding properties similar to those of Sobolev spaces. In particular,

$$W^{\lambda,p}(\Omega) \subset L^{np/(n-\lambda p)}(\Omega), \text{ with continuous injection.} \tag{1.19}$$

Also, similar to the Miranda–Nirenberg interpolation result,

$$W^{1+\lambda,r}(\Omega) \cap C^{0,\alpha}(\overline{\Omega}) \subset W^{1,q}(\Omega), \quad \forall q < \frac{rn(1+\lambda)}{n - r\lambda\alpha} \tag{1.20}$$

with continuous injection. For this last result, see [8].

1.1.2 Regularity of Domains

We shall need some classical notions related to the smoothness of domains.

• Local Maps

Let Ω be a bounded domain of R^n, and let $\Gamma = \partial\Omega$. We say that Ω is of class $C^{k,\delta}$, $0 \leq \delta \leq 1$, if at each point $x_0 \in \Gamma$ there exists a ball $B = B_{x_0}$ and a one-to-one mapping ψ of B onto $D \subset R^n$, such that

$$\Omega^+ = \psi(B \cap \Omega) \subset \{y \in R^n | y_n > 0\},$$
$$\Gamma' = \psi(B \cap \Gamma) \subset \{y \in R^n | y_n = 0\},$$
$$\psi \in (C^{k,\delta}(B))^n, \quad \psi^{-1} \in (C^{k,\delta}(D))^n.$$

We say that the diffeomorphism ψ straightens the boundary in B. The case $k = 0, \delta = 1$ corresponds to a Lipschitz boundary.

It will be convenient to introduce the open set obtained by reflection of Ω^+, namely,

$$\Omega^- = \{y | y_n < 0, (y_1, \ldots, y_{n-1}, -y_n) \in \Omega^+\},$$

and set

$$\Omega' = \Omega^+ \cup \Omega^- \cup \Gamma'.$$

Then Ω' is an open bounded set of R^n. By compactness we can pick a finite number of the above balls, say $B^1, \ldots, B^m$, such that

$$\Gamma \subset \bigcup_{i=1}^{m} B^i,$$

and we can complete by finding an open set ω such that $\overline{\omega} \subset \Omega$ and

$$\overline{\Omega} \subset \omega \bigcup_{i=1}^{m} B^i.$$

We say that $\omega \cup_{i=1}^m B^i$ forms a system of local maps for $\overline{\Omega}$. Then it is possible to find a partition of unity related to this system of local maps , namely functions $\theta_0, \theta_1, \ldots, \theta_n$, such that

- for $\theta_0 \in C^\infty(R^n)$, support $\theta_0 \subset \omega$;
- for $\theta_j \in C^\infty(R^n)$, support $\theta_j \subset B^j$, $j = 1, \ldots, m$, $0 \le \theta_j \le 1$,
  $\sum_{j=0}^{n} \theta_j = 1$ in $\overline{\Omega}$.

This procedure permits us to reduce regularity on $\overline{\Omega}$ to interior regularity and regularity of functions on $y_n > 0$. For instance, suppose the domain is Lipschitz, and $u \in C^{0,\delta}(\Omega)$, with

$$u(\psi^{-1}(y)) \in C^{0,\delta}(\Omega^+),$$

where B is any ball centered on the boundary, as mentioned above. Then

$$u \in C^{0,\delta}(\overline{\Omega}).$$

An important property of boundaries we shall be considering is the "sphere condition":

$$\forall x_0 \in \Gamma, \quad |B_R(x_0) \cap (R^n - \Omega)| \geq c R^n. \tag{1.21}$$

It is satisfied, for instance, when the boundary is Lipschitz.

We shall also consider a splitting (D for Dirichlet, and N for Neumann)

$$\Gamma = \Gamma_D \cup \Gamma_N$$

and assume that

$$\Gamma_D \text{ is open in } \Gamma.$$

Suppose we have a system of local maps . Pick one of the balls B, perform the transformation ψ, and then set

$$\Gamma'_D = \psi(B \cap \Gamma_D),$$

$$\Gamma'_N = \psi(B \cap \Gamma_N).$$

Obviously,

$$\Gamma' = \Gamma'_D \cup \Gamma'_N.$$

We shall assume the following properties related to the splitting Γ_D, Γ_N:

$$\forall x_0 \in \Gamma_D, \text{ then } |B_R(x_0) \cap \Gamma_D| \geq c R^{n-1};$$

$$\forall x_0 \in \Gamma_N, \text{ if } B_R(x_0) \cap \Gamma_D = \emptyset, \text{ then } |B_R(x_0) \cap \Omega| \geq c R^n. \tag{1.22}$$

The same properties hold for the images Γ'_D, Γ'_N.

1.1.3 Poincaré Inequality

There are two types of Poincaré inequality. One occurs when the function does not vanish on the domain, in which case we must consider the function minus its average over the domain; the other occurs when the function vanishes on a set of sufficiently large measure. We present here some combinations of those, which we will need often.

Let Ω be a domain of R^n, and set $\Gamma = \partial\Omega$. We then state the Poincaré (or Sobolev–Poincaré) estimates that we need. We shall consider only balls such that

$$|B_R(x_0) \cap \Omega| > 0. \tag{1.23}$$

a. Standard form

Consider first a function $u \in H_0^1(\Omega)$. We extend it by 0 outside Ω. We assume that

$$\Omega \text{ is a Lipschitz domain,} \tag{1.24}$$

in which case the sphere condition (1.21) is satisfied. Define a constant associated to u and $B_R(x_0)$ by

$$
c_R = \begin{cases} \dfrac{1}{|B_R|} \displaystyle\int_{B_R} u \, dx & \text{if } B_R \subset \Omega, \\[3mm] 0 & \text{if } B_R \cap (R^n - \Omega) \neq \emptyset. \end{cases} \tag{1.25}
$$

Then we have

$$
\left(\int_{B_R \cap \Omega} |u - c_R|^\lambda \, dx \right)^{1/\lambda} \leq c R^{n\left(\frac{1}{\lambda} - \frac{1}{\mu}\right)+1} \left(\int_{B_{3R} \cap \Omega} |Du|^\mu \, dx \right)^{1/\mu}, \tag{1.26}
$$

where

$$
1 \leq \mu \leq 2, \quad n\left(\frac{1}{\lambda} - \frac{1}{\mu}\right) + 1 \geq 0. \tag{1.27}
$$

In fact, (1.26) summarizes two Poincaré inequalities. First, if $B_R \subset \Omega$, then the left-hand side of (1.26) can be improved as

$$
\left(\int_{B_R} |u - c_R|^\lambda \, dx \right)^{1/\lambda} \leq c R^{n\left(\frac{1}{\lambda} - \frac{1}{\mu}\right)+1} \left(\int_{B_R} |Du|^\mu \, dx \right)^{1/\mu}
$$

from the Poincaré inequality in a ball, and (1.26) holds a fortiori. Whenever $B_R \cap (R^n - \Omega) \neq \emptyset$, then from (1.24) and (1.23) there exists a point $x_0' \in \Gamma \cap B_R(x_0)$. Since $B_R(x_0) \subset B_{2R}(x_0')$, then

$$
\leq \left(\int_{B_{2R}(x_0') \cap \Omega} |u|^\lambda \, dx \right)^{1/\lambda} = \left(\int_{B_{2R}(x_0')} |u|^\lambda \, dx \right)^{1/\lambda},
$$

and from the assumption (1.21) and Poincaré's inequality,

$$
\left(\int_{B_R \cap \Omega} |u|^\lambda \, dx \right)^{1/\lambda} \leq c R^{n\left(\frac{1}{\lambda} - \frac{1}{\mu}\right)+1} \left(\int_{B_{2R}(x_0')} |Du|^\mu \, dx \right)^{1/\mu}
$$

$$
\leq c R^{n\left(\frac{1}{\lambda} - \frac{1}{\mu}\right)+1} \left(\int_{B_{3R}(x_0)} |Du|^\mu \, dx \right)^{1/\mu}.
$$

Hence (1.26) holds.

b. Function vanishing on a part of the boundary

Consider now a function

$$
u \in H^1(\Omega), \quad u = 0 \text{ on } \Gamma_D. \tag{1.28}
$$

We assume in this case (1.24) and (1.22). Then the definition of c_R is more involved.

$$c_R = \begin{cases} \dfrac{1}{|B_R|} \displaystyle\int_{B_R} u\,dx & \text{if } B_R \subset \Omega \\[3ex] \dfrac{1}{|B_{2R}(x_0') \cap \Omega|} \displaystyle\int_{B_{2R}(x_0')\cap\Omega} u\,dx & \text{if } B_{2R}(x_0') \cap \Gamma_D = \emptyset \\[3ex] 0 & \text{if } B_{2R}(x_0') \cap \Gamma_D \neq \emptyset, \end{cases} \quad (1.29)$$

where if $B_R \cap (R^n - \Omega) \neq \emptyset$ we have set $x_0' \in \Gamma \cap B_R(x_0)$. Then we have

$$\left(\int_{B_R \cap \Omega} |u - c_R|^\lambda \, dx \right)^{1/\lambda} \leq c R^{n\left(\frac{1}{\lambda} - \frac{1}{\mu}\right)+1} \left(\int_{B_{7R}\cap\Omega} |Du|^\mu \, dx \right)^{1/\mu}. \quad (1.30)$$

Consider the cases successively. If $B_R \subset \Omega$, then the values of u on the boundary are irrelevant. If not, from (1.24) and (1.23) B_R contains points of Γ. Pick $x_0' \in \Gamma \cap B_R(x_0)$. If $B_{2R}(x_0') \cap \Gamma_D = \emptyset$, then first

$$\left(\int_{B_R(x_0)\cap\Omega} |u - c_R|^\lambda \, dx \right)^{1/\lambda} \leq \left(\int_{B_{2R}(x_0')\cap\Omega} |u - c_R|^\lambda \, dx \right)^{1/\lambda},$$

and thanks to the second part of (1.22) we have

$$|B_{2R}(x_0') \cap \Omega| \geq c R^n.$$

Hence we can apply the Poincaré inequality to obtain

$$\left(\int_{B_{2R}(x_0')\cap\Omega} |u - c_R|^\lambda \, dx \right)^{1/\lambda} \leq c R^{n\left(\frac{1}{\lambda} - \frac{1}{\mu}\right)+1} \left(\int_{B_{2R}(x_0')\cap\Omega} |Du|^\mu \, dx \right)^{1/\mu},$$

and we have again (1.30).

Finally, when

$$B_{2R}(x_0') \cap \Gamma_D \neq \emptyset,$$

either x_0' itself belongs to Γ_D or it does not. Recall assumption (1.22) and suppose, for instance, that x_0' belongs to Γ_N; then $B_{2R}(x_0')$ contains a point $x_0'' \in \Gamma_D$. Consequently,

$$\left(\int_{B_{2R}(x_0')\cap\Omega} |u|^\lambda \, dx \right)^{1/\lambda} \leq \left(\int_{B_{4R}(x_0'')\cap\Omega} |u|^\lambda \, dx \right)^{1/\lambda},$$

and from the first part of (1.22) we can apply the Poincaré inequality to assert that

$$\left(\int_{B_{4R}(x_0'')\cap\Omega} |u|^\lambda \, dx\right)^{1/\lambda} \leq cR^{n\left(\frac{1}{\lambda}-\frac{1}{\mu}\right)+1} \left(\int_{B_{4R}(x_0'')\cap\Omega} |Du|^\mu \, dx\right)^{1/\mu}.$$

Since $B_{4R}(x_0'') \subset B_{7R}(x_0)$, we have (1.30) again. The case where x_0' belongs to Γ_D is simpler and also relies on assumption (1.22).

c. Poincaré inequality on an annulus

We shall also have to use Poincaré inequalities on an annulus of the form $(B_{2R}(x_0) - B_{R/2}(x_0)) \cap \Omega$. We pick m (depending on n) such that

$$c(4m)^n > \frac{\varpi_n}{2^n}, \tag{1.31}$$

where ϖ_n is the Lebesgue measure of B_1, and c is the constant entering in the sphere condition (1.21).

Take a function $u \in H_0^1(\Omega)$, and assume (1.24), (1.21). We define the constant c_R as follows:

$$c_R = \begin{cases} \dfrac{1}{|B_{2R} - B_{R/2}|} \displaystyle\int_{B_{2R}-B_{R/2}} u \, dx & \text{if} \quad B_{2R} \subset \Omega, \\[4mm] 0 & \text{if} \quad B_{2R} \cap (R^n - \Omega) \neq \emptyset. \end{cases} \tag{1.32}$$

Then we have the following Poincaré inequality:

$$\left(\int_{(B_{2R}-B_{R/2})\cap\Omega} |u - c_R|^\lambda \, dx\right)^{1/\lambda}$$

$$\leq cR^{n\left(\frac{1}{\lambda}-\frac{1}{\mu}\right)+1} \left(\int_{(B_{2(2m+1)R}-B_{R/2})\cap\Omega} |Du|^\mu \, dx\right)^{1/\mu}. \tag{1.33}$$

The only case that is not immediately clear is that with $B_{2R} \not\subset \Omega$. Then $c_R = 0$. There exists a point $x' \in \Gamma \cap (B_{2R}(x_0) - B_{R/2}(x_0))$. We first state

$$\left(\int_{(B_{2R}(x_0)-B_{R/2}(x_0))\cap\Omega} |u|^\lambda \, dx\right)^{1/\lambda} \leq \left(\int_{(B_{4R}(x')-B_{R/2}(x_0))\cap\Omega} |u|^\lambda \, dx\right)^{1/\lambda}$$

$$\leq \left(\int_{(B_{4mR}(x')-B_{R/2}(x_0))\cap\Omega} |u|^\lambda \, dx\right)^{1/\lambda}.$$

Next we check that

$$
\begin{aligned}
|(B_{4mR}(x') - B_{R/2}(x_0)) &\cap (R^n - \Omega)| \\
&= |B_{4mR}(x') \cap (R^n - \Omega)| - |B_{R/2}(x_0) \cap \Gamma_D| \\
&\geq c(4mR)^n - \varpi_n \left(\frac{R}{2}\right)^n
\end{aligned}
$$

from the "sphere condition." From the choice of m, it follows that the Poincaré inequality applies, and thus

$$
\begin{aligned}
\left(\int_{(B_{4mR}(x') - B_{R/2}(x_0)) \cap \Omega} |u|^\lambda \, dx\right)^{1/\lambda} \\
\leq cR^{n\left(\frac{1}{\lambda} - \frac{1}{\mu}\right) + 1} \left(\int_{(B_{4mR}(x') - B_{R/2}(x_0)) \cap \Omega} |Du|^\mu \, dx\right)^{1/\mu} \\
\leq cR^{n\left(\frac{1}{\lambda} - \frac{1}{\mu}\right) + 1} \left(\int_{(B_{2(2m+1)R}(x_0) - B_{R/2}(x_0)) \cap \Omega} |Du|^\mu \, dx\right)^{1/\mu}.
\end{aligned}
$$

Hence we obtain (1.33).

d. Function vanishing inside the domain

The final situation is the following: Suppose $u \in H^1(\Omega) \cap L^\infty(\Omega)$ and u vanishes not on the boundary but on an open surface Γ_D, with

$$
\Gamma_D \subset \Omega, \quad \Gamma_D \text{ singly connected}, \quad \Gamma_D \text{ lying in a plane.} \tag{1.34}
$$

We also assume that

$$
\text{if } x_0 \in \Gamma_D, \quad |B_R(x_0) \cap \Gamma_D| \geq c_0 R^{n-1}. \tag{1.35}
$$

We can always extend u beyond Ω so that it has compact support and remains in $H^1(R^n) \cap L^\infty(R^n)$. We define

$$
c_R = \begin{cases} \dfrac{1}{|(B_{2R} - B_R) \cap \Omega|} \displaystyle\int_{(B_{2R} - B_R) \cap \Omega} u \, dx & \text{if } B_{2R} \text{ contains no point of } \Gamma_D, \\[2mm] 0 & \text{if } B_{2R} \text{ contains a point of } \Gamma_D. \end{cases} \tag{1.36}
$$

Define m this time such that

$$
c_0(4m)^{n-1} > \varpi_{n-1}, \tag{1.37}
$$

where c_0 is the constant defined in (1.35). Then we have the following Poincaré inequality:

$$\left(\int_{(B_{2R}-B_R)\cap\Omega}|u-c_R|^\lambda\,dx\right)^{1/\lambda}$$

(1.38)

$$\leq cR^{n\left(\frac{1}{\lambda}-\frac{1}{\mu}\right)+1}\left(\int_{(B_{2(2m+1)R}-B_R)\cap\Omega}|Du|^\mu\,dx\right)^{1/\mu}+c_1R^{n/\lambda+1},$$

where the constant c_1 depends on the L^∞ norm of u and the diameter of Γ_D, but not on R.

Let us check (1.38). The only difficult case is where B_{2R} contains a point of Γ_D, say x', and $c_R = 0$. If $2R <$ diameter of Γ_D, then $B_R(x_0)$ cannot contain all of Γ_D. Hence if x' is not already in $B_{2R} - B_R$, then since Γ_D is singly connected, there must be a point $x'' \in (B_{2R} - B_R) \cap \Gamma_D$. But then

$$\left(\int_{(B_{2R}(x_0)-B_R(x_0))\cap\Omega}|u|^\lambda\,dx\right)^{1/\lambda}\leq\left(\int_{(B_{4R}(x'')-B_R(x_0))\cap\Omega}|u|^\lambda\,dx\right)^{1/\lambda}$$

$$\leq\left(\int_{(B_{4mR}(x'')-B_R(x_0))\cap\Omega}|u|^\lambda\,dx\right)^{1/\lambda}.$$

We see that

$$\begin{aligned}|(B_{4mR}(x'')-B_R(x_0))\cap\Omega\cap\Gamma_D| &= |(B_{4mR}(x'')-B_R(x_0))\cap\Gamma_D|\\ &= |B_{4mR}(x'')\cap\Gamma_D|-|B_R(x_0)\cap\Gamma_D|\\ &\geq |B_{4mR}(x'')\cap\Gamma_D|-\varpi_{n-1}R^{n-1},\end{aligned}$$

since Γ_D lies in a plane. From the assumption (1.35) and the choice of m, it follows that the Poincaré inequality applies, and

$$\left(\int_{(B_{4mR}(x'')-B_R(x_0))\cap\Omega}|u|^\lambda\,dx\right)^{1/\lambda}$$

$$\leq cR^{n\left(\frac{1}{\lambda}-\frac{1}{\mu}\right)+1}\left(\int_{(B_{4mR}(x'')-B_R(x_0))\cap\Omega}|Du|^\mu\,dx\right)^{1/\mu}$$

$$\leq cR^{n\left(\frac{1}{\lambda}-\frac{1}{\mu}\right)+1}\left(\int_{(B_{2(2m+1)R}(x_0)-B_R(x_0))\cap\Omega}|Du|^\mu\,dx\right)^{1/\mu},$$

and we obtain (1.38). It remains to consider the case where $2R \geq$ diameter of Γ_D. In this case,

$$\left(\int_{(B_{2R}-B_R)\cap\Omega} |u|^\lambda \, dx\right)^{1/\lambda} \leq c_1 R^{n/\lambda+1},$$

with c_1 depending only on the L^∞ norm of u and on the diameter of Γ_D. So in both cases we obtain (1.38).

Remark 1.1. The result extends to a situation where Γ_D is composed of a finite number of pieces, all satisfying (1.34) and (1.35).

1.1.4 Covering of Domains

We present here some covering lemmas and Whitney decomposition results, following the presentation of E.M. Stein [95], [96].

Lemma 1.2. *Let E be a measurable subset of R^n covered by a sequence B^j of balls of bounded diameter. There exists a disjoint subsequence $B_k = B^{j_k}$ such that*

$$|E| \leq 5^n \sum_k |B_k|. \tag{1.39}$$

Proof. We pick our subsequence by induction. First choose $B_1 = B^{j_1}$ such that

$$\text{diam } B_1 \geq \frac{1}{2} \sup_j \{\text{diam } B^j\}.$$

Then suppose we have defined the balls $(B_1, \ldots, B_k)$; we choose $B_{k+1} = B^{j_{k+1}}$ disjoint from the preceding ones and such that

$$\text{diam } B_{k+1} \geq \frac{1}{2} \sup_{\{j| \, B^j \cap B_l = \emptyset, l=1,\ldots,k\}} \{\text{diam } B^j\}. \tag{1.40}$$

If one cannot find a ball disjoint from $B_1, \ldots, B_k$, then one stops. Suppose first that our subsequence is infinite, and satisfies

$$\sum_k |B_k| = \infty.$$

Then there is nothing to prove. Suppose it is infinite and

$$\sum_k |B_k| < \infty.$$

Therefore, also

$$|B_k| \to 0, \text{ as } k \to \infty.$$

Define $B_k^* =$ ball of same center as B_k with $\text{diam } B_k^* = 5 \text{ diam } B_k$.
    Our claim is, that

$$E \subset \cup_k B_k^*. \tag{1.41}$$

It is, of course, enough to check that

$$B^j \subset \cup_k B_k^*, \ \forall j, \tag{1.42}$$

referring to the initial covering. We may, of course, assume that

$$B^j \neq B_1, \ldots, B_k, \ldots.$$

Otherwise, the assertion is immediate. Let $k + 1$ be the first index greater than or equal to 2 such that

$$\text{diam } B_{k+1} < \frac{1}{2} \text{ diam } B^j.$$

This index necessarily exists by our assumptions and the fact that the sequence is infinite. We claim that there exists $j_0 \in \{1, \ldots, k\}$ such that

$$B^j \cap B_{j_0} \neq \emptyset. \tag{1.43}$$

Indeed, otherwise, we would have by the construction of B_{k+1} (see (1.40)) that

$$\text{diam } B_{k+1} \geq \frac{1}{2} \text{ diam } B^j,$$

which is impossible by the definition of the index $k+1$. We can always assume j_0 to be the smallest integer. Thus

$$\text{diam } B_{j_0} \geq \frac{1}{2} \text{ diam } B^j,$$

which together with (1.43) implies, obviously,

$$B^j \subset B_{j_0}^*,$$

and thus (1.42), (1.41) have been obtained. Finally,

$$|E| \leq \sum_k |B_k^*| \leq 5^n \sum_k |B_k|.$$

When the sequence is finite and stops at k, the property (1.43) is obvious, and the same reasoning applies. The proof is complete. $\diamondsuit$

We proceed with the Whitney decomposition. We consider the lattice of points of R^n whose coordinates are positive or negative integers. This defines a mesh, denoted by $\mathcal{M}_0$, of cubes of unit length. To this initial mesh we associate

$$\mathcal{M}_k = 2^{-k} \mathcal{M}_0, \tag{1.44}$$

where k runs from $-\infty$ to $+\infty$. The cubes of $\mathcal{M}_k$ have length 2^{-k}, and a cube of $\mathcal{M}_k$ gives rise to 2^n cubes of $\mathcal{M}_{k+1}$. The diameter of a cube of $\mathcal{M}_k$ is, of course, $\sqrt{n}2^{-k}$.

Consider now

$$\mathcal{O} \text{ an open set of } R^n, \quad F = \mathcal{O}^c \text{ (the complement of } \mathcal{O}).$$

Define

$$\mathcal{O}_k = \left\{ x \,|\, \sqrt{n}2^{-k+1} < \text{ dist } (x, F) \leq \sqrt{n}2^{-k+2} \right\}.$$

Clearly,

$$\mathcal{O} = \bigcup_{k=-\infty}^{+\infty} \mathcal{O}_k.$$

We next define

$$\mathcal{F}_0 = \bigcup_k \{ Q \in \mathcal{M}_k \,|\, Q \cap \mathcal{O}_k \neq \emptyset \}. \tag{1.45}$$

We have the following lemma.

Lemma 1.3.

$$Q \in \mathcal{F}_0 \Rightarrow \text{diam } Q \leq \text{dist } (Q, F) \leq 4 \text{ diam } Q. \tag{1.46}$$

Proof. Suppose

$$Q \in \mathcal{M}_k.$$

Hence

$$\text{diam } Q = \sqrt{n}2^{-k}.$$

Pick some $x \in Q \cap \mathcal{O}_k$. Then

$$\text{dist } (Q, F) \leq \text{ dist } (x, F) \leq \sqrt{n}2^{-k+2}.$$

On the other hand,

$$\text{dist } (Q, F) \geq \text{ dist } (x, F) - \text{ diam } Q \geq 2\sqrt{n}2^{-k} - \sqrt{n}2^{-k} = \sqrt{n}2^{-k}.$$

Thus we obtain (1.46). $\diamond$

Since

$$Q \in \mathcal{F}_0 \Rightarrow Q \subset \mathcal{O},$$

it is clear that

$$\bigcup_{Q \subset \mathcal{F}_0} Q = \mathcal{O}. \tag{1.47}$$

One problem with the collection $\mathcal{F}_0$ is that the cubes of this family are not necessarily disjoint. We see now that we can extract a subfamily of cubes with disjoint interiors while preserving the covering property (1.47).

A first remark concerns the meshes. It is the following:

$$Q_1 \in \mathcal{M}_{k_1}, Q_2 \in \mathcal{M}_{k_2}, k_1 > k_2,$$

$$\text{interior } (Q_1) \cap \text{ interior } (Q_2) \neq \emptyset \Rightarrow Q_1 \subset Q_2 \tag{1.48}$$

This implies that given any cube $Q \in \mathcal{F}_0$ there is a *unique maximal* cube of the collection containing it. Indeed, from (1.46),

$$Q, Q' \in \mathcal{F}_0, Q \subset Q' \Rightarrow \text{dist}\,(Q', F) \leq 4 \text{ diam } Q.$$

Hence
$$\text{diam } Q' \leq 4 \text{ diam } Q.$$

On the other hand, from (1.48), if Q', Q'' contain Q, one of the two is contained in the other. Hence the assertion. Clearly, these maximal cubes have disjoint interiors and form a covering of $\mathcal{O}$. We make the following definition:

$$\mathcal{F} = \text{ subfamily of } \mathcal{F}_0 \text{ composed of the unique maximal extensions,} \quad (1.49)$$

and we have proven the following property:

Theorem 1.4. *To each*

$$\mathcal{O} \text{ open set of } R^n, \quad F = \mathcal{O}^c,$$

there corresponds a family of cubes $\mathcal{F}$ such that:

1. $\cup_{Q \subset \mathcal{F}} Q = \mathcal{O}$.
2. *The cubes of $\mathcal{F}$ have disjoint interiors.* $\qquad\qquad\qquad\qquad$ (1.50)
3. $\text{diam } Q \leq \text{ dist}\,(Q, F) \leq 4 \text{ diam } Q, \; \forall Q \in \mathcal{F}$.

We proceed with some additional useful properties. We begin with a lemma.

Lemma 1.5. *Suppose $Q_1, Q_2 \in \mathcal{F}, Q_1 \cap Q_2 \neq \emptyset$. Then*

$$\frac{1}{4} \text{ diam } Q_2 \leq \text{ diam } Q_1 \leq 4 \text{ diam } Q_2. \qquad (1.51)$$

Proof. We recall that the two cubes have disjoint interiors. Hence their common points belong to their respective boundaries. Using

$$\text{dist}\,(Q_1, F) \leq 4 \text{ diam } Q_1$$

and, because the cubes touch (boundaries have common points),

$$\text{dist}\,(Q_2, F) \leq \text{ dist}\,(Q_1, F) + \text{ diam } Q_1,$$

it follows that
$$\text{diam } Q_2 \leq 5 \text{ diam } Q_1.$$

But the ratio between the diameters being necessarily a power of 2, we have, in fact,
$$\text{diam } Q_2 \leq 4 \text{ diam } Q_1.$$

From the symmetry of the roles of the two cubes, the assertion follows. $\Diamond$

Lemma 1.6. *If $Q \in \mathcal{F}$, then the number of cubes of $\mathcal{F}$ that touch Q cannot exceed $N = 12^n$.*

Proof. This upper bound is certainly not optimal, but the remarkable feature is that it does not depend on $\mathcal{F}$.

Now we remark that if a cube that touches Q has a length larger than that of Q, then since the factor is a multiple of 2, we can decompose it into cubes of $\mathcal{F}_0$ that touch Q, have disjoint interiors, and belong to the same mesh as Q. So we obtain an upper bound for the number of cubes touching Q by picking all cubes of the same (or finer) mesh as Q that touch Q but do not overlap. Because of (1.51), the length of the cubes that touch Q is also larger than $\frac{1}{4}$ length (Q). Hence we majorize the number of cubes of $\mathcal{F}$ that touch Q by taking the total number of cubes of $\mathcal{F}_0$ that touch Q (but do not overlap) and whose size is between $\frac{1}{4}$ length Q and length Q. Since there are 3^n cubes that belong to the same mesh as Q and that touch Q (including Q), and since each of these cubes contains at most 4^n cubes of length $\frac{1}{4}$ smaller than them, we obtain the number N.   ◇

Now to any Q of $\mathcal{F}$ we associate

$$Q^* = (1 + \epsilon)Q,$$

which is the cube of same center as Q whose length is multiplied by $1 + \epsilon$.

We first note that

$$\left(1 - \frac{\epsilon}{2}\right) \operatorname{diam} Q \leq \operatorname{dist}(Q^*, F) \leq 4 \operatorname{diam} Q,$$

which is due to the observation

$$\operatorname{dist}(x, F) \geq \operatorname{dist}(\operatorname{Pr} x, F) - |x - \operatorname{Pr} x|, \ \forall x \in Q^*,$$

where $\operatorname{Pr} x$ denotes the projection of x onto Q. Then also

$$|x - \operatorname{Pr} x| \leq \frac{\epsilon}{2} \operatorname{diam} Q.$$

Since Q^* is contained in $\mathcal{O}$, we have also

$$\bigcup_{Q \subset \mathcal{F}} Q^* = \mathcal{O}.$$

Of course, the cubes Q^* may have nondisjoint interiors. Nevertheless, interesting properties still occur.

Lemma 1.7. *Assume $\epsilon < \frac{1}{4}$. Let $Q, Q' \in \mathcal{F}$. Then Q', Q^* intersect if and only if Q, Q' touch each other.*

Proof. The first thing to remark is that if Q', Q^* intersect, then we have a size condition similar to (1.51). Indeed, one first shows that

$$\frac{1}{5+\epsilon} \text{ diam } Q' \leq \text{ diam } Q \leq \frac{5}{1-\epsilon/2} \text{diam } Q',$$

and because of the condition on ϵ, and since the ratio can be only a power of 2, we have, in fact,

$$\frac{1}{4} \text{ diam } Q' \leq \text{ diam } Q \leq 4 \text{ diam } Q'.$$

Consider the union of Q and all the cubes that touch Q whose diameter is larger than $\frac{1}{4}$ diam Q. From the choice of ϵ, Q^* is contained in the interior of this union. So if Q' intersects Q^*, it also intersects the interior of the union and thus the interior of one of the cubes that touch Q. Thus it is a subset of one these cubes, since it cannot strictly include any of them. Because of the size condition it coincides with one of them. Hence the result.   ◇

Lemma 1.8. *The cubes Q^* have the finite intersection property, namely, each point of $\mathcal{O}$ is contained in at most N cubes Q^*.*

Proof. If $x \in \mathcal{O}$, there exists a $Q_0 \in \mathcal{F}$ that contains x. If x also belongs to some Q^*, then Q touches Q_0. There can be at most N of these cubes. Hence the result.   ◇

We associate a partition of unity with the Whitney decomposition. We index the cubes Q_k of the collection $\mathcal{F}$, and denote by x_k, ρ_k the center and side length of Q_k. Let Q_0 be the cube of center 0 and side length 1. Let θ be a function that is $W^{1,\infty}(R^n)$ and

$$0 \leq \theta \leq 1, \quad \theta = 1 \text{ on } Q_0, \quad \text{supp } \theta \subset \left(1 + \frac{\epsilon}{2}\right) Q_0.$$

We introduce

$$\tilde{Q}_k = \left(1 + \frac{\epsilon}{2}\right) Q_k.$$

Thus we have the strict inclusion

$$Q_k \subset \tilde{Q}_k \subset Q_k^*.$$

We set

$$\theta_k(x) = \theta\left(\frac{x - x_k}{\rho_k}\right). \tag{1.52}$$

Then

$$\theta_k = 1 \text{ on } Q_k, \quad \text{supp } \theta_k \subset \tilde{Q}_k,$$

and

$$\|\theta_k\|_\infty = \frac{\|\theta\|_\infty}{\rho_k}. \tag{1.53}$$

Define

$$\eta_k(x) = \frac{\theta_k}{\sum_l \theta_l}.$$

(1.54)

Then we have the partition of unity property

$$\sum_k \eta_k = \mathbb{1}_{\mathcal{O}}.$$

If we compute the gradient of η_k, we see easily that

$$\|\eta_k\|_\infty \le \|\theta_k\|_\infty + \sum_{\{l | Q_l \text{ touches } Q_k\}} \|\theta_l\|_\infty,$$

and from the results of lemmas 1.6 and 1.5 we can assert, that

$$\|\eta_k\|_\infty \le (1+4N)\frac{\|\theta\|_\infty}{\rho_k}.$$

(1.55)

$\diamond$

It will be helpful in some applications to consider an open subset of $\mathcal{O}$, say $\mathcal{O}'$, and the Whitney decomposition associated with it. We call it $\mathcal{F}'$. We state the following result.

Lemma 1.9. *Let $Q \in \mathcal{F}$, $Q' \in \mathcal{F}'$, and assume $Q^* \cap Q'^* \ne \emptyset$. Then one has*

$$\operatorname{diam} Q' \le 4 \operatorname{diam} Q.$$

Proof. Note that $F \subset F'$. Then

$$\operatorname{dist}(Q^*, F') \le \operatorname{dist}(Q^*, F) \le 4 \operatorname{diam} Q.$$

Now, since $Q^* \cap Q'^* \ne \emptyset$, we have

$$\operatorname{dist}(Q'^*, F') \le \operatorname{dist}(Q^*, F') + \operatorname{diam} Q^*,$$

and combining the inequalities yields

$$\operatorname{dist}(Q'^*, F') \le (5 + \epsilon) \operatorname{diam} Q.$$

Since

$$\left(1 - \frac{\epsilon}{2}\right) \operatorname{diam} Q' \le \operatorname{dist}(Q'^*, F'),$$

it follows that

$$\left(1 - \frac{\epsilon}{2}\right) \operatorname{diam} Q' \le (5 + \epsilon) \operatorname{diam} Q.$$

In view of the size of ϵ and the fact that the ratio of diameters is a power of 2, the claim is proven. $\diamond$

1.2 Useful Techniques

1.2.1 Reverse Hölder's Inequality

We present here a basic estimate initiated by F.G. Gehring [37], extended by
N.G. Meyers and A. Elcrat [77], a local version of which has been given by
M. Giaquinta and M. Modica [44]. We present the latest version of the result,
which is the most useful one for our purpose.
We shall use the notation

$$Q_R(x_0) = \{x \mid x_i - x_{0i} \leq R, \forall i\}$$

for the cube in R^n of center x_0 and side $2R$. Let also

$$\fint_{Q_R} \phi \, dx = \frac{1}{|Q_R|} \int_{Q_R} \phi \, dx.$$

Let Q_1 be a bounded cube in R^n. We consider

$$g \geq 0, \quad f \geq 0, \quad g \in L^q(Q_1), \quad f \in L^r(Q_1), \quad 1 < q < r, \tag{1.56}$$

and we make the assumption

$$\fint_{Q_{R/2}} g^q \, dx \leq b \left[\left(\fint_{Q_R} g \, dx \right)^q + \fint_{Q_R} f^q \, dx \right] + \theta \fint_{Q_R} g^q \, dx \tag{1.57}$$

for each $Q_R = Q_R(x_0)$ contained in Q_1. We then state the following result.

Theorem 1.10. *There exists a constant $\theta_0(q, n)$, and for $\theta < \min(b, \theta_0)$, a
constant $\epsilon(q, n, \theta, b)$ such that $\forall p$ with $q \leq p < q + \min(r - q, \epsilon)$, one has*

$$\left(\fint_{Q_{R/2}} g^p \, dx \right)^{1/p} \leq c_p \left[\left(\fint_{Q_R} g^q \, dx \right)^{1/q} + \left(\fint_{Q_R} f^p \, dx \right)^{1/p} \right] \tag{1.58}$$

for each $Q_R = Q_R(x_0)$ contained in Q_1, where $c_p = c_p(q, n, \theta, b)$.

We begin with the following result on Stieltjes integrals.

Lemma 1.11. *Let $q > 1$, $a > 1$, $h : [1, \infty) \rightarrow [0, \infty)$ nonincreasing, and
$\lim_{t \to \infty} h(t) = 0$. Suppose that*

$$-\int_t^\infty s^{q-1} dh(s) \leq at^{q-1}(h(t) + H(t)), \quad \forall t \geq 1, \tag{1.59}$$

where $H(t) : [1, \infty) \rightarrow [0, \infty)$. Let

$$\epsilon = \frac{q-1}{a-1}. \tag{1.60}$$

Then for $q \leq p < q + \epsilon$ one has

$$-\int_1^\infty s^{p-1} dh(s) \leq \mu_p \left[-\int_1^\infty s^{q-1} dh(s) + a(p-q) \int_1^\infty t^{p-2} H(t)\, dt \right], \tag{1.61}$$

where we have set

$$\mu_p = \frac{p-1}{p-1-a(p-q)}. \tag{1.62}$$

Proof. Note first that μ_p is a positive number.
Since $H(t) \geq 0$, we may assume without loss of generality that

$$\int_1^\infty t^{p-2} H(t)\, dt < \infty.$$

Otherwise, (1.61) is trivially satisfied.
    Let us prove the result under the additional assumption

$$h(t) = 0, \quad t \in [j, \infty). \tag{1.63}$$

Define

$$I(r) = -\int_1^\infty t^r\, dh(t).$$

We compute easily, for $p \geq q$,

$$I(p-1) = I(q-1) + (p-q)J, \tag{1.64}$$

where

$$J = -\int_1^j t^{p-q-1} \left(\int_t^j s^{q-1}\, dh(s) \right) dt.$$

Using the assumption (1.59) yields

$$J \leq a \left[\int_1^j t^{p-2} (h(t) + H(t))\, dt \right].$$

Hence also, by an easy integration by parts,

$$J \leq a \frac{I(p-1)}{p-1} + a \int_1^\infty t^{p-2} H(t)\, dt.$$

Using this estimate in (1.64) we obtain

$$I(p-1) \leq \mu_p \left(I(q-1) + a(p-q) \int_1^\infty t^{p-2} H(t)\, dt \right),$$

which is the desired result (1.61).

We now drop the assumption (1.63). Define then

$$h_j(t) = h(t)\, \mathrm{I\!I}_{[1,j)},$$

which is also defined on $[1, \infty)$ and is nonincreasing and satisfies (1.63). Let us check that

$$-\int_t^\infty s^{q-1} dh_j(s) \le at^{q-1}(h_j(t) + H(t)), \quad \forall t \ge 1. \tag{1.65}$$

The only case to consider is $t < j$, in which case the left-hand side of (1.65) becomes

$$-\int_t^\infty s^{q-1} dh_j(s) = -\int_t^j s^{q-1} dh(s) + j^{q-1} h(j)$$

$$= -\int_t^\infty s^{q-1} dh(s) + j^{q-1} h(j) + \int_j^\infty s^{q-1} dh(s).$$

Note that

$$j^{q-1} h(j) + \int_j^\infty s^{q-1} dh(s) \le 0.$$

Therefore,

$$-\int_t^\infty s^{q-1} dh_j(s) \le -\int_t^\infty s^{q-1} dh(s)$$

$$\le at^{q-1}(h(t) + H(t))$$

$$= at^{q-1}(h_j(t) + H(t)),$$

and (1.65) has been verified. Applying the first part of the proof we get

$$-\int_1^\infty s^{p-1} dh_j(s) \le \mu_p \left[-\int_1^\infty s^{q-1} dh_j(s) + a(p-q) \int_1^\infty t^{p-2} H(t)\, dt \right],$$

hence also

$$-\int_1^\infty s^{p-1} dh_j(s) \le \mu_p \left[-\int_1^\infty s^{q-1} dh(s) + a(p-q) \int_1^\infty t^{p-2} H(t)\, dt \right].$$

This means that

$$-\int_1^j s^{p-1} dh(s) \le \mu_p \left[-\int_1^\infty s^{q-1} dh(s) + a(p-q) \int_1^\infty t^{p-2} H(t)\, dt \right],$$

and letting $j \to \infty$, we get the desired result. $\diamond$

We proceed with an application. Consider a domain Q of R^n and a function $\phi \ge 0$. We write

$$E(\phi, t) = \{x \in Q | \phi(x) > t\}, \quad t \ge 1.$$

Lemma 1.12. *Let F, G be two positive functions. Let $q > 1$ and assume that*

$$F, G \in L^q(Q)$$

and

$$\int_{E(G,t)} G^q \, dx \leq a \left[t^{q-1} \left(\int_{E(G,t)} G \, dx \right) + \int_{E(F,t)} F^q \, dx \right], \quad \forall t \geq 1, \ a > 1.$$

$$(1.66)$$

Using the definitions of ϵ and $\mu(p)$, (1.60), and (1.62), we have

$$\int_Q G^p \, dx \leq \mu(p) \left[\int_Q G^q \, dx + a \int_Q F^p \, dx \right] \qquad (1.67)$$

with

$$q \leq p < q + \epsilon.$$

Proof. Without loss of generality, we may assume that

$$F \in L^p(Q).$$

Otherwise, (1.67) is trivially verified. We set

$$h(t) = \int_{E(G,t)} G \, dx,$$

$$H(t) = \frac{\displaystyle\int_{E(F,t)} F^q \, dx}{t^{q-1}},$$

and we notice that

$$\int_{E(G,t)} G^q \, dx = - \int_t^\infty s^{q-1} \, dh(s),$$

which follows from

$$(t + \theta)^{q-1}(h(t + \theta) - h(t)) \leq \left[\int_{E(G,t+\theta)} G^q \, dx - \int_{E(G,t)} G^q \, dx \right]$$

$$\leq t^{q-1}(h(t + \theta) - h(t)), \quad \theta > 0.$$

We then see from (1.66) that all the assumptions of Lemma 1.11 are satisfied. Therefore, (1.61) implies

$$\int_{E(G,1)} G^p \, dx \leq \mu(p) \left[\int_{E(G,1)} G^q \, dx + a(p-q) \int_1^\infty t^{p-q-1} \left(\int_{E(F,t)} F^q \, dx \right) dt \right].$$

It is easy to check that

$$(p-q)\int_1^\infty t^{p-q-1}\left(\int_{E(F,t)} F^q \, dx\right) dt \le \int_{E(F,1)} F^p \, dx.$$

Noting also that

$$\int_{\{x\in Q|G(x)<1\}} G^p \, dx \le \int_{\{x\in Q|G(x)\le 1\}} G^q \, dx,$$

we easily obtain (1.67).   ◇

We proceed with some technical results, which constitute a variant of the Whitney decomposition (see Section 1.1.4).

Let $Q = Q_{R_0}(a_0)$ be some cube of R^n, of center a_0 and side-length $2R_0$. We define the following subsets:

$$C_0 = \left\{ x \in Q \middle| \, \mathrm{dist}(x,\partial Q) > \frac{2}{3}R_0 \right\},$$

$$C_k = \left\{ x \in Q \middle| \frac{2^{-k+1}}{3} R_0 \le \mathrm{dist}(x,\partial Q) < \frac{2^{-k+2}}{3} R_0 \right\}, \quad k = 1,\dots .$$

Clearly,

$$Q = \bigcup_{k=0,1,\dots} C_k.$$

One next splits C_k into cubes of equal size $(2^{-k+1}/3)R_0$, for $k = 0,1,\dots$. Note that there are

$$\nu_k = (3 \times 2^k - 2)^n - (3 \times 2^k - 4)^{+n}$$

such cubes. We denote them by $Q^1_{j,k}, j = 1,\dots,\nu_k$. We next split each of these cubes into 2^n cubes of equal size, creating a new generation of subcubes denoted by $Q^2_{j,k}, j = 1\dots\nu_k 2^n$, of size $(2^{-k}/3)R_0$. We proceed by splitting this new generation into a new one, by dividing the side-length by 2, called $Q^3_{j,k}, j = 1\dots\nu_k 2^{2n}$, and so on.

Lemma 1.13. *Let $\gamma(x) \ge 0$ and $\gamma \in L^1(Q)$. Let ξ be any number such that*

$$\fint_Q \gamma(x) \, dx < \xi.$$

Then considering the sets C_k defined above, there exists a sequence of cubes $Q_{j,k} \subset C_k$ that do not overlap, have parallel sides, satisfy the condition

$$\xi 2^{kn} < \fint_{Q_{j,k}} \gamma(x) \, dx \le \xi 2^{kn} \times 3^n, \tag{1.68}$$

and satisfy

$$\gamma(x) \le \xi 2^{kn} \quad a.e. \ in \ C_k - \bigcup_j Q_{j,k}.$$

Proof. Consider the first generation cubes $Q^1_{j,k}$. Clearly,

$$\fint_{Q^1_{j,k}} \gamma(x)\,dx \le \fint_Q \gamma(x)\,dx \frac{|Q|}{|Q^1_{j,k}|} < \xi 2^{kn} \times 3^n.$$

If we pick those $Q^1_{j',k}$ such that

$$\xi 2^{kn} < \fint_{Q^1_{j',k}} \gamma(x)\,dx,$$

then we have obtained all the first generation cubes for which the condition (1.68) holds. By construction, the remaining $Q^1_{j'',k}$ satisfy the condition

$$\fint_{Q^1_{j'',k}} \gamma(x)\,dx \le \xi 2^{kn}.$$

We may consider the cubes of the second generation obtained by splitting the $Q^1_{j'',k}$. Each of these cubes $Q^2_{j,k}$ satisfies

$$\fint_{Q^2_{j,k}} \gamma(x)\,dx \le \fint_{Q^1_{j'',k}} \gamma(x)\,dx \frac{|Q^1_{j'',k}|}{|Q^2_{j,k}|} \le \xi 2^{kn} \times 2^n,$$

and hence also

$$\fint_{Q^2_{j,k}} \gamma(x)\,dx \le \xi 2^{kn} \times 3^n.$$

Therefore, if we pick those $Q^2_{j',k}$ such that

$$\xi 2^{kn} < \fint_{Q^2_{j',k}} \gamma(x)\,dx,$$

we have obtained the second generation cubes for which the relation (1.68) holds. Thus, successively, we obtain cubes of all generations for which (1.68) holds. They form the list $Q_{j,k}$. For points $y \in C_k - \cup_j Q_{j,k}$ we can find cubes of any generation $h, Q^h_{j_h,k}$ containing them, and such that

$$\fint_{Q^h_{j_h,k}} \gamma(x)\,dx \le \xi 2^{kn}.$$

Since by the theory of Lebesgue points, as $h \to \infty$

$$\fint_{Q_{j_h,k}^h} \gamma(x)\, dx \to \gamma(y) \text{ a.e.,}$$

the desired result is proven.   ◇

Proof of Theorem 1.10. Choose a cube $Q \subset Q_1$ with sides parallel to those of Q_1. Consider the sequence C_k associated with it, as constructed in Lemma 1.13. Set

$$\gamma(x) = g^q(x)\left(\fint_Q g^q\right)^{-1}.$$

Let $s > t \geq 1$. Set $\xi = s^q$. Since

$$\fint_Q \gamma(x)\, dx = 1 < \xi,$$

we can apply Lemma 1.13 to assert that there exists a sequence of disjoint cubes $Q_{j,k} \subset C_k$ with parallel sides such that (1.68) holds and such that

$$\gamma(x) \leq \xi 2^{kn} \text{ a.e. on } C_k - \bigcup_j Q_{j,k}.$$

Next set

$$G(x) = g(x)\left(\fint_Q g^q\right)^{-1/q} 2^{-nk/q}, \text{ for } x \in C_k.$$

Then we can assert that

$$s^q < \fint_{Q_{j,k}} G^q(x)\, dx \leq s^q 3^n$$

and

$$G(x) \leq s \text{ a.e. on } Q - \bigcup_{j,k} Q_{j,k}. \tag{1.69}$$

If we consider

$$E(G, s) = \{x \in Q | G(x) > s\},$$

then by (1.69), the set $E(G, s) - E(G, s) \cap (\cup_{j,k} Q_{j,k})$ has measure 0, and therefore

$$\int_{E(G,s)} G^q\, dx = \int_{E(G,s) \cap (\cup_{j,k} Q_{j,k})} G^q\, dx$$

$$\leq \sum_{j,k} \int_{Q_{j,k}} G^q\, dx \leq 3^n s^q \sum_{j,k} |Q_{j,k}|. \tag{1.70}$$

Consider now the cube $\overline{Q}_{j,k}$, which has the same center as $Q_{j,k}$ with double the side-length and parallel sides. Applying the assumption (1.57) we can write

$$\fint_{Q_{j,k}} g^q \, dx \leq b \left[\left(\fint_{\overline{Q}_{j,k}} g \, dx \right)^q + \fint_{\overline{Q}_{j,k}} f^q \, dx \right] + \theta \fint_{\overline{Q}_{j,k}} g^q \, dx. \qquad (1.71)$$

Set

$$F(x) = f(x) \left(\fint_Q g^q \right)^{-1/q} 2^{-nk/q}, \text{ for } x \in C_k.$$

Since

$$\overline{Q}_{j,k} \subset \bigcup_{i=(k-1)^+}^{k+1} C_i, \qquad (1.72)$$

we can assert that

$$F(x) \geq \frac{f(x)}{\left(\fint_Q g^q \right)^{1/q} 2^{n(k+1)/q}}, \text{ for } x \in \overline{Q}_{j,k},$$

$$G(x) \geq \frac{g(x)}{\left(\fint_Q g^q \right)^{1/q} 2^{n(k+1)/q}}, \text{ for } x \in \overline{Q}_{j,k}, \qquad (1.73)$$

and therefore, from the definition of F, G and (1.73) we can assert that

$$s^q < \fint_{Q_{j,k}} G^q \, dx \leq b2^n \left[\left(\fint_{\overline{Q}_{j,k}} G \, dx \right)^q + \fint_{\overline{Q}_{j,k}} F^q \, dx \right] + \theta 2^n \fint_{\overline{Q}_{j,k}} G^q \, dx. \qquad (1.74)$$

We now make the following choice for s:

$$s^q = 2^n b \left(\frac{5q}{q-1} t \right)^q.$$

It then follows easily from (1.74) that

$$\frac{5q}{q-1} t |\overline{Q}_{j,k}| \leq \int_{\overline{Q}_{j,k}} G \, dx + |\overline{Q}_{j,k}|^{1-1/q} \left(\int_{\overline{Q}_{j,k}} F^q \, dx \right)^{1/q}$$

$$+ \left(\frac{\theta}{b} \right)^{1/q} |\overline{Q}_{j,k}|^{1-1/q} \left(\int_{\overline{Q}_{j,k}} G^q \, dx \right)^{1/q}. \qquad (1.75)$$

We notice that

$$
\int_{\overline{Q}_{j,k}} G\, dx \leq \int_{\overline{Q}_{j,k}\cap E(G,t)} G\, dx + t|\overline{Q}_{j,k}|,
$$
$$
\int_{\overline{Q}_{j,k}} F\, dx \leq \int_{\overline{Q}_{j,k}\cap E(F,t)} F\, dx + t|\overline{Q}_{j,k}|,
\tag{1.76}
$$

from which we can deduce easily, provided that $\theta < b$, that

$$
|\overline{Q}_{j,k}|^{1-1/q}\left(\int_{\overline{Q}_{j,k}} F^q\, dx\right)^{1/q}
$$
$$
\leq t^{1-q}\left(\int_{\overline{Q}_{j,k}\cap E(F,t)} F^q\, dx\right) + 2t|\overline{Q}_{j,k}|,
$$
$$
\left(\frac{\theta}{b}\right)^{1/q}|\overline{Q}_{j,k}|^{1-1/q}\left(\int_{\overline{Q}_{j,k}} G^q\, dx\right)^{1/q}
$$
$$
\leq \frac{\theta}{b}t^{1-q}\left(\int_{\overline{Q}_{j,k}\cap E(G,t)} G^q\, dx\right) + 2t\,|\overline{Q}_{j,k}|.
\tag{1.77}
$$

Using (1.77) and the first inequality (1.76) in (1.75), we obtain the estimate

$$
\frac{5}{q-1}|\overline{Q}_{j,k}| \leq t^{-1}\int_{\overline{Q}_{j,k}\cap E(G,t)} G\, dx + t^{-q}\int_{\overline{Q}_{j,k}\cap E(F,t)} F^q\, dx
$$
$$
+ \frac{\theta}{b}t^{-q}\int_{\overline{Q}_{j,k}\cap E(G,t)} G^q\, dx.
\tag{1.78}
$$

Set

$$
D_k = \bigcup_j \overline{Q}_{j,k}.
$$

Since D_k is bounded, we use the classical covering result of Lemma 1.2 (see [95]) to assert that there exists a sequence of disjoint sets $\overline{Q}_{j(h),k}$ such that

$$
|D_k| \leq 5^n \sum_h \left|\overline{Q}_{j(h),k}\right|.
$$

Therefore, using (1.78) we have

$$
|D_k| \leq 5^{n-1}(q-1)\sum_h \left[t^{-1}\int_{\overline{Q}_{j(h),k}\cap E(G,t)} G\, dx + t^{-q}\int_{\overline{Q}_{j(h),k}\cap E(F,t)} F^q\, dx \right.
$$
$$
\left. + \frac{\theta}{b}t^{-q}\int_{\overline{Q}_{j(h),k}\cap E(G,t)} G^q\, dx \right].
\tag{1.79}
$$

Recalling (1.72),

$$\bigcup_h \overline{Q}_{j(h),k} \subset \bigcup_{i=(k-1)+}^{k+1} C_i,$$

and noting that the left- and right-hand sides are disjoint unions, we deduce from (1.79) that

$$|D_k| \leq 5^{n-1}(q-1) \sum_{i=(k-1)+}^{k+1} \left[t^{-1} \int_{C_i \cap E(G,t)} G\,dx \right.$$

$$\left. + t^{-q} \int_{C_i \cap E(F,t)} F^q\,dx + \frac{\theta}{b} t^{-q} \int_{C_i \cap E(G,t)} G^q\,dx \right].$$

Therefore, we also have that

$$\sum_k |D_k| \leq 5^{n-1}(q-1) \sum_{i=0}^{\infty} \sum_{k=(i-1)+}^{i+1} \left[t^{-1} \int_{C_i \cap E(G,t)} G\,dx + t^{-q} \int_{C_i \cap E(F,t)} F^q\,dx \right.$$

$$\left. + \frac{\theta}{b} t^{-q} \int_{C_i \cap E(G,t)} G^q\,dx \right]$$

$$\leq 5^n(q-1) \sum_{i=0}^{\infty} \left[t^{-1} \int_{C_i \cap E(G,t)} G\,dx + t^{-q} \int_{C_i \cap E(F,t)} F^q\,dx \right.$$

$$\left. + \frac{\theta}{b} t^{-q} \int_{C_i \cap E(G,t)} G^q\,dx \right].$$

Hence finally, we get

$$\sum_k |D_k| \leq 5^n(q-1) \left[t^{-1} \int_{E(G,t)} G\,dx + t^{-q} \right.$$

$$\left. \int_{E(F,t)} F^q\,dx + \frac{\theta}{b} t^{-q} \int_{E(G,t)} G^q\,dx \right]. \tag{1.80}$$

Since $Q_{j,k} \subset D_k$ and are disjoint sets, it is easily seen that

$$\sum_{j,k} |Q_{j,k}| \leq \sum_k |D_k|.$$

Thus going back to (1.70), we deduce from (1.80), recalling the value of s^q, that

$$\int_{E(G,s)} G^q\,dx \leq a_1 b \left[t^{q-1} \int_{E(G,t)} G\,dx + \int_{E(F,t)} F^q\,dx \right] + \theta a_1 \int_{E(G,t)} G^q\,dx, \tag{1.81}$$

where we have set

$$a_1 = \left(\frac{5q}{q-1}\right)^q (q-1)30^n.$$

On the other hand,

$$\int_{E(G,t)-E(G,s)} G^q \, dx \le s^{q-1} \int_{E(G,t)} G \, dx.$$

Hence

$$\int_{E(G,t)-E(G,s)} G^q \, dx \le a_2 b t^{q-1} \int_{E(G,t)} G \, dx, \qquad (1.82)$$

where we have set

$$a_2 = \left(\frac{5q}{q-1}\right)^{q-1} 2^{n(q-1)/q}. \qquad (1.83)$$

Adding up (1.81) and (1.82) and performing some easy calculations we obtain exactly (1.66), where we have set

$$a = \frac{(a_1 + a_2)b}{1 - \theta a_1},$$

provided that

$$\theta < \theta_0(q,n) = \left(\frac{q-1}{5q}\right)^{q-1} \frac{1}{5q \, 30^n}.$$

We can then use Lemma 1.12 to assert that

$$\int_Q G^p \, dx \le \mu(p) \left[\int_Q G^q \, dx + a \int_Q F^p \, dx \right].$$

Using the definitions of G and of F, we obtain

$$\sum_k \frac{\int_{C_k} g^p \, dx}{\left(\fint_Q g^q \, dx \right)^{\frac{p}{q}} 2^{npk/q}} \le \mu_p \sum_k \frac{\int_{C_k} g^q \, dx}{\left(\fint_Q g^q \, dx \right) 2^{npk/q}}$$

$$+\mu_p a \sum_k \frac{\int_{C_k} f^p \, dx}{\left(\fint_Q g^q \, dx \right)^{\frac{p}{q}} 2^{npk/q}}. \qquad (1.84)$$

Noting that $Q_{R_0/2}(a_0) \subset C_0 \cup C_1$, we obtain from (1.84) (after some easy calculations) that

$$\left(\fint_{Q_{\frac{R_0}{2}}} g^p \, dx \right)^{1/p} \le 2^{n\left(\frac{1}{p}+\frac{1}{q}\right)} (\mu_p)^{1/p} \left[\left(\fint_{Q_{R_0}} g^q \, dx \right)^{1/q} + a^{1/p} \left(\fint_{Q_{R_0}} f^p \, dx \right)^{1/p} \right],$$

and if we set, for instance,

$$c_p = 2^{2n/q}(a\mu_p)^{1/q},$$

we have proven the estimate (1.58), with R replaced by R_0. Since $Q_{R_0}(a_0)$ is any cube in Q_1 with parallel sides, the proof of the theorem is complete. $\diamondsuit$

Remark 1.14. Suppose we make an assumption that is slightly weaker than (1.57), namely,

$$\fint_{Q_{R/4}} g^q \, dx \le b \left[\left(\fint_{Q_R} g \, dx \right)^q + \fint_{Q_R} f^q \, dx \right] + \theta \fint_{Q_R} g^q \, dx \tag{1.85}$$

for any $Q_R = Q_R(x_0)$ contained in Q_1. Then we get a correspondingly slightly weaker result than Theorem 1.10:

There exist a constant $\theta_0(q, n)$ and for $\theta < \min(b, \theta_0)$ a constant $\epsilon(q, n, \theta, b)$ such that $\forall q \le p < q + \min(r - q, \epsilon)$, one has

$$\left(\fint_{Q_{R/4}} g^p \, dx \right)^{1/p} \le c_p \left[\left(\fint_{Q_R} g^q \, dx \right)^{1/q} + \left(\fint_{Q_R} f^p \, dx \right)^{1/p} \right] \tag{1.86}$$

for each $Q_R = Q_R(x_0)$ contained in Q_1, where $c_p = c_p(q, N, \theta, b)$.

Of course, the constants take slightly different values from those of Theorem 1.10, but they retain exactly the same properties. This is clear from the proof of Theorem 1.10. So the ratio of sizes of the cubes appearing on the left- and right-hand sides of assumption (1.57) is arbitrary, provided that it is fixed (not depending on R).

1.2.2 Gehring's Result

We shall apply Theorem 1.10 in the following way. For any x_0, let $B_R(x_0)$ denote the ball of center x_0 and radius R,

$$\{x \mid |x - x_0| \le R\}.$$

Note the obvious inclusions

$$B_R(x_0) \subset Q_R(x_0) \subset B_{R\sqrt{n}}(x_0). \tag{1.87}$$

Let Ω be a bounded domain of R^n. We consider a function $u \in H^1(\Omega)$. Let us assume that for each $x_0 \in R^n$ (writing B_R for $B_R(x_0)$),

$$\int_{B_R \cap \Omega} |Du|^2 \, dx \le k \left[\int_{B_{2R} \cap \Omega} |f|^2 \, dx \right.$$

$$+ \frac{1}{R^2} \left(\int_{B_{2R} \cap \Omega} |Du|^{2n/(n+2)} \, dx \right)^{(n+2)/n}$$

$$\left. + \frac{1}{R} \left(\int_{B_{2R} \cap \Omega} |Du|^{2n/(n+1)} \, dx \right)^{(n+1)/n} \right] \tag{1.88}$$

with

$$f \in L^r(\Omega), \quad r > 2. \tag{1.89}$$

Then we assert the following:

Theorem 1.15. *If we make the assumptions (1.88), (1.89), then there exists ϵ (depending only on k, n, Ω) such that $\forall 2 \leq p < 2 + \min(r - 2, \epsilon)$, $Du \in (L^p(\Omega))^n$ and*

$$\|Du\|_{L^p} \leq c_p |f|_{L^p}.$$

Proof. Let R_0 be such that $\Omega \subset B_{R_0}(0)$. Clearly, we may assume without loss of generality that $R \leq R_0$. We also check that

$$\frac{1}{R^2} \left(\int_{B_{2R} \cap \Omega} |Du|^{2n/(n+2)} \, dx \right)^{(n+2)/n}$$
$$\leq c(n) \frac{1}{R} \left(\int_{B_{2R} \cap \Omega} |Du|^{2n/(n+1)} \, dx \right)^{(n+1)/n}, \tag{1.90}$$

where the constant $c(n)$ depends only on n. Substituting (1.90) into (1.88) and dividing by R^n, we reduce the assumption to

$$\fint_{B_R \cap \Omega} |Du|^2 \, dx \leq b \left[\fint_{B_{2R} \cap \Omega} |f|^2 \, dx + \left(\fint_{B_{2R} \cap \Omega} |Du|^{2n/(n+1)} \, dx \right)^{(n+1)/n} \right], \tag{1.91}$$

where the constant b depends on k, n, R_0. Now define

$$z = |Du|^{2n/(n+1)} \mathbb{1}_\Omega,$$
$$\phi = |f|^{2n/(n+1)} \mathbb{1}_\Omega.$$

Then (1.91) becomes

$$\fint_{B_R} z^{(n+1)/n} \, dx \leq b \left[\fint_{B_{2R}} \phi^{(n+1)/n} dx + \left(\fint_{B_{2R}} z \, dx \right)^{(n+1)/n} \right], \tag{1.92}$$

but this is exactly like the basic assumption (1.57), except that balls replace cubes. We can take a cube Q_1 containing Ω and rewrite (1.92) with cubes on the left- and right-hand sides whose size ratio can be chosen fixed, thanks to (1.87). The desired result follows at once from Theorem 1.10. ◇

Remark 1.16. Just as in Remark 1.14, we can have in the statement of the assumption (1.88) balls with a different ratio of sizes, provided, of course, that it is fixed.

Remark 1.17. As we shall see later, assumption (1.88) arises naturally in applications.

Remark 1.18. We can extend Theorem 1.15 naturally to a vector function

$$u(x) = (u^1(x), \dots, u^N(x))$$

with

$$|Du|^2 = \sum_{i=1}^{N} |Du^i|^2.$$

This will be helpful for systems of equations.

Remark 1.19. Suppose $u \in H_0^1(\Omega)$, and we extend u outside Ω by 0. We assume that Ω has Lipschitz boundary (see (1.21), (1.24)). To any ball $B_R(x_0), x_0 \in \Omega$, we associate the constant c_R defined by (1.25). One often obtains "Cacciopoli type" inequalities such as

$$\int_{B_R} |Du|^2 \, dx \leq k \left[\frac{1}{R^2} \int_{B_{2R}} |u - c_R|^2 \, dx + \int_{B_{2R} \cap \Omega} |f|^2 \, dx \right]. \tag{1.93}$$

Then the result of Theorem 1.15 is valid for u. This is an immediate consequence of the Poincaré inequality (see (1.26))

$$\int_{B_{2R}} |u - c_R|^2 \, dx \leq c \left(\int_{B_{6R}} |Du|^{2n/(n+2)} \, dx \right)^{(n+2)/n}.$$

1.2.3 Hole-Filling Technique of Widman

We describe here another important technique in the regularity theory of functions [105].

Let $u \in H_0^1(\Omega)$. We extend it by 0 outside Ω. We assume that for any ball $B_R(x_0), R \leq R_0$, one has ($n > 2$)

$$\int_{B_R} |Du|^2 |x - x_0|^{2-n} \, dx \leq k_0 \int_{B_{2R} - B_R} |Du|^2 |x - x_0|^{2-n} \, dx + k_1 R^\beta, \tag{1.94}$$

where

$$k_0, k_1 \text{ constants}, \quad 0 < \beta, \tag{1.95}$$

and also

$$\int_{B_{R_0}} |Du|^2 |x - x_0|^{2-n} \, dx \leq C_0. \tag{1.96}$$

We then state the following result.

Theorem 1.20. *Assume (1.94), (1.95), (1.96). Then u belongs to $C^\delta(\overline{\Omega})$ for all*

$$\delta < \delta_0 = \frac{1}{2} \min \left(\beta, \log_2 \frac{1 + k_0}{k_0} \right). \tag{1.97}$$

Proof. By "filling" the hole in (1.94) we get

$$(1 + k_0) \int_{B_R} |Du|^2 |x - x_0|^{2-n} \, dx \leq k_0 \int_{B_{2R}} |Du|^2 |x - x_0|^{2-n} \, dx + k_1 R^\beta.$$

Let us set

$$\theta = \frac{k_0}{k_0 + 1}, \quad K = \frac{k_1}{k_0 + 1}.$$

we get

$$\int_{B_R} |Du|^2 |x - x_0|^{2-n} \, dx \leq \theta \int_{B_{2R}} |Du|^2 |x - x_0|^{2-n} \, dx + K R^\beta.$$

Dividing by $R^{2\delta}$, where δ is chosen as in (1.97), and noting that (1.97) implies $\beta - 2\delta > 0$, so

$$K R^{\beta - 2\delta} \leq K R_0^{\beta - 2\delta} = C R_0^{-2\delta},$$

we can write

$$R^{-2\delta} \int_{B_R} |Du|^2 |x - x_0|^{2-n} \, dx \leq \theta R^{-2\delta} \int_{B_{2R}} |Du|^2 |x - x_0|^{2-n} \, dx + C R_0^{-2\delta}.$$

$$(1.98)$$

We apply (1.98) with $R = R_0/2^{j+1}$, where j is a positive integer:

$$\left(\frac{R_0}{2^{j+1}} \right)^{-2\delta} \int_{B_{R_0 2^{-j-1}}} |Du|^2 |x - x_0|^{2-n} \, dx$$

$$\leq \theta \left(\frac{R_0}{2^j} \right)^{-2\delta} 2^{2\delta} \int_{B_{R_0 2^{-j}}} |Du|^2 |x - x_0|^{2-n} \, dx + C R_0^{-2\delta}.$$

Next set

$$\mu = \theta 2^{2\delta},$$

and by the condition on δ, we have $\mu < 1$. Therefore, setting

$$\phi_j = 2^{2j\delta} \int_{B_{R_0/2^j}} |Du|^2 |x - x_0|^{2-n} \, dx$$

we have the property

$$\phi_{j+1} \leq \mu \phi_j + C.$$

Hence

$$\phi_j \leq \phi_0 + \frac{C}{1 - \mu},$$

and from the assumption (1.96) we can assert that

$$\left(\frac{R_0}{2^j} \right)^{-2\delta} \int_{B_{R_0/2^j}} |Du|^2 |x - x_0|^{2-n} \, dx \leq C_1 \quad \forall j,$$

from which it easily follows that

$$R^{-2\delta} \int_{B_R} |Du|^2 |x - x_0|^{2-n} \, dx \leq C_2 \quad \forall R \leq R_0.$$

Hence also

$$R^{2-n-2\delta} \int_{B_R} |Du|^2 \, dx \leq C_2.$$

Using Hölder's inequality, we obtain

$$\int_{B_R} |Du| \, dx \leq C_3 R^{n-1+\delta},$$

and the result follows from Morrey's theorem (see (1.6)). ◇

Remark 1.21. In the right-hand side of (1.94) one may have

$$\int_{B_{2mR} - B_R} |Du|^2 |x - x_0|^{2-n} \, dx$$

instead of

$$\int_{B_{2R} - B_R} |Du|^2 |x - x_0|^{2-n} \, dx$$

without changing the result. We just have to change δ_0 into

$$\delta_0 = \frac{1}{2} \min \left(\beta, \log_{2m} \frac{1 + k_0}{k_0} \right).$$

1.2.4 Inhomogeneous Hole-Filling

The following technique, due to the second author [24], is of a similar spirit. It is called "inhomogeneous hole-filling." It is of particular use in the case $n = 2$. We first assume that the function $u \in H_0^1(\Omega)$ satisfies, for any ball $B_R(x_0)$,

$$\int_{B_R} |Du|^2 \, dx \leq k_0 \left(\int_{B_{2R} - B_R} |Du|^2 \, dx \right)^{1/2} + k_1 R^\beta, \quad \beta > 0. \qquad (1.99)$$

Then we have the following result.

Theorem 1.22. *If $u \in H_0^1(\Omega)$ satisfies (1.99), then*

$$\int_{B_R} |Du|^2 \, dx \leq C_0 \frac{1 + \log \log_2 \frac{R_0}{2R}}{1 + \log_2 \frac{R_0}{2R}} \quad \forall R \leq \frac{R_0}{2}.$$

Proof. We first notice that

$$R^\beta \leq k_\beta \frac{1}{\log_2 \frac{R_0}{R}} \quad \forall R < R_0$$

with

$$k_\beta = \frac{R_0^\beta e^{-1}}{\beta \log(2)}.$$

Hence the "inhomogeneous hole-filling inequality" (1.99) also implies

$$\int_{B_R} |Du|^2 \, dx \leq k_0 \left(\int_{B_{2R} - B_R} |Du|^2 \, dx \right)^{1/2} + k_1 k_\beta \frac{1}{\log_2 \frac{R_0}{R}} \quad \forall R < R_0. \tag{1.100}$$

We will need the following algebraic result.

Lemma 1.23. *Let s_i be a sequence of positive real numbers such that*

$$s_{i+1} \leq s_i \leq a\sqrt{s_{i-1} - s_i} + bi^{-1}, \tag{1.101}$$

where $a, b > 0$, $i \geq 1$.
  Then

$$s_i \leq \frac{1}{i+1} s_0 + \left(b + \frac{a^2}{4} \right) \frac{1}{i+1} (1 + \log i), \quad i \geq 1. \tag{1.102}$$

Proof. Note that (1.101) implies

$$s_i \leq i(s_{i-1} - s_i) + \left(b + \frac{a^2}{4} \right) i^{-1},$$

and hence, setting $b_0 = b + a^2/4$,

$$s_i \leq \frac{i}{i+1} s_{i-1} + b_0 \frac{1}{i(i+1)}.$$

By induction, one checks that

$$s_{i+j} \leq \frac{i}{i+j+1} s_{i-1} + b_0 \frac{1}{i+j+1} \sum_{l=i}^{j+i} \frac{1}{l}.$$

In the previous relation we let $i = 1, j = j - 1$, and we obtain

$$s_j \leq \frac{1}{j+1} s_0 + b_0 \frac{1}{j+1} \sum_{l=1}^{j} \frac{1}{l}$$

with $j \geq 1$.

Now we use the property

$$1 + \log j \geq \sum_{l=1}^{j} \frac{1}{l} \quad \forall j \geq 1$$

to obtain (1.102).   ◇

Proof of Theorem 1.22. For $j \geq 1$ we set

$$R_j = R_0 \, 2^{-j}, \quad s_j = \int_{B_{R_j}} |Du|^2 \, dx.$$

By definition,

$$j = \log_2 \frac{R_0}{R_j},$$

and the condition $j \geq 1$ implies $R_j \leq R_0/2$. Applying (1.100) we deduce that

$$s_j \leq \sqrt{s_{j-1} - s_j} + k_1 k_\beta \frac{1}{j},$$

which coincides with (1.101). Therefore, applying Lemma 1.23, we obtain

$$\int_{B_{R_j}} |Du|^2 \, dx \leq C_0 \frac{1 + \log \log_2 \frac{R_0}{R_j}}{1 + \log_2 \frac{R_0}{R_j}},$$

where C_0 is a suitable constant (we use the fact that the function

$$(1 + \log \log_2 x)/(1 + \log_2 x)$$

is decreasing for $x \geq 2$). Then we take $R \leq R_0/2$. It is possible to find $j \geq 1$ such that

$$R_{j+1} \leq R \leq R_j;$$

hence $2R \geq R_j$. The expected result follows easily.   ◇

Let us now show how to apply the "inhomogeneous hole-filling technique" to obtaining C^δ regularity results in the case $n = 2$.

We assume (1.24) and thus (1.21). We then begin by stating Cacciopoli's inequality

$$\int_{B_R} |Du|^2 \, dx \leq K_1 \int_{B_{2qR} - B_R} |Du|^2 \, dx + K_2 \int_{B_{2qR}} |Du|^2 |u - c_R| \, dx$$

$$+ K_3 \int_{B_{2qR}} |u - c_R|^2 \, dx + K_4 R^\beta, \quad q \geq 1, \ \beta > 0,$$

(1.103)

where c_R is an arbitrary positive constant, possibly depending on R.

We state the following result.

Theorem 1.24. *Assume that $u \in H_0^1(\Omega; R^N) \cap C^\gamma(\Omega; R^N)$ satisfies (1.99) and (1.103)(without loss of generality we can take the same value of β in both assumptions). Then if $n = 2$, we have*

$$u \in C^\alpha(\overline{\Omega}), \text{ where } \alpha \leq \min\left(\alpha_0, \frac{\beta}{2}, \gamma\right),$$

α_0 depending only on K_1, q. Moreover,

$$\|u\|_\alpha \leq C\left(\|u\|_{H_0^1}\right),$$

where C is bounded when $\|u\|_{H_0^1}$ is bounded and depends only on the constants entering into the assumptions (1.99) and (1.103) and on Ω. In particular, it does not depend on γ.

Remark 1.25. The interest of this theorem is to obtain an a priori estimate and a value of the Hölder exponent that depends only on the constants entering into the assumptions, and as far as the estimate is concerned only on the H_0^1 norm of u. But we have to know a priori that the function u is Hölder, however the exponent may depend on u. Otherwise, the result is wrong. For instance, the pair

$$u^1 = \cos\log\log\frac{1}{|x|}, \quad u^2 = \sin\log\log\frac{1}{|x|}$$

would be a counterexample.

Proof of Theorem 1.24. To simplify, we restrict ourselves to the case of a scalar u. Since u is $C^\gamma(\Omega)$, it is $C^\alpha(\Omega)$ for any $\alpha < \gamma$. In particular,

$$|u(x) - c_R| \leq 2(2qR)^\alpha[u]_\alpha, \quad \forall x \in B_{2qR}(x_0).$$

We deduce from Caccioppoli's inequality, by filling the hole, that

$$\int_{B_R} |Du|^2 \, dx \leq \frac{K_1}{K_1 + 1} \int_{B_{2qR}} |Du|^2 \, dx + k_2 \int_{B_{2qR}} |Du|^2 |u - c_R| \, dx$$

$$+ k_3 \int_{B_{2qR}} |u - c_R|^2 \, dx + k_4 R^\beta.$$

Multiplying by $R^{-n+2-2\alpha}$ $\left(\alpha \leq \min\left(\frac{\beta}{2}, \gamma\right)\right)$ yields

$$R^{-n+2-2\alpha} \int_{B_R} |Du|^2 \, dx$$

$$\leq \frac{K_1}{K_1 + 1} R^{-n+2-2\alpha} \int_{B_{2qR}} |Du|^2 \, dx$$

$$+ k_2(2q)^\alpha[u]_\alpha R^{-n+2-\alpha} \int_{B_{2qR}} |Du|^2 \, dx$$

$$+ k_3(2q)^{2\alpha}[u]_\alpha^2 R^2 + k_4 R^{-n+2-2\alpha+\beta}.$$

$$(1.104)$$

If $n = 2$, then (1.104) yields, if we set

$$\phi_q(R)^2 = C_0 \frac{1 + \log \log_2 \frac{R_0}{4qR}}{1 + \log_2 \frac{R_0}{4qR}}$$

and use Theorem 1.22,

$$R^{-2\alpha} \int_{B_R} |Du|^2 \, dx \leq k_4 R^{\beta - 2\alpha} + \frac{K_1}{K_1 + 1} (2q)^{2\alpha} (2qR)^{-2\alpha} \int_{B_{2qR}} |Du|^2 \, dx$$

$$+ k_2 (2q)^{2\alpha} [u]_\alpha \phi_q(R) \left((2qR)^{-2\alpha} \int_{B_{2qR}} |Du|^2 \, dx \right)^{1/2}$$

$$+ k_3 (2q)^{2\alpha} R^2 [u]_\alpha^2,$$

where the constants k_2, k_3, k_4 are generic, independent of α. We now need to choose $\alpha \leq \alpha_0$ with $\frac{K_1}{K_1+1}(2q)^{2\alpha_0} < 1$. Then we find θ such that

$$\frac{K_1}{K_1 + 1} (2q)^{2\alpha_0} < \theta < 1,$$

and we can write

$$R^{-2\alpha} \int_{B_R} |Du|^2 \, dx \leq \theta (2qR)^{-2\alpha} \int_{B_{2qR}} |Du|^2 \, dx + \psi_q(R)[u]_\alpha^2 + C,$$

where $\psi_q(R) \to 0$ as $R \to 0$. From this inequality it is easy to convince oneself that $\|u\|_\alpha$ is finite, and

$$\|u\|_\alpha^2 \leq \frac{1}{1 - \theta} \sup_{R \leq R_\delta} \psi_q(R) \, [u]_\alpha^2 + C_\delta,$$

where $R_\delta \to 0$ as $\delta \to 0$. Recalling Morrey's theorem, we have also

$$\|u\|_\alpha^2 \leq C \frac{1}{1 - \theta} \sup_{R \leq R_\delta} \psi_q(R) \, \|u\|_\alpha^2 + C_\delta.$$

Thus, we can pick δ sufficiently small so that

$$\|u\|_\alpha^2 \leq C_1,$$

where the constant is characterized as in the statement of the theorem. The desired result follows. $\diamond$

1.3 Green Function

1.3.1 Statement of Results

Consider a matrix $a(x) \in \mathcal{L}(R^n; R^n)$ such that

$$\alpha |\xi|^2 \leq a(x)\xi.\xi \leq M|\xi|^2, \quad \forall \, \xi \in R^n, \, x \in R^n. \tag{1.105}$$

Let Q be a ball such that $\overline{\Omega} \subset Q$, and let $x_0 \in \Omega$. We consider the Green Function $G = G^{x_0}$ relative to x_0; see [52] for details. It is the solution of

$$\int_Q aD\phi.DG\,dx = \phi(x_0), \quad \forall\,\phi \in C_0^\infty(Q). \tag{1.106}$$

We have the following properties.

Theorem 1.26. *Assume (1.105). Then*

$$G \in W_0^{1,\mu}(Q), \quad 1 \le \mu < \frac{n}{n-1}, \quad G \in L^\nu(Q), \quad 1 \le \nu < \frac{n}{n-2}. \tag{1.107}$$

Theorem 1.27. *Assume (1.105). Then*

$$c_0|x - x_0|^{2-n} \le G^{x_0}(x) \le c_1|x - x_0|^{2-n} \tag{1.108}$$

if $n \ge 3$.[1] The constants c_0, c_1 depend only on α, M. The estimate (1.108) holds in a neighborhood of x_0 whose closure is contained in Q. In particular, it will hold for any $x \in \overline{\Omega}$.

We shall use the following approximation to G, called G^h (where we shall let $h \to 0$), and where for clarity of notation we suppress the x_0 dependence,

$$-D_i(a_{i\,k}D_kG^h) = \delta^h, G^h = 0 \text{ on } \partial Q, \tag{1.109}$$

with

$$\delta^h = \frac{\mathbb{1}_{B_h}}{|B_h|},$$

where the ball B_h is centered at the singularity x_0.

1.3.2 Proof of Theorem 1.26

We test equation (1.109) with

$$\frac{G^h}{(1 + (G^h)^s)^{\frac{1}{s}}}, \quad 0 < s < 1,$$

which is less than one. Hence we get

$$\int_Q a_{i\,k}D_kG^h D_i[G^h(1 + (G^h)^s)^{-1/s}]\,dx \le 1.$$

To simplify the notation in the following calculations we suppress the parameter h. So

$$\int_Q \frac{a_{i\,k}D_kGD_iG}{(1 + G^s)^{(1+s)/s}}\,dx \le 1. \tag{1.110}$$

---

[1] for $n = 2$, one should replace $|x - x_0|^{2-n}$ by $-\log|x - x_0|$.

Note that for $0 < s < 1$,

$$\frac{1}{1+G} \le \frac{2^{(1-s)/s}}{(1+G^s)^{1/s}}.$$

Combining this inequality with (1.110) yields

$$\int_Q \frac{|DG|^2}{(1+G)^{1+s}}\, dx \le K_s = \frac{1}{\alpha} 2^{(1-s^2)/s},$$

hence also

$$\int_Q \left| D\left(1+G\right)^{(1-s)/2} \right|^2 dx \le L_s = \left(\frac{1-s}{2}\right)^2 K_s. \tag{1.111}$$

From the estimate (1.111) it follows that

$$\left\| (1+G^h)^{(1-s)/2} - 1 \right\|_{H_0^1(Q)} \le C_s,$$

hence from Sobolev embedding

$$\left\| (1+G^h)^{(1-s)/2} \right\|_{L^{2n/(n-2)}} (Q) \le C_s,$$

and this implies the second property (1.107). Next set

$$\beta = \frac{n(1-s)}{2(n-1-s)}.$$

Then writing

$$\int_Q |DG|^{2\beta}\, dx = \int_Q \left| \frac{DG}{(1+G)^{(1+s)/2}} \right|^{2\beta} (1+G)^{1+s\beta}\, dx$$

and using Hölder's inequality, the first property (1.107) follows from the second property and the above. The proof is complete. ◇

1.3.3 Estimates on $\log G$

In this section we obtain estimates on $\log G$, for $n \ge 3$. We consider a ball B_{2R_0} such that

$$\overline{B_{2R_0}} \subset Q.$$

We then have the following result.

Lemma 1.28.

$$\int_{B_R} |D \log G|^2\, dx \le C R^{n-2} \quad \forall R \le R_0. \tag{1.112}$$

Proof. We consider a cutoff function τ_R such that

$$\tau_R = 1 \text{ on } B_R, \quad \text{supp } \tau_R \subset B_{2R}, \quad |D\tau_R| \leq \frac{C}{R}.$$

We test (1.109) with

$$\frac{\tau_R^2}{G},$$

which is a positive function. It is then easy to derive

$$\int_Q |D \log G|^2 \tau_R^2 \, dx \leq C \int_Q |D\tau_R|^2 \, dx,$$

and the result (1.112) follows immediately.   ◇

Consider now a constant c_R satisfying

$$|\{x \in B_R, \ \log G \geq c_R\}| \geq \frac{1}{2}|B_R|,$$

$$|\{x \in B_R, \ \log G \leq c_R\}| \geq \frac{1}{2}|B_R|. \tag{1.113}$$

Then we state the following result.

Lemma 1.29.

$$\|(\log G - c_R)^-\|_{L^\infty(B_R)} \leq C, \quad \forall R \leq R_0. \tag{1.114}$$

In order to perform a local version of Moser's iteration technique (a general presentation will be made in Section 3.3.1) we shall introduce

$$\rho \leq \frac{R}{2}$$

and a cutoff function $\tau_{R,\rho}$ such that

$$\tau_{R,\rho} = \begin{cases} 1 & \text{on } B_{R+\rho}, \\ 0 & \text{outside } B_{R+2\rho}, \end{cases}$$

$$|D\tau_{R,\rho}| \leq \frac{3}{2}\rho.$$

The function $\tau_{R,\rho}$ is explicitly defined by

$$\tau_{R,\rho}(x) = \frac{1}{\rho^3} \left[2|x - x_0|^3 - (9\rho + 6R)|x - x_0|^2 \right.$$

$$\left. + 6(\rho + R)(R + 2\rho)|x - x_0| - (R + 2\rho)^2(2R + \rho) \right]$$

for $R + \rho \leq |x - x_0| \leq R + 2\rho$. To simplify the notation we write

$$\tau_{R,\rho} = \tau_\rho.$$

Proof of Lemma 1.29. Let be $s > 1$, and test (1.109) with

$$\frac{\tau_\rho^2}{G}((\log G - c)^-)^{(s-1)}.$$

We obtain the inequality

$$\int_Q |D \log G|^2 \tau_\rho^2 ((\log G - c)^-)^{(s-2)} \, dx$$

$$\leq \frac{C}{(s-1)^2} \int_Q |D\tau_\rho|^2 ((\log G - c)^-)^s \, dx$$

and also

$$\int_Q |D \tau_\rho((\log G - c)^-)^{\frac{s}{2}}|^2) \, dx \leq C \frac{s}{(s-1)}^2 \int_Q |D\tau_\rho|^2 ((\log G - c)^-)^s \, dx.$$

We perform Moser's iteration technique. Using the Poincaré inequality yields

$$\left(\int_{B_{R+\rho}} ((\log G - c)^-)^{\frac{sn}{n-2}} \, dx \right)^{(n-2)/n}$$

$$\leq \frac{C}{\rho^2} \left(\frac{s}{s-1} \right)^2 \int_{B_{R+2\rho}} ((\log G - c)^-)^s \, dx,$$

or again

$$\|(\log G - c)^-\|_{L^{sn/(n-2)}(B_{R+\rho})}$$

$$\leq \left(\frac{C}{\rho^2} \left(\frac{s}{s-1} \right)^2 \right)^{1/s} \|(\log G - c)^-\|_{L^s(B_{R+2\rho})}. \tag{1.115}$$

We apply this inequality with

$$\rho = R_{j+1} = \frac{R}{2^{j+1}}, \quad s = s_j = s_0 a^j, \quad a = \frac{n}{n-2}$$

and set

$$z_j = \|(\log G - c)^-\|_{L^{s_j}(B_{R+R_j})}.$$

Then (1.115) reads

$$z_{j+1} \leq \left(\frac{C}{R_{j+1}^2} \left(\frac{s_j}{s_j - 1} \right)^2 \right)^{1/s_j} z_j.$$

From this iteration one obtains

$$z_j \leq \frac{C_{s_0}}{R^{2a/(a-1)s_0}} z_0 \qquad (1.116)$$

with

$$C_{s_0} = \left(C \left(\frac{s_0}{s_0 - 1} \right)^2 \right)^{\frac{1}{s_0}\left(\frac{a}{a-1}^2\right)}.$$

We apply (1.116) with $s_0 = 2$ and $c = c_R$. We note that by the Poincaré inequality

$$z_0^2 \leq CR^2 \int_{B_{2R}} |D\,G|^2 \, dx \leq CR^n,$$

where we have made use of Lemma 1.28. Letting $j \to \infty$ in (1.116), the result (1.114) follows. $\diamondsuit$

1.3.4 Estimates on Positive and Negative Powers of G

Let us set

$$T_R = B_{2R} - B_R,$$

$$T_R' = B_{4R} - B_{R/2}.$$

We assume

$$R \leq R_0, \quad \overline{B_{6R_0}} \subset Q.$$

Our objective is to prove the following result.

Lemma 1.30.

$$\|G^\beta\|_{L^\infty(T_R)} \leq C_\beta \fint_{T_R'} G^\beta \, dx \qquad (1.117)$$

for all real values of β.

We prove the estimate (1.117) for G^h. We can let h tend to 0, provided that the right-hand side of (1.117) is finite. The fact that it is indeed finite for any β will follow from later results.

We shall need a cutoff function similar to $\tau_{R,\rho}$ defined above. We use the same notation ($\rho \leq R$):

$$\tau_{R,\rho}(x)$$

$$= \begin{cases} 0, & \text{if } |x - x_0| \leq R - \rho, \\ -\dfrac{4}{\rho^3}[4|x - x_0|^3 - 3(4R - 3\rho)|x - x_0|^2 \\ \quad + 6(2R - \rho)(R - \rho)|x - x_0| \\ \quad - (R - \rho)^2(4R - \rho)] & \text{if } R - \rho \leq |x - x_0| \leq R - \frac{\rho}{2}, \end{cases}$$

$$\tau_{R,\rho}(x)$$

$$
= \begin{cases}
1, & \text{if } R - \frac{\rho}{2} \le |x - x_0| \le 2R + 2\rho \\
\begin{aligned}
\frac{1}{4\rho^3}[&|x - x_0|^3 - 3(2R + 3\rho)|x - x_0|^2 \\
&+12(R + 2\rho)(R + \rho)|x - x_0| \\
&-4(R + 2\rho)^2(2R + \rho)]
\end{aligned} & \text{if } 2R + 2\rho \le |x - x_0| \le 2R + 4\rho \\
0 & \text{if } |x - x_0| \ge 2R + 4\rho.
\end{cases}
$$

We can check again that

$$|D\tau_{R,\rho}| \le \frac{3}{\rho}.$$

Proof of Lemma 1.30. We denote $\tau_{R,\rho}$ by τ_ρ, and test (1.109) with $\tau_\rho^{q+2}G^{s-1}$, $s \ge 2$. Let

$$T_{R,\rho} = B_{2R+2\rho} - B_{R-\rho/2},$$

$$T'_{R,\rho} = B_{2R+4\rho} - B_{R-\rho}.$$

By construction

$$\tau_\rho = 1, \text{ on } T_{R,\rho}, \operatorname{supp}\tau_\rho \subset T'_{R,\rho}.$$

After standard calculations one arrives at

$$
\int_Q aD(\tau_\rho^{\frac{q}{2}+1}G^{\frac{s}{2}})D(\tau_\rho^{\frac{q}{2}+1}G^{\frac{s}{2}})\,dx
$$
$$
= \frac{(q + 2)(s - 2)}{2(s - 1)} \int_Q \tau_\rho^{\frac{q}{2}}G^{\frac{s}{2}} aD\left(\tau_\rho^{\frac{q}{2}+1}G^{\frac{s}{2}}\right) D\tau_\rho\,dx
$$
$$
+ \frac{(q + 2)^2}{4(s - 1)} \int_Q \tau_\rho^q G^s a D\tau_\rho D\tau_\rho\,dx.
$$

Applying Hölder's inequality to the first integral on the right-hand side, we majorize it by

$$
\frac{s - 2}{2(s - 1)} \int_Q aD\left(\tau_\rho^{\frac{q}{2}+1}G^{\frac{s}{2}}\right) D\left(\tau_\rho^{\frac{q}{2}+1}G^{\frac{s}{2}}\right)\,dx
$$
$$
+ \frac{(q + 2)^2(s - 2)}{8(s - 1)} \int_Q \tau_\rho^q G^s a D\tau_\rho D\tau_\rho\,dx,
$$

and thus we obtain, collecting results,

$$\int_Q aD\left(\tau_\rho^{\frac{q}{2}+1}G^{\frac{s}{2}}\right)D\left(\tau_\rho^{\frac{q}{2}+1}G^{\frac{s}{2}}\right)dx \le \frac{(q+2)^2}{4}\int_Q \tau_\rho^q G^s aD\tau_\rho D\tau_\rho\,dx.$$

Therefore, we deduce

$$\int_Q \left|D\left(\tau_\rho^{\frac{q}{2}+1}G^{\frac{s}{2}}\right)\right|^2 dx \le C\left(\frac{q+2}{\rho}\right)^2 \int_Q \tau_\rho^q G^s\,dx.$$

Applying the Poincaré inequality yields

$$\int_Q \tau_\rho^{(q+2)n/(n-2)}G^{sn/(n-2)}\,dx \le C\left(\frac{q+2}{\rho}\right)^{2n/(n-2)}\left(\int_Q \tau_\rho^q G^s\,dx\right)^{n/(n-2)}.$$

On the right-hand side we write

$$\tau_\rho^q G^s G^s = \tau_\rho^q G^s G^{qs/(q+2)}G^{2s/(q+2)}.$$

So we may apply Hölder's inequality with exponents

$$\frac{(q+2)n}{q(n-2)},\quad \frac{n(q+2)}{2(n+q)},$$

and we obtain on the right-hand side the integral of the left-hand side to the power $q/(q+2)$. After cancellation, we finally obtain the following estimate:

$$\|G\|_{L^{sn/(n-2)}(T_{R,\rho})}^{2s/(q+2)} \le C\left(\frac{q+2}{\rho}\right)^2 \|G\|_{L^{sn/(n+q)}(T'_{R,\rho})}^{2s/(q+2)}. \tag{1.118}$$

Then we apply (1.118) with

$$\rho = R_{j+1} = \frac{R}{2^{j+1}},\quad s = s_j = s_0 a^j,\quad a = \frac{n+q}{n-2}.$$

We note that

$$T'_{R,R_{j+1}} = T_{R,R_j}.$$

Setting

$$z_j = \|G\|_{L^{s_j n/(n+q)}(T_{R,R_j})},$$

we obtain from (1.118) a Moser iteration (see Section 3.3.1 for a general presentation), from which it follows that

$$\|G\|_{L^\infty(T_R)} \le \frac{C_{s_0,q}}{R^{(n+q)/s_0}}\|G\|_{L^{s_0 n/(n+q)}(T'_R)}.$$

We have used the notation

$$C_{s_0,q} = (C(q+2))^{(n+q)^2/s_0(q+2)}.$$

Thus playing with the two parameters $s_0 \geq 2, q \geq 0$, we can achieve any positive value for

$$\beta = \frac{ns_0}{n+q}.$$

Thus the estimate (1.117) has been proven for $\beta > 0$.

We next test (1.109) with $\tau_\rho^{q+2} G^{-s-1}$, $s > 0$. We now obtain

$$\|G^{-1}\|^{2s/(q+2)}_{L^{sn/(n-2)}(T_{R,\rho})} \leq C \left(\frac{q+2}{\rho}\right)^2 \|G^{-1}\|^{2s/(q+2)}_{L^{sn/(n+q)}(T'_{R,\rho})}. \tag{1.119}$$

Then we apply (1.119) with

$$\rho = R_{j+1} = \frac{R}{2^{j+1}}, \quad s = s_j = s_0 a^j, \quad a = \frac{n+q}{n-2}.$$

We note that

$$T'_{R,R_{j+1}} = T_{R,R_j}.$$

Setting

$$z_j = \|G^{-1}\|_{L^{s_j n/(n+q)}(T_{R,R_j})}$$

and performing the Moser iteration we obtain, letting $j \to \infty$,

$$\|G^{-1}\|_{L^\infty(T_R)} \leq \frac{C_{s_0,q}}{R^{(n+q)/s_0}} \|G^{-1}\|_{L^{s_0 n/(n+q)}(T'_R)},$$

where $C_{s_0,q}$ is of the same form as above.

Again, playing with the two parameters $s_0 > 0, q \geq 0$, we can achieve any negative value of

$$\beta = -\frac{ns_0}{n+q}.$$

Thus (1.117) is proven for any value of β.   ◇

1.3.5 Harnack's Inequality

We first make the comment that the equivalent of property (1.114) for the positive part is not true. We shall give a weaker result, which will be sufficient in order to obtain later the important Harnack's inequality. We state a lemma.

Lemma 1.31. *For $\beta \leq \beta_0$, where $\beta_0 < 1$ is determined by the Poincaré inequality constant and α, M (see (1.105)), we have*

$$\fint_{T_R} \exp \beta (\log G - c_{2R})^+ \, dx \leq C, \tag{1.120}$$

where the definition of T_R has been given in Section 1.3.4, and that of c_R in (1.113). In fact, this is rigourous only for G^h; the constant C is independent of h.

Proof. We shall use the sets $T_{R,\rho}, T'_{R,\rho}$ and the cutoff functions $\tau_{R,\rho}$ defined in the proof of Lemma 1.30. We write τ_ρ for $\tau_{R,\rho}$ to simplify the notation, and we recall that

$$T_R = T_{R,0}, \quad T'_R = T'_{R,R/2}, \quad T_{R,2\rho} = T'_{R,\rho}.$$

Step 1: preliminary calculations

We test (1.109) with

$$\frac{\tau_\rho^2}{G} \exp \beta (\log G - c)^+,$$

where $\beta < 1$, and c is a constant to be defined later (in the calculation we proceed formally, replacing G^h by G). We obtain

$$(1 - \beta) \int_Q \tau_\rho^2 a \, D \log G \, D \log G \exp \beta (\log G - c)^+ \, dx$$

$$\leq 2 \int_Q a \, D \log G \, D\tau_\rho \, \tau_\rho \exp \beta (\log G - c)^+ \, dx.$$

One deduces easily

$$\int_{T_{R,\rho}} |D \log G|^2 \exp \beta (\log G - c)^+ \, dx$$

$$\leq \frac{c_0}{(1 - \beta)^2} \frac{1}{\rho^2} \int_{T'_{R,\rho} - T_{R,\rho}} \exp \beta (\log G - c)^+ \, dx, \tag{1.121}$$

where c_0 depends only on α, M.

We also note that

$$C_1 R^{n-1} \rho \leq |T'_{R,\rho} - T_{R,\rho}| \leq C_2 R^{n-1} \rho$$

with

$$C_1 = |S| n \, 2^n, \quad C_2 = |S|((3^n - 2^n) 2^n + n).$$

Let us now choose the constant c. We shall use the general notation $c_{R,\rho}, c'_{R,\rho}, \bar{c}_{R,\rho}$ for the constants satisfying

$$|\{x \in T_{R,\rho}, \ \log G \geq c_{R,\rho}\}| \geq \frac{1}{2} |T_{R,\rho}|,$$

$$|\{x \in T'_{R,\rho}, \ \log G \leq c'_{R,\rho}\}| \geq \frac{1}{2} |T'_{R,\rho}|, \tag{1.122}$$

$$|\{x \in T'_{R,\rho} - T_{R,\rho}, \ \log G \leq \bar{c}_{R,\rho}\}| \geq \frac{1}{2} |T'_{R,\rho} - T_{R,\rho}|.$$

Of course, with our notation, we have

$$c'_{R,\rho} = c_{R,2\rho}.$$

We will compare these constants later on. For the time being, we choose $c = \bar{c}_{R,\rho}$ in (1.121), and use the Poincaré inequality to assert that

$$\int_{T'_{R,\rho} - T_{R,\rho}} (\exp \beta (\log G - \bar{c}_{R,\rho})^+ - 1) \, dx$$

$$\leq c_1 \beta R^{1 - \frac{1}{n}} \rho^{\frac{1}{n}} \int_{T'_{R,\rho} - T_{R,\rho}} \exp \beta (\log G - \bar{c}_{R,\rho})^+ |D \log G| \, dx$$

$$\leq c_1 \beta \int_{T'_{R,\rho} - T_{R,\rho}} \exp \beta (\log G - \bar{c}_{R,\rho})^+ \, dx$$

$$+ 4 c_1 \beta R^{2 - \frac{2}{n}} \rho^{\frac{2}{n}} \int_{T'_{R,\rho} - T_{R,\rho}} \exp \beta (\log G - \bar{c}_{R,\rho})^+ |D \log G|^2 \, dx.$$

We pick β such that $1 - c_1 \beta > 0$. We use the last estimate in (1.121) to obtain

$$\int_{T_{R,\rho}} |D \log G|^2 \exp \beta (\log G - \bar{c}_{R,\rho})^+ \, dx$$

$$\leq \frac{C c_0}{(1 - c_1 \beta)(1 - \beta)^2} \frac{R^{n-1}}{\rho} + \frac{4 c_0 c_1 \beta}{(1 - \beta)^2 (1 - c_1 \beta)} \left(\frac{R}{\rho} \right)^{2 - \frac{2}{n}}$$

$$\int_{T'_{R,\rho} - T_{R,\rho}} |D \log G|^2 \exp \beta (\log G - \bar{c}_{R,\rho})^+ \, dx.$$

Filling the hole, we obtain

$$\int_{T_{R,\rho}} |D \log G|^2 \exp \beta (\log G - \bar{c}_{R,\rho})^+ \, dx$$

$$\leq \frac{4 c_0 c_1 \beta}{(1 - \beta)^2 (1 - c_1 \beta)} \left(\frac{R}{\rho} \right)^{2 - \frac{2}{n}} \tag{1.123}$$

$$\int_{T'_{R,\rho}} |D \log G|^2 \exp \beta (\log G - \bar{c}_{R,\rho})^+ \, dx + \frac{C}{\beta} R^{n-2}.$$

Step 2: comparing constants
We prove that

$$|\bar{c}_{R,\rho} - c_{2R+4\rho}| \leq C, \tag{1.124}$$

where the constant does not depend on R, ρ.

We estimate only the positive part $(\bar{c}_{R,\rho} - c_{2R+4\rho})^+$. To estimate the negative part is done in the same way. We write, as is easily seen,

$$(\bar{c}_{R,\rho} - c_{2R+4\rho})^+ |T'_{R,\rho}| \leq \int_{T'_{R,\rho}} (\bar{c}_{R,\rho} - \log G)^+ \, dx$$

$$+ \int_{B_{2R+4\rho}} (\log G - c_{2R+4\rho})^+ \, dx.$$

Using the definition of the numbers $\bar{c}_{R,\rho}, c_{2R+4\rho}$, we may apply the Poincaré inequality to both integrals. We can then refer to Lemma 1.28 to get the desired result. Of course, (1.124) is just an example of an estimate that can be obtained. For example, we also have

$$|c_{2R+4\rho} - c_{2R+2\rho}| \le C.$$

Step 3: obtaining the estimate
We apply (1.123) with $\rho = R/2$. Taking account of (1.124) we obtain

$$\int_{B_{3R}-B_{3R/4}} |D \log G|^2 \exp \beta (\log G - c_{3R})^+ \, dx$$

$$\le C\beta \int_{B_{4R}-B_{R/2}} |D \log G|^2 \exp \beta (\log G - c_{4R})^+ \, dx + C R^{n-2}$$

and also, by splitting the integral on the right-hand side and majorizing,

$$\int_{B_{3R}-B_{3R/4}} |D \log G|^2 \exp \beta (\log G - c_{3R})^+ \, dx$$

$$\le C\beta \int_{B_{4R}-B_R} |D \log G|^2 \exp \beta (\log G - c_{4R})^+ \, dx$$

$$+ C\beta \int_{B_R-B_{R/4}} |D \log G|^2 \exp \beta (\log G - c_R)^+ \, dx + C R^{n-2}.$$

Thus it follows that

$$\frac{1}{(3R)^{n-2}} \int_{B_{3R}-B_{3R/4}} |D \log G|^2 \exp \beta (\log G - c_{3R})^+ \, dx$$

$$\le C\beta \frac{1}{(4R)^{n-2}} \int_{B_{4R}-B_R} |D \log G|^2 \exp \beta (\log G - c_{4R})^+ \, dx$$

$$+ C\beta \frac{1}{R^{n-2}} \int_{B_R-B_{R/4}} |D \log G|^2 \exp \beta (\log G - c_R)^+ \, dx + C.$$

At this stage it is important to recall that in reality, we are working with G^h and not directly with G. Therefore, the quantity

$$\sup_{0<R<R_0} \left(\frac{1}{R^{n-2}} \int_{B_R-B_{R/4}} |D \log G^h|^2 \exp \beta (\log G^h - c_R^h)^+ \, dx \right)$$

is finite, since G^h is in L^∞ and Lemma 1.28 holds. From the previous estimate, picking β_0 sufficiently small it is then easy to conclude that

$$\sup_{0<R<R_0} \left(\frac{1}{R^{n-2}} \int_{B_R-B_{R/4}} |D \log G^h|^2 \exp \beta (\log G^h - c_R^h)^+ \, dx \right) < C.$$

In particular, we have

$$\int_{B_{2R}-B_R} |D \log G^h|^2 \exp \beta (\log G^h - c_{2R}^h)^+ \, dx \leq CR^{n-2}.$$

We write formally, dropping the h,

$$\int_{B_{2R}-B_R} |D \log G|^2 \exp \beta (\log G - c_{2R})^+ \, dx \leq CR^{n-2}.$$

We next replace the constant c_{2R} by $c_{R,0}$ (see (1.122)), and use the Poincaré inequality to get

$$\int_{B_{2R}-B_R} \left(\exp \frac{\beta}{2} (\log G - c_{R,0})^+ - 1 \right)^2 \, dx \leq CR^n,$$

from which we deduce

$$\int_{B_{2R}-B_R} \exp \beta (\log G - c_{R,0})^+ \, dx \leq CR^n$$

and also

$$\int_{B_{2R}-B_R} \exp \beta (\log G - c_{2R})^+ \, dx \leq CR^n,$$

which is the desired result.   ◇

We deduce Harnack's inequality from Lemma 1.30 as follows.

Theorem 1.32. *If (1.105) holds, then*

$$\sup_{T_R} G \leq C \inf_{T_R} G. \tag{1.125}$$

Proof. From Lemma 1.31 we deduce that for β sufficiently small,

$$\fint_{T_R} \exp \beta (\log G - c_{2R})^+ \, dx \leq C.$$

We have also, as is easily seen by splitting integrals and changing c_{2R} into c_{4R},

$$\fint_{T'_R} \exp \beta (\log G - c_{4R})^+ \, dx \leq C.$$

From Lemma 1.29 we can assert that

$$\fint_{T'_R} \exp \beta (\log G - c_{4R})^- \, dx \leq C.$$

Hence for $\beta > 0$ sufficiently small,

$$\fint_{T'_R} \exp \beta (\log G - c_{4R}) \, dx \fint_{T'_R} \exp -\beta (\log G - c_{4R}) \, dx \leq K_\beta,$$

and since clearly the constants cancel,

$$\fint_{T'_R} G^\beta \, dx \fint_{T'_R} G^{-\beta} \, dx \leq K_\beta.$$

We can then write the sequence of inequalities

$$\|G\|^\beta_{L^\infty(T_R)} \leq C_\beta \fint_{T'_R} G^\beta \, dx \leq \frac{C'_\beta}{\fint_{T'_R} G^{-\beta} \, dx} \leq \frac{C''_\beta}{\|G^{-1}\|^\beta_{L^\infty(T_R)}}.$$

Hence

$$\sup_{T_R} G \leq \frac{C}{\sup_{T_R} G^{-1}} = C \inf_{T_R} G.$$

Thus we obtain (1.125).   ◇

1.3.6 Proof of Theorem 1.27

We begin with a consequence of the weak maximum principle.

Lemma 1.33. *The function* $\sup_{\partial B_R} G$ *decreases with* R.

Proof. Again we shall consider only G^h, for h small, and proceed formally with G. Take $R < R'$, and assume that

$$\sup_{\partial B_R} G < \sup_{\partial B_{R'}} G.$$

Then considering the domain $Q - B_R$, it follows that

$$\sup_{Q - B_R} G$$

cannot be attained on the boundary, but this contradicts the maximum principle . Hence the result.   ◇

• Estimate from Above

We need here to work explicitly with G^h. Even formally, the reasoning that we shall perform does not make any sense for G directly. We shall extend G^h by 0 outside Q, and we assume

$$R \geq 2h.$$

Recall the definitions

$$T_R = B_{2R} - B_R,$$
$$T'_R = B_{4R} - B_{R/2}.$$

By Moser's iteration technique we can prove, for any $c \geq 0$,

$$\|(G^h - c)^+\|^2_{L^\infty(T_R)} \leq \frac{C}{R^n} \|(G^h - c)^+\|^2_{L^2(T'_R)}.$$

Thus taking

$$c = \max_{\partial B_{2R}} G$$

and noting that thanks to Lemma 1.33,

$$\text{supp} \left(G^h - \max_{\partial B_{2R}} G^h \right)^+ \subset B_{2R} - B_{R/2},$$

we obtain

$$\left(\max_{\partial B_R} G^h - \max_{\partial B_{2R}} G^h \right)^2 \leq \frac{C}{R^{n-2}} \int_{B_{2R} - B_{R/2}} |DG^h|^2 \, \text{1I}_{G^h > \max_{\partial B_{2R}} G^h} \, dx. \tag{1.126}$$

On the other hand, consider the function

$$\left(G^h - \max_{\partial B_{2R}} G^h \right)^+ \vee \left(\max_{\partial B_{R/2}} G^h - \max_{\partial B_{2R}} G^h \right),$$

whose support is in Q, and hence is a possible test function for (1.109). Note that

$$D \left[\left(G^h - \max_{\partial B_{2R}} G^h \right)^+ \vee \left(\max_{\partial B_{R/2}} G^h - \max_{\partial B_{2R}} G^h \right) \right] = DG^h \, \text{1I}_{G^h > \max_{\partial B_{2R}} G^h}.$$

Therefore, we obtain

$$\int_{B_{2R} - B_{R/2}} |DG^h|^2 \, \text{1I}_{G^h > \max_{\partial B_{2R}} G^h} \, dx \leq C \left(\max_{\partial B_{R/2}} G^h - \max_{\partial B_{2R}} G^h \right),$$

which combined with (1.126) yields the property

$$\left(R^{n-2} \left(\max_{\partial B_R} G^h - \max_{\partial B_{2R}} G^h \right) \right)^2 \leq C R^{n-2} \left(\max_{\partial B_{R/2}} G^h - \max_{\partial B_{2R}} G^h \right). \tag{1.127}$$

By elementary arguments, we can assert that

$$\max_{\partial B_R} G^h - \max_{\partial B_{2R}} G^h \leq \frac{K_\epsilon}{R^{n-2}} + \epsilon \left(\max_{\partial B_{R/2}} G^h - \max_{\partial B_{2R}} G^h \right), \quad \forall \epsilon. \tag{1.128}$$

We apply this relation with

$$R = R_i = h\,2^i, \quad i \geq 1.$$

Setting

$$z_i = R_i^{n-2}\left(\max_{\partial B_{R_i}} G^h - \max_{\partial B_{R_{i+1}}} G^h\right)$$

and picking

$$\epsilon 2^{n-2} = a < 1$$

we deduce from (1.128) the relation

$$z_i \leq K + a z_{i-1}.$$

Therefore, we obtain

$$z_i \leq \frac{K}{1-a} + z_0.$$

To finish we need to estimate

$$z_0 = h^{n-2}\left(\max_{\partial B_h} G^h - \max_{\partial B_{2h}} G^h\right).$$

However, by testing (1.109) with

$$\left(G^h - \max_{\partial B_{2h}} G^h\right)^+$$

we obtain

$$\int_{B_{2h}} |DG^h|^2 1\!\!1_{G^h > \max_{\partial B_{2h}} G^h}\, dx \leq \frac{C}{|B_h|} \int_{B_h} \left(G^h - \max_{\partial B_{2h}} G^h\right)^+ dx.$$

Using Hölder's inequality, then the Poincaré inequality, we easily deduce that z_0 is bounded independently of h.

Hence we have obtained

$$(2^i h)^{n-2}\left(\max_{\partial B_{2^i h}} G^h - \max_{\partial B_{2^{i+1} h}} G^h\right) \leq K, \quad \forall h, i \geq 0.$$

Now setting

$$\zeta_i = (2^i h)^{n-2} \max_{\partial B_{2^i h}} G^h$$

we get

$$\zeta_i \leq K + \frac{1}{2^{n-2}} \zeta_{i+1}.$$

But for i sufficiently large,

$$\zeta_i = 0.$$

Therefore, we have obtained

$$(2^i h)^{n-2} \max_{\partial B_{2^i h}} G^h \leq K, \quad \forall h, i \geq 0.$$

It easily follows that

$$(R)^{n-2} \max_{\partial B_R} G^h \leq K, \quad \forall h, R \geq h.$$

The estimate from above is easily deduced.

• Estimate from Below
Of course, the estimate from below holds only outside a neighborhood of the boundary of Q. To fix ideas, we shall consider balls as in Section 1.3.4:

$$R \leq R_0, \quad \overline{B_{6R_0}} \subset Q.$$

We begin with a lemma.

Lemma 1.34.
$$\int_{T_R} |DG|^2 \, dx \geq cR^{2-n}. \tag{1.129}$$

Proof. We use the test function

$$\theta_R = \begin{cases} 1 \text{ on } B_R, \\ 0 \text{ on } Q - B_{2R}, \end{cases}$$

with

$$|D\theta_R| \leq \frac{C}{R}.$$

We immediately obtain

$$1 \leq C \|DG\|_{L^2(T_R)} R^{\frac{n}{2}-1}.$$

Hence the result. ◇

We proceed with another lemma.

Lemma 1.35.
$$\max_{\partial B_R} G \geq cR^{2-n}. \tag{1.130}$$

Proof. We test with

$$\psi = G \vee \max_{\partial B_R} G$$

and we note that

$$\psi = G, \text{ on } Q - B_R.$$

Therefore,

$$\max_{\partial B_R} G \geq \int_Q a_{ik} D_k G\, D_i \psi\, dx = \int_Q a_{ik} D_k G\, D_i G \mathbb{1}_{\psi = G}\, dx$$

$$\geq C \int_{Q-B_R} |DG|^2\, dx,$$

and from Lemma 1.34 the result follows immediately. $\diamondsuit$

Since

$$\sup_{T_R} G = \max_{\partial B_R} G,$$

we also have

$$\sup_{T_R} G \geq cR^{2-n}.$$

Using Harnack's inequality (1.125) we obtain

$$\inf_{T_R} G \geq cR^{2-n}.$$

This implies the estimate from below. $\diamondsuit$

2. General Regularity Results

2.1 Introduction

We show in this chapter how to obtain general regularity results for solutions of PDEs or systems of PDEs. We shall stress the techniques for obtaining $W^{1,p}$ and C^δ regularity. We shall then derive additional regularity. In the sequel everything is self-contained, except that we shall make use of the $W^{2,p}$ regularity results for the linear Dirichlet problem. For equations we rely heavily on O.A. Ladyzhenskaya, N.N. Ural'tseva [66].

2.2 Obtaining $W^{1,p}$ Regularity

2.2.1 Linear Equations

We consider a domain Ω satisfying

$$\Omega \text{ is a Lipschitz domain.} \tag{2.1}$$

Then the "sphere condition" holds:

$$\forall x_0 \in \Gamma = \partial\Omega, \quad |B_R(x_0) \cap (R^n - \Omega)| \geq cR^n, \quad c > 0. \tag{2.2}$$

Consider a matrix $a(x) \in \mathcal{L}(R^n; R^n)$, which is a measurable function such that

$$\alpha|\xi|^2 \leq a(x)\xi.\xi \leq M|\xi|^2, \quad \forall \xi \in R^n, \, x \in \Omega. \tag{2.3}$$

To the matrix a, we associate the second order differential operator

$$A = -\frac{\partial}{\partial x_i}\left(a_{ij}(x)\frac{\partial}{\partial x_j}\right). \tag{2.4}$$

Let us introduce f_0, f such that

$$f_0 \in L^2(\Omega), \quad f \in (L^r(\Omega))^n, \quad r > 2, \tag{2.5}$$

and consider the variational problem

$$\int_\Omega a(x)Du.D\phi\,dx = \int_\Omega f_0\phi\,dx + \int_\Omega f.D\phi\,dx \quad \forall \phi \in H_0^1(\Omega), u \in H_0^1(\Omega). \tag{2.6}$$

We want to prove the following theorem.

Theorem 2.1. *Assume (2.1), (2.2), (2.3), (2.5). Then there exists ϵ depending only on α, M, Ω, n such that for $2 \leq p < 2 + \min(r - 2, \frac{4}{n-2}, \epsilon)$,*

$$\|u\|_{W_0^{1,p}(\Omega)} \leq c_p(|f_0|_{L^2} + \|f\|_{L^r}).$$

Proof. It is important to use the representation

$$\int_\Omega f_0 \phi \, dx = \int_\Omega h_0 \phi \, dx + \sum_{i=1}^n \int_\Omega h_i D_i \phi \, dx$$

with

$$h_0, h_1, \ldots, h_n \in L^{2^*}(\Omega),$$

where

$$2^* = \frac{2n}{n-2}.$$

We may set

$$\tilde{f}_0 = h_0, \quad \tilde{f}_i = h_i + f_i.$$

The new functions $\tilde{f}_0, \tilde{f}_i$ belong to $L^{\bar{r}}(\Omega)$ with

$$\bar{r} = \min(r, 2^*),$$

and clearly,

$$\|\tilde{f}_0\|_{L^{\bar{r}}}, \quad \|\tilde{f}_i\|_{L^{\bar{r}}} \leq c(|f_0|_{L^2} + \|f\|_{L^r}).$$

It is therefore sufficient to prove the theorem under the additional assumption $f_0 \in L^r$, for it will then hold for the functions $\tilde{f}_0, \tilde{f}_i$, and $\bar{r}$, which will imply the result, since

$$2 + \min\left(r - 2, \frac{4}{n-2}\right) = \bar{r}.$$

Let us consider the balls $B_R(x_0), R \leq R_0$, and $\Omega \subset B_{R_0}(0)$. We consider only those such that

$$|B_{2R}(x_0) \cap \Omega| > 0.$$

We consider a cutoff function τ such that

$$\tau = 1 \text{ if } |x| \leq 1, \quad \tau = 0 \text{ if } |x| \geq 2, \quad 0 \leq \tau \leq 1,$$

and τ is smooth. We set

$$\tau_R(x) = \tau\left(\frac{x - x_0}{R}\right).$$

In (2.6) we take

$$\phi = (u - c_R)\tau_R^2,$$

where c_R has been defined in (1.25), with R replaced by $2R$. Note that this choice is possible, since $\phi \in H_0^1(\Omega)$. We easily deduce

$$\alpha \int_{B_{2R} \cap \Omega} |Du|^2 \tau_R^2 \, dx \leq c \int_{B_{2R} \cap \Omega} |Du| \frac{|u - c_R|}{R} \tau_R \, dx$$

$$+ c \int_{B_{2R} \cap \Omega} |f_0||u - c_R|\tau_R^2 \, dx + c \int_{B_{2R} \cap \Omega} |f| \left(|Du|\tau_R^2 + \frac{|u - c_R|}{R}\tau_R \right) dx$$

We use Hölder's inequality and the Poincaré inequality (1.26) with

$$\lambda = \frac{2n}{n - 1}, \quad \mu = \frac{2n}{n + 1}$$

to obtain

$$\int_{B_{2R} \cap \Omega} |Du||u - c_R| \, dx \leq c \left(\int_{B_{6R} \cap \Omega} |Du|^{2n/(n+1)} \, dx \right)^{(n+1)/n}.$$

We majorize

$$\int_{B_{2R} \cap \Omega} |f_0||u - c_R| \, dx \leq c \int_{B_{2R} \cap \Omega} |f_0|^2 \, dx + c \int_{B_{2R} \cap \Omega} |u - c_R|^2 \, dx$$

and apply the Poincaré inequality (1.26) with

$$\lambda = 2, \quad \mu = 2n/(n + 2)$$

to get

$$\int_{B_{2R} \cap \Omega} |u - c_R|^2 \, dx \leq \left(\int_{B_{6R} \cap \Omega} |Du|^{2n/(n+2)} \, dx \right)^{(n+2)/n}.$$

Similarly,

$$\int_{B_{2R} \cap \Omega} |f||u - c_R| \, dx \leq c \int_{B_{2R} \cap \Omega} |f|^2 \, dx + c \int_{B_{2R} \cap \Omega} |u - c_R|^2 \, dx.$$

The term

$$\int_{B_{2R} \cap \Omega} |f||Du|\tau_R^2 \, dx$$

is majorized using Young's inequality ($ab \leq \eta a^2 + b^2/\eta$), and the term $|Du|^2$ is absorbed by the left-hand side. Collecting results, and noting that $\tau_R = 1$ on $B_R(x_0)$, we can write

$$\int_{B_R \cap \Omega} |Du|^2 \, dx \leq c \left[\int_{B_{6R} \cap \Omega} (|f|^2 + |f_0|^2) \, dx \right.$$

$$+ \frac{1}{R^2} \left(\int_{B_{6R} \cap \Omega} |Du|^{2n/(n+2)} \, dx \right)^{(n+2)/2}$$

$$\left. + \frac{1}{R} \left(\int_{B_{6R} \cap \Omega} |Du|^{2n/(n+1)} \, dx \right)^{(n+1)/n} \right].$$

Hence (cf. Remark 1.16) the conditions of applicability of Theorem 1.15 are valid, and thus the proof is complete.   ◊

We proceed with the mixed boundary value problem. We assume

$$\forall x_0 \in \Gamma_D, |B_R(x_0) \cap \Gamma_D| \geq c R^{n-1},$$

$$\forall x_0 \in \Gamma_N, \text{ if } B_R(x_0) \cap \Gamma_D = \emptyset, \text{ then } |B_R(x_0) \cap \Omega| \geq c R^n.$$

(2.7)

We write

$$H^1_{\Gamma_D}(\Omega) = \{u \in H^1(\Omega) | u = 0, \text{ on } \Gamma_D\}$$

and consider the variational problem

$$\int_\Omega a(x) Du.D\phi \, dx = \int_\Omega f_0 \phi \, dx + \int_\Omega f.D\phi \, dx \quad \forall \phi \in H^1_{\Gamma_D}(\Omega), u \in H^1_{\Gamma_D}(\Omega),$$

(2.8)

and we want to prove the following theorem.

Theorem 2.2. *Assume (2.1), (2.7), (2.3), (2.5). Then there exists ϵ depending only on α, M, Ω, n such that for $2 \leq p < 2 + \min(r - 2, \frac{4}{n-2}, \epsilon)$,*

$$\|u\|_{W^{1,p}_D(\Omega)} \leq c_p(|f_0|_{L^2} + \|f\|_{L^r}).$$

Proof. The proof is exactly similar to that of Theorem 2.1, testing (2.8) with

$$\phi = (u - c_R)\tau_R^2,$$

where c_R is this time defined by (1.29), with R replaced by $2R$. Note that this choice is possible, since $\phi \in H^1_{\Gamma_D}(\Omega)$. We proceed as in the proof of Theorem 2.1, this time using the Poincaré inequality (1.30). We apply Theorem 1.15 again, taking account of Remark 1.16. Thus the result has been proven.   ◊

2.2.2 Nonlinear Problems

We describe here a technique that will be applicable to nonlinear systems. Let us consider a vector function

$$u(x) = (u^1(x), \ldots, u^N(x)).$$

We assume that

$$u \in (H^1_0(\Omega))^N \cap (L^\infty(\Omega))^N.$$

(2.9)

Consider a vector

$$s = (s^1, \ldots, s^N)$$

in R^N and denote by $|s|$ the usual Euclidean norm. We shall also use the notation

$$\|s\| = \max(|s^1|, \ldots, |s^N|).$$

Let us introduce a function on R^N, denoted by $X_0(s)$, that has the following properties:

$$X_0(s) \geq 0, \quad X_0(0) = 0, \tag{2.10}$$

$$\left| \frac{\partial X_0}{\partial s^\nu}(s) \right| \leq |s^\nu| \beta(\|s\|), \tag{2.11}$$

where

$$\beta(\rho) \text{ is positive monotone increasing for } \rho \geq 0. \tag{2.12}$$

Note that (2.12) implies

$$X_0(s) \leq \frac{1}{2} |s|^2 \beta(\|s\|), \tag{2.13}$$

which can be readily seen by writing

$$X_0(s) = \int_0^1 \frac{d}{d\lambda} \frac{\partial X_0}{\partial s^\nu}(\lambda s) s^\nu \, d\lambda$$

and using (2.11), (2.12).

A trivial example of such a function $X_0(s)$ is, of course,

$$X_0(s) = \frac{|s|^2}{2}.$$

To state our fundamental condition, we need to associate with u a vector of constants

$$c = (c^1, \ldots, c^N),$$

which is arbitrary, except for the restriction

$$\|c\| \leq \sup_x \|u(x)\|. \tag{2.14}$$

Let also introduce a function $\psi(x)$ that is subject only to the restrictions

$$\begin{aligned} &\psi(x) \geq 0, \psi \in H^1 \cap L^\infty(\Omega), \\ &\psi(x) = 0 \text{ on } \Gamma \text{ if } c \neq 0. \end{aligned} \tag{2.15}$$

The main condition is the following:

$$\begin{aligned} &\exists k_0, \ K_0 > 0 \text{ depending only on the } L^\infty \text{ norm of } u \text{ such that} \\ &\forall c, \ \psi, \text{ satisfying (2.14), (2.15) one has} \\ &k_0 \int_\Omega |Du|^2 \psi \, dx + \int_\Omega a D X_0(u - c) . D\psi \, dx \leq K_0 \int_\Omega \psi \, dx, \end{aligned} \tag{2.16}$$

where of course,

$$DX_0(u - c)(x) = \sum_\nu \frac{\partial X_0}{\partial s^\nu}(u(x) - c) Du^\nu(x). \tag{2.17}$$

We are going to prove the following theorem.

Theorem 2.3. *If we assume (2.1), (2.2), (2.9), (2.3), (2.10), (2.11), (2.12), and the main condition (2.16), then there exists ϵ, depending only on α, M, Ω, and the L^∞ norm of u, such that for $2 \le p < 2 + \epsilon$, $u \in (W_0^{1,p}(\Omega))^N$, and the $W^{1,p}$ norm depends only on the same constants and on p.*

Proof. We consider, as in the linear case, the balls $B_{2R}(x_0)$ and the cutoff function τ_R. We define c_R as in the linear case with Dirichlet boundary conditions, namely by (1.25), with R replaced by $2R$. Note that c_R is a vector, since u is a vector. We apply the main condition (2.16) with $c = c_R$ and

$$\psi = \tau_R^2.$$

Considering the terms in (2.16), we have

$$k_0 \int |Du|^2 \tau_R^2 \, dx \ge k_0 \int_{Q_R} |Du|^2 \, dx \tag{2.18}$$

and

$$K_0 \int \tau_R^2 \, dx \le CR^n. \tag{2.19}$$

Then we have to estimate

$$\left| \int aDX_0 . D\tau_R \tau_R \, dx \right| \le C \int_{B_{2R} \cap \Omega} |Du| \frac{|u - c_R|}{R} \, dx.$$

As in the linear case, we use Hölder's inequality and the Poincaré inequality (1.26) with

$$\lambda = \frac{2n}{n-1}, \quad \mu = \frac{2n}{n+1}$$

to obtain

$$\int_{B_{2R} \cap \Omega} |Du||u - c_R| \, dx \le c \left(\int_{B_{6R} \cap \Omega} |Du|^{2n/(n+1)} \, dx \right)^{(n+1)/n}.$$

Therefore we have proven, that

$$\left| \int aDX_0 . D\tau_R \tau_R \, dx \right| \le \frac{C}{R} \left(\int_{B_{6R}} |Du|^{2n/(n+1)} \, dx \right)^{(n+1)/n}. \tag{2.20}$$

If we recall (2.18), (2.19), then (2.20) in (2.16) yields

$$\int_{B_R} |Du|^2 \, dx \le CR^n + \frac{C}{R} \left(\int_{B_{6R}} |Du|^{2n/(n+1)} \, dx \right)^{(n+1)/n}, \tag{2.21}$$

and thus we satisfy the conditions of applicability of Theorem 1.15, and the proof is complete.   ◇

Remark 2.4. The main condition is used in the proof of the previous theorem in a relatively trivial way. However, this form will appear naturally in the applications to stochastic games (see Section 3.3).

● **A Variant**

We can state a variant of Theorem 2.3 when we know a priori that the function u is continuous, and not just bounded. This allows us to get rid of the main condition (2.16).

Theorem 2.5. *We assume (2.1), (2.2), (2.3), and let u be such that*

$$u \in (H_0^1(\Omega))^N \cap (C^0(\overline{\Omega}))^N \tag{2.22}$$

and

$$Au^\nu \in L^1(\Omega), \quad |Au^\nu| \le b^\nu + B^\nu |Du(x)|^2, \\ b^\nu, B^\nu \text{ depend only on the } L^\infty \text{ norm of } u. \tag{2.23}$$

Then there exists ϵ, depending only on α, M, Ω, and the modulus of continuity of u, such that for $2 \le p < 2+\epsilon$, $u \in (W_0^{1,p}(\Omega))^N$, and the $W^{1,p}$ norm depends only on the same constants and on p.

Proof. We notice that in obtaining the crucial estimate (2.21) we can restrict ourselves to $R \le R_0$, fixed but arbitrarily small, since for $R > R_0$ the estimate is obvious. We then proceed as in the proof of Theorem 2.3, using the same quantities c_R and

$$\psi = \tau_R^2.$$

We test Au^ν with $(u^\nu - c_R^\nu)\psi$, which yields

$$\int_\Omega Au^\nu (u^\nu - c_R^\nu)\psi \, dx \le \int_\Omega (b^\nu + B^\nu |Du|^2)|u^\nu - c_R^\nu|\psi \, dx.$$

Then we have

$$\int_\Omega a \, Du^\nu . Du^\nu \psi \, dx + \int_\Omega a \, Du^\nu . D\psi (u^\nu - c_R^\nu) \, dx$$
$$\le \int_\Omega (b^\nu + B^\nu |Du|^2)|u^\nu - c_R^\nu|\psi \, dx.$$

Now clearly, from the continuity of u, calling $\delta(.)$ the modulus of continuity of u, we have

$$|u^\nu - c_R^\nu| \le \delta(2R)$$

in $B_{2R}(x_0)$. Then taking R_0 sufficiently small we deduce after easy cancellations the estimate

$$k_0 \int_\Omega |Du|^2 \psi \, dx + \int_\Omega aD \frac{|u - c|^2}{2} . D\psi \, dx \le K_0 \int_\Omega \psi \, dx,$$

and we recover, in fact, the inequality of (2.16). We can then complete the proof as in Theorem 2.3. $\diamond$

Of course, it is not easy to prove continuity a priori. Nevertheless, we shall see that this is the case for harmonic mappings; see Chapter 6.

2.3 Obtaining C^δ Regularity

2.3.1 L^∞ Bounds for Linear Problems

We begin by describing Moser's technique [81]. Consider the Dirichlet problem (2.6), with data

$$f_0 \in L^p, \quad f \in (L^q)^n, \quad p > \frac{n}{2}, \quad q > n. \tag{2.24}$$

Then we state the following theorem.

Theorem 2.6. *If (2.1), (2.2), (2.3), (2.24) hold, then the solution of (2.6) satisfies*

$$\|u\|_{L^\infty} \le c(\|f_0\|_{L^p} + \|f\|_{(L^q)^n}).$$

Proof. In (2.6) take $\phi = |u|^{s-2}u, s \ge 2$. We easily obtain (assuming for the moment that the integrals that we obtain are finite)

$$\frac{\alpha}{2}(s-1)\int_\Omega |Du|^2|u|^{s-2}\,dx \le \int_\Omega |f_0||u|^{s-1}\,dx + \frac{s-1}{2\alpha}\int_\Omega |f|^2|u|^{s-2}\,dx.$$

To simplify the notation, we set

$$v = |u|.$$

Note that from the Poincaré inequality we have

$$\left(\int_\Omega v^{ns/(n-2)}\,dx\right)^{(n-2)/n} \le c\int_\Omega |D\,v^{s/2}|^2\,dx$$

$$\le cs^2\int_\Omega |Du|^2 u^{s-2}\,dx.$$

Collecting results we obtain

$$\|v\|_{ns/(n-2)} \le (c|f_0|_p)^{1/s}s^{1/s}\|v\|_{(p/(p-1))(s-1)}^{(s-1)/s} + (c\|f\|_q)^{1/s}s^{2/s}\|v\|_{(q/(q-2))(s-2)}^{(s-2)/s}. \tag{2.25}$$

Let

$$r = \max\left(\frac{p}{(p-1)}, \frac{q}{(q-2)}\right).$$

Then we deduce easily from (2.25) that

$$\max(1, \|v\|_{ns/(n-2)}) \le (c\max(|f_0|_p, \|f\|_q))^{1/s}s^{2/s}\max(1, \|v\|_{rs}). \tag{2.26}$$

Call $K = c\max(|f_0|_p, \||f|^2|_q)$, and define the sequence

$$s_j = 2a^j$$

with
$$a = \frac{n}{(n-2)r}.$$

We also use the notation
$$z_j = \max(1, \|v\|_{rs_j}).$$

Then we deduce from (2.26) with
$$s = s_j$$

that
$$z_{j+1} \le K^{1/s_j}(s_j)^{2/s_j} z_j,$$

with
$$z_0 = \max(1, \|v\|_{2r}).$$

We know that $a > 1$ from assumption (2.24); in particular, s_j increases and tends to ∞ as $j \to \infty$. Since $v \in H_0^1$ (and hence is in $L^{2n/(n-2)}$) the assumptions on p, q imply $z_0 < \infty$. Therefore, the sequence z_j is by induction well-defined. Moreover, we have
$$z_j \le z_0 \prod_{h=0}^{j-1}(s_h)^{2/s_h} \times K^{1/s_h}.$$

Thanks again to the property $a > 1$, the product is convergent, and hence
$$\max(1, \|v\|_\infty) = \lim_{j \to \infty} z_j$$

is finite. To assert that the integrals considered above are finite, we consider as a test function $|u_l|^{s-2}u_l$, where u_l is the function u truncated when its absolute value is larger than l, namely,
$$u_l = u \mathbb{I}_{\{|u|<l\}}.$$

The proof is then identical. The proof is complete. ◇

Remark 2.7. For the inhomogeneous Dirichlet problem, i.e.,
$$u = \psi \text{ on } \Gamma, \quad \|\psi\|_{L^\infty(\Gamma)} < \infty,$$

the preceding proof has to be adapted using
$$\phi = ((u - \|\psi\|_\infty)^+)^{s-1},$$

since this function vanishes on the boundary.

Remark 2.8. For the mixed boundary value problem $u = 0$ on Γ_D, the proof extends, provided that Γ_D has a positive $(n-1)$-dimensional Lebesgue measure, which is the case in particular when we make assumption (2.7).

There is also a useful trick to obtain an $L^\infty(\Omega)$ bound. See [66], [104]. Consider a function $u \in H^1(\Omega)$. Assume

$$u \leq k_0 \text{ on } \Gamma_D; k_0 > 0 \text{ on } \Gamma_D. \tag{2.27}$$

Set

$$A_k = \{x \in \Omega | u(x) > k\}$$

and

$$\omega_k = |A_k|.$$

Assume also that

$$\int_\Omega |D(u-k)^+|^2 \, dx)^{1/2} \leq C(\omega_k)^{(p-2)/2p}, \quad k \geq k_0, \quad p > n. \tag{2.28}$$

Then there exists λ such that

$$u(x) \leq k_0 + \lambda, \text{ a.e.} \tag{2.29}$$

Indeed, the function ω_k is positive and nonincreasing for $k \geq k_0$. Moreover, for $h \geq k$, we have

$$(h-k)\omega_h \leq \int_\Omega (u-k)^+ \, dx,$$

and by Hölder's inequality,

$$(h-k)\omega_h \leq (\omega_k)^{(n+2)/2n} \left(\int_\Omega ((u-k)^+)^{2n/(n-2)} \, dx \right)^{(n-2)/2n}.$$

Then from the Poincaré inequality,

$$(h-k)\omega_h \leq (\omega_k)^{(n+2)/2n} \left(\int_\Omega |D(u-k)^+|^2 \, dx \right)^{1/2}.$$

Using (2.28) we obtain

$$(h-k)\omega_h \leq C\omega_k^{1+\delta}, \quad \delta > 0, \quad h \geq k \geq k_0 \tag{2.30}$$

with

$$\delta = \frac{1}{n} - \frac{1}{p}.$$

From the following lemma we have

$$\omega_k = 0, \quad k \geq k_0 + 2C(\omega_{k_0})^\delta 2^{1/\delta}, \tag{2.31}$$

and hence the result follows.

Lemma 2.9. *Consider a positive nonincreasing function ω_k such that (2.30) holds. Then property (2.31) holds.*

Proof. Consider the sequence

$$k_{n+1} = k_n + \frac{\lambda}{2^n}$$

(where λ will be chosen later). Suppose we have, for some n,

$$\omega_{k_n} \leq \frac{\omega_{k_0}}{2^{n/\delta}}. \tag{2.32}$$

Then from assumption (2.30) it follows that

$$\omega_{k_{n+1}} \leq \frac{C}{\lambda} \frac{(\omega_{k_0})^{1+\delta}}{2^{n/\delta}},$$

and if we choose

$$\lambda = C(\omega_{k_0})^\delta 2^{1/\delta},$$

we see that (2.32) is satisfied at step $n + 1$. Since clearly, (2.32) holds for $n = 0$, the property holds for any n. But

$$\omega_{k_0 + 2\lambda} \leq \omega_{k_n} \to 0, \quad n \to \infty.$$

The proof is complete.   ◇

2.3.2 C^δ Regularity for Dirichlet Problems

We shall consider a framework that will be applicable to linear as well as nonlinear problems. A major role will be played by the Green function. We begin by recalling the fundamental properties that will be used.

Let Q be a ball such that $\overline{\Omega} \subset Q$, and let $x_0 \in \Omega$. We consider the Green Function $G = G^{x_0}$ relative to x_0; see Section 1.3 or [52] for details. It is the solution of

$$\int_Q aD\phi.DG \, dx = \phi(x_0), \quad \forall \, \phi \in C_0^\infty(Q). \tag{2.33}$$

We have (see Section 1.3)

$$G \in W_0^{1,\mu}(Q), \quad 1 \leq \mu < \frac{n}{n-1}, \quad G \in L^\nu(Q), \quad 1 \leq \nu < \frac{n}{n-2}, \tag{2.34}$$

and if $n \geq 3$, then[2]

$$c_0|x - x_0|^{2-n} \leq G^{x_0}(x) \leq c_1|x - x_0|^{2-n}, \tag{2.35}$$

where c_0, c_1 depend only on α, M. The estimate (2.35) holds in a neighborhood of x_0 whose closure is contained in Q. In particular, it will hold for any $x \in \overline{\Omega}$.

---

[2] for $n = 2$, one should replace $|x - x_0|^{2-n}$ by $-\log|x - x_0|$.

Note that because of the singularity, G does not belong to H^1 or L^∞. This is why it is convenient in dealing with Green functions to use mollifiers obtained as follows. Let $\rho > 0$ and replace (2.33) by

$$\int_Q aD\phi.DG_\rho \, dx = \fint_{B_\rho(x_0)} \phi \, dx, \quad \forall \, \phi \in C_0^\infty(Q). \tag{2.36}$$

Then G^ρ is in $H_0^1 \cap L^\infty(\Omega)$ and converges as $\rho \to \infty$ to G in the various spaces where G makes sense (see (2.34)).

Let us consider a vector function

$$u(x) = (u^1(x), \ldots, u^N(x)),$$

and we assume that

$$u \in (H_0^1(\Omega))^N \cap (L^\infty(\Omega))^N \tag{2.37}$$

as well as

$$Au^\nu - f_0^\nu + \sum_i D_i f_i^\nu \in L^1(\Omega),$$

$$|Au^\nu - f_0^\nu + \sum_i D_i f_i^\nu| \le b^\nu + B^\nu |Du(x)|^2,$$
$$b^\nu, B^\nu \text{ depend only on the } L^\infty \text{ norm of } u, \tag{2.38}$$

$$f_0^\nu \in L^p, \quad f_i^\nu \in L^q, \quad p > \tfrac{n}{2}, \quad q > n.$$

We shall also assume the condition (2.16), or more precisely, a slight extension of it:

$$\exists k_0, K_0 > 0 \text{ depending only on the } L^\infty \text{ norm of } u \text{ such that}$$
$$\forall c, \, \psi, \text{ satisfying (2.14), (2.15),}$$

$$k_0 \int_\Omega |Du|^2 \psi \, dx + \int_\Omega aDX_0(u - c).D\psi \, dx \tag{2.39}$$
$$\le K_0 \int_\Omega ((1 + |f_0| + |f|^2)\psi + |f| \|D\psi|) \, dx,$$

where f_0 represents the vector f_0^ν, and f the matrix f_i^ν. By replacing p by

$$\tilde{p} = \min\left(p, \frac{q}{2}\right)$$

we may assume, without loss of generality, that

$$|f_0| + |f|^2 \in L^p.$$

We are going to prove the following theorem.

Theorem 2.10. *If we assume (2.1), (2.2), (2.3), (2.37), (2.10), (2.11), (2.12), (2.38), and the main condition (2.39), then u belongs to $C^{0,\delta}(\overline{\Omega})$, with $\delta \le \delta_0 < 1$, where δ_0 as well as the $C^{0,\delta}$ norm depend only on the constants α, M, the norms of f_0, f, and the L^∞ norm of u.*

Remark 2.11. This includes the de Giorgi–Nash theorem. See "Application 1" after the proof of Theorem 2.10.

Remark 2.12. In fact, in nonlinear problems the H_0^1 norm will also be estimated by the L^∞ norm. The assumption that u belongs to H_0^1 is nevertheless necessary to give a meaning to (2.6), (2.39). However, the a priori estimate is extremely useful in practice when one uses a method of approximation.

Proof of Theorem 2.10. Let $\tau(x)$ again be a smooth cutoff function such that

$$\tau = 1 \text{ if } |x| \leq 1, \quad \tau = 0 \text{ if } |x| \geq 2, \quad 0 \leq \tau \leq 1.$$

We set

$$\tau_R(x) = \tau\left(\frac{x - x_0}{R}\right).$$

Let us consider balls $B_R(x_0)$ of center x_0 and fixed radius R. We suppose $0 < R \leq R_0$. We shall apply (2.39) with

$$\psi = G\tau_R^2 \tag{2.40}$$

and

$$c = c_R = 0 \text{ if } B_{2R} \cap (R^n - \Omega) \neq \emptyset,$$

$$c_R = \frac{1}{|B_{2R} - B_{R/2}|} \int_{B_{2R}-B_{R/2}} u \, dx \text{ if } B_{2R} \subset \Omega. \tag{2.41}$$

Clearly, all of the conditions on ψ and c as required by (2.14), (2.15) are satisfied, except the regularity. So in fact, one should first use

$$\psi = G^\rho \tau_R^2$$

and then let $\rho \to 0$. To simplify the presentation we shall omit this preliminary step. This does not affect the estimates we achieve, and these are our sole objective.

Consider the various terms in (2.39). To simplify the notation, we do not write Ω in the integrals, by extending u and ψ to 0 outside the domain. We also shall denote all constants by the generic letter C. These constants depend only on the L^∞ norm of u.

We first note that

$$k_0 \int |Du|^2 \psi \, dx \geq C \int_{B_R} |Du|^2 |x - x_0|^{2-n} \, dx$$

$$\geq C \int_{B_{R/2}} |Du|^2 |x - x_0|^{2-n} \, dx. \tag{2.42}$$

Next we show that

$$K_0 \int \left((|f_0| + |f|^2)\psi + |f| \|D\psi\| \right) dx \leq C R^\beta, \quad 0 < \beta < 2. \tag{2.43}$$

In fact, $\beta < 1$ when $f \neq 0$. Indeed, using the properties of Green functions, we have that

$$\int (|f_0| + |f|^2)\psi \, dx \leq C |B_{2R}|^{\frac{1}{\mu'} - \frac{1}{p}} \leq C R^{n\left(\frac{1}{\mu'} - \frac{1}{p}\right)}, \tag{2.44}$$

where μ' is the Hölder conjugate of a number μ, which we must pick such that

$$\frac{n}{n-2} > \mu > \frac{p}{p-1},$$

which is possible, thanks to the assumption on p. Similarly,

$$\int |f| |G_{\tau R}| |D_{\tau R}| \, dx \leq C R^{n\left(\frac{1}{\mu'} - \frac{1}{q}\right) - 1}, \tag{2.45}$$

and μ must satisfy

$$\frac{n}{n-2} > \mu > \frac{qn}{qn - n - q},$$

which is possible, thanks to the assumption on q. Finally,

$$\int |f| \|DG| \tau_R^2 \, dx \leq C R^{n\left(\frac{1}{\nu'} - \frac{1}{q}\right)}, \tag{2.46}$$

and ν must satisfy

$$\frac{n}{n-1} > \nu > \frac{q}{q-1},$$

which is again possible because of the assumption on q. Using (2.44), (2.45), (2.46) in estimating the left-hand side of (2.43), and setting

$$\beta = \min \left[n \left(\frac{1}{\mu'} - \frac{1}{p} \right), n \left(\frac{1}{\mu'} - \frac{1}{q} \right) - 1, n \left(\frac{1}{\nu'} - \frac{1}{q} \right) \right],$$

we obtain (2.43).

We then turn to the term involving X_0. Since

$$D\psi = DG\tau_R^2 + 2G\tau_R D\tau_R,$$

we have two terms. Note also that it readily follows from (2.17), (2.11) that

$$|DX_0(u - c_R)(x)| \leq C|Du| |u - c_R|.$$

Hence

$$\left| \int a DX_0 . D_{\tau R} G_{\tau R} \, dx \right| \leq C \int_{B_{2R} - B_{R/2}} |Du| \frac{|u - c_R|}{R} G \, dx$$

$$\leq C \int_{B_{2R} - B_{R/2}} |Du|^2 |x - x_0|^{2-n} \, dx$$

$$+ C \int_{(B_{2R} - B_{R/2}) \cap \Omega} \frac{|u - c_R|^2}{R^2} |x - x_0|^{2-n} \, dx.$$

But

$$\int_{(B_{2R}-B_{R/2})\cap\Omega} \frac{|u-c_R|^2}{R^2}|x-x_0|^{2-n}\,dx \le \frac{C}{R^n}\int_{(B_{2R}-B_{R/2})\cap\Omega}|u-c_R|^2\,dx.$$

Using the Poincaré inequality (see (1.33)) we have

$$\int_{(B_{2R}-B_{R/2})\cap\Omega}|u-c_R|^2\,dx \le CR^2\int_{B_{2(2m+1)R}-B_{R/2}}|Du|^2\,dx.$$

Hence collecting results, we have

$$\left|\int aDX_0.D\tau_R G\tau_R\,dx\right| \le C\int_{B_{2(2m+1)R}-B_{R/2}}|Du|^2|x-x_0|^{2-n}\,dx. \quad (2.47)$$

We then turn to the most important term (since it involves DG):

$$Z = \int aDG.DX_0\tau_R^2\,dx.$$

We write it as

$$Z = \int aDG.D(X_0\tau_R^2)\,dx - 2\int aDG.D\tau_R X_0\tau_R\,dx,$$

and we can assert that the first term is positive, from the equation giving the Green Function (2.33). Therefore,

$$Z \ge -\frac{C}{R}\int_{(B_{2R}-B_R)\cap\Omega}|DG\|u-c_R|^2\tau_R\,dx,$$

where we have made use of (2.13) in estimating X_0.

We then write this inequality as follows:

$$Z \ge -C\int_{(B_{2R}-B_{R/2})\cap\Omega}G\frac{|u-c_R|^2}{R^2}\,dx$$

$$-C\int_{(B_{2R}-B_R)\cap\Omega}G^{-1}|DG|^2|u-c_R|^2(\tau_R)^2\,dx.$$

The first integral has already been estimated by a term similar to the right-hand side of (2.47). It remains then to estimate the term

$$Y = \int_{(B_{2R}-B_R)\cap\Omega}G^{-1}|DG|^2|u-c_R|^2(\tau_R)^2\,dx.$$

We now introduce a new cutoff function satisfying

$$\xi = \begin{cases} 0 & \text{for } |x| \leq \frac{1}{2}, \\ \tau & \text{for } |x| \geq 1, \end{cases}$$

and we set

$$\xi_R(x) = \xi\left(\frac{x - x_0}{R}\right).$$

We may assume that $\xi_R = \tau_R$ on $B_{2R} - B_R$.

Consider (2.33), the equation defining the Green Function, and take[3]

$$\phi = G^{-1/2}|u - c_R|^2 \xi_R^2.$$

Noting that $\phi(x_0) = 0$, we get

$$\frac{1}{2} \int aDG.DG\, G^{-3/2}|u - c_R|^2 \xi_R^2\, dx$$

$$= \int aD(|u - c_R|^2 \xi_R^2).DG\, G^{-1/2}\, dx. \tag{2.48}$$

We now make use of assumption (2.38), and after easy transformations we have

$$\int Au^\nu(u^\nu - c_R^\nu)G^{1/2}\xi_R^2\, dx = \int aDu^\nu.Du^\nu G^{1/2}\xi_R^2\, dx$$

$$+ \frac{1}{4} \int aD(|u - c_R|^2 \xi_R^2).DGG^{-1/2}\, dx$$

$$+ 2 \int aDu^\nu.D\xi_R(u^\nu - c_R^\nu)G^{1/2}\xi_R\, dx$$

$$- \frac{1}{2} \int aD\xi_R.DG|u - c_R|^2 G^{-1/2}\xi_R\, dx.$$

Using the second part of assumption (2.38), we can check that

$$\int aD(|u - c_R|^2 \xi_R^2).DGG^{-1/2}\, dx \leq 2 \int |f|\|u - c_R\| DG|G^{-1/2}\xi_R^2\, dx$$

$$+ 2 \int aD\xi_R.DG|u - c_R|^2 G^{-1/2}\xi_R\, dx$$

$$+ CR^{(2-n)/2} \int_{B_{2R}-B_{R/2}} |Du|^2\, dx$$

$$+ CR^{1+\frac{n}{2}-\frac{n}{p}} + CR^{\frac{n}{2}-\frac{n}{q}}.$$

But we may write

---

[3] This is formal, since ϕ is not C^∞, but one may proceed with mollifiers.

$$2 \int aD\xi_R.DG|u - c_R|^2 G^{-1/2}\xi_R \, dx \leq C\delta \int |DG|^2 G^{-3/2}|u - c_R|^2\xi_R^2 \, dx$$

$$+\frac{C}{\delta} \int_{(B_{2R}-B_{R/2})\cap\Omega} \frac{|u - c_R|^2}{R^2}G^{1/2} \, dx,$$

where δ is arbitrary. Similarly, we have

$$2 \int |f||u - c_R||DG|G^{-1/2}\xi_R^2 \, dx$$

$$\leq C\delta \int |DG|^2 G^{-3/2}|u - c_R|^2\xi_R^2 \, dx + \frac{C}{\delta}CR^{1+\frac{n}{2}-\frac{n}{2q}}.$$

Combining these three last estimates in (2.48) and using the Poincaré inequality (1.33) we obtain

$$\int |DG|^2 G^{-3/2}|u - c_R|^2\xi_R^2 \, dx$$

$$\leq CR^{(2-n)/2} \int_{B_{2(2m+1)R}-B_{R/2}} |Du|^2 \, dx + CR^{1+\frac{n}{2}-\frac{n}{p}} + CR^{\frac{n}{2}-\frac{n}{q}}.$$

Going back to the definition of Y and recalling that $\xi_R = \tau_R$ on $B_{2R} - B_R$ we have

$$Y \leq CR^{(2-n)/2} \int |DG|^2 G^{-3/2}|u - c_R|^2\xi_R^2 \, dx$$

$$\leq CR^{2-n} \int_{B_{2(2m+1)R}-B_{R/2}} |Du|^2 \, dx + CR^{2-\frac{n}{p}} + CR^{1-\frac{n}{q}}$$

$$\leq C \int_{B_{2(2m+1)R}-B_{R/2}} |Du|^2|x - x_0|^{2-n} \, dx + CR^{2-\frac{n}{p}} + CR^{1-\frac{n}{q}},$$

and turning back to the definition of Z we have

$$Z \geq -C \int_{B_{2(2m+1)R}-B_{R/2}} |Du|^2|x - x_0|^{2-n} \, dx - CR^{2-\frac{n}{p}} - CR^{1-\frac{n}{q}}. \quad (2.49)$$

Using (2.42), (2.43), (2.47), (2.49) in the relation (2.39), with our choice of ψ, and replacing R with $2R$, we obtain the basic estimate

$$\int_{B_R} |Du|^2|x - x_0|^{2-n} \, dx \leq C \int_{B_{4(2m+1)R}-B_R} |Du|^2|x - x_0|^{2-n} \, dx + CR^\beta$$

for any x_0 and any $R \leq R_0$, with constants depending only on the L^∞ norm of u. Now proceeding in the same way, taking $\psi = G$ in (2.39) one also proves that

$$\int |Du|^2|x - x_0|^{2-n} \, dx \leq C,$$

and using the hole-filling technique we complete the proof of the result.   ◇

• Application 1: De Giorgi–Nash Theorem

Returning to the linear problem (2.6), note first that from Theorem 2.6 the property (2.37) is satisfied. The property (2.38) is trivially satisfied with $b, B = 0$ (recall that we are in the scalar case). It remains to check the main assumption (2.39). We define

$$X_0(s) = \frac{s^2}{2}.$$

Taking

$$\phi = (u - c_R)\psi$$

we easily deduce from (2.6) that

$$\int_\Omega aDu.Du\,\psi\,dx + \int_\Omega aDX_0(u-c).D\psi\,dx = \int_\Omega \left(f_0\phi + \sum_i f_i D_i\phi \right) dx,$$

from which (2.39) follows at once, if we recall that u is bounded and $c = c_R$ is bounded in R. Theorem 2.10 reduces to the de Giorgi–Nash theorem.   ◇

• Application 2: The Scalar Case

In the scalar case the condition (2.39) is also automatically satisfied, provided that the solution u is known to be bounded a priori. Indeed, one may pick the function

$$X_0(s) = \frac{\exp\frac{\lambda s^2}{2} - 1}{\lambda}$$

and

$$\phi = X_0'(u - c_R)\psi.$$

We can compute

$$\int_\Omega a(x)Du.D\phi\,dx = \int_\Omega f_0\phi\,dx + \int_\Omega f.D\phi\,dx$$
$$+ \int_\Omega \left(Au - f_0 + \sum_i D_i f_i \right) \phi\,dx,$$

and for λ large enough one can deduce the property (2.39). So the C^δ regularity follows from the $H_0^1 \cap L^\infty(\Omega)$ regularity in the scalar case. This property does not carry over to the case of systems. This result is used in the proof of Theorem 2.25 below   ◇

• A Variant

As in Section 2.2.2 we can state a variant of Theorem 2.10, where we know a priori that the functions are uniformly continuous, and not just bounded. In that case we do not need the condition (2.39).

Theorem 2.13. *If (2.1), (2.2), (2.3), (2.38) hold, and if*

$$u \in (H_0^1(\Omega))^N \cap (C^0(\overline{\Omega}))^N, \tag{2.50}$$

then u also belongs to $C^{0,\delta}(\overline{\Omega})$ for all $\delta \le \delta_0 < 1$, where δ_0 as well as the $C^{0,\delta}$ depend only on the constants α, M, the norms of f_0, f, and as far as the norm is concerned on the modulus of continuity of u.

Proof of Theorem 2.13.
We note that in applying Morrey's theorem (1.6) we can restrict ourselves to $R \le R_0$, fixed but arbitrarily small. Therefore, we proceed as in the proof of Theorem 2.10, where the size of R_0 will be made precise later. In fact, testing

$$Au^\nu - f_0^\nu + \sum_i D_i f_i^\nu$$

with $(u^\nu - c_R^\nu)G\tau_R^2$ yields

$$\int_\Omega (Au^\nu - f_0^\nu + \sum_i D_i f_i^\nu)(u^\nu - c_R^\nu)G\tau_R^2\, dx$$

$$\le \int_\Omega (b^\nu + B^\nu |Du|^2)|u^\nu - c_R^\nu|G\tau_R^2\, dx.$$

Letting

$$\psi = G\tau_R^2$$

we have

$$\int_\Omega a\, Du^\nu . Du^\nu \psi\, dx + \int_\Omega a\, Du^\nu . D\psi(u^\nu - c_R^\nu)\, dx$$

$$\le \int_\Omega (b^\nu + B^\nu |Du|^2)(u^\nu - c_R^\nu)\psi\, dx + \int_\Omega \left(f_0^\nu - \sum_i D_i f_i^\nu \right)(u^\nu - c_R^\nu)\psi\, dx.$$

Now we use continuity as in Theorem 2.5. Recalling that $\delta(.)$ is the modulus of continuity of u, we have

$$|u^\nu - c_R^\nu| \le \delta(2R)$$

in $B_{2R}(x_0)$. Then taking R_0 sufficiently small, we deduce after easy cancellations the estimate

$$k_0 \int_\Omega |Du|^2 \psi\, dx + \int_\Omega aD\frac{|u - c|^2}{2}.D\psi\, dx$$

$$\le K_0 \int_\Omega ((1 + |f_0|)\psi + |f\|D\psi|)\, dx,$$

and we recover the inequality of (2.39). Note that the modulus of continuity plays a role only in the choice of R_0, and not in the constants of the above inequality. We can then proceed as in the proof of Theorem 2.10. To obtain

$$\int_{B_{R_0}} |Du|^2 |x - x_0|^{2-n}\, dx \le C$$

we just test with

$$(u^\nu - c_{R_0}^\nu) G \tau_{R_0}^2$$

and proceed with simplified calculations. This completes the proof.   ◇

2.3.3 C^δ Regularity for Linear Mixed Boundary Value Problems

Assume that

$$\Omega \text{ is a Lipschitz domain,} \tag{2.51}$$

$$\Gamma_D \text{ is made up of a finite number of simply connected pieces,} \tag{2.52}$$

and also assume (2.7). Consider the variational problem

$$\int_\Omega a(x) Du . D\phi\, dx = \int_\Omega f_0 \phi\, dx + \int_\Omega f . D\phi\, dx$$

$$\forall\, \phi \in H^1_{\Gamma_D}(\Omega), \quad u \in H^1_{\Gamma_D}(\Omega), \tag{2.53}$$

and we assume (2.24). We want to prove the following result.

Theorem 2.14. *If (2.7), (2.51), (2.52), (2.3), and (2.24) hold, then the solution of (2.53) belongs to $C^{0,\delta}(\overline{\Omega})$, with $\delta \le \delta_0 < 1$, where δ_0 as well as the $C^{0,\delta}$ norm depend only on the constants α, M, the domain, and the norms of f_0, f.*

Proof. We know that $u \in L^\infty(\Omega)$ (see remark 2.8). Moreover, the interior regularity follows from the Dirichlet case. So in fact, it is sufficient to prove that for any sufficiently small ball B centered on the boundary, we have

$$u'(y) = u(\psi^{-1}(y)) \in C^{0,\delta}(\overline{\Omega^+}),$$

where ψ is the diffeomorphism that straightens the boundary in B.

(a) Reduce the problem to the domain $B \cap \Omega$
We first notice that from the variational problem (2.53) we have

$$\int_{\Omega \cap B} a(x) Du . D\phi\, dx = \int_{\Omega \cap B} f_0 \phi\, dx + \int_{\Omega \cap B} f . D\phi\, dx$$

$$\forall \phi \in H^1_{(\partial B \cap \Omega) \cup (\Gamma_D \cap B)}(B \cap \Omega), \quad u \in H^1_{\Gamma_D \cap B}(B \cap \Omega), \tag{2.54}$$

that is, for all test functions that vanish on $(\partial B \cap \Omega) \cup (\Gamma_D \cap B)$. Of course, now the symmetry between the boundary conditions of the test function ϕ and those of u has been broken.

(b) Change of coordinates
In (2.54) we perform the change of variables

$$x = \psi^{-1}(y).$$

Let

$$J_\psi(x) = \text{matrix } \left(\frac{\partial \psi_i}{\partial x_j}\right),$$

and set

$$u'(y) = u(\psi^{-1}(y)),$$
$$a'(y) = \frac{J_\psi(\psi^{-1}(y))a(\psi^{-1}(y))(J_\psi)^*(\psi^{-1}(y))}{|\det J_\psi(\psi^{-1}(y))|},$$
$$f_0'(y) = f_0(\psi^{-1}(y)), \quad f'(y) = J_\psi(\psi^{-1}(y))f(\psi^{-1}(y)).$$

Notice that

$$u'(y) \in H^1_{\Gamma_D'}.$$

Now if we pick a function $\phi'(y)$ such that

$$\phi'(y) \in H^1_{\psi(\partial B \cap \Omega) \cup \Gamma_D'},$$

then it is easy to check that (2.20) implies

$$\int_{\Omega^+} a'(y)Du'.D\phi' \, dy = \int_{\Omega^+} f_0'\phi' \, dy + \int_{\Omega^+} f'.D\phi' \, dy,$$

$$\forall \phi' \in H^1_{\psi(\partial B \cap \Omega) \cup \Gamma_D'}(\Omega^+), \quad u' \in H^1_{\Gamma_D'}(\Omega^+),$$

(2.55)

where of course, Du' means the gradient with respect to y. Note also that

$$a'(y)\xi.\xi \geq \alpha \min_{\psi(B)} |\det J_\psi(\psi^{-1}(y))| \, |\xi|^2, \quad \forall \xi \in R^n.$$

(c) Reflection procedure
Using the specific feature of Γ', we produce from (2.55) a variational problem in Ω' (for the definition of Γ', Ω' cf. Section 1.3). Writing $y = (y', y_n)$, we define

$$u'(y', y_n) = u'(y', -y_n) \text{ if } y_n < 0,$$
$$\phi'(y', y_n) = \phi'(y', -y_n) \text{ if } y_n < 0.$$

Then we see that

$$u'(y) \in H^1_{\Gamma_D'}(\Omega')$$

and

$$\phi'(y) \in H^1_0(\Omega'), \quad \phi'(y) = 0 \text{ on } \Gamma_D'.$$

Similarly, if $y_n < 0$, define

$$a'_{ii}(y', y_n) = a'_{ii}(y', -y_n), \quad \forall i,$$

$$a'_{ij}(y', y_n) = a'_{ij}(y', -y_n), \quad \forall i \neq j, \quad i, j \neq n,$$

$$a'_{in}(y', y_n) = -a'_{in}(y', -y_n), \quad \forall i \neq n,$$

$$f'_i(y', y_n) = f'_i(y', -y_n), \quad \forall i \neq n,$$

$$f'_n(y', y_n) = -f'_n(y', -y_n).$$

Then from the relation (2.55) we can write

$$\int_{\Omega'} a'(y) Du'.D\phi'\, dy = \int_{\Omega'} f'_0 \phi'\, dy + \int_{\Omega'} f'.D\phi'\, dy,$$

(2.56)

$$\forall \phi' \in H^1_0(\Omega'), \phi' = 0 \text{ on } \Gamma'_D, \quad u' \in H^1(\Omega'), \quad u' = 0 \text{ on } \Gamma'_D.$$

(d) Final step
With (2.56) we are essentially in the Dirichlet case, with some additional features. First, to simplify the notation we drop the prime symbol, and write (2.56) as

$$\int_{\Omega} a(y) Du.D\phi\, dy = \int_{\Omega} f_0 \phi\, dy + \int_{\Omega} f.D\phi\, dy,$$

(2.57)

$$\forall \phi \in H^1_0(\Omega), \phi = 0 \text{ on } \Gamma_D, \quad u \in H^1(\Omega), \quad u = 0 \text{ on } \Gamma_D.$$

Note that Γ_D does not lie on $\partial\Omega$, but in the interior (of course, recall that we are dealing with Ω' and not the original Ω). Moreover, thanks to the assumptions (2.7), (2.52) we can assert that all assumptions of (1.34), (1.35) are satisfied, provided that we keep in mind that we may allow a finite number of pieces. Moreover, by extending u outside Ω, we can assume that u has compact support, and that $u \in H^1(R^n) \cap L^\infty(R^n)$. We next consider the Green Function defined on Ω itself, namely,

$$\int_{\Omega} aD\phi.DG\, dx = \phi(x_0), \quad \forall \phi \in C^\infty_0(\Omega).$$

Here, we notice that we do not take a bigger set Q such that $\overline{\Omega} \subset Q$, as we did in the pure Dirichlet case (see (2.33)). The reason for this will become apparent. The consequence is that the hole-filling technique will give us only

$$u \in C^{0,\delta}(\Omega)$$

(rather than on all of $\overline{\Omega}$). But this is sufficient for our needs, since we are really interested in the Hölder regularity on Γ' (which is a part of Ω', going back to the previous notation). We should also first consider the regularized version of G, namely G_ρ, but we skip this step as usual.

Define the constant c_R as follows:

$$c_R = \begin{cases} \dfrac{1}{|(B_{2R} - B_R) \cap \Omega|} \displaystyle\int_{(B_{2R} - B_R) \cap \Omega} u \, dx & \text{if } B_{2R} \text{ contains no point of } \Gamma_D, \\ 0 & \text{if } B_{2R} \text{ contains a point of } \Gamma_D. \end{cases}$$

Set

$$\psi = G\tau_R^2,$$

where τ_R denotes the usual cutoff function, which is equal to 1 on $B_R(x_0)$, vanishes outside $B_{2R}(x_0)$, and satisfies

$$|D\tau_R| \le \frac{C}{R}.$$

We also take

$$X_0(s) = \frac{s^2}{2},$$

and in (2.57) let

$$\phi = (u - c_R)\psi.$$

Since G vanishes on $\partial\Omega$, so does this function. Thanks to the choice of the constant, this function is admissible for (2.57) (leaving aside the issue of regularity, which we dealt with by taking G_ρ instead of G). We deduce immediately that

$$\int_\Omega aDu.Du\psi \, dx + \int_\Omega aDX_0(u - c).D\psi \, dx = \int_\Omega (f_0\psi + f.D\psi) \, dx,$$

and thus we can proceed as in Theorem 2.10, where we make use of the special Poincaré inequality (1.38). We obtain $C^{0,\delta}(\Omega)$ regularity, and the proof is complete.   ◇

2.3.4 C^δ Regularity in the Case $n = 2$

We explain here how to make use of the results of Section 1.2.4 to prove $C^{0,\delta}$ properties in the case $n = 2$.

Let us consider a vector function

$$u(x) = (u^1(x), \dots, u^N(x))$$

such that

$$u \in (H_0^1(\Omega))^N. \tag{2.58}$$

We assume also

$$\sum_{\nu=1}^N Au^\nu \ge k_0|Du|^2 - K, \quad Au^\nu \le b^\nu + B^\nu|Du(x)|^2. \tag{2.59}$$

Then we can state the following theorem.

Theorem 2.15. *Assume (2.1), (2.2), (2.58), (2.59), and $n = 2$. Assume also that the functions u^ν are Hölder continuous. Then the functions u^ν belong to $C^{0,\delta}(\overline{\Omega})$, with $\delta \leq \delta_0 < 1$, where δ_0 as well as the $C^{0,\delta}$ norm depend only on the constants, α, M, k_0, K, b, B, the domain, and the H_0^1 norm of u. The number δ_0 does not depend on the initial Hölder exponent.*

Proof. Consider again the balls $B_R(x_0)$, $\forall x_0 \in R^n$. Let, as usual, τ_R be a smooth function that equals 1 in $B_R(x_0)$ and whose support is $B_{2R}(x_0)$, such that $|D\tau_R| \leq C/R$. We begin by testing the first inequality (2.59) with τ_R. Using Hölder's inequality,

$$\int_{B_R} |Du|^2 \, dx \leq K \left(\int_{B_{2R} - B_R} |Du|^2 \, dx \right)^{1/2} + KR^n. \qquad (2.60)$$

Now considering $c_R = (c_R^\nu)$ defined by

$$c_R = \begin{cases} \dfrac{1}{|B_{2R} - B_R|} \displaystyle\int_{B_{2R} - B_R} u \, dx & \text{if} \quad B_{2R} \subset \Omega, \\[2ex] 0 & \text{if} \quad B_{2R} \cap (R^n - \Omega) \neq \emptyset, \end{cases}$$

we test the second inequality (2.59) with $(u^\nu - c_R^\nu)\tau_R^2$. We then obtain (we use a generic constant C)

$$\alpha \int_{B_{2R}} |Du^\nu|^2 \tau_R^2 \, dx \leq K_1 \int_{B_{2R}} |u^\nu - c_R^\nu|^2 |D\tau_R|^2 \, dx$$
$$+ K_2 \int_{B_{2R}} |Du|^2 |u^\nu - c_R^\nu| \tau_R^2 \, dx$$
$$+ K_3 \int_{B_{2R}} |u^\nu - c_R^\nu| \tau_R^2 \, dx.$$

We notice that

$$\int_{B_{2R}} |u^\nu - c_R^\nu|^2 |D\tau_R|^2 \, dx \leq \frac{1}{R^2} \int_{B_{2R} - B_R} |u^\nu - c_R^\nu|^2 \, dx$$
$$\leq C \int_{B_{2(2m+1)R} - B_R} |Du^\nu|^2 \, dx,$$

where we have used the Poincaré inequality (1.33).

Now we have

$$\int_{B_{2R}} |u^\nu - c_R^\nu| \tau_R^2 \, dx \leq \frac{1}{2} \int_{B_{2R}} |u^\nu - c_R^\nu|^2 \, dx + CR^n.$$

Collecting results and summing over ν, we derive Cacciopoli's inequality

$$\int_{B_R} |Du|^2\, dx \le K_1 \int_{B_{2qR}-B_R} |Du|^2\, dx + K_2 \int_{B_{2qR}} |Du|^2 |u - c_R|\, dx$$

$$+ K_3 \int_{B_{2qR}} |u - c_R|^2\, dx + K_4 R^n, \quad q \ge 1, \quad K_1 \ge 1.$$

(2.61)

Thanks to (2.60) and (2.61), the conditions of applicability of Theorem 1.24 are satisfied. Thus we obtain the result.   ◇

2.4 Maximum Principle

2.4.1 Assumptions

We consider a domain Ω as described in Section 1.1.3 and the operator A defined in (2.4). Let u be a scalar function such that

$$u \in H_0^1(\Omega) \cup L^\infty(\Omega) \tag{2.62}$$

and

$$Au \le \lambda + \lambda_0(\rho)|Du|^2 - cu, \tag{2.63}$$

where

$$\begin{aligned}
&\lambda,\ c \text{ are positive constants,}\\
&\lambda_0(\rho) \text{ is positive monotone increasing,}\\
&\rho = \sup_x |u(x)|.
\end{aligned} \tag{2.64}$$

The meaning of the inequality (2.63) must be taken in the sense of two elements of $H^{-1}(\Omega)$ (dual of $H_0^1(\Omega)$). The weak maximum principle is the following result.

Theorem 2.16. *If we assume (2.3), (2.4) and (2.63), (2.64), then we have*

$$u \le \frac{\lambda}{c}. \tag{2.65}$$

Before we present the proof we can recall that in the case of more regularity, namely $u \in C^2(\Omega)$, we can refer to the strong maximum principle. Indeed, (2.63) holds for any point in Ω. Suppose there exists a point $x \in \Omega$ with $u(x) > 0$ (otherwise, the property (2.65) is automatically satisfied). Then, take x_0 to be a maximum point of u in Ω. At such a point, $Du(x_0) = 0$, and $Au(x_0) \ge 0$; hence $u(x_0) \le \lambda/c$, and since $u(x_0) = \max u(x)$, the result follows.

Remark 2.17. Since the value of λ_0 does not matter, we have the same result when assumption (2.63) is replaced with

$$Au \le \lambda + \lambda_0(\rho)|Du|^2 + \lambda_1(\rho)|Du|^\delta - cu \tag{2.66}$$

where

$$\delta < 2. \tag{2.67}$$

This is not a more general result. Indeed, from (2.66) and Young's inequality, it follows that

$$Au \le \lambda + 2 - \delta/2\lambda_1 \epsilon^{\delta/(2-\delta)} + \left(\lambda_0 + \lambda_1\frac{\delta}{2\epsilon}\right)|Du|^2 - cu,$$

and thus we may infer that

$$u \le \frac{1}{c}\left(\lambda + \frac{2-\delta}{2}\lambda_1\epsilon^{\delta/(2-\delta)}\right).$$

Since ϵ is arbitrary, we obtain the same result.

2.4.2 Proof of Theorem 2.16

Consider the function

$$E = \exp\frac{\lambda_0}{\alpha}\left(u - \frac{\lambda}{c}\right),$$

which belongs to $H^1 \cap L^\infty(\Omega)$. We can compute

$$AE = \frac{\lambda_0}{\alpha}E\,Au - \left(\frac{\lambda_0}{\alpha}\right)^2 E\,aDu.Du.$$

The function AE is defined as an element of the dual of $H_0^1 \cap L^\infty(\Omega)$.
From the assumption (2.63) it follows that

$$AE \le \frac{\lambda_0}{\alpha}E(\lambda - cu).$$

We test this relation with $(E-1)^+$. This last function belongs to $H_0^1 \cap L^\infty(\Omega)$, and thus the test is possible. We deduce immediately

$$\int_\Omega aD(E-1)^+.D(E-1)^+\,dx + \frac{\lambda_0}{\alpha}\int_\Omega E(E-1)^+(cu-\lambda)^+\,dx,$$

and thus necessarily

$$(E-1)^+(cu-\lambda)^+ = 0 \text{ a.e.,}$$

which means also that

$$(cu-\lambda)^+ = 0 \text{ a.e.,}$$

and thus the result is proven. $\diamond$

2.5 More Regularity

2.5.1 From C^δ and W^{1,p_0}, $p_0 > 2$, to H^2_{loc}

Let us consider functions $a_i(x, s, \xi)$ defined on $R^n \times R \times R^n$ satisfying the assumptions

$$|D_x a_i(x, s, \xi)| \leq a_i^0(x) + a_i^1|\xi|^2, \quad a_i^0 \in L^2(\Omega), \quad a_i > 0 \text{ constant},$$

$$|D_s a_i(x, s, \xi)| \leq b_i^0(x) + b_1|\xi|, \quad b_i^0 \in L^4(\Omega), \quad b_i^1 > 0, \tag{2.68}$$

$$|D_\xi a_i(x, s, \xi)| \leq c_i \tag{2.69}$$

$(a(x, s, \xi) - a(x, s, \eta)).(\xi - \eta) \geq \alpha|\xi - \eta|^2, \alpha > 0.$
Suppose

$$f \in L^2(\Omega). \tag{2.70}$$

We consider a function u satisfying

$$u \in H^1(\Omega) \cap C^{0,\delta}(\Omega), \quad \delta > 0, \tag{2.71}$$

such that
$$\begin{aligned} &\operatorname{div} a(x, u, Du) + f \in L^1(\Omega), \\ &|\operatorname{div} a(x, u, Du) + f| \leq k_0 + k_1|Du|^2, \text{ a.e.} \end{aligned} \tag{2.72}$$

Our first objective is to prove an a priori estimate.

Theorem 2.18. *If (2.68), (2.69), (2.70), (2.71), and (2.72) hold and if the function u satisfies*

$$u \in H^2_{\text{loc}}(\Omega), \tag{2.73}$$

then for any subset ω with $\overline{\omega} \subset \Omega$,

$$\|u\|_{H^2_\omega} \leq c_\omega(\|u\|_{H^1}, [u]_\delta), \tag{2.74}$$

where c_ω is a function of the arguments, bounded when the arguments are bounded, and depending only on the constants entering into the assumptions and on the data.

Proof. First we consider the subdomain Ω_ρ. For $x_0 \in \Omega_\rho$, define the cube $Q_R(x_0)$, of center x_0 and side-length $2R$. We pick

$$R \leq \frac{\rho}{2\sqrt{n}}$$

Such that $Q_{4R}(x_0) \subset \Omega$. Let τ_R be a smooth cutoff function such that

$$\tau_R = 1 \text{ on } Q_{2R}, \quad \tau_R = 0 \text{ outside } Q_{4R}, \quad |D\tau_R| \leq \frac{C}{R}.$$

We first notice that from the assumption (2.72), we have

$$\left| \int_\Omega (D_i a_i + f) D_k(\tau_R^2 D_k u) \, dx \right| \le \int_\Omega (k_0 + k_1 |Du|^2) |D_k(\tau_R^2 D_k u)| \, dx. \quad (2.75)$$

Now, since τ_R and $D\tau_R$ vanish on $\partial\Omega$, we may perform two integrations by parts to obtain

$$\int_\Omega (D_i a_i + f) D_k(\tau_R^2 D_k u) \, dx$$

$$= \int_\Omega (D_k a_i \, D_i(\tau_R^2 D_k u) + f D_k(\tau_R^2 D_k u)) \, dx. \quad (2.76)$$

We make use of the monotonicity assumption (2.69) to obtain, by classical arguments,

$$\sum_{ik} D_k(a_i(x, u, Du)) D_i D_k u \ge \alpha \sum_{ij} |D_i D_k u|^2 + \sum_{ik} (D_{x_k} a_i(x, u, Du)) D_i D_k u$$

$$+ D_s a_i(x, u, Du) D_k u D_i D_k u). \quad (2.77)$$

From (2.77) and (2.75), (2.76), it then follows that

$$\alpha \sum_{ik} \int_\Omega \tau_R^2 |D_i D_k u|^2 \, dx + \int_\Omega \tau_R^2 (D_{x_k} a_i + D_s a_i D_k u) D_i D_k u \, dx$$

$$+ \sum_{ik} \int_\Omega \left(D_{x_k} a_i + D_s a_i D_k u + \sum_l D_{\xi_l} a_i D_k D_l u \right) D_k u D_i \tau_R^2 \, dx \quad (2.78)$$

$$+ \int_\Omega f D_k(\tau_R^2 D_k u) \, dx \le \int_\Omega (k_0 + k_1 |Du|^2) |D_k(\tau_R^2 D_k u)| \, dx.$$

We make use of the assumptions (2.68), (2.70) in inequality (2.78) to obtain after some easy steps

$$\int_\Omega \tau_R^2 |D^2 u|^2 \, dx \le C \int_\Omega \tau_R^2 |Du|^4 \, dx + \frac{C}{R^2} \int_{\Omega \cap Q_{4R}} \phi \, dx,$$

in which the constants depend on the parameters, and

$$\phi = |Du|^2 + |f|^2 + \sum_i ((a_i^0)^2 + (b_i^0)^4) + 1,$$

hence also

$$\int_{\Omega \cap Q_{2R}} |D^2 u|^2 \, dx \le C \int_{\Omega \cap Q_{4R}} |Du|^4 \, dx + \frac{C}{R^2} \int_{\Omega \cap Q_{4R}} \phi \, dx. \quad (2.79)$$

Now we make use of the Miranda–Nirenberg interpolation result, see (1.18), with $q = 4, \theta = \frac{1}{2}$ to obtain

$$\int_{\Omega \cap Q_{4R}} |Du|^4 \, dx \leq C \left(\int_{\Omega \cap Q_{4R}} |D^2u|^{2n/(n+2\delta)} \, dx \right)^{(n+2\delta)/n} [u]^2_{\delta, Q_{4R}} + C[u]^4_{\delta, Q_{4R}},$$

and using Hölder's inequality yields

$$\int_{\Omega \cap Q_{4R}} |Du|^4 \, dx \leq CR^{2\delta} \int_{\Omega \cap Q_{4R}} |D^2u|^2 \, dx \times [u]^2_{\delta, Q_{4R}} + C[u]^4_{\delta, Q_{4R}}. \quad (2.80)$$

So using (2.80) in (2.79) yields

$$\int_{\Omega \cap Q_{2R}} |D^2u|^2 \, dx \leq C_0 R^{2\delta} \int_{\Omega \cap Q_{4R}} |D^2u|^2 \, dx + \frac{C}{R^2}, \quad (2.81)$$

where the constants depend on the H^1 and $C^{0,\delta}$ norms of u, and on the data, but not on R or x_0. We now take the sup in $x_0 \in \Omega_\rho$, R fixed, and we notice that

$$\sup_{x_0 \in \Omega_\rho} \int_{\Omega \cap Q_{4R}(x_0)} |D^2u|^2 \, dx \leq 4^n \sup_{x_0 \in \Omega_\rho} \int_{\Omega \cap Q_R(x_0)} |D^2u|^2 \, dx,$$

where K is independent of R. We now choose R sufficiently small that

$$C_0 R^{2\delta} 4^n < 1.$$

Then it follows from (2.81) that

$$\sup_{x_0 \in \Omega_\rho} \int_{\Omega \cap Q_R(x_0)} |D^2u|^2 \, dx \leq K_R.$$

Since we can cover Ω_ρ with a finite number of such cubes, the result follows.
$\diamond$

Remark 2.19. There is an elementary way to use $C^{0,\delta}$ continuity, avoiding the use of the Miranda–Nirenberg interpolation result; see [25]. Consider the quantity

$$\int_\Omega \tau_R^2 |Du|^4 \, dx.$$

Recall that τ_R vanishes on the boundary of Ω. So we may write, by an integration by parts,

$$\int_\Omega \tau_R^2 |Du|^4 \, dx \leq \frac{3}{2} \int_\Omega |u - u(x_0)| \tau_R^2 |D^2u|^2 \, dx$$

$$+ \frac{5}{2} \int_\Omega |u - u(x_0)| \tau_R^2 |Du|^4 \, dx \quad (2.82)$$

$$+ \int_\Omega |u - u(x_0)| |D\tau_R|^2 |Du|^2 \, dx.$$

Note that from $C^{0,\delta}$,

$$\tau_R^2 |u(x) - u(x_0)| \le (2R)^\delta \sqrt{n} [u]_\delta \tau_R^2.$$

Imposing first $\frac{5}{2}(2R)^\delta \sqrt{n}[u]_\delta < \frac{1}{2}$, we deduce from (2.82) that

$$\frac{1}{2} \int_\Omega \tau_R^2 |Du|^4 \, dx \le \frac{3}{2}(2R)^\delta \sqrt{n}[u]_\delta \int_\Omega \tau_R^2 |D^2 u|^2 \, dx$$
$$+ 2R\sqrt{n}[u]_\delta \int_\Omega |D\tau_R|^2 |Du|^2 \, dx,$$

and thus we recover (2.81) and we can proceed.

Our next objective is to prove the regularity in

$$H_{\text{loc}}^2(\Omega)$$

in a particular case. Namely, we shall assume that

$$k_1 = 0 \text{ in } (2.72) \tag{2.83}$$

and in (2.68) we strengthen the first part as follows:

$$|D_x a_i(x, s, \xi)| \le a_i^0(x) + a_i^1|\xi|, a_i^0 \in L^2(\Omega), a_i^1 \in L^4(\Omega) \tag{2.84}$$

(clearly, a stronger assumption). Moreover, we assume

$$|a_i(x, s, \xi) - a_i(x', s, \xi)| \le \ell_i |x - x'|^\delta (1 + |\xi|). \tag{2.85}$$

Furthermore,

$$u \in W^{1,p_0}(\Omega), \quad p_0 > 2. \tag{2.86}$$

We then state the following theorem.

Theorem 2.20. *We make the assumptions of Theorem (2.18) and in addition, (2.83), (2.84), (2.85), (2.86). Then we have*

$$u \in H_{\text{loc}}^2(\Omega). \tag{2.87}$$

The point is to find an adequate approximation for the solution. It will be obtained through finite differences.

We proceed with intermediate results. Consider, in addition to Ω_ρ, the additional subdomain $\Omega_{2\rho}$, and let ζ be a cutoff function that is equal to 1 on $\Omega_{2\rho}$, 0 on $R^n - \Omega_\rho$, and is smooth and takes values in $[0, 1]$. We shall consider the finite difference operators Δ_{te_k}, e_k being the kth unit coordinate vector, and $0 \le t \le \rho$. We are going to prove the following lemma.

Lemma 2.21. *Assume in addition to all the assumptions of Theorem 2.20 that*

$$u \in W^{1,p}, \quad p \ge p_0.$$

Then $\forall \theta$, $0 < \theta < (p-2)/2$, one has

$$u \in W^{1+(\theta+(1-\theta)\delta),2}(\Omega_{2\rho}). \tag{2.88}$$

Proof. Computations similar to those done in the proof of the a priori estimate can be done with finite differences. So take $\Delta_{-te_k}(\zeta^2 \Delta_{te_k} u)$. From the assumption (2.72) completed with (2.83), we have

$$\left| \int_\Omega (D_i a_i + f) \Delta_{-te_k}(\zeta^2 \Delta_{te_k} u)\, dx \right| \le \int_\Omega k_0 |\Delta_{-te_k}(\zeta^2 \Delta_{te_k} u)|\, dx.$$

Now since $\Delta_{-te_k}(\zeta^2 \Delta_{te_k} u)$ vanishes on $\partial\Omega$, we may perform an integration by parts and a discrete integration by parts to obtain

$$\int_\Omega (D_i a_i + f) \Delta_{-te_k}(\zeta^2 \Delta_{te_k} u)\, dx$$
$$= \int_\Omega (-\Delta_{te_k} a_i\, D_i(\zeta^2 \Delta_{te_k} u) + f\Delta_{-te_k}(\zeta^2 \Delta_{te_k} u))\, dx.$$

We make use of the monotonicity assumption (2.69) to obtain

$$\sum_{ik} \Delta_{te_k} a_i \Delta_{te_k} D_i u \ge \alpha \sum_k |\Delta_{te_k} Du|^2$$
$$+ \sum_{ik} (a_i(x + te_k, u(x + te_k), Du) - a_i(x, u(x + te_k), Du))\Delta_{te_k} D_i u$$
$$+ \sum_{ik} \int_0^1 D_s a_i(x, u + \lambda\Delta_{te_k} u, Du)\Delta_{te_k} u \Delta_{te_k} D_i u\, d\lambda.$$

From (2.77) and (2.75), (2.76), it then follows that

$$\alpha \sum_k \int_\Omega \zeta^2 |\Delta_{te_k} Du|^2\, dx + \sum_{ik} \int_\Omega \zeta^2 (a_i(x + te_k, u(x + te_k), Du)$$
$$- a_i(x, u(x + te_k), Du))\, dx$$
$$+ \sum_{ik} \int_\Omega \zeta^2 \int_0^1 D_s a_i(x, u + \lambda\Delta_{te_k} u, Du)\Delta_{te_k} u\, \Delta_{te_k} D_i u\, d\lambda\, dx \quad (2.89)$$
$$+ \sum_{ik} \int_\Omega \Delta_{te_k} a_i\, D_i(\zeta^2)\Delta_{te_k} u\, dx$$
$$\le \int_\Omega (|f| + k_0)|\Delta_{-te_k}(\zeta^2 \Delta_{te_k} u)|\, dx.$$

We notice the formula

$$\int_\Omega (k_0 + |f|)|\Delta_{-te_k}(\zeta^2 \Delta_{te_k} u)|\, dx$$
$$\le t \int_{\Omega_\rho} \int_0^1 (k_0 + |f|(x + \lambda te_k))|D_k(\zeta^2 \Delta_{te_k} u)|\, dx\, d\lambda.$$

We then make use of the assumptions (2.68), (2.70) and the above formula in inequality (2.89) to obtain

$$\sum_k \int_\Omega \zeta^2 |\Delta_{te_k} Du|^2 \, dx \le C \sum_{ik} \int_{\Omega_\rho} (a_i(x + te_k, u(x + te_k), Du)$$

$$- a_i(x, u(x + te_k), Du))^2 \, dx$$

$$+ C \int_{\Omega_\rho} |Du|^2 |\Delta_{te_k} u|^2 \, dx \qquad (2.90)$$

$$+ C \left(\int_{\Omega_\rho} |\Delta_{te_k} u|^4 \, dx \right)^{1/2} + Ct^2,$$

where the constants depend only on

$$\|u\|_{H^1 \cap L^\infty}, \quad |f|_{L^2}, \quad \|b_i^0\|_{L^4}, \quad \|a_i^0\|_{L^2}, \quad \|a_i^1\|_{L^2}, \quad k_0, \quad \alpha,$$

and not on p, nor on $\|a_i^1\|_{L^4}$.

Consider, to begin with,

$$\int_{\Omega_\rho} |Du|^2 |\Delta_{te_k} u|^2 \, dx \le (\|Du\|_{L^p(\Omega_\rho)})^2 \left(\int_{\Omega_\rho} |\Delta_{te_k} u|^{2p/(p-2)} \, dx \right)^{(p-2)/p}.$$

Introduce $0 < \theta < 1$, and write

$$\left(\int_{\Omega_\rho} |\Delta_{te_k} u|^{2p/(p-2)} \, dx \right)^{(p-2)/p}$$

$$= \left(\int_{\Omega_\rho} |\Delta_{te_k} u|^{2p\theta/(p-2)} |\Delta_{te_k} u|^{\frac{2p(1-\theta)}{p-2}} \, dx \right)^{(p-2)/p}.$$

Now pick $\theta < (p-2)/2$. We can use Hölder's inequality and the $C^{0,\delta}$ regularity to deduce

$$\left(\int_{\Omega_\rho} |\Delta_{te_k} u|^{2p/(p-2)} \right)^{(p-2)/p} \le t^{2\delta(1-\theta)} [u]_\delta^{2(1-\theta)} \left(\int_{\Omega_\rho} |\Delta_{te_k} u|^p \, dx \right)^{2\theta/p}.$$

Since

$$\left(\int_{\Omega_\rho} |\Delta_{te_k} u|^p \, dx \right)^{2\theta/p} \le t^{2\theta} (\|Du\|_{L^p(\Omega)})^{2\theta},$$

we can conclude that

$$\int_{\Omega_\rho} |Du|^2 |\Delta_{te_k} u|^2 \, dx \le t^{2\theta + 2\delta(1-\theta)} (\|Du\|_{L^p(\Omega)})^{2+2\theta} ([u]_\delta)^{2(1-\theta)}.$$

The other terms on the right-hand side of (2.90) are estimated in a similar way. Note that in estimating the first term on the right-hand side the assumptions (2.84), (2.85) are used. It follows that

$$\sum_k \int_{\Omega_{2\rho}} |\Delta_{le_k} Du|^2 \, dx \leq C t^{2\theta + 2\delta(1-\theta)},$$

and this is the same as (2.88). The lemma has been proven.   ◇

Proof of Theorem 2.20. The key point is to prove that

$$u \in W_{\text{loc}}^{1,4}(\Omega). \tag{2.91}$$

Indeed, if (2.91) is done, then the right-hand side of (2.90) is estimated by t^2, and the estimate (2.90), after dividing by t^2 and letting t tend to 0, yields the expected result.

Now from Lemma 2.21 and assumption (2.86), we know that

$$Du \in W^{\theta_0 + \delta(1-\theta_0),2}(\Omega_\rho), \quad \text{with } \theta_0 < \frac{p_0 - 2}{2}.$$

Set

$$\theta_1 = \theta_0 + \delta(1 - \theta_0).$$

From the inclusion property (2.86), we know that

$$Du \in L^{p_1}(\Omega_\rho) \quad \forall p_1 < \frac{2n(1 + \theta_1)}{n - 2\delta\theta_1},$$

so if

$$\frac{2n(1 + \theta_1)}{n - 2\delta\theta_1} > 4,$$

we are done. Otherwise, pick p_1 with

$$2(1 + \theta_1) < p_1 < \frac{2n(1 + \theta_1)}{n - 2\delta\theta_1},$$

which is possible. We can reapply the lemma, changing Ω into Ω_ρ, p_0 into p_1, and θ_1 for θ_0. Setting

$$\theta_2 = \theta_1 + \delta(1 - \theta_1)$$

we obtain that

$$Du \in W^{\theta_2}(\Omega_{2\rho}).$$

We can proceed, defining a sequence p_k, θ_k with

$$\theta_k = \theta_{k-1} + \delta(1 - \theta_{k-1}),$$

$$2(1 + \theta_k) < p_k < \frac{2n(1 + \theta_k)}{n - 2\delta\theta_k},$$

and

$$Du \in W^{\theta_k}(\Omega_{k\rho}).$$

Since $\theta_k \to 1$ and

$$\frac{2n(1+\theta_k)}{n-2\delta\theta_k} \to \frac{4n}{n-2\delta} > 4,$$

we can, after a finite number of steps, find a $p_k \geq 4$, and hence (2.91) is obtained. This completes the proof.   ◇

Remark 2.22. The bootstrap argument used in the proof is due to J. Naumann [83].

• **Application**
Pick

$$a_i(x, s, \xi) = \sum_j a_{ij}(x)\xi_j$$

with

$$\sum_{ij} a_{ij}\xi_j\xi_i \geq \alpha|\xi|^2, \forall\xi, \quad a_{ij} \in C^{0,\delta} \cap W^{1,4}(\Omega).$$

Then the assumptions are satisfied.

2.5.2 Using the Linear Theory of Regularity

Consider the linear problem of Section 2.2.1. We recall briefly that

$$\alpha|\xi|^2 \leq a(x)\xi.\xi \leq M|\xi|^2, \quad \forall\,\xi \in R^n, \; x \in \Omega, \tag{2.92}$$

and

$$A = -\frac{\partial}{\partial x_i}\left(a_{ij}(x)\frac{\partial}{\partial x_j}\right). \tag{2.93}$$

We next assume that

$$\Omega \text{ is of class } C^{2,\delta}, \quad \delta \in (0,1), \tag{2.94}$$

$$a(x) \in W^{1,\infty}(\Omega). \tag{2.95}$$

Also, let

$$f \in L^p(\Omega), \quad p > 1. \tag{2.96}$$

Consider the variational problem

$$\int_\Omega a(x)Du.D\phi\,dx = \int_\Omega f.D\phi\,dx,$$

$$\forall\phi \in W_0^{1,p'}(\Omega), \quad u \in W_0^{1,p}(\Omega). \tag{2.97}$$

Then one has the fundamental result

$$u \in W^{2,p}(\Omega). \tag{2.98}$$

In the case of nonvariational problems written as

$$-a_{i,j}\frac{\partial^2 u}{\partial x_i \partial x_j} = f, \quad u \in W^{2,p}(\Omega) \cap W_0^{1,p}(\Omega), \tag{2.99}$$

the assumption (2.95) can be replaced by

$$a_{i,j} \in C^0(\overline{\Omega}), \tag{2.100}$$

and there exists one and only one solution of (2.99).

Let us see how we can make use of this result to derive more regularity results for the vector functions u considered in Theorem 2.10.

Theorem 2.23. *We assume (2.92), (2.93), (2.94), (2.95), (2.37), (2.38) with*

$$f^\nu = 0, \quad f_0^\nu \in L^\infty(\Omega).$$

We also assume (2.39) with X_0 satisfying (2.10), (2.11), (2.12). Then one has

$$u^\nu \in W^{2,p}(\Omega), \quad \forall 1 \le p < \infty.$$

Proof. We note first that Theorems 2.3 and 2.10 apply, so

$$\exists p_0 > 2 \text{ such that } u \in (W_0^{1,p_0}(\Omega))^N$$

and

$$u \in (C^\delta(\overline{\Omega}))^N, \quad 0 < \delta < 1.$$

Now using (2.38) we see that

$$A u^\nu \in L^{p_0/2}(\Omega).$$

From the linear theory, see (2.98), we can assert that

$$u^\nu \in W^{2,p_0/2}(\Omega).$$

We can then make use of the Miranda–Nirenberg interpolation result (1.17) applied with $\theta = \frac{1}{2}$ to assert that

$$u^\nu \in W_0^{1,p_1}(\Omega), \quad \frac{1}{p_1} = \frac{1}{p_0} - \frac{\delta}{2n},$$

and hence

$$p_1 > p_0.$$

Of course, this supposes that

$$p_0 < \frac{2n}{\delta}.$$

We can then proceed with a bootstrap argument, showing that

$$u^\nu \in W_0^{1,p_i}(\Omega), \quad \frac{1}{p_i} = \frac{1}{p_{i-1}} - \frac{\delta}{2n},$$

as long as the sequence is smaller than $2n/\delta$. Necessarily there is an i such that $p_i \geq 2n/\delta$. Therefore, it follows that

$$u^\nu \in W_0^{1,p}(\Omega), \quad p > 2n.$$

But then from the linear theory

$$u^\nu \in W^{2,s}(\Omega), \quad s > n.$$

From Sobolev embedding we have $u^\nu \in W^{1,r}$, $\forall r$, and from the linear theory again the result follows. $\diamond$

We can again give a variant of Theorem 2.23 when we know that u is continuous and not just bounded.

Theorem 2.24. *If (2.92), (2.93), (2.94), (2.95), (2.22), and (2.23) hold, then one has*

$$u^\nu \in W^{2,p}(\Omega), \quad \forall 1 \leq p < \infty.$$

Proof. The proof is identical to that of Theorem 2.23, since we know that Theorems 2.5 and 2.13 apply. We can then proceed, using the Miranda–Nirenberg interpolation result and a bootstrap argument. $\diamond$

2.5.3 Full Regularity for a General Quasilinear Scalar Equation

We consider as in Section 2.5.1 functions $a_i(x, s, \xi)$ defined on $R^n \times R \times R^n$, and we use the notation

$$a_{i,x_k}(x, s, \xi) = \frac{\partial a_i}{\partial x_k}(x, s, \xi),$$

$$a_{i,s}(x, s, \xi) = \frac{\partial a_i}{\partial s}(x, s, \xi),$$

$$a_{i,k}(x, s, \xi) = \frac{\partial a_i}{\partial \xi_k}(x, s, \xi).$$

We make the assumptions

$$a_{i,x_k}, a_{i,s}, a_{i,k} \text{ are measurable,}$$
$$a_{i,k} \text{ is globally continuous in all arguments,}$$

$$|a_{i,x_k}(x, s, \xi)| \leq K|\xi|^2 + K,$$

$$|a_{i,s}(x, s, \xi)| \leq K|\xi| + K, \tag{2.101}$$

$$|a_{i,k}(x, s, \xi)| \leq K, \quad \forall x, s, \xi, \text{ with } |s| \leq C,$$

$$a_{i,k}(x, s, \xi)\lambda_i \lambda_k \geq \alpha|\lambda|^2, \quad \forall \lambda \in R^n, \alpha > 0.$$

We assume that there exists u such that

$$u \in H_0^1(\Omega) \cap L^\infty(\Omega),$$

$$\text{div } a(x, u, Du) \in L^1(\Omega), \quad |\text{div } a(x, u, Du)| \le K|Du|^2 + K, \tag{2.102}$$

where a denotes the vector a_i.

Our objective is to prove the following full regularity property.

Theorem 2.25. *We assume that Ω satisfies (2.94) and that (2.101), (2.102) are satisfied. Then*

$$u \in W^{2,p}(\Omega), \quad \forall 1 \le p < \infty.$$

Proof.
1. Preliminaries
It will be convenient to use the notation

$$a_{i,k}(x) = \int_0^1 a_{i,k}(x, u, \theta Du) \, d\theta.$$

Then thanks to the assumptions (2.101) we have

$$\alpha I \le a(x) \le MI.$$

Of course, no confusion should be made between this matrix $a(x)$ and the vector $a(x, u, Du)$ previously considered.

Defining the differential operator

$$A = -\text{div } a(x)D,$$

we can write

$$Au = -\text{div } a(x, u, Du) + a_{i,x_i}(x, u, 0) + a_{i,s}(x, u, 0)D_i u,$$

and thus from the third assumption (2.101) it follows immediately that

$$|Au| \le K|Du|^2 + K \tag{2.103}$$

with, of course, constants K different from those of (2.102). Note that we also have (see "Application 2" after the proof of Theorem 2.10)

$$\alpha \int_\Omega |Du|^2 \psi \, dx + \int_\Omega a \, DX_0(u - c)D\psi \, dx \le K_0 \int_\Omega \psi \, dx, \tag{2.104}$$

where

$$X_0(s) = \frac{\exp \gamma s^2 - 1}{\gamma}, \quad \gamma = \frac{K^2}{4\alpha^2},$$

and c is an arbitrary constant, $|c| \le \|u\|_{L^\infty}$, and ψ is arbitrary, $\psi(x) \ge 0, \psi \in H^1 \cap L^\infty(\Omega), \psi = 0$ on $\partial\Omega$ if $c \ne 0$.

Therefore, all the conditions of applicability of Theorems 2.3 and 2.10 hold, so

$$\exists p_0 > 2 \text{ such that } u \in W_0^{1,p_0}(\Omega)$$

and

$$u \in C^\delta(\overline{\Omega}), \quad 0 < \delta < 1.$$

It will be helpful to give a variant of (2.104) as follows. We allow a constant $c \neq 0$ with a ψ which does not vanish on the boundary, provided that c is not too large. Namely, suppose

$$|c| \leq \frac{\alpha^2}{2K^2\|u\|_\infty}.$$

Then we have

$$\frac{\alpha}{2}\int_\Omega |Du|^2\psi\,dx + 2\int_\Omega a\,DuD\psi u\exp\gamma(u-c)^2\,dx \leq K_0'\int_\Omega \psi\,dx \quad (2.105)$$

for any $\psi(x) \geq 0, \psi \in H^1 \cap L^\infty(\Omega)$. The proof is similar to that of (2.104) and consists in testing Au with $2u\exp\gamma(u-c)^2$.

2. Additional estimates

We prove here some useful additional estimates. Let x_0 be any point of Ω, and consider the Green Function $G = G^{x_0}$ associated with the matrix

$$a_{i,k}(x) = a_{i,k}(x, u, Du),$$

namely, as usual,

$$\int_Q aD\phi.DG\,dx = \phi(x_0), \quad \forall\,\phi \in C_0^\infty(Q),$$

where Q is a ball and $Q \supset \overline{\Omega}$.

We recall that

$$G \in W_0^{1,\mu}(Q), \quad 1 \leq \mu < \frac{n}{n-1}, \quad G \in L^\nu(Q), \quad 1 \leq \nu < \frac{n}{n-2},$$

$$c_0|x - x_0|^{2-n} \leq G^{x_0}(x) \leq c_1|x - x_0|^{2-n}, \quad x \in \Omega.$$

We also note the property

$$\int_\Omega |DG|^2 G^{-1-\epsilon}\,dx \leq C_{\epsilon_0}, \quad \forall\epsilon \geq \epsilon_0 > 0. \quad (2.106)$$

Indeed, we just test the Green Function equation with

$$\phi = G^{-\epsilon}\zeta^2,$$

where

$\zeta = 1$ on Ω, $\zeta = 0$ on the boundary of Q, $0 \leq \zeta \leq 1$, ζ smooth,

and note that this function vanishes at the singularity x_0.

Moreover, we have a more precise estimate of local character. Let $B_R(\xi)$ be a ball of radius R and let

$$S_R = \Omega \cap B_R.$$

Suppose that $x_0 \in S_R$. Then we have the estimate

$$\int_{S_R} |DG|^2 G^{-1-\epsilon} \, dx \leq \frac{C_\mu}{\epsilon^2} R^{n-2-\frac{n}{\mu}(1-\epsilon)}, \quad \forall \mu, \quad \frac{n}{n-2} > \mu > \frac{n(1-\epsilon)}{n-2}, \tag{2.107}$$

where C_μ depends only on μ.

We want next to prove that (see also Remark 2.26)

$$\int_\Omega |Du|^2 (G^{x_0})^{1+\beta} \, dx \leq C_\beta, \quad \forall x_0 \in \Omega, \quad \beta < \frac{\delta}{n-2}, \tag{2.108}$$

where δ is the Hölder's coefficient related to the function u,

$$|u(x) - u(x_0)| \leq c|x - x_0|^\delta.$$

We note the property

$$(G^{x_0})^{\delta/(n-2)} |u(x) - u(x_0)| \leq C.$$

The proof will be split in two parts, according to whether x_0 is close to the boundary of Ω or not. Suppose first

$$|u(x_0)| \leq \frac{\alpha^2}{2K^2 \|u\|_\infty}.$$

Then we apply (2.105) with $c = u(x_0)$ and

$$\psi = (G^{x_0})^{1+\beta}.$$

The right-hand side of (2.105) is clearly finite, for

$$\beta < \frac{2}{n-2}.$$

The main term is

$$J = 2 \int_\Omega a \, Du D(G^{1+\beta}) u \exp \gamma (u - u(x_0))^2 \, dx.$$

Now

$$J = J_1 + J_2$$

with

$$J_1 = \int_\Omega a\, D(u^2) D(G^{1+\beta})\, dx,$$

and as is easily seen,

$$|J_2| \le C \int |Du| |DG| G^\beta |x - x_0|^{2\delta}\, dx.$$

Using Hölder's inequality and absorbing the $|Du|^2 G^{1+\beta}$ term into the left-hand side of (2.105) we reduce the problem to that of estimating the integral

$$\int_\Omega |DG|^2 G^{-1-\frac{4\delta}{n-2}+\beta},$$

which is finite, provided that

$$\beta < \frac{4\delta}{n-2}.$$

We next write

$$J_1 = \int_Q a D(u^2 - u^2(x_0)) D(G^{1+\beta}) \zeta^2\, dx,$$

where ζ is the function defined above (recall that it is 1 on Ω). Using the equation of the Green function as well as (2.106) we check easily that J_1 is finite, provided that

$$\beta < \frac{\delta}{n-2}.$$

Suppose next that

$$|u(x_0)| > \frac{\alpha^2}{2K^2 \|u\|_\infty}.$$

Then $x_0 \in \tilde\Omega$, with $\overline{\tilde\Omega} \subset \Omega$. We take a new ζ such that

$$\zeta = 1 \text{ on } \tilde\Omega, \quad \zeta = 0 \text{ on the boundary of } \Omega, \quad 0 \le \zeta \le 1, \quad \zeta \text{ smooth},$$

and we apply (2.104) with

$$\psi = (G^{x_0})^{1+\beta} \zeta^2, \quad c = u(x_0).$$

We proceed as in the previous case. In fact, we have here a weaker restriction on β, namely

$$\beta < \frac{2\delta}{n-2}.$$

So (2.108) has been obtained.

We can next state a local estimate as in (2.107). Namely, we have

$$\int_{S_R} |Du|^2 (G^{x_0})^{1+\beta}\, dx \leq C_{\mu,\beta} R^{n-2-\frac{n}{\mu}(1+\beta-\frac{\delta}{n-2})} \quad \forall x_0 \in S_R,$$

(2.109)

$$\left(\frac{\delta}{n-2} - \frac{1}{2}\right)^+ \leq \beta < \frac{\delta}{n-2},$$

$$\frac{n}{n-2}\left(1 + \beta - \frac{\delta}{n-2}\right) < \mu < \frac{n}{n-2}.$$

The proof is an adaptation of (2.108). Let us indicate the main points. There will again be two parts, but with a choice depending on R. Consider a function ζ_R smooth, $0 \leq \zeta_R \leq 1$, $\zeta_R = 1$, on B_R, with support in B_{2R} and $|D\zeta_R| \leq C/R$. We consider

$$\psi = (G^{x_0})^{1+\beta}\zeta_R^2, \quad c = u(x_0).$$

Now, if $B_{2R} \subset \Omega$, we can apply (2.104). If not, we must apply (2.105), but then we must have

$$|u(x_0)| \leq \frac{\alpha^2}{2K^2\|u\|_\infty}.$$

However, since there exists a point of Γ that also belongs to B_{2R}, we can assert that

$$|u(x_0)| \leq c(4R)^\delta,$$

and thus the condition will be fulfilled, provided that

$$R \leq R_0 = \frac{1}{4}\left(\frac{\alpha^2}{2cK^2\|u\|_\infty}\right)^{1/\delta}.$$

This can be assumed without loss of generality.

Suppose, for instance, that we are in the case where B_{2R} crosses the boundary and we apply (2.105). The right-hand side is clearly majorized by

$$C_\mu R^{n\left(1-\frac{1+\beta}{\mu}\right)}.$$

The two other terms are

$$J_1 = \int_\Omega a\, D(u^2) D(G^{1+\beta})\zeta_R^2 \exp \gamma (u - u(x_0))^2\, dx$$

and

$$J_2 = \int_\Omega a\, D(u^2) D\zeta_R^2\, G^{1+\beta} \exp \gamma (u - u(x_0))^2\, dx.$$

For J_1 we use the Green function equation, and after some easy steps using Hölder's inequality and absorbing the $|Du|^2 G^{1+\beta}$ term into the left-hand side of (2.105), we reduce the problem to that of estimating the integrals

$$I_1 = \int_\Omega |DG|^2 G^{-1+\beta} |x - x_0|^\delta \zeta_R^2 \, dx,$$

$$I_2 = \int_\Omega G^{1+\beta} |u|^2 |D\zeta_R|^2 \, dx,$$

$$I_3 = \int_\Omega G^{1+\beta} |x - x_0|^\delta |D\zeta_R|^2 \, dx,$$

and thanks to the conditions on β, μ, we easily obtain (2.109).

3. Local L^∞ bound for the gradient
We shall cover Ω with a finite number of balls of size R, denoted by B_R for simplicity. The size R will be chosen small but fixed. It is thus sufficient to prove that

$$\| \, |Du| \, \|_{L^\infty(\Omega \cap B_R)} < \infty.$$

Set $S_R = \Omega \cap B_R$. We introduce a function $\tau \in C_0^\infty(S_R)$, and we shall prove that

$$\| \, \tau |Du| \, \|_{L^\infty(S_R)} < \infty.$$

We take $x_0 \in S_R$, and we test div $a(x, u, Du)$ with $D_k(G^{x_0} \tau^2 D_k u)$, and we obtain after easy calculations

$$\begin{aligned}
\int & a_{i,l} D_k D_l u D_i D_k u \, G\tau^2 \, dx + \tau^2 |Du|^2(x_0) \\
&= \int \, \text{div} \, a(x, u, Du) D_k(G\tau^2 D_k u) \\
&\quad - \frac{1}{2} \int a_{i,l} D_l(|Du|^2) G D_i \tau^2 \, dx \\
&\quad + \frac{1}{2} \int a_{i,l} D_l \tau^2 \, |Du|^2 D_i G \, dx \\
&\quad - \int (a_{i,x_k} + a_{i,s} D_k u) D_k u \, D_i(G\tau^2) \, dx \\
&\quad - \int (a_{i,x_k} + a_{i,s} D_k u) \, D_i D_k u G\tau^2 \, dx.
\end{aligned} \tag{2.110}$$

We indicate how to estimate the terms. First we make obvious majorizations in (2.110) leading to

$$\begin{aligned}
\alpha \int & |D^2 u|^2 \, G\tau^2 \, dx + \tau^2 |Du|^2(x_0) \\
&\leq C \left(\int |D^2 u| |Du|^2 G\tau^2 \, dx \right. \\
&\quad \left. + \int |D^2 u| G\tau^2 \, dx + \int |Du|^3 |D\tau| G\tau \, dx \right. \tag{2.111}
\end{aligned}$$

$$+ \int |DG||Du|^3 \tau^2 \, dx + \int |D^2u||Du||D\tau|G\tau \, dx$$

$$+ \int |Du|^2|DG||D\tau|\tau \, dx \Big) + C.$$

Considering the right-hand side of (2.111), we see that the second integral can be absorbed into the first integral of the left-hand side. The third integral is majorized by

$$\int |Du|^3|D\tau|G\tau \, dx \le C_R \|\tau \, Du\|_\infty \int_\Omega |Du|^2 G \, dx \le C_R \|\tau \, Du\|_\infty.$$

The fifth integral can be estimated by using Young's inequality and absorbing the $|D^2u|$ term into the left-hand side, leaving a constant C_R. The last integral is majorized by

$$\int |Du|^2|DG||D\tau|\tau \, dx \le \|\tau \, Du\|_\infty \int |Du||DG||D\tau| \, dx,$$

and since

$$\int_\Omega |Du||DG| \, dx \le C,$$

thanks to (2.106), (2.108) it is also majorized by $C_R \|\tau \, Du\|_\infty$. Collecting results, we have shown that

$$\frac{\alpha}{2} \int |D^2u|^2 \, G\tau^2 \, dx + \tau^2|Du|^2(x_0)$$

$$\le C \Big(\int |D^2u||Du|^2 G\tau^2 \, dx \tag{2.112}$$

$$+ \int |DG||Du|^3\tau^2 \, dx \Big) + C_R\|\tau \, Du\|_\infty + C_R.$$

So far, R is arbitrary. It is in estimating the two remaining integrals on the right-hand side of (2.112) that we shall need to choose R. Indeed, the first integral is majorized by

$$\|\tau \, Du\|_\infty \Big(\int |D^2u||Du|^2 G\tau^2 \, dx \Big)^{1/2} \Big(\int_{S_R} |Du|^2 G \, dx \Big)^{1/2},$$

and the second by

$$(\|\tau \, Du\|_\infty)^2 \int_{S_R} |Du||DG| \, dx.$$

It is sufficient to choose R such that

$$\int_{S_R} |Du||DG| \, dx \le \frac{1}{4C},$$

$$\int_{S_R} |Du|^2 G\, dx \le \frac{\alpha}{2C^2},$$

to obtain

$$\|\tau\, Du\|_\infty + \int |D^2 u|^2\, G\tau^2\, dx \le C_R.$$

4. L^∞ bound for the gradient on the boundary and Hölder property

Near the boundary, the L^∞ bound for the gradient will be obtained at the same time as the Hölder property. After straightening the boundary, we reduce the problem to the following. The domain is

$$\Omega = \tilde\Omega \cap \{x_n > 0\},$$
$$\tilde\Omega - \Omega = (\tilde\Omega \cap \{x_n = 0\}) \cup \{(x_t, -x_n)| x = (x_t, x_n) \in \Omega\},$$

where $t = 1, \ldots, n-1$ and where $\tilde\Omega$ is a bounded smooth domain of R^n, whose size will be fixed but sufficiently small. We can assume that

$$\tilde\Omega \subset B_{R_0}(0),$$

where R_0 will be chosen conveniently. We set

$$\Gamma = \tilde\Omega \cap \{x_n = 0\}$$

and we consider a function,

$$u \in H^1(\Omega), \quad u = 0 \text{ on } \Gamma,$$

such that

$$u \in C^\delta(\overline{\Omega}), \quad |\operatorname{div} a(x, u, Du)| \le K|Du|^2 + K, \quad x \in \Omega.$$

It will be convenient to assume that u is extended (with value 0) to $\tilde\Omega$. Then

$$u \in H^1(\tilde\Omega).$$

Our objective is to show that

$$Du \in C^\gamma(\Omega \cup \Gamma), \quad \gamma > 0.$$

Of course, the situation for the components of the gradient depends on whether we take the normal derivative x_n or any of the tangential derivatives $x_i, i = 1, \ldots, n-1$. We write for simplicity

$$D_t, \quad t = 1, \ldots, n-1,$$

whereas the notation D_i refers to any value $i = 1, \ldots, n$. The main difference is that

$$D_t u = 0 \text{ on } \Gamma.$$

Thus we can extend $D_t u$ to $\tilde{\Omega}$ by the value 0. Since we want to establish a priori estimates, we view $D_t u$ as a function in $H^1(\tilde{\Omega})$.

Consider points

$$x_0 \in \tilde{\Omega}.$$

We shall associate the Green function on a domain

$$Q \supset \overline{\tilde{\Omega}},$$

with x_0 and with the matrix $a(x)$, defined again by

$$a_{i,k}(x) = a_{i,k}(x, u, Du).$$

Of course, we have to define $a(x)$ outside Ω and preserve its properties. We do this in an arbitrary way.

We next consider balls $B_R(x_0)$ such that

$$B_{2R}(x_0) \subset \tilde{\Omega}$$

and the usual localization function τ_R, which is equal 1 on B_R and has support in B_{2R}. As usual, to obtain Hölder estimates we shall consider constants

$$c_R = c_{t,R} = \begin{cases} \frac{1}{|B_{2R}-B_{R/2}|} \int_{B_{2R}-B_{R/2}} D_t u \, dx & \text{if } B_{2R} \subset \Omega, \\ 0 & \text{if } B_{2R} \cap (\tilde{\Omega} - \Omega) \neq \emptyset. \end{cases}$$

Consider the function $\tau_R^2 G(D_t u - c_R)$, which vanishes on the boundary of Ω. Hence testing div $a(x, u, Du)$ with $D_t(\tau_R^2 G(D_t u - c_R))$, we obtain the relation

$$\int_\Omega D_i a_i D_t(\tau_R^2 G(D_t u - c_R)) \, dx = \int_\Omega D_t a_i D_i(\tau_R^2 G(D_t u - c_R)) \, dx.$$

Computing $D_t a_i$ yields

$$\int_\Omega D_i a_i D_t(\tau_R^2 G(D_t u - c_R)) \, dx$$

$$= \int_\Omega a_{i,k} D_k D_t u \, D_i D_t u \, \tau_R^2 G \, dx$$

$$+ \int_\Omega a_{i,k} D_k \left(\frac{(D_t u - c_R)^2}{2} \right) D_i(\tau_R^2 G) \, dx$$

$$+ \int_\Omega (a_{i,x_t} + a_{i,s} D_t u) D_i D_t u \, \tau_R^2 G \, dx$$

$$+ \int_\Omega (a_{i,x_t} + a_{i,s} D_t u)(D_t u - c_R) D_i(\tau_R^2 G) \, dx.$$

We then test the Green function equation with $1/2(D_t u - c_R)^2 \tau_R^2$. Collecting results, it follows that

$$\int_\Omega a_{i,k} D_k D_t u \, D_i D_t u \, \tau_R^2 G \, dx + \frac{(D_t u - c_R)^2}{2} \tau_R^2(x_0)$$

$$= \int_\Omega D_i a_i D_t(\tau_R^2 G(D_t u - c_R)) \, dx + \int_\Omega a_{i,k} \frac{(D_t u - c_R)^2}{2} D_k \, \tau_R^2 D_i G \, dx$$

$$- \int_\Omega a_{i,k} D_k \frac{(D_t u - c_R)^2}{2} D_i \, \tau_R^2 G \, dx$$

$$- \int_\Omega (a_{i,x_t} + a_{i,s} D_t u) D_i D_t u \, \tau_R^2 G \, dx$$

$$- \int_\Omega (a_{i,x_t} + a_{i,s} D_t u)(D_t u - c_R) \, D_i(\tau_R^2 G) \, dx.$$

Performing majorizations and using the Hölder inequality, we obtain

$$\int_\Omega |DD_t u|^2 \, \tau_R^2 G \, dx + \frac{(D_t u - c_R)^2}{2} \tau_R^2(x_0)$$

$$\leq \int_\Omega (K|Du|^4 + K)\tau_R^2 G \, dx$$

$$+ K \int_\Omega |D_t u - c_R|^2 |D\tau_R|^2 G \, dx + \int_\Omega (K|Du|^2 + K)|D_t u$$

$$- c_R|\tau_R^2|DG| \, dx + K \int_\Omega |D_t u - c_R|^2 \tau_R |D\tau_R| \, |DG| \, dx,$$

hence also

$$\int_\Omega |DD_t u|^2 \, \tau_R^2 G \, dx + \frac{(D_t u - c_R)^2}{2} \tau_R^2(x_0)$$

$$\leq \int_\Omega (K|Du|^4 + K)\tau_R^2 G^{1+\beta} \, dx + K \int_\Omega |D_t u - c_R|^2 |D\tau_R|^2 G \, dx$$

$$+ K \int_\Omega |D_t u - c_R|^2 |DG|^2 G^{-1-\beta} \tau_R^2 \, dx$$

$$+ K \int_\Omega |D_t u - c_R|^2 \tau_R |D\tau_R| \, |DG| \, dx.$$

The last integral requires a treatment similar to that in the proof of Theorem 2.10 (see in particular the treatment of integrals Z and Y). With the same calculations one checks that

$$\int_\Omega |D_t u - c_R|^2 \tau_R |D\tau_R| \, |DG| \, dx \leq K_0 \int_{\Omega \cap (B_{2(2m+1)R} - B_{R/2})} |D \, D_t u|^2 G \, dx$$

$$+ \int_\Omega (K|Du|^4 + K)\tau_R^2 G \, dx.$$

Similarly one obtains

$$\int_\Omega |D_t u - c_R|^2 |D\tau_R|^2 G \, dx \leq K_0 \int_{\Omega \cap (B_{2(2m+1)R} - B_{R/2})} |D \, D_t u|^2 G \, dx.$$

Hence collecting results, we have obtained the following estimate:

$$
\int_\Omega |DD_t u|^2 \, \tau_R^2 G \, dx + \frac{(D_t u - c_R)^2}{2} \tau_R^2(x_0)
$$
$$
\leq \int_\Omega (K|Du|^4 + K)\tau_R^2 G^{1+\beta} \, dx
$$
$$
+ K \int_\Omega |D_t u - c_R|^2 |DG|^2 G^{-1-\beta} \tau_R^2 \, dx
$$
$$
+ K_0 \int_{\Omega \cap (B_{2(2m+1)R} - B_{R/2})} |D \, D_t u|^2 G \, dx. \tag{2.113}
$$

Let

$$
\lambda_R = \int_{\Omega \cap B_{2R}} (|Du|^2 G^{1+\beta} + |DG|^2 G^{-1-\beta}) \, dx.
$$

Then from the estimates (2.107) and (2.109), choosing

$$
\beta = \frac{\delta}{2(n-2)}, \quad \mu = \frac{n}{n-2} \frac{1 - \delta/2(n-2)}{1 - \delta/4(n-2)},
$$

we have

$$
\lambda_R \leq CR^{\delta/4} \leq CR^{\delta/8} R_0^{\delta/8},
$$

and we deduce from (2.113) that

$$
\int_{B_R \cap \Omega} |DD_t u|^2 |x - x_0|^{2-n} \, dx \leq k \int_{\Omega \cap (B_{2(2m+1)R} - B_{R/2})} |D \, D_t u|^2 |x - x_0|^{2-n} \, dx
$$
$$
+ KR_0^{\delta/8}(1 + \|Du\|_\infty^2) R^{\delta/8},
$$

where

$$
\|Du\|_\infty = \|Du\|_\infty = \|Du\|_{L^\infty(\Omega)}.
$$

Similarly, we can assert

$$
\int_\Omega |DD_t u|^2 |x - x_0|^{2-n} \, dx \leq KR_0^{\delta/8}(1 + \|Du\|_\infty^2).
$$

Using the hole-filling technique we arrive at

$$
R^{2-n-2\epsilon} \int_{B_R \cap \Omega} |DD_t u|^2 \, dx \leq KR_0^{\delta/8}(1 + \|Du\|_\infty^2), \quad R \leq R_0, \tag{2.114}
$$

for some positive ϵ. Going back to the expression of div a, we notice the inequality

$$
\left| \sum_{i,k} a_{i,k} D_i D_k u \right| \leq K|Du|^2 + K.
$$

In particular, this implies

$$|D_n^2 u| \leq K|D\, D_t u| + K|Du|^2 + K.$$

Using this, the estimate (2.114), and (2.109) again, we can complete (2.114) and write

$$R^{2-n-2\epsilon} \int_{B_R \cap \Omega} |DD_n u|^2\, dx \leq KR_0^{\delta/8}(1 + \|Du\|_\infty^2), \quad R \leq R_0, \quad (2.115)$$

possibly with a smaller ϵ.

To apply Morrey's result in $\tilde{\Omega}$ we have to define $D_t u$ and $D_n u$ outside Ω so that they are $H^1(\tilde{\Omega})$. This we did for $D_t u$ by extending it by 0, but the same cannot be done for $D_n u$, since it is not 0 on the boundary. To avoid confusion set

$$z(x) = D_n u(x) \text{ if } x \in \Omega,$$
$$z(x_t, x_n) = z(x_t, -x_n) \text{ if } x \in \tilde{\Omega} - \Omega.$$

Hence we have extended $D_n u(x)$ by reflection. It is clear from (2.115) that

$$R^{2-n-2\epsilon} \int_{B_R} |Dz|^2\, dx \leq KR_0^{\delta/8}(1 + \|Du\|_\infty^2), \quad R \leq R_0.$$

From Morrey's result we can then assert that

$$[D_t u]_{\epsilon, \tilde{\Omega}} \leq KR_0^{\delta/8}(1 + \|Du\|_\infty^2),$$
$$[z]_{\epsilon, \tilde{\Omega}} \leq KR_0^{\delta/8}(1 + \|Du\|_\infty^2),$$

and thus combining these two results, we have

$$[Du]_{\epsilon, \Omega \cup \Gamma} \leq KR_0^{\delta/8}(1 + \|Du\|_\infty^2).$$

Since R_0 can be chosen sufficiently small, this implies necessarily

$$\|Du\|_\infty^2(\Omega) \leq C,$$

and of course also the Hölder property.

5. End of proof of full regularity

The situation now is easy. From the second assumption (2.101) we deduce that the matrix

$$a_{j,k}(x) = a_{j,k}(x, u, Du)$$

belongs to $C(\overline{\Omega})$. Since u satisfies

$$|a_{j,k}(x)D_j D_k u(x)| \leq K|Du|^2 + K,$$

we can now rely on the regularity results of the linear theory, in nonvariational form, see (2.99), (2.100), and a bootstrap argument to assert that $u \in W^{2,p}(\Omega), 1 \leq p < \infty$. The proof of Theorem 2.25 is complete. $\diamond$

Remark 2.26. To prove property (2.108) we can also use the results deduced from the hole-filling method to obtain the C^δ property. Let us sketch the local argument. Suppose we have the estimate

$$\int_{B_R(x_0)} \frac{|Du|^2}{|x - x_0|^{n-2}} \, dx \leq CR^{2\delta}, \quad R \leq R_0.$$

Then we notice that for $\delta' < \delta$ we have

$$\int_{B_R(x_0)} \frac{|Du|^2}{|x - x_0|^{n-2+2\delta'}} \, dx \leq C, \quad R \leq R_0. \tag{2.116}$$

Indeed, clearly,

$$\int_{B_R - B_{R/2}} \frac{|Du|^2}{|x - x_0|^{n-2+2\delta'}} \, dx \leq CR^{2\delta - 2\delta'}.$$

Applying this relation with $R = R_0 2^{-j}$, $j = 0, \ldots,$ and summing up, we obtain (2.116). Now using the estimates on the Green function, it follows that

$$\int_{B_R(x_0)} |Du|^2 (G^{x_0})^{1+2\frac{\delta'}{n-2}} \, dx \leq C, \quad R \leq R_0,$$

and we recover the result in the case where x_0 is far from the boundary.

3. Nonlinear Elliptic Systems Arising from Stochastic Games

3.1 Stochastic Games Background

The objective of this section is to present some results concerning a class of stochastic games for N players, obtained from the study of nonlinear elliptic systems, the Bellman system. The method consists in applying the dynamic programming approach. The value functions are the solutions of a system of partial differential equations. From regularity results, one is able to construct continuous feedbacks, which will represent optimal controls for each player. The standard verification approach will show the optimality. The optimality is to be taken in the sense of Nash [82]; see also J.P. Aubin [3].

3.1.1 Statement of the Problem and Results

Description of a Stochastic Game. Let

$$\Omega = C^0([0, +\infty); R^n), \quad \mathcal{A} = \text{ Borel } \sigma\text{-algebra on } \Omega. \tag{3.1}$$

The elements of Ω are denoted by $\omega \equiv \omega(t)$, and we provide Ω with a probability law P such that

$$\omega(t) \text{ is a standardized } n\text{-dimensional Wiener process.} \tag{3.2}$$

We then set

$$\mathcal{F}^t = \sigma\{\omega(s), s \leq t, \omega \in \Omega\}. \tag{3.3}$$

A trajectory starting at x is simply

$$x(t; \omega) = x + \omega(t). \tag{3.4}$$

We now consider N players, each of whom acts through a control $v_\nu(t)$, $\nu = 1, \ldots, N$. We assume

$$v(t) = (v_1(t), \ldots, v_N(t)), \text{ adapted process with bounded values in } R^{nN}, \tag{3.5}$$

which we call an admissible control.

Also, let

$g(x)$ be a measurable bounded function with values in R^n. $\qquad$ (3.6)

To a pair $x, v(t)$, where $v(t)$ is a control vector as above, we associate the process

$$\beta_{x,v}(t) = g(x(t)) + \sum_\mu v_\mu(t)$$

and the probability $P_{x,v}$ such that

$$\frac{dP_{x,v}}{dP}|_{\mathcal{F}^t} = \exp\left\{ \int_0^t \beta_{x,v}(s)\, d\omega(s) - \frac{1}{2}\int_0^t |\beta_{x,v}(s)|^2\, ds \right\}. \qquad (3.7)$$

From the Girsanov theorem, if we introduce the process

$$w_{x,v}(t) = \omega(t) - \int_0^t \beta_{x,v}(s)\, ds, \qquad (3.8)$$

then the system $\Omega, \mathcal{A}, \mathcal{F}^t, P_{x,v}, w_{x,v}(t)$ forms a probability system in which $w_{x,v}(t)$ is an $\mathcal{F}^t$ standardized Wiener process. Note that from (3.8) one has

$$dx = \left(g(x(t)) + \sum_\mu v_\mu(t) \right) dt + dw_{x,v}(t), \quad x(0) = x. \qquad (3.9)$$

Now let

$\mathcal{O}$ be an open smooth bounded domain of R^n $\qquad$ (3.10)

and let

$$\tau_x = \inf\{t | x(t) \notin \mathcal{O}\}.$$

We shall stop the process $x(t)$ at the exit of the domain $\mathcal{O}$, and to save the notation, we shall still denote by $x(t)$ the stopped process. Also, let

$f_\nu(x)$ be a scalar measurable bounded function. $\qquad$ (3.11)

We set

$$J_\nu(x, v) = E_{x,v} \int_0^{\tau_x} e^{-ct} (f_\nu(x(t)) + \frac{1}{2}|v_\nu(t)|^2 + \theta v_\nu(t).\bar{v}_\nu(t)) dt \qquad (3.12)$$

with the notation

$$\bar{v}_\nu = \sum_{\mu \neq \nu} v_\mu \qquad (3.13)$$

and where

$$c > 0, \quad \theta \text{ a real parameter.} \qquad (3.14)$$

We shall also use the notation

$$v = (v_\nu, \bar{v}^\nu),$$ (3.15)

where of course, the part $\bar{v}^\nu$ represents all components that are different from v_ν.

A Nash point for the game defined by the functionals (3.12) is a control $\hat{v}(\cdot)$ such that

$$J_\nu(x, \hat{v}_\nu, \bar{\hat{v}}^\nu) \le J_\nu(x, v_\nu, \bar{\hat{v}}^\nu), \quad \forall \nu,$$ (3.16)

for any admissible control v.

Existence of a Nash Point. Our objective is to prove the following theorem.

Theorem 3.1. *We make the assumptions (3.5), (3.6), (3.10), (3.11),(3.14). Suppose also*

$$\theta \le \frac{1}{2}, \quad \theta \ne -\frac{1}{N-1}; \ or \ \theta > 1.$$ (3.17)

Then there exists a Nash point for the game defined by the functionals (3.12).

As indicated in the introduction, the method consists in considering a system of Bellman equations for the value functions of the game. This means here that for a convenient control $\hat{v}$, possibly depending on x, the functions

$$u_\nu(x) = J_\nu(x, \hat{v})$$ (3.18)

are the solutions of a system of partial differential equations. This system will permit us to characterize optimal feedbacks for the N players. The proof of optimality will be performed by a verification argument. A key point is to obtain sufficient regularity properties for the value functions. Otherwise, it is not possible to obtain feedbacks. Techniques of partial differential equations are instrumental in obtaining these necessary regularity properties.

3.1.2 Bellman Equations

Notation. Here we introduce the Lagrangians (see (3.76) below) for a more general form)

$$L_\nu(v, p) = \frac{1}{2}|v_\nu|^2 + \theta v_\nu \cdot \bar{v}_\nu + p_\nu \cdot \sum_\mu v_\mu,$$ (3.19)

where

$$p = (p_1, \ldots, p_N).$$

The first point is to consider, for a given p, a Nash point in v for the functions $L_\nu(v, p)$. Clearly, the following conditions must hold (by differentiation) for such a Nash point $v(p)$:

$$v_\nu(p) + \theta \bar{v}_\nu(p) + p_\nu = 0.$$ (3.20)

Provided, that

$$\theta \neq 1, \quad \theta \neq -\frac{1}{N-1}, \tag{3.21}$$

it is easy to check that the system (3.20) has a unique solution given by the formulas

$$v_\nu(p) = \frac{\theta \sum_\mu p_\mu}{(1-\theta)(1+(N-1)\theta)} - \frac{p_\nu}{1-\theta}. \tag{3.22}$$

We note also the complementary formulas

$$\bar{v}_\nu(p) = \frac{-\sum_\mu p_\mu}{(1-\theta)(1+(N-1)\theta)} + \frac{p_\nu}{1-\theta}. \tag{3.23}$$

We can then define the quantities

$$L_\nu(p) = L_\nu(v(p), p). \tag{3.24}$$

It is also useful to express, from (3.22) and (3.23), the vectors p_ν in terms of $v(p)$ as

$$p_\nu = -v_\nu(p) - \theta \bar{v}_\nu(p), \tag{3.25}$$

and also

$$\bar{p}_\nu = -(N-1)\theta v_\nu(p) + (N\theta - 1)\bar{v}_\nu(p). \tag{3.26}$$

We can, in particular, write

$$L_\nu(p) = -\frac{1}{2}|v_\nu(p)|^2 - \frac{1}{\theta}p_\nu.(p_\nu + v_\nu(p)). \tag{3.27}$$

The Bellman equations are written as follows (see (3.89) for the Hamiltonian notation):

$$-\frac{1}{2}\Delta u_\nu - g(x).Du_\nu + cu_\nu = f_\nu(x) + L_\nu(Du),$$

$$u_\nu = 0, \text{ on } \partial\mathcal{O}. \tag{3.28}$$

3.1.3 Verification Property

We begin by stating the following result concerning the system of Bellman equations, which will be proved later.

Theorem 3.2. *We make the assumptions of Theorem 3.1. Then there exists a solution of the system (3.28) such that*

$$u_\nu \in W^{2,s}(\mathcal{O}), \quad \forall\, 2 \leq s < \infty. \tag{3.29}$$

In particular, the functions u_ν are continuously differentiable, with second derivatives in $L^s(\mathcal{O})$.

From this regularity result we can prove Theorem 3.1

Proof of Theorem 3.1. Consider

$$\hat{v}_\nu(x) = v_\nu(Du(x)).$$

It will correspond to an optimal feedback for the player ν. Next, we set

$$\hat{v}_\nu(t) = \hat{v}_\nu(x(t)), \tag{3.30}$$

which defines a stochastic process, depending on x, the initial value of $x(t)$. Since from the regularity of u the functions $\hat{v}_\nu(x)$ are continuous, and if we recall that $x(t)$ refers to the process stopped at the exit of $\mathcal{O}$, we get that $\hat{v}_\nu(t)$ is a bounded adapted process. Define $\beta_{x,\hat{v}}(t)$, $P_{x,\hat{v}}$, $w_{x,\hat{v}}(t)$, by formulas (3.7), (3.8). Then the system $\Omega, \mathcal{A}, \mathcal{F}^t, P_{x,\hat{v}}, w_{x,\hat{v}}(t)$ forms a probability system in which $w_{x,\hat{v}}(t)$ is an $\mathcal{F}^t$ standardized Wiener process. Note that from (3.8) one has

$$dx = \left(g(x(t)) + \sum_\mu \hat{v}_\mu(t) \right) dt + dw_{x,\hat{v}}(t), \quad x(0) = x. \tag{3.31}$$

Furthermore, from (3.19), one has

$$L_\nu(Du(x)) = \frac{1}{2}|\hat{v}_\nu(x)|^2 + \theta\hat{v}_\nu(x).\bar{\hat{v}}_\nu(x) + Du_\nu(x).\sum_\mu \hat{v}_\mu(x).$$

Hence, from (3.28),

$$-\frac{1}{2}\Delta u_\nu - g(x).Du_\nu + cu_\nu = f_\nu(x) + \frac{1}{2}|\hat{v}_\nu(x)|^2 + \theta\hat{v}_\nu(x).\bar{\hat{v}}_\nu(x)$$

$$+ Du_\nu(x).\sum_\mu \hat{v}_\mu(x), \tag{3.32}$$

$$u_\nu = 0, \text{ on } \partial\mathcal{O}.$$

From the regularity of u_ν one may use Ito's formula to assert, taking account of (3.31), (3.32) and notation (3.30),

$$d(u_\nu(x(t))e^{-ct}) = -e^{-ct}\left[f_\nu(x(t)) + \frac{1}{2}|\hat{v}_\nu(t)|^2 + \theta\hat{v}_\nu(t).\bar{\hat{v}}_\nu(t) \right] dt$$

$$+ e^{-ct} Du_\nu(x(t)).dw_{x,\hat{v}}(t),$$

and thus, integrating between 0 and τ_x, then taking the mathematical expectation with respect to $P_{x,\hat{v}}$, we obtain easily

$$u_\nu(x) = J_\nu(x,\hat{v}) = J_\nu\left(x, \hat{v}_\nu, \bar{\hat{v}}^\nu\right). \tag{3.33}$$

Next, we notice that from the definition of Nash points

$$L_\nu(p) \leq L_\nu(v_\nu, \bar{v}(p)^\nu), \quad \forall v_\nu.$$

Hence also, from (3.28) and (3.32), we can state

$$-\frac{1}{2}\Delta u_\nu - g(x).Du_\nu + cu_\nu \leq f_\nu(x) + \frac{1}{2}|v_\nu|^2 + \theta v_\nu.\bar{\hat{v}}_\nu(x)$$
$$+Du_\nu(x).(v_\nu + \bar{\hat{v}}_\nu(x)), \quad \forall v_\nu, \quad (3.34)$$
$$u_\nu = 0, \text{ on } \partial\mathcal{O}.$$

Consider then the control $(v_\nu(t), \bar{\hat{v}}^\nu(t))$, where $v_\nu(t)$ is any bounded, adaptive process with values in R^n, and $\bar{\hat{v}}^\nu(t)$ is the part of $\hat{v}(t)$ without the component ν, as above. Then define $\beta_{x,v_\nu,\bar{\hat{v}}^\nu}(t)$, $P_{x,v_\nu,\bar{\hat{v}}^\nu}$, $w_{x,v_\nu,\bar{\hat{v}}^\nu}(t)$, by formulas (3.7), (3.8). Then the system $\Omega, \mathcal{A}, \mathcal{F}^t, P_{x,v_\nu,\bar{\hat{v}}^\nu}, w_{x,v_\nu,\bar{\hat{v}}^\nu}(t)$ forms a probability system in which $w_{x,v_\nu,\bar{\hat{v}}^\nu}(t)$ is an $\mathcal{F}^t$ standardized Wiener process. Note again that from (3.8) one has

$$dx = (g(x(t)) + v_\nu(t) + \bar{\hat{v}}_\nu(t)) \, dt + dw_{x,v_\nu,\bar{\hat{v}}^\nu}(t), \quad x(0) = x. \quad (3.35)$$

It follows from (3.34) and Ito's formula that

$$d(u_\nu(x(t))e^{-ct}) \geq -e^{-ct}\left[f_\nu(x(t)) + \frac{1}{2}|v_\nu(t)|^2 + \theta v_\nu(t).\bar{\hat{v}}_\nu(t)\right] dt$$

$$+ e^{-ct}Du_\nu(x(t)).dw_{x,v_\nu,\bar{\hat{v}}^\nu}(t).$$

Thus, integrating between 0 and τ_x, then taking the mathematical expectation with respect to $P_{x,v_\nu,\bar{\hat{v}}^\nu}$, we obtain easily

$$u_\nu(x) \leq J_\nu\left(x, \hat{v}_\nu, \bar{\hat{v}}^\nu\right), \quad (3.36)$$

which implies that (3.16) is satisfied, and thus the proof of Theorem 3.1 has been completed. $\diamond$

3.2 Introduction to the Analytic Part

We concentrate now on the analytic part of the theory. We refer to A. Bensoussan, J. Frehse [7] for more details on motivation. We here need to know only that the theory of stochastic games leads to an important class of systems of nonlinear elliptic (or parabolic) PDEs (partial differential equations) called Bellman systems. They are of the form

$$Au^\nu = H^\nu(x, u, Du), \quad (3.37)$$

where $\nu = 1, \ldots, N$ indexes the equations. We write

$$u = (u^1, \ldots, u^\nu, \ldots, u^N),$$

and Du is a matrix,

$$H = (H^1, \ldots, H^\nu, \ldots, H^N).$$

So in vector form we have

$$Au = H(x, u, Du). \tag{3.38}$$

The operator A is a second-order linear differential operator, and the same holds for all equations. It will be defined on a smooth bounded domain Ω of R^n, over which the space variable x runs. The operator A is given in the divergence form

$$A = -\sum_{ij=1}^{n} \frac{\partial}{\partial x_i} a_{i,j} \frac{\partial}{\partial x_j}, \tag{3.39}$$

where the matrix $a = a_{ij}$ satisfies

$$a(x) \text{ bounded}, a(x)\xi.\xi \geq \alpha|\xi|^2, \forall \xi \in R^n, x \in \Omega; \ \alpha > 0. \tag{3.40}$$

The operators H^ν are called Hamiltonians. They are of quadratic growth in the gradient for bounded values of u. Unfortunately, in contrast to the scalar case, one cannot consider a general quadratic growth. The dependence of the ν equation on Du^μ (for $\mu \neq \nu$) is of drastic importance. However, we can take advantage of an available degree of freedom in the writing of the system.

Indeed, consider a matrix Γ that is invertible. Consider the change of unknown functions

$$z = \Gamma u.$$

Clearly, we can rewrite (3.38) as

$$Az = \Gamma H(x, \Gamma^{-1}z, D\Gamma^{-1}z).$$

So the Hamiltonians have been transformed into

$$H(x, s, p) \to \Gamma H(x, \Gamma^{-1}s, \Gamma^{-1}p). \tag{3.41}$$

But then the new Hamiltonians defined in (3.41) may enjoy an obvious structure with respect to the arguments that the original ones do not have. This is done without changing anything in the problem. So we introduce the following definition.

Definition 3.3. *We say that a "structure" holds for the Hamiltonians H up to a linear transformation if there exists a matrix Γ such that when the transformation (3.41) is performed, the new Hamiltonians enjoy the desired structure.*

In fact, it will be very useful to play with several linear transformations such as Γ, so that different forms of structures appear. Each structure will provide information on the solution z, which will carry over to u (see Remark 3.13 for the application to stochastic games).

At this stage, let us just give a trivial example of what we have in mind, as far as using the trick of the linear transformation is concerned. Suppose we consider the following system:

$$-\Delta u^1 + g.Du^1 = -|Du^1 + Du^2|^2 + f^1,$$

$$-\Delta u^2 + g.Du^2 = -|Du^1 + Du^2|^2 + f^2.$$

Trivially, if we define

$$v^1 = u^1 - u^2, \quad v^2 = u^2,$$

then we have, by subtraction,

$$-\Delta v^1 + g.Dv^1 = f^1 - f^2,$$
$$-\Delta v^2 + g.Dv^2 = -|Dv^1 + 2Dv^2|^2 + f^2.$$

The new structure is obviously better, since we no longer have a coupled system, and v^1 is the solution of a linear equation.

In studying (3.38), two types of developments will be needed: one to obtain L^∞ estimates and one to obtain estimates in Sobolev spaces and in C^δ. Different linear transformations may yield structures for which the desired properties will be derived. We shall consider (3.38) with Dirichlet data, namely

$$u_{|\partial\Omega} = 0. \tag{3.42}$$

3.3 Estimates in Sobolev spaces and in C^δ

3.3.1 Assumptions and Statement of Results

The main assumption will be the following. Up to a linear transformation the following structure holds:

$$H^\nu(x, s, p) = Q(x, s, p).p^\nu + H_0^\nu(x, s, p),$$

$$|H_0^\nu(x, s, p)| \leq k^\nu(\|s\|) + \sum_\mu K_\mu^\nu(\|s\|)|p^\mu|^2, \quad \|s\| = \max_\mu |s^\mu|, \tag{3.43}$$

where $k^\nu(\rho)$ and $K_\mu^\nu(\rho)$ are monotone increasing, $K_\mu^\nu(\rho)$ satisfies a smallness condition for $\mu > \nu$, $|Q(x, s, p)|^2 \leq k(\|s\|) + K(\|s\|)|p|^2$, $k(\rho), K(\rho)$ are monotone increasing.

Remark 3.4. In fact, the last Hamiltonian H^N can have general quadratic growth, since there is no smallness condition for $\nu = N$. If there is a function Q such that the structure described in the first part of 3.43 holds for all the other Hamiltonians, then we may just define

$$H_0^N(x,s,p) = H^N(x,s,p) - Q(x,s,p).p^N,$$

and this completes the conditions of (3.43).

Let us make the smallness condition explicit. Of course, at this stage we just state it. Its rationale will become clear from the forthcoming proofs.

We are going to define functions $\gamma^\nu(\rho), \lambda^\nu(\rho)$ backwards successively, as follows:

$$\alpha(\lambda^N(\rho))^2 - \lambda^N(\rho)K_N^N(\rho) - \frac{K(\rho)}{4\alpha} > 0 , \quad \gamma^N(\rho) = 1,$$

$$\alpha\lambda^\nu(\rho) > K_\nu^\nu(\rho),$$

$$\gamma^\nu(\rho)\lambda^\nu(\rho)(\alpha\lambda^\nu(\rho) - K_\nu^\nu(\rho)) - \frac{K(\rho)}{4\alpha} \tag{3.44}$$

$$- \sum_{\mu > \nu} \gamma^\mu(\rho)\lambda^\mu(\rho)K_\nu^\mu(\rho) \exp 2\rho\lambda^\mu(\rho) > 0,$$

$$\nu = 1, \ldots, N-1.$$

The smallness condition is then

$$\sum_{\mu < \nu} \gamma^\mu(\rho)\lambda^\mu(\rho)K_\nu^\mu(\rho) \exp 2\rho\lambda^\mu(\rho)$$

$$< \gamma^\nu(\rho)\lambda^\nu(\rho)(\alpha\lambda^\nu(\rho) - K_\nu^\nu(\rho)) \tag{3.45}$$

$$- \frac{K(\rho)}{4\alpha} - \sum_{\mu > \nu} \gamma^\mu(\rho)\lambda^\mu(\rho)K_\nu^\mu(\rho) \exp 2\rho\lambda^\mu(\rho), \quad \forall 2 \leq \nu \leq N.$$

Of course, the natural case in which the smallness condition is automatically satisfied is

$$K_\nu^\mu(\rho) = 0, \forall \mu < \nu.$$

Moreover, the smallness condition will require only one value of ρ, which will be made explicit later on.

Remark 3.5. The statement of the smallness condition clarifies why the use of the linear transformation to discover structures is crucial.

Clearly, the smallness condition implies the property

$$\alpha\gamma^\nu(\rho)(\lambda^\nu(\rho))^2 - \gamma^\nu(\rho)\lambda^\nu(\rho)K_\nu^\nu(\rho)$$

$$- \frac{K(\rho)}{4\alpha} - \sum_{\mu \neq \nu} \gamma^\mu(\rho)\lambda^\mu(\rho)K_\nu^\mu(\rho) \exp 2\rho\lambda^\mu(\rho) > 0, \tag{3.46}$$

$$\forall \nu, \forall \rho > 0,$$

and in fact, this is the property we shall use.

Remark 3.6. When $N = 2$, the assumption can be simplified. Indeed, it is sufficient to assume that (up to a linear transformation)

$$|H^1(x, s, p)| \leq \bar{k}(\|s\|) + \bar{K}(\|s\|)|p||p^1|,$$

and of course, H^2 has general quadratic growth. Indeed, if we define

$$Q(x, s, p) = \frac{H^1(x, s, p)p^1(1 + |p|^2)}{1 + |p^1|^2(1 + |p|^2)}$$

and

$$H_0^1(x, s, p) = H^1(x, s, p) - Q(x, s, p).p^1,$$

then one checks that

$$|H_0^1(x, s, p)| \leq \bar{k}(\|s\|) + \frac{\bar{K}(\|s\|)}{2} = k^1(\|s\|)$$

and

$$|Q(x, s, p)| \leq \frac{\bar{k}(\|s\|)}{2} + \left(\frac{\bar{k}(\|s\|)}{2} + \bar{K}\right)|p|.$$

Thus the assumptions (3.43) are clearly satisfied.

We can then state the main result.

Theorem 3.7. *We assume (3.40), (3.43), (3.45). Suppose we have an $H_0^1(\Omega) \cap L^\infty(\Omega)$ solution of (3.38), (3.42); then this solution is also in $C^\delta(\Omega) \forall \delta, 0 \leq \delta < \delta_0 < 1$, and in $W_0^{1,p}(\Omega), \forall p \in [2, 2 + p_0)$. The exponents δ_0, p_0 as well as the bounds of the norms depend only on the values $\alpha, k(\rho), K(\rho), k^\nu(\rho), K_\mu^\nu(\rho)$, where*

$$\rho = \sup_x \|u(x)\|.$$

3.3.2 Preliminaries

Introduce

$$\theta(x) = e^x - x - 1$$

and for $\gamma > 0$,

$$\sigma_\gamma(x) = \exp \gamma \, \theta(x) + \exp \gamma \, \theta(-x).$$

We note that

$$\theta'(x) = e^x - 1,$$
$$\sigma_\gamma'(x) = \gamma(\theta'(x) \exp \gamma\theta(x) - \theta'(-x) \exp \gamma\theta(-x)),$$
$$\sigma_\gamma''(x) = \gamma \left((\gamma(\theta'(x))^2 + \theta''(x)) \exp \gamma\theta(x) \right.$$
$$\left. + (\gamma(\theta'(-x))^2 + \theta''(-x)) \exp \gamma\theta(-x)\right).$$

Hence

$$\theta(x) \geq 0, \quad \theta(0) = 0,$$
$$\sigma_\gamma(x) \geq 2, \quad \sigma_\gamma'(-x) = -\sigma_\gamma'(x), \quad \sigma_\gamma''(x) \geq 0,$$
$$\sigma_\gamma(0) = 2, \quad \sigma_\gamma'(0) = 0.$$

We shall make use of the relations

$$\sigma_\gamma(x) \leq e^x \exp \gamma\, \theta(x) + e^{-x} \exp \gamma\, \theta(-x) \leq \sigma_\gamma(x)\, e^{|x|}. \qquad (3.47)$$

It is sufficient to check them for $x > 0$, by symmetry. In this case, set

$$a = \exp \gamma\, \theta(x), \quad b = \exp \gamma\, \theta(-x).$$

Then $a > b$ and $ae^x + be^{-x} > a + b$. Hence the left part of (3.47) holds. The second part is obvious.

Next,

$$\frac{1}{\gamma}|\sigma_\gamma'(x)| \leq e^x \exp \gamma\, \theta(x) + e^{-x} \exp \gamma\, \theta(-x). \qquad (3.48)$$

It is again sufficient to check it for $x > 0$. Then introducing a and b as above, we have to check that

$$(e^x - 1)a + (1 - e^{-x})b \leq ae^x + be^{-x},$$

or

$$b \leq a + 2be^{-x},$$

which is true, since $b \leq a$. We also have

$$\frac{1}{\gamma}|\sigma_\gamma'(x)| \leq |x|e^{|x|}\sigma_\gamma(x), \qquad (3.49)$$

and if we check it for $x > 0$, we notice that

$$e^x - 1 \leq xe^x, \quad 1 - e^{-x} \leq x \leq xe^x.$$

Moreover, (3.49) implies

$$\frac{1}{\gamma}(\sigma_\gamma(x) - 2) \leq |x|^2 e^{|x|}\sigma_\gamma(x).$$

Now consider the function of N arguments

$$X(s) = \prod_{\nu=1}^{N} \sigma_{\gamma^\nu}(s^\nu). \qquad (3.50)$$

We then set

$$X_{s^\nu}(s) = X(s)\frac{(\sigma_{\gamma^\nu})'(s^\nu)}{\sigma_{\gamma^\nu}(s^\nu)} \tag{3.51}$$

and also

$$X^\nu(s) = X(s)\frac{e^{s^\nu}\exp\gamma^\nu\,\theta(s^\nu) + e^{-s^\nu}\exp\gamma^\nu\,\theta(-s^\nu)}{\sigma_{\gamma^\nu}(s^\nu)}. \tag{3.52}$$

We then derive from the inequalities (3.47), (3.48), (3.49) the following ones:

$$X(s) \le X^\nu(s) \le X(s)e^{|s^\nu|} \tag{3.53}$$

and

$$\frac{1}{\gamma^\nu}|X_{s^\nu}(s)| \le X^\nu(s) \le X(s)e^{|s^\nu|}, \tag{3.54}$$

$$\frac{1}{\gamma^\nu}|X_{s^\nu}(s)| \le X(s)|s^\nu|e^{|s^\nu|}. \tag{3.55}$$

Finally, considering $X(\lambda s)$ as a function of λ between 0 and 1, and computing

$$\frac{d}{d\lambda}X(\lambda s) = \sum_\nu s^\nu X_{s^\nu}(\lambda s),$$

we see from (3.52) that it is positive for $\lambda > 0$. Hence

$$2^N \le X(\lambda s) \le X(s).$$

Moreover, from (3.55) we derive

$$0 \le X(s) - 2^N \le \frac{1}{2}X(s)\sum_\nu \gamma^\nu(s^\nu)^2 e^{|s^\nu|}.$$

We proceed with a basic inequality. We consider our solution u of (3.38) and recall that

$$\rho = \sup\|u(x)\|.$$

We associate with each component $u^\nu(x)$ a constant called c^ν, which can be chosen arbitrarily except that we require

$$|c^\nu| \le \rho.$$

We consider arbitrary real $\gamma^\nu > 0$ and λ^ν constants and define

$$F = \prod_{\nu=1}^N \exp\gamma^\nu\,\theta(\lambda^\nu(u^\nu - c^\nu)). \tag{3.56}$$

We also introduce a function

$$\psi(x) \ge 0, \quad \psi \in C^1(\overline{\Omega}), \quad \psi_{|\partial\Omega} = 0$$

$$\tag{3.57}$$

if and only if one of the constants satisfies $c^\nu \ne 0$.

Lemma 3.8. *We have the inequality*

$$\sum_\nu \gamma^\nu (\lambda^\nu)^2 \int_\Omega \theta''(\lambda^\nu(u^\nu - c^\nu))aDu^\nu.Du^\nu \, F\psi \, dx + \int_\Omega aDF.D\psi \, dx$$

$$\text{(3.58)}$$

$$\leq \frac{1}{4} \int_\Omega a^{-1}Q.QF\psi \, dx + \sum_\nu \gamma^\nu \lambda^\nu \int_\Omega \theta'(\lambda^\nu(u^\nu - c^\nu))H_0^\nu(x, u, Du) \, F\psi \, dx.$$

Proof. We test equation (3.38) with

$$\gamma^\nu \lambda^\nu \theta'(\lambda^\nu(u^\nu - c^\nu))F\psi,$$

which by the assumption on ψ vanishes on the boundary of Ω. Note also that

$$DF(x) = F\sum_\nu \gamma^\nu \lambda^\nu \theta'(\lambda^\nu(u^\nu - c^\nu))Du_\nu(x).$$

We obtain

$$\sum_\nu \gamma^\nu (\lambda^\nu)^2 \int_\Omega \theta''(\lambda^\nu(u^\nu - c^\nu))aDu^\nu.Du^\nu \, F\psi \, dx$$

$$+ \int_\Omega aDF.D\psi \, dx + \int_\Omega \frac{aDF.DF}{F}\psi \, dx = \int_\Omega Q.DF\psi \, dx \quad \text{(3.59)}$$

$$+ \sum_\nu \gamma^\nu \lambda^\nu \int_\Omega \theta'(\lambda^\nu(u^\nu - c^\nu))H_0^\nu(x, u, Du) \, F\psi \, dx.$$

If we write

$$\frac{aDF.DF}{F} - Q.DF$$

$$= a\left(F^{-1/2}DF - \frac{1}{2}F^{1/2}a^{-1}Q\right).\left(F^{-1/2}DF - \frac{1}{2}F^{1/2}a^{-1}Q\right)$$

$$- \frac{1}{4}Fa^{-1}Q.Q \geq -\frac{1}{4}Fa^{-1}Q.Q$$

and use it in (3.59), we obtain (3.58).   ◇

3.3.3 Proof of Theorem 3.7

Define Λ to be the diagonal matrix whose diagonal elements are the λ^ν. We shall use the functional $X(\Lambda(u - c))$, where c represents the vector whose components are the c^ν. More explicitly,

$$X(\Lambda(u - c)) = \prod_{\nu=1}^N \sigma_{\gamma^\nu}(\lambda^\nu(u^\nu - c^\nu)).$$

Note that

$$DX(\Lambda(u-c)) = \sum_\nu \lambda^\nu X_{s^\nu}(\Lambda(u-c))Du^\nu.$$

We also note that $X(\Lambda(u-c))$ can be written as the sum of 2^N terms similar to F, defined in (3.56). These terms differ only by the fact that we use the 2^N vectors

$$(\pm\lambda^1, \ldots, \pm\lambda^N).$$

We then apply our basic inequality (3.58) to these 2^N different F, and we add them up. Collecting results, we obtain

$$\sum_\nu \gamma^\nu(\lambda^\nu)^2 \int_\Omega X^\nu(\Lambda(u-c))aDu^\nu.Du^\nu\,\psi\,dx$$

$$+ \int_\Omega aDX(\Lambda(u-c)).D\psi\,dx$$

$$\leq \frac{1}{4}\int_\Omega a^{-1}Q.QX(\Lambda(u-c))\psi\,dx$$

$$+ \sum_\nu \lambda^\nu \int_\Omega X_{s^\nu}(\Lambda(u-c))H_0^\nu(x,u,Du)\psi\,dx,$$

and using (3.54), the left part of (3.53), and the assumption (3.43), we obtain

$$\sum_\nu \gamma^\nu(\lambda^\nu)^2 \int_\Omega X^\nu aDu^\nu.Du^\nu\,\psi\,dx + \int_\Omega aDX.D\psi\,dx$$

$$\leq \frac{1}{4\alpha}\int_\Omega X(k+K|Du|^2)\psi\,dx$$

$$+ \sum_\nu \lambda^\nu\gamma^\nu \int_\Omega X^\nu\left(k^\nu + \sum_\mu K_\nu^\mu|Du^\mu|^2\right)\psi\,dx,$$

or

$$\sum_\nu \gamma^\nu(\lambda^\nu)^2\alpha \int_\Omega X^\nu|Du^\nu|^2\,\psi\,dx + \int_\Omega aDX.D\psi\,dx$$

$$\leq \frac{1}{4\alpha}\int_\Omega X(k+K|Du|^2)\psi\,dx$$

$$+ \sum_\nu \lambda^\nu\gamma^\nu \int_\Omega X^\nu\left(k^\nu + \sum_\mu K_\mu^\nu|Du^\mu|^2\right)\psi\,dx,$$

or again,

$$\sum_\nu \int_\Omega |Du^\nu|^2 \left[X^\nu(\alpha\gamma^\nu(\lambda^\nu)^2 - \frac{K}{4\alpha} - \lambda^\nu\gamma^\nu K_\nu^\nu) - \sum_{\mu\neq\nu}\lambda^\mu\gamma^\mu K_\nu^\mu X^\mu\right]\psi\,dx$$

$$+ \int_\Omega aDX.D\psi\,dx \leq \int_\Omega \left(\frac{k}{4\alpha}X + \sum_\nu \lambda^\nu\gamma^\nu k^\nu X^\nu\right)\psi\,dx.$$

Now from (3.53) we can assert that for any pair μ, ν,

$$X^\mu(\Lambda(u-c)(x)) \le X^\nu(\Lambda(u-c)(x)) \exp \lambda^\mu |u^\mu(x) - c^\mu|.$$

Hence also, by the definition of ρ and the restriction on c^μ, we have

$$X^\mu \le X^\nu \exp 2\rho\lambda^\mu.$$

Recall that all constants k, K, k^ν, K_μ^ν depend on ρ. Then, if we pick the $\gamma^\nu(\rho), \lambda^\nu(\rho)$ as indicated above (in order that (3.46) holds), recalling that

$$2^N \le X \le X^\nu$$

and using the smallness condition, we find that there exist 2 positive constants k_0, K_0 depending on ρ such that

$$k_0 \int_\Omega |Du|^2 \psi \, dx + \int_\Omega aDX.D\psi \, dx \le K_0 \int_\Omega \psi \, dx.$$

But if we set

$$X_0(s) = X(\Lambda s) - 2^N,$$

we see that the general condition for deriving C^δ and $W^{1,p}$ properties that we gave in Chapter 3 (see Theorems 2.10 and 2.3) holds for u. Thus we obtain the result. $\diamond$

3.4 Estimates in L^∞

3.4.1 Assumptions

We shall consider linear transformations that satisfy the "maximum principle," that is,

$$\Gamma u \ge 0 \Rightarrow u \ge 0.$$

In practice it is sufficient that all the elements of Γ^{-1} be nonnegative. It is well known that this is by no means a necessary condition. In particular, positive definite matrices such that their nondiagonal terms are less than or equal to 0 satisfy the maximum principle (M matrices). We shall first assume that there exists a linear transformation Γ such that

$$\Gamma \text{ or } -\Gamma \text{ satisfies the maximum principle.} \tag{3.60}$$

Then also

$$\sum_\mu \Gamma_{\nu,\mu} H^\mu(x, s, p) \le \lambda^\nu + \lambda_0^\nu(\|s\|) \left| \sum_\mu \Gamma_{\nu,\mu} p^\mu \right|^2 - c^\nu \sum_\mu \Gamma_{\nu,\mu} s^\mu, \tag{3.61}$$

where

$$\lambda^\nu, \quad c^\nu > 0, \text{ constants,}$$
$$\lambda_0^\nu(\rho) \text{ positive monotone increasing.} \tag{3.62}$$

Finally, we assume that

There exist strictly positive constants α_μ, c, λ, and $\lambda_0(\rho)$positive monotone increasing such that either Γ satisfies the maximum principle and

$$\sum_\mu \alpha_\mu H^\mu(x, s, p) \geq -\lambda - \lambda_0(\|s\|) \left| \sum_\mu \alpha_\mu p^\mu \right|^2 - c \sum_\mu \alpha_\mu s^\mu \tag{3.63}$$

or $-\Gamma$ satisfies the maximum principle and

$$\sum_\mu \alpha_\mu H^\mu(x, s, p) \leq \lambda + \lambda_0(\|s\|) \left| \sum_\mu \alpha_\mu p^\mu \right|^2 - c \sum_\mu \alpha_\mu s^\mu.$$

3.4.2 Statement of Results

We now state the L^∞ estimate result.

Theorem 3.9. *We assume (3.40) and (3.60), (3.61), (3.62), (3.63). Any solution of (3.38), (3.42) that belongs to $\left(H_0^1(\Omega) \cap L^\infty(\Omega)\right)$ satisfies the following estimates: If Γ satisfies the maximum principle , then*

$$u^\nu(x) \leq \sum_\mu (\Gamma^{-1})_{\nu,\mu} \frac{\lambda^\mu}{c^\mu},$$
$$u^\nu(x) \geq -\frac{\lambda}{c\alpha^\nu} - \sum_{\substack{\mu \neq \nu \\ \varpi}} \frac{\alpha_\mu}{\alpha_\nu}(\Gamma^{-1})_{\mu,\varpi} \frac{\lambda^\varpi}{c^\varpi}; \tag{3.64}$$

if $-\Gamma$ satisfies the maximum principle , then

$$u^\nu(x) \geq -\sum_\mu (\Gamma^{-1})_{\nu,\mu} \frac{\lambda^\mu}{c^\mu},$$
$$u^\nu(x) \leq \frac{\lambda}{c\alpha^\nu} + \sum_{\substack{\mu \neq \nu \\ \varpi}} \frac{\alpha_\mu}{\alpha_\nu}(\Gamma^{-1})_{\mu,\varpi} \frac{\lambda^\varpi}{c^\varpi}. \tag{3.65}$$

Proof. If we consider the system obtained after the linear transformation Γ, we have (setting $z = \Gamma u$)

$$Az = \Gamma H(x, \Gamma^{-1}z, D\Gamma^{-1}z).$$

From the assumption (3.61) it follows that

$$Az^\nu \leq \lambda^\nu + \lambda_0^\nu(\rho)|Dz^\nu|^2 - c^\nu z^\nu, \tag{3.66}$$

where

$$\rho = \sup \|u(x)\|,$$

and from the weak maximum principle (Section 1.2.4), it follows that

$$z^\nu \leq \frac{\lambda^\nu}{c^\nu}.$$

Suppose now that Γ satisfies the maximum principle ; then by definition we obtain the first estimate in (3.64). Next consider the function

$$\overline{u}(x) = \sum_\mu \alpha_\mu u^\mu(x).$$

Then from assumption (3.63) we have

$$A\overline{u} \geq -\lambda - \lambda_0(\rho)|D\overline{u}|^2 - c\overline{u},$$

and thus again from the weak maximum principle we deduce

$$\overline{u} \geq -\frac{\lambda}{c}. \tag{3.67}$$

The second part of (3.64) follows from (3.67) and the first estimate (3.64). The proof of (3.65) is similar. Hence the proof is complete. $\diamond$

3.5 Existence of Solutions

3.5.1 Setting of the Problem and Assumptions

We are now going to give an existence result for solutions of (3.38), (3.42). In addition to the preceding assumptions, we shall need the following:

$$H(x,s,p) \text{ is continuous in } s,p \tag{3.68}$$

and

$$a_{ij}(x) \in W^{1,\infty}. \tag{3.69}$$

We now state our main result.

Theorem 3.10. *We assume (3.40), (3.43), (3.45), (3.60), (3.61), (3.62), (3.63), with all c^ν identical and equal to c. Finally, we assume (3.68), (3.69). Then there exist solutions of (3.38), (3.42) in $W^{2,p}(\Omega)$ for every $2 \leq p < \infty$.*

3.5.2 Proof of Existence

We shall construct an approximation as follows. Let us define

$$H^\epsilon(x, s, p) = \frac{H(x, s, p) + cs}{1 + \epsilon |H(x, s, p) + cs|} - cs.$$

Note that here the c refers explicitly to the one corresponding to the linear transformation Γ introduced in the assumptions (3.61), (3.63).
By construction,

$$|H^\epsilon(x, s, p) + cs| \le \frac{1}{\epsilon}.$$

A linear transformation $\tilde{\Gamma}$ (not necessarily Γ) on H induces a corresponding one on H^ϵ as follows:

$$\tilde{\Gamma} H^\epsilon(x, \tilde{\Gamma}^{-1}s, \tilde{\Gamma}^{-1}p) = \frac{\tilde{\Gamma} H(x, \tilde{\Gamma}^{-1}s, \tilde{\Gamma}^{-1}p) + cs}{1 + \epsilon |H(x, \tilde{\Gamma}^{-1}s, \tilde{\Gamma}^{-1}p) + c\tilde{\Gamma}^{-1}s|} - cs. \qquad (3.70)$$

Consider assumption (3.43). Then from (3.70) it follows that the same formula holds for H^ϵ, with the same functions

$$k(\rho), \quad K(\rho), \quad K_\nu^\mu(\rho)$$

and with

$$k^\nu(\rho) \to k^\nu(\rho) + c\rho.$$

In particular, it is very important to notice that these functions of ρ do not depend on ϵ.

Next notice that the assumptions (3.60), (3.61), (3.62), (3.63) will hold for H^ϵ with the same $\alpha_\mu, \lambda^\nu, \lambda_0^\nu(\rho)$ and of course the same c. Also, (3.68) holds for H^ϵ. We now consider the approximating problem

$$Au^\epsilon = H^\epsilon(x, u^\epsilon, Du^\epsilon), \qquad (3.71)$$

which of course can be written as

$$Au^\epsilon + cu^\epsilon = H^\epsilon(x, u^\epsilon, Du^\epsilon) + cu^\epsilon, \qquad (3.72)$$

and the right-hand side of (3.72) is bounded by the number $1/\epsilon$.

We are going to check that there exists a solution of (3.71) such that

$$u^\epsilon \in (W^{2,p} \cap W_0^{1,p}(\Omega))^N. \qquad (3.73)$$

This is obtained by a fixed point argument. Indeed, consider the map

$$T^\epsilon : (W_0^{1,p}(\Omega))^N \Rightarrow (W_0^{1,p}(\Omega))^N$$

defined by the equation

$$Av + cv = H^\epsilon(x, z, Dz) + cz, \quad v \in (W_0^{1,p}(\Omega))^N.$$

The right-hand side is in L^∞. From (3.69) we may refer to the classical regularity theory of elliptic PDEs [46], [66] to assert that

$$\|v\|_{(W^{2,p})^N} \leq C_\epsilon.$$

Consider the set

$$K_\epsilon = \{z \in (W_0^{1,p}(\Omega))^N \,|\, \|z\|_{(W^{2,p})^N} \leq C_\epsilon\}.$$

Then K_ϵ is a compact subset of $(W_0^{1,p}(\Omega))^N$, and T^ϵ maps K_ϵ into itself. From the assumption (3.68) and the corresponding continuity property for H^ϵ, we check immediately that T^ϵ is continuous. From Schauder's fixed point theorem, the map T^ϵ has a fixed point. We can then apply to u^ϵ the theory developed in Sections 3.3, 3.4 and in particular the results of Theorems 3.7, 3.9. In view of the properties of H^ϵ discussed above, we get the following estimates:

$$\|u^\epsilon\|_{(L^\infty)^N} \leq C, \quad \|u^\epsilon\|_{(H_0^1)^N} \leq C,$$

and

$$\|u^\epsilon\|_{(C^\delta)^N} \leq C, \quad \|u^\epsilon\|_{(W^{1,p})^N} \leq C,$$

with

$$0 < \delta \leq \delta_0 < 1, \quad 2 \leq p < 2 + \sigma_0.$$

Now also from equation (3.71), we have

$$\|Au^\epsilon\|_{(L^1)^N} \leq C.$$

If we consider a subsequence (still denoted by u^ϵ to save notation) that converges to u in $(H_0^1)^N$ weakly and in $(L^\infty)^N$ in the weak-star topology, then the above estimates and standard compactness properties are sufficient to guarantee that there is strong convergence in $(H_0^1)^N$ and in $(L^\infty)^N$. From this and Vitali's theorem, it follows that

$$H^\epsilon(x, u^\epsilon, Du^\epsilon) \to H(x, u, Du)$$

in $(L^1)^N$. Hence u is a solution.

Once we have a solution in $(H_0^1)^N \cap (C^\delta)^N$ or in $(W^{1,p})^N$, then the regularity results of Chapter 2 (see the proof of Theorem 2.23) imply that

$$u \in (W^{2,p})^N.$$

The proof of Theorem 3.10 is complete. ◇

3.5.3 Existence of a Weak Solution

Assumption (3.69) was used in the proof of Theorem 3.10 only to define the approximation u^ϵ through a Schauder fixed-point type argument (and of course at the end to obtain the $W^{2,p}$ regularity). None of the estimates in the spaces L^∞, C^δ, or $W^{1,p}$ make use of it; they require only the assumption (3.40) on the matrix a. We can show that it is possible to delete the assumption (3.69) and prove the existence of a weak solution of (3.38), (3.42). By weak solution we mean a solution in

$$(L^\infty(\Omega))^N \cap (H_0^1(\Omega))^N.$$

In fact, it will automatically be in

$$(C^\delta(\Omega))^N \cap (W_0^{1,p}(\Omega))^N,$$

where $0 < \delta \leq \delta_0 < 1$, $2 \leq p < 2 + \sigma_0$. We state the following result:

Theorem 3.11. *We assume (3.40), (3.43), (3.45), (3.60), (3.61), (3.62), (3.63), with all c^ν identical and equal to c. We also assume (3.68). Then there exist weak solutions of (3.38), (3.42), that is, solutions in*

$$(L^\infty(\Omega))^N \cap (H_0^1(\Omega))^N$$

that are also

$$(C^\delta(\Omega))^N \cap (W_0^{1,p}(\Omega))^N, \quad 0 < \delta \leq \delta_0 < 1, \; 2 \leq p < 2 + \sigma_0.$$

Proof. Using mollifiers, we can build a sequence of matrices $a^\epsilon(x)$ that satisfy (3.40) uniformly in ϵ, and that are smooth and converge pointwise to a. Denote by A^ϵ the corresponding second-order differential operator. Consider then a solution u^ϵ of

$$A^\epsilon u^\epsilon = H(x, u^\epsilon, Du^\epsilon)$$

(which exists, by virtue of Theorem 3.10). We have the estimates

$$\|u^\epsilon\|_{(L^\infty)^N} \leq C, \quad \|u^\epsilon\|_{(H_0^1)^N} \leq C,$$

and

$$\|u^\epsilon\|_{(C^\delta)^N} \leq C, \quad \|u^\epsilon\|_{(W^{1,p})^N} \leq C,$$

with

$$0 < \delta \leq \delta_0 < 1, \quad 2 \leq p < 2 + \sigma_0.$$

None of these estimates make use of the fact that a^ϵ is in $W^{1,\infty}$. Moreover,

$$\|A^\epsilon u^\epsilon\|_{L^1} \leq C, \tag{3.74}$$

so we may extract a subsequence (still denoted by u^ϵ) that converges weakly in $(H_0^1)^N$ and weak-star in $(L^\infty)^N$ to u. From the compactness of the injection of C^δ into L^∞, we may assert that

$$u^\epsilon \to u \text{ in } (L^\infty)^N.$$

But from the estimate (3.74) it then follows that

$$\int A^\epsilon u^\epsilon (u^\epsilon - u)\, dx \to 0.$$

This means that

$$\int a^\epsilon Du^\epsilon.Du^\epsilon\, dx - \int a^\epsilon Du^\epsilon.Du\, dx \to 0. \tag{3.75}$$

Now from the weak convergence of u^ϵ in H_0^1 and the pointwise convergence of a^ϵ we deduce easily that

$$\int a^\epsilon Du^\epsilon.Du\, dx \to \int a Du.Du\, dx,$$

and in view of (3.75) we have in particular,

$$\int a^\epsilon (Du^\epsilon - Du).(Du^\epsilon - Du)\, dx \to 0.$$

Hence

$$u^\epsilon \to u \text{ in } (H_0^1)^N.$$

It then follows that

$$H(x, u^\epsilon, Du^\epsilon) \to H(x, u, Du) \text{ in } (L^1)^N$$

and also, for instance,

$$A^\epsilon u^\epsilon \to Au \text{ weakly in } (H^{-1})^N.$$

Hence u is a solution, and the proof is complete. $\diamond$

3.6 Hamiltonians Arising from Games

3.6.1 Notation

We shall use vectors

$$p = (p^1, \dots, p^N),$$

where each p^ν is in R^n (as already introduced as arguments of Hamiltonians; see Section 3.2). We shall need also other vectors of the same size called "controls" and denoted by

$$v = (v^1, \dots, v^N).$$

We shall make use of the notation

$$\bar{v}^\nu = \sum_{\mu \neq \nu} v^\mu, \quad \bar{p}^\nu = \sum_{\mu \neq \nu} p^\mu,$$

and of course,

$$\bar{v}^\nu + v^\nu = \sum_\mu v^\mu.$$

We introduce the functions (called Lagrangians)

$$L^\nu(v,p) = \frac{1}{2}v^{\nu*}Mv^\nu + v^{\nu*}\Theta\bar{v}^\nu + p^{\nu*}(\bar{v}^\nu + v^\nu), \tag{3.76}$$

where

$$M, \Theta \text{ are } n \times n \text{ symmetric matrices, and } M > 0. \tag{3.77}$$

We shall assume

$$M - \Theta \text{ and } M + (N-1)\Theta \text{ are invertible.} \tag{3.78}$$

We shall next consider the Nash point of the Lagrangians with respect to the controls, considering the arguments p as parameters. This means that we look for a stationary point of each Lagrangian L^ν in the variable v^ν, freezing all other controls. A Nash point is thus defined by the relations

$$Mv^\nu + \Theta\bar{v}^\nu + p^\nu = 0. \tag{3.79}$$

To save notation, we shall keep the notation $v^\nu, \bar{v}^\nu$ for the solution of (3.79), as well as for general arguments of the Lagrangians. In practice we shall only use the solution of (3.79). Note that thanks to (3.78), we can solve (3.79). Indeed, it is easy to check the formulas

$$v^\nu = (M - \Theta)^{-1}[\Theta(M + (N-1)\Theta)^{-1}\bar{p}^\nu$$
$$-(M + (N-2)\Theta)(M + (N-1)\Theta)^{-1}p^\nu], \tag{3.80}$$

$$\bar{v}^\nu = (M - \Theta)^{-1}[-M(M + (N-1)\Theta)^{-1}\bar{p}^\nu$$
$$+(N-1)\Theta(M + (N-1)\Theta)^{-1}p^\nu], \tag{3.81}$$

and

$$v^\nu + \bar{v}^\nu = -(M + (N-1)\Theta)^{-1}(\bar{p}^\nu + p^\nu). \tag{3.82}$$

It is also useful to invert these relations, expressing the p^ν in terms of the v^ν. Note that

$$\sum_\nu \bar{v}^\nu = (N-1)\sum_\nu v^\nu, \tag{3.83}$$

$$\sum_\nu \bar{p}^\nu = (N-1)\sum_\nu p^\nu, \tag{3.84}$$

and

$$p^\nu = -Mv^\nu - \Theta\bar{v}^\nu, \tag{3.85}$$

$$\bar{p}^\nu = -(N-1)\Theta v^\nu - (M+(N-2)\Theta)\bar{v}^\nu. \tag{3.86}$$

We shall write the resulting Lagrangians $L^\nu(p)$, where the notation stresses the fact that the last remaining arguments are the p, in the form

$$L^\nu(p) = -\frac{1}{2}v^{\nu*}Mv^\nu - (Mv^\nu + \Theta\bar{v}^\nu)^*\bar{v}^\nu, \tag{3.87}$$

or

$$L^\nu(p) = -\frac{1}{2}(v^\nu + \bar{v}^\nu)^*M(v^\nu + \bar{v}^\nu) + (\bar{v}^\nu)^*\left(\frac{M}{2} - \Theta\right)\bar{v}^\nu. \tag{3.88}$$

Of course, one has to keep in mind that the v^ν and the $\bar{v}^\nu$ have to be replaced by their values in terms of the p^ν, given by (3.80), (3.81). We shall consider the following Hamiltonians:

$$H^\nu(x,s,p) = L^\nu(p) + g(x).p^\nu - cs^\nu + f^\nu(x), \tag{3.89}$$

where

$$g,\, f^\nu \text{ measurable and bounded, } c > 0. \tag{3.90}$$

The objective now is to check whether these Hamiltonians satisfy the conditions of validity of Theorems 3.10, 3.11.

3.6.2 Verification of the Assumptions for Hölder Regularity

We shall consider the following linear transformation $\tilde{\Gamma}$:

$$\tilde{\Gamma}_{\nu,\mu} = \begin{cases} 1 & \text{if } \nu \neq N,\ \mu = \nu, \\ 0 & \text{if } \nu \neq N,\ \mu \neq \nu, \neq N, \\ -1 & \text{if } \nu \neq N,\ \mu = N, \end{cases} \tag{3.91}$$

$$\tilde{\Gamma}_{N,\mu} = \begin{cases} 0 & \text{if } \mu \neq N, \\ 1 & \text{if } \mu = N. \end{cases}$$

To simplify notation, we introduce

$$\tilde{p}^\nu = \sum_\mu \tilde{\Gamma}_{\nu,\mu}p^\mu,$$

$$\tilde{s}^\nu = \sum_\mu \tilde{\Gamma}_{\nu,\mu}s^\mu,$$

that is,

$$\tilde{p}^\nu = p^\nu - p^N \text{ if } \nu \neq N; \quad \tilde{p}^N = p^N,$$

and
$$\tilde{s}^\nu = s^\nu - s^N \text{ if } \nu \neq N; \quad \tilde{s}^N = s^N.$$

The corresponding Hamiltonians are
$$\tilde{H}^\nu(x, \tilde{s}, \tilde{p}) = H^\nu - H^N \text{ if } \nu \neq N; \quad \tilde{H}^N = H^N.$$

Using formulas (3.89), (3.88), (3.81), it is easy to check the relation
$$\tilde{H}^\nu = (\bar{v}^\nu + \bar{v}^N)^* \left(\frac{M}{2} - \Theta \right) (M - \Theta)^{-1} \tilde{p}^\nu + g.\tilde{p}^\nu - c\tilde{s}^\nu$$

for $\nu \neq N$. Let us define
$$Q(x, \tilde{s}, \tilde{p}) = (M - \Theta)^{-1} \left(\frac{M}{2} - \Theta \right) (\bar{v}^1 + \bar{v}^N) + g$$

and
$$\tilde{H}_0^1 = -c\tilde{s}^1.$$

We then define successively for $\nu = 2, \ldots, N$,
$$\tilde{H}_0^\nu = \tilde{H}^\nu - Q.\tilde{p}^\nu,$$

so in fact, after an easy check, for $\nu = 1, \ldots, N - 1$,
$$\tilde{H}_0^\nu = (\tilde{p}^\nu - \tilde{p}^1)^*(M - \Theta)^{-1} \left(\frac{M}{2} - \Theta \right) (M - \Theta)^{-1} \tilde{p}^\nu - c\tilde{s}^\nu.$$

We thus see that the correlation terms involve only $\tilde{p}^\nu$ and $\tilde{p}^1$. Hence the smallness condition is trivially satisfied, since we are in the case of $K_\mu^\nu = 0$, when $\mu > \nu$. We can thus assert the following:

Proposition 3.12. *Assuming (3.78), the Hamiltonians defined by the relations (3.88), (3.89) satisfy assumption (3.43) and trivially (3.45), using the linear transformation defined by (3.91).*

Remark 3.13. The original Hamiltonians do not satisfy the assumptions (3.43), (3.45), and thus it would have been impossible to derive the C^δ or $W^{1,p}$ regularity without using the linear transformation on the system. This clearly emphasizes the importance of this degree of freedom that the linear transformations represent.

3.6.3 Verification of the Assumptions for the L^∞ Bound

The next important point is to check whether the Hamiltonians introduced in (3.88), (3.89) satisfy the conditions leading to the L^∞ bounds. As we shall see, these conditions will not be automatically satisfied, and restrictions on the correlation matrix Θ will be introduced. To what extent these restrictions are necessary is an open problem.

Case $\Theta \leq \frac{M}{2}$

We assume here that

$$\Theta \leq \frac{M}{2}, \quad \Theta \text{ invertible.} \tag{3.92}$$

Note that with this assumption, the first condition (3.78) is satisfied.

This is an easy case. Indeed, we have the following result.

Proposition 3.14. *We assume (3.92) and the second condition of (3.78). Then the Hamiltonians defined by the relations (3.88), (3.89) satisfy the assumptions (3.60), (3.61), (3.62), (3.63) with $c^\nu = c$ and without using a linear transformation.*

Proof. In other words, we take $\Gamma = I$. Using

$$\bar{v}^\nu = -\Theta^{-1}(Mv^\nu + p^\nu)$$

we can write the Hamiltonians as follows (see (3.87)):

$$H^\nu(x, s, p) = -\frac{1}{2}v^{\nu*}Mv^\nu - p^{\nu*}\Theta^{-1}(Mv^\nu + p^\nu)$$
$$+ g(x).p^\nu - cs^\nu + f^\nu(x). \tag{3.93}$$

Thus for a convenient constant λ_0, we have

$$H^\nu(x, s, p) \leq \|f^\nu\| + \|g\|\|p^\nu\| + \lambda_0|p^\nu|^2 - cs^\nu.$$

The assumptions (3.61), (3.62) are satisfied, with

$$\lambda^\nu = \|f^\nu\|, \quad c^\nu = c$$

(where we use the fact that we may add the term $\|g\|\|p^\nu\|$ on the right-hand side of (3.61) without changing the result (maximum principle , Chapter 1, and Remark 2.17)).

Let us check (3.63). We take $\alpha_\mu = 1$. Using (3.88), (3.82), and assumption (3.92) we have

$$H^\nu(x, s, p) \geq -\frac{1}{2}(p^\nu + \bar{p}^\nu)^*(M + (N-1)\Theta)^{-1}M(M + (N-1)\Theta)(p^\nu + \bar{p}^\nu)$$
$$+ g.p^\nu - cs^\nu + f^\nu.$$

Since

$$\sum_\mu \alpha_\mu p^\mu = p^\nu + \bar{p}^\nu,$$

if we add up the preceding estimates, we find that (3.63) is satisfied with

$$\lambda = \left\| \sum_\nu f^\nu \right\|.$$

The proof is complete.   ◇

Case of a scalar correlation
We assume here
$$M = I, \quad \Theta = \theta I. \tag{3.94}$$

Our objective is to prove the following statement.

Proposition 3.15. *We assume (3.94) with $\theta > 1$. Then the Hamiltonians defined by the relations (3.88), (3.89) satisfy the assumptions (3.60), (3.61), (3.62), (3.63), with $c^\nu = c$.*

Remark 3.16. With $\theta > 1$, the conditions (3.78) are automatically satisfied.

Proof of Proposition 3.15: Considering (3.87), we have

$$H^\nu(x, s, p) = \left(\frac{1}{2} - \theta \right) |\bar{v}^\nu|^2 - \frac{1}{2} |v^\nu + \bar{v}^\nu|^2 + g.p^\nu - cs^\nu + f^\nu, \tag{3.95}$$

hence also

$$H^\nu(x, s, p) = \left(\frac{1}{2} - \theta \right) |\bar{v}^\nu|^2 - \frac{1}{2(N-1)^2} \left| \sum_\mu \bar{v}^\mu \right|^2 + g.p^\nu - cs^\nu + f^\nu,$$

and from (3.80) we have

$$\bar{v}^\nu = \frac{1}{(1 - \theta)(1 + (N - 1)\theta)} [-\bar{p}^\nu + (N - 1)\theta p^\nu]. \tag{3.96}$$

In view of formula (3.96), it is natural to consider the following linear transformation Γ:
$$\Gamma_{\nu\,\nu} = -(N - 1)\theta \quad \Gamma_{\nu\,\mu} = 1, \quad \forall \mu \neq \nu,$$

and from (3.96) we see that

$$\bar{v}^\nu = \frac{1}{(\theta - 1)(1 + (N - 1)\theta)} \sum_\mu \Gamma_{\nu\,\mu} p^\mu. \tag{3.97}$$

Let us check that if $\theta > 1$, then $-\Gamma$ satisfies the maximum principle . Indeed, its inverse is given by

$$\Gamma^{-1}_{\nu\,\nu} = -\frac{(N - 1)\theta - (N - 2)}{(N - 1)(\theta - 1)(1 + (N - 1)\theta)},$$

$$\Gamma^{-1}_{\nu\,\mu} = -\frac{1}{(N - 1)(\theta - 1)(1 + (N - 1)\theta)}, \quad \forall \mu \neq \nu,$$

and thus all coefficients are negative when $\theta > 1$.

We now compute

$$\sum_{\mu} \Gamma_{\nu\,\mu} L^{\mu} = -\theta(N-1)L^{\nu} + \sum_{\mu \neq \nu} L^{\mu}$$

$$= (N-1)\theta\left(\theta - \frac{1}{2}\right)|\bar{v}^{\nu}|^2 - \sum_{\mu \neq \nu}\left(\theta - \frac{1}{2}\right)|\bar{\eta}^{\mu}|^2 + \frac{\theta - 1}{2(N-1)}\left|\sum_{\mu} \bar{\eta}^{\mu}\right|^2$$

$$= |\bar{v}^{\nu}|^2\left[(N-1)\theta\left(\theta - \frac{1}{2}\right) + \frac{\theta - 1}{2(N-1)}\right] + \frac{\theta - 1}{N-1}\bar{v}^{\nu} \cdot \sum_{\mu \neq \nu} \bar{v}^{\mu}$$

$$+ \frac{\theta - 1}{2(N-1)}\left|\sum_{\mu \neq \nu} \bar{v}^{\mu}\right|^2 - \sum_{\mu \neq \nu}\left(\theta - \frac{1}{2}\right)|v_{\mu}|^2.$$

Now we have

$$\left|\sum_{\mu \neq \nu} \bar{v}^{\mu}\right|^2 \leq (N-1)\sum_{\mu \neq \nu} |\bar{v}^{\mu}|^2,$$

and thus

$$\frac{\theta - 1}{2(N-1)}\left|\sum_{\mu \neq \nu} \bar{v}^{\mu}\right|^2 - \sum_{\mu \neq \nu}\left(\theta - \frac{1}{2}\right)|v_{\mu}|^2 \leq -\sum_{\mu \neq \nu} \frac{\theta}{2}|v^{\mu}|^2.$$

Therefore, collecting results, we have

$$\sum_{\mu} \Gamma_{\nu\,\mu} L^{\mu} \leq |\bar{v}^{\nu}|^2\left[(N-1)\theta\left(\theta - \frac{1}{2}\right) + \frac{\theta - 1}{2(N-1)}\right]$$

$$+ \frac{\theta - 1}{N-1}\bar{v}^{\nu} \cdot \sum_{\mu \neq \nu} \bar{v}^{\mu} - \sum_{\mu \neq \nu} \frac{\theta}{2}|v^{\mu}|^2$$

$$\leq \lambda_0|\bar{v}^{\nu}|^2$$

for a convenient constant λ_0. It follows that

$$\sum_{\mu} \Gamma_{\nu\,\mu} H^{\mu} \leq \lambda_0|\bar{v}^{\nu}|^2 + \|g\|\left|\sum_{\mu} \Gamma_{\nu\,\mu} p^{\mu}\right| - \sum_{\mu} \Gamma_{\nu\,\mu} s^{\mu} + \left\|\sum_{\mu} \Gamma_{\nu\,\mu} f^{\mu}\right\|,$$

and in view of (3.97), we can assert that the assumptions (3.61), (3.62) are satisfied, with

$$\lambda^{\nu} = \left\|\sum_{\mu} \Gamma_{\nu\,\mu} f^{\mu}\right\|, \quad c^{\nu} = c.$$

As far (3.63) is concerned, we again take $\alpha_{\mu} = 1$. This time, since $-\Gamma$ satisfies the maximum principle , we must majorize $\sum_{\mu} H^{\mu}$. But from (3.95),

$$\sum_{\mu} H^{\mu} \le g. \sum_{\mu} p^{\mu} - c \sum_{\mu} s^{\mu} + \sum_{\mu} f^{\mu},$$

and thus (3.63) is immediate. The proof of Proposition 3.15 is complete. ◊

General Case

The method used for the scalar correlation does not carry over to the general case, since we cannot use $\bar{v}^{\nu}$ as an intermediary. We begin with a general calculation. Consider a positive number $\rho > 1$. Consider a linear transformation Γ such that

$$\Gamma_{\nu\,\nu} = -\rho(N-1), \quad \Gamma_{\nu\,\mu} = 1 \quad \forall \mu \ne \nu,$$

Then, as we have seen in the scalar correlation case, $-\Gamma$ satifies the maximum principle .

Note that from (3.81), (3.82) we also have

$$\bar{v}^{\nu} = -(\Theta - M)^{-1}p^{\nu} + (\Theta - M)^{-1}M(M + (N-1)\Theta)^{-1}(v^{\nu} + \bar{v}^{\nu}).$$

We use this formula in the Lagrangian (3.88), as well as (3.82). Reducing terms, we obtain the formula

$$L^{\nu} = -(p^{\nu} + \bar{p}^{\nu})^{*}(M + (N-1)\Theta)^{-1} \left(\frac{M}{2} + MBM \right)(M + (N-1)\Theta)^{-1}$$
$$(p^{\nu} + \bar{p}^{\nu}) - p^{\nu*}Bp^{\nu} + 2p^{\nu*}BM(M + (N-1)\Theta)^{-1}(p^{\nu} + \bar{p}^{\nu}),$$

where we have set

$$B = (\Theta - M)^{-1} \left(\Theta - \frac{M}{2} \right)(\Theta - M)^{-1}.$$

As is easily checked,

$$\frac{M}{2} + MBM = \frac{1}{2}M(\Theta - M)^{-1}\Theta M^{-1}\Theta(\Theta - M)^{-1}M. \qquad (3.98)$$

Let

$$\tilde{L}^{\nu} = -\rho(N-1)L^{\nu} + \sum_{\mu \ne \nu} L^{\mu},$$

which corresponds to performing the linear transformation on the Hamiltonians. We obtain

$$\tilde{L}^{\nu} = (\rho - 1)(N - 1)(p^{\nu} + \bar{p}^{\nu})^{*}(M + (N-1)\Theta)^{-1} \left(\frac{M}{2} + MBM \right)$$
$$(M + (N-1)\Theta)^{-1}(p^{\nu} + \bar{p}^{\nu}) + \rho(N-1)p^{\nu*}Bp^{\nu} \qquad (3.99)$$
$$- \sum_{\mu \ne \nu} p^{\mu*}Bp^{\mu} + 2(\bar{p}^{\nu} - \rho(N-1)p^{\nu})^{*}BM(M + (N-1)\Theta)^{-1}(p^{\nu} + \bar{p}^{\nu}).$$

Let us set

$$\tilde{p}^\nu = \bar{p}^\nu - \rho(N-1)p^\nu.$$

We shall use this new quantity to eliminate p^ν in the expression (3.99) and express $\tilde{L}^\nu$ in terms of $\tilde{p}^\nu$ and p^μ, $\mu \neq \nu$. The important term for our purpose is the term that does not involve $\tilde{p}^\nu$, which is obtained by taking

$$\tilde{p}^\nu = 0.$$

We get

$$\tilde{L}^\nu|_{\tilde{p}^\nu=0} = (\rho-1)\frac{(1+\rho(N-1))^2}{\rho^2(N-1)}(\bar{p}^\nu)^*(M+(N-1)\Theta)^{-1}\left(\frac{M}{2}+MBM\right)$$

$$(M+(N-1)\Theta)^{-1}\bar{p}^\nu + \frac{1}{\rho(N-1)}p^{\nu*}Bp^\nu - \sum_{\mu\neq\nu}p^{\mu*}Bp^\mu. \tag{3.100}$$

We begin by making the assumption

$$\Theta - \frac{M}{2} > 0. \tag{3.101}$$

Hence $B > 0$. This implies $M + (N-1)\Theta > 0$, but we must still assume

$$\Theta - M \text{ invertible.} \tag{3.102}$$

We now majorize (3.100) as follows (taking account of (3.101)):

$$\tilde{L}^\nu|_{\tilde{p}^\nu=0} \leq \frac{(\rho-1)(2N-3)}{N-1}\frac{(1+\rho(N-1))^2}{\rho^2}\sum_{\mu\neq\nu}p^{\mu*}(M+(N-1)\Theta)^{-1}$$

$$\left(\frac{M}{2}+MBM\right)(M+(N-1)\Theta)^{-1}p^\mu \tag{3.103}$$

$$+\left(\frac{2N-3}{\rho(N-1)}-1\right)\sum_{\mu\neq\nu}p^{\mu*}Bp^\mu.$$

The conditions to be introduced become clear from (3.103). Firstly, we need to pick a ρ that is not just larger than 1, but also that satisfies

$$\rho > \frac{2N-3}{N-1}.$$

Furthermore, we must have

$$\frac{(\rho-1)(2N-3)}{N-1}\frac{(1+\rho(N-1))^2}{\rho^2}\xi^*\left(\frac{M}{2}+MBM\right)\xi$$

$$< \left(1-\frac{2N-3}{\rho(N-1)}\right)\xi^*(M+(N-1)\Theta)B(M+(N-1)\Theta)\xi$$

for every $\xi \neq 0$. Recalling (3.98) and noting that

$$(M + (N - 1)\Theta)B(M + (N - 1)\Theta)$$
$$= (N\Theta(\Theta - M)^{-1} - I)\left(\Theta - \frac{M}{2}\right)(N\Theta(\Theta - M)^{-1} - I),$$

we can write the condition as follows:

$$\left(N - \frac{3}{2}\right) \min_{\rho > \frac{2N-3}{N-1}} \frac{(\rho - 1)(1 + \rho(N - 1))^2}{\rho((N - 1)\rho - (2N - 3))}$$

$$< \inf_{\xi} \frac{\xi^*(N\Theta(\Theta - M)^{-1} - I)(\Theta - \frac{M}{2})(N\Theta(\Theta - M)^{-1} - I)\xi}{\xi^* M(\Theta - M)^{-1}\Theta M^{-1}\Theta(\Theta - M)^{-1}M\xi}. \tag{3.104}$$

If this condition is satisfied, then we may find a convenient ρ such that

$$\tilde{L}^\nu|_{\tilde{p}^\nu = 0} \leq -c_0 \sum_{\mu \neq \nu} |p^\mu|^2,$$

where c_0 is an appropriate positive constant. Therefore, we also have

$$\tilde{L}^\nu \leq \lambda_0 |\tilde{p}^\nu|^2,$$

and thus we may assert that the assumptions (3.61), (3.62) are satisfied for the linear transformation obtained from this choice of ρ.

The verification of (3.63) is immediate, thanks to (3.101), as in the case of a scalar correlation. Finally, we can state the following result.

Proposition 3.17. *If we assume (3.101), (3.102), (3.104), then the Hamiltonians defined by the relations (3.88), (3.89) satisfy the assumptions (3.60), (3.61), (3.62), (3.63), with $c^\nu = c$*

For $N = 2$, the condition (3.104) reduces to

$$2 < \inf_{\xi} \frac{\xi^*(2\Theta(\Theta - M)^{-1} - I)(\Theta - \frac{M}{2})(2\Theta(\Theta - M)^{-1} - I)\xi}{\xi^* M(\Theta - M)^{-1}\Theta M^{-1}\Theta(\Theta - M)^{-1}M\xi}.$$

The condition (3.104) does not lead to the best result when applied to the scalar correlation case. We can take $\rho = \theta$ only when θ is large enough. One can check that θ must satisfy

$$\theta^2 - \theta(3N - 4) + 2N - 3 > 0.$$

When $N = 2$, however, this restriction holds for $\theta > 1$.

3.7 The Case of Two Players with Different Coupling Terms in the Payoffs

When $N = 2$ the Lagrangians (3.19) take the form

$$L_1(v,p) = \frac{1}{2}|v_1|^2 + \theta v_1.v_2 + p_1.(v_1 + v_2),$$

$$L_2(v,p) = \frac{1}{2}|v_2|^2 + \theta v_1.v_2 + p_2.(v_1 + v_2),$$

and Theorem 3.1 amounts to the following: If

$$\theta \leq \frac{1}{2}, \quad \Theta \neq -1; \text{ or } \theta > 1,$$

then there exists a Nash point for the game defined by the functionals (3.12). If we examine carefully the proof, we needed the following combinations of equations. In the case $\theta \leq \frac{1}{2}, \Theta \neq -1$, then we simply added up the equations to derive the L^∞ bound from below. When $\theta > 1$, we computed

$$\theta L_1(p) - L_2(p),$$

$$\theta L_2(p) - L_1(p),$$

to derive bounds from below for

$$\tilde{u}_1 = \theta u_1 - u_2,$$

$$\tilde{u}_2 = \theta u_2 - u_1,$$

and since

$$u_1 = \frac{\theta \tilde{u}_1 + \tilde{u}_2}{\theta^2 - 1},$$

$$u_2 = \frac{\theta \tilde{u}_2 + \tilde{u}_1}{\theta^2 - 1},$$

the same type of bound from below holds for u_1, u_2. To realize the special structure (see Section 3.6.2) we just subtract the equations and consider the new variables

$$\tilde{u}_1 = u_1 - u_2, \quad \tilde{u}_2 = u_2.$$

In this section we shall consider different coupling terms in the two Lagrangians, and see how our existence results extend.

3.7.1 Description of the Model and Statement of Results

We consider the following payoffs, for two players:

$$J_1(x,v) = E_{x,v} \int_0^{T_x} e^{-ct} \left(f_1(x(t)) + \frac{1}{2}|v_1(t)|^2 + \theta v_1(t).v_2(t) \right) dt,$$

$$J_2(x,v) = E_{x,v} \int_0^{T_x} e^{-ct} \left(f_2(x(t)) + \frac{1}{2}|v_2(t)|^2 + \sigma v_1(t).v_2(t) \right) dt. \tag{3.105}$$

This model leads to the following Lagrangians:

$$L_1(v,p) = \frac{1}{2}|v_1|^2 + \theta v_1.v_2 + p_1.(v_1 + v_2),$$

$$L_2(v,p) = \frac{1}{2}|v_2|^2 + \sigma v_1.v_2 + p_2.(v_1 + v_2). \tag{3.106}$$

We shall need the condition

$$\sigma\theta \neq 1. \tag{3.107}$$

We then state the following conditions

$$\sigma > 0, \quad \theta > 0, \quad \sigma\theta > 1, \tag{3.108}$$

or

$$0 < \theta \leq \frac{1}{2},$$

$$\frac{1-\theta-(1+\theta)\sqrt{1-2\theta}}{2\theta} \leq \sigma \leq -(\theta^2+\theta+1)+(\theta+1)\sqrt{\theta^2+2} \tag{3.109}$$

or

$$-1 < \theta < 0,$$

$$\frac{1-\theta-(1+\theta)\sqrt{1-2\theta}}{2\theta} \leq \sigma \leq -(\theta^2+\theta+1)+(\theta+1)\sqrt{\theta^2+2} \tag{3.110}$$

or

$$\theta < -1,$$

$$-(\theta^2+\theta+1)+(\theta+1)\sqrt{\theta^2+2} \leq \sigma \leq \frac{1-\theta-(1+\theta)\sqrt{1-2\theta}}{2\theta}. \tag{3.111}$$

The case $\theta = 0$ can be obtained as a limit case in (3.110), which yields

$$\theta = 0, \quad -\frac{1}{2} \leq \sigma \leq -1 + \sqrt{2}. \tag{3.112}$$

The case $\theta = -1$ can be obtained as a limit case in (3.110) and (3.111), which yields

$$\theta = -1, \quad \sigma = -1, \tag{3.113}$$

but this is forbidden by the condition (3.107). We then state the following theorem.

Theorem 3.18. *We make the assumptions (3.5), (3.6), (3.10), (3.11), (3.14), with $N = 2$. Suppose also that the parameters θ, σ satisfy (3.107) and one of the conditions (3.108), (3.109), (3.110), (3.111), (3.112). Then there exists a Nash point for the game defined by the functionals (3.105).*

Note that except in the case (3.108), we have $\sigma \leq \frac{1}{2}$. This can be checked by verifying that in all cases (3.109), (3.110), (3.111), (3.112) the upper bound on σ is always smaller than or equal to $\frac{1}{2}$. We can also check that the case $\sigma = \theta$ reduces to results already obtained.

We shall not give a full proof of Theorem 3.18, since many of the steps are in common with those of Theorem 3.1. The main differences concern the way we arrive at the L^∞ bounds, and at the special structure that leads to the H_0^1 and C^δ bounds. These two aspects are dealt with in the next two sections, respectively.

3.7.2 L^∞ Bounds

We present here the equivalent of Section 3.6.3. First, we notice that a Nash point for the Lagrangians $L_1(v, p), L_2(v, p)$ yields

$$v_1(p) + \theta v_2(p) + p_1 = 0,$$
$$v_2(p) + \sigma v_1(p) + p_2 = 0. \tag{3.114}$$

Hence

$$L_1(p) = L_1(v(p), p) = -\frac{1}{2}|v_1(p)|^2 + p_1 . v_2(p),$$

$$L_2(p) = L_2(v(p), p) = -\frac{1}{2}|v_2(p)|^2 + p_2 . v_1(p).$$

From this form we get the equivalent of (3.93). So we only need bounds from below. Note that provided that $\sigma\theta \neq 1$ we can solve the system (3.114), and obtain

$$v_1(p) = \frac{-p_1 + \theta p_2}{1 - \sigma\theta},$$

$$v_2(p) = \frac{-p_2 + \sigma p_1}{1 - \sigma\theta}.$$

We now consider the case (3.108) and set

$$\tilde{u}_1 = \sigma u_1 - u_2,$$
$$\tilde{u}_2 = \theta u_2 - u_1,$$

from which we deduce

$$u_1 = \frac{\theta \tilde{u}_1 + \tilde{u}_2}{\sigma \theta - 1},$$
$$u_2 = \frac{\sigma \tilde{u}_2 + \tilde{u}_1}{\sigma \theta - 1}.$$

Thanks to conditions (3.108), it is thus sufficient to obtain bounds from below on $\tilde{u}_1, \tilde{u}_2$ to derive similar ones on u_1, u_2. After easy computations, one can derive the following expressions:

$$\sigma L_1(p) - L_2(p) = \frac{\sigma}{2}|v_1(p) + v_2(p)|^2$$
$$+ |v_2(p)|^2 \left(-\frac{\sigma}{2} - \sigma\theta + \frac{1}{2} \right) + (1 - 2\sigma)v_1(p).v_2(p),$$

$$\theta L_2(p) - L_1(p) = \frac{\theta}{2}|v_1(p) + v_2(p)|^2$$
$$+ |v_1(p)|^2 \left(-\frac{\theta}{2} - \sigma\theta + \frac{1}{2} \right) + (1 - 2\theta)v_1(p).v_2(p).$$

Combining Bellman equations, we arrive at

$$-\frac{1}{2}\Delta\tilde{u}_1 - g(x).D\tilde{u}_1 + c\tilde{u}_1 \geq \sigma f_1 - f_2$$
$$+ |D\tilde{u}_1|^2 \frac{-\frac{\sigma}{2} - \sigma\theta + \frac{1}{2}}{(1 - \sigma\theta)^2} + \frac{1 - 2\sigma}{(1 - \sigma\theta)^2}D\tilde{u}_1.D\tilde{u}_2,$$

$$-\frac{1}{2}\Delta\tilde{u}_2 - g(x).D\tilde{u}_2 + c\tilde{u}_2 \geq \theta f_2 - f_1$$
$$+ |D\tilde{u}_2|^2 \frac{-\frac{\sigma}{2} - \sigma\theta + \frac{1}{2}}{(1 - \sigma\theta)^2} + \frac{1 - 2\sigma}{(1 - \sigma\theta)^2}D\tilde{u}_1.D\tilde{u}_2,$$

from which bounds from below on $\tilde{u}_1, \tilde{u}_2$ are easily obtained.

We want now to consider the sum

$$\tilde{u} = u_1 + u_2,$$

which implies computing the sum

$$L_1(p) + L_2(p).$$

After easy computations, whose details are left to the reader, we can check the following expression:

$$L_1(p) + L_2(p) = -\frac{1}{2}\frac{\theta^2 + \sigma^2}{(1 - \sigma\theta)^2}|p_1 + p_2|^2$$

$$+ \frac{1}{(1 - \sigma\theta)^2}(p_1 + p_2)((\sigma + \theta + 1)(\theta p_1 + \sigma p_2) - (p_1 + p_2))$$

$$+ \frac{1}{2(1 - \sigma\theta)^2}[p_1^2(2(\sigma + 1)(1 - \sigma\theta) - (\theta + 1)^2)$$

$$+ p_2^2(2(\theta + 1)(1 - \sigma\theta) - (\sigma + 1)^2)].$$

In order to realize

$$L_1(p) + L_2(p) \geq 0, \text{ when } p_1 + p_2 = 0,$$

we must assume the inequalities

$$2(\sigma + 1)(1 - \sigma\theta) - (\theta + 1)^2 \geq 0,$$

$$2(\theta + 1)(1 - \sigma\theta) - (\sigma + 1)^2 \geq 0,$$

which rewritten as second-order polynomials in σ yields

$$2\sigma^2\theta + 2\sigma(\theta - 1) + \theta^2 + 2\theta - 1 \leq 0,$$
$$\sigma^2 + 2\sigma(\theta^2 + \theta + 1) - (2\theta + 1) \leq 0. \qquad (3.115)$$

Note that if $\theta > 0$, the first condition implies that there must exist roots for the second-order polynomial in σ, and writing the discriminant, we obtain easily

$$\theta \leq \frac{1}{2},$$

and similarly

$$\sigma \leq \frac{1}{2}.$$

So we necessarily limit ourselves to the possibilities concerning θ described in (3.109), (3.110), (3.111), (3.112). Note that the second polynomial always has roots in σ. We first consider the case (3.112). In this case, $\theta = 0$, the conditions (3.115) reduce to (3.112). So, there remain the three possibilities expressed in (3.109), (3.110), (3.111). Note that since the second polynomial has roots in σ, we derive the conditions

$$-(\theta^2 + \theta + 1) - |\theta + 1|\sqrt{\theta^2 + 2} \leq \sigma$$

$$\leq -(\theta^2 + \theta + 1) + |\theta + 1|\sqrt{\theta^2 + 2}. \qquad (3.116)$$

The roots of the first polynomial are

$$\frac{-(\theta - 1) \pm |\theta + 1|\sqrt{1 - 2\theta}}{2\theta},$$

but to state the conditions on σ, we have to distinguish the cases $\theta > 0$ from $\theta < 0$. So in the case (3.109), we must write

$$\frac{1 - \theta - (\theta + 1)\sqrt{1 - 2\theta}}{2\theta} \leq \sigma \leq \frac{1 - \theta + (\theta + 1)\sqrt{1 - 2\theta}}{2\theta}, \qquad (3.117)$$

and (3.116) reduces to

$$-(\theta^2 + \theta + 1) - (\theta + 1)\sqrt{\theta^2 + 2} \leq \sigma$$

$$\leq -(\theta^2 + \theta + 1) + (\theta + 1)\sqrt{\theta^2 + 2}. \qquad (3.118)$$

To proceed we shall need to study the signs of the following functions:

$$\chi(\theta) = 2\theta(\sqrt{\theta^2 + 2} - \theta) - 1 - \sqrt{1 - 2\theta},$$

$$\phi(\theta) = 2\theta(\sqrt{\theta^2 + 2} - \theta) - 1 + \sqrt{1 - 2\theta},$$

$$\psi(\theta) = 2\theta(\sqrt{\theta^2 + 2} + \theta) + 1 + \sqrt{1 - 2\theta}.$$

A simple calculation shows that

$$\chi'(\theta) = 2\frac{(\sqrt{\theta^2 + 2} - \theta)^2}{\sqrt{\theta^2 + 2}} + (1 - 2\theta)^{-1/2} > 0.$$

Hence $\chi'(\theta) > 0$, and, since $\chi(\frac{1}{2}) = 0$, we can assert that

$$\chi(\theta) \leq 0, \quad \text{for } \theta \leq \frac{1}{2}. \qquad (3.119)$$

Next,

$$\phi'(\theta) = 2\frac{(\sqrt{\theta^2 + 2} - \theta)^2}{\sqrt{\theta^2 + 2}} - (1 - 2\theta)^{-1/2} > 0$$

and

$$\phi''(\theta) = -2\frac{(\sqrt{\theta^2 + 2} - \theta)^2}{\sqrt{\theta^2 + 2}}\left[2 + \frac{\theta}{\sqrt{\theta^2}}\right] - (1 - 2\theta)^{-3/2} > 0.$$

Therefore, $\phi'(\theta)$ is decreasing, and since

$$\phi'(-\infty) = +\infty, \quad \phi'(0) = 2\sqrt{2} - 1, \quad \phi'\left(\frac{1}{2}\right) = -\infty,$$

it follows that the equation $\phi'(\theta) = 0$ has only one root θ_0, with

$$0 < \theta_0 < \frac{1}{2}.$$

Noting that

$$\phi(0) = \phi\left(\frac{1}{2}\right) = 0,$$

we can assert that

$$\phi(\theta) \leq 0, \quad \text{for } \theta \leq 0,$$
$$\phi(\theta) \geq 0, \quad \text{for } 0 \leq \theta \leq \frac{1}{2}. \tag{3.120}$$

Finally, we have

$$\psi(\theta) \geq 0, \quad \text{for } \theta \leq \frac{1}{2}. \tag{3.121}$$

Indeed, the positivity being obvious when $\theta \geq 0$, it is sufficient to consider the case $\theta \leq 0$. But then writing

$$\psi(\theta) = -2\theta^2\sqrt{1 + \frac{2}{\theta^2}} + 2\theta^2 + 1 + \sqrt{1 - 2\theta}$$

and using

$$\sqrt{1 + \frac{2}{\theta^2}} \leq 1 + \frac{1}{\theta^2},$$

we check easily that

$$\psi(\theta) \geq -1 + \sqrt{1 - 2\theta} \geq 0.$$

With the properties (3.119), (3.120), (3.121) in mind, we first analyze the case $0 \leq \theta \leq \frac{1}{2}$, in which (3.117), (3.118) apply. We check easily that

$$-(\theta^2 + \theta + 1) - (\theta + 1)\sqrt{\theta^2 + 2} \leq \frac{1 - \theta - (\theta + 1)\sqrt{1 - 2\theta}}{2\theta}$$

and

$$-(\theta^2 + \theta + 1) + (\theta + 1)\sqrt{\theta^2 + 2} \leq \frac{1 - \theta + (\theta + 1)\sqrt{1 - 2\theta}}{2\theta},$$

from which (3.109) follows, checking also that the interval in which σ lies is not empty.

Turning now to the situation where $\theta < 0$, we first notice that the conditions on σ derived from the first polynomial are expressed as

$$\sigma \geq \frac{1 - \theta - |\theta + 1|\sqrt{1 - 2\theta}}{2\theta}$$

or

$$\sigma \leq \frac{1 - \theta + |\theta + 1|\sqrt{1 - 2\theta}}{2\theta}.$$

So we have to distinguish two cases. First

$$-1 < \theta < 0,$$

on which we have the conditions

$$-(\theta^2 + \theta + 1) - (\theta+1)\sqrt{\theta^2+2} \le \sigma \le -(\theta^2+\theta+1) + (\theta+1)\sqrt{\theta^2+2}$$

and

$$\sigma \ge \frac{1 - \theta - (\theta+1)\sqrt{1-2\theta}}{2\theta}$$

or

$$\sigma \le \frac{1 - \theta + (\theta+1)\sqrt{1-2\theta}}{2\theta}.$$

Using the signs of the functions ϕ and ψ, one can check that

$$-(\theta^2 + \theta + 1) - (\theta+1)\sqrt{\theta^2+2} \le \frac{1 - \theta - (\theta+1)\sqrt{1-2\theta}}{2\theta}$$

$$\le -(\theta^2+\theta+1) + (\theta+1)\sqrt{\theta^2+2}$$

and

$$\frac{1 - \theta + (\theta+1)\sqrt{1-2\theta}}{2\theta} \le -(\theta^2+\theta+1) - (\theta+1)\sqrt{\theta^2+2}.$$

Comparing intervals, we conclude easily that (3.110) holds. The last case is

$$\theta < -1,$$

on which we have the conditions

$$-(\theta^2+\theta+1) + (\theta+1)\sqrt{\theta^2+2} \le \sigma \le -(\theta^2+\theta+1) - (\theta+1)\sqrt{\theta^2+2}$$

and

$$\sigma \ge \frac{1 - \theta + (\theta+1)\sqrt{1-2\theta}}{2\theta}$$

or

$$\sigma \le \frac{1 - \theta - (\theta+1)\sqrt{1-2\theta}}{2\theta}.$$

We proceed in a way similar to (3.110) to obtain (3.111).

3.7.3 H_0^1 Bound

We shall derive the H_0^1 bound once the L^∞ bound is available. Similar methods, although more involved, permit us to derive the C^δ estimates, and will not be detailed here. Recalling the method of Section 3.6.2 (so we consider here temporarily the case $\sigma = \theta$), for $N = 2$, we set

$$\tilde{u}_1 = u_1 - u_2,$$
$$\tilde{u}_2 = u_2,$$
$$\tilde{p}_1 = p_1 - p_2,$$
$$\tilde{p}_2 = p_2,$$

and we get

$$\tilde{L}_1(p) = L_1(p) - L_2(p) = \frac{\frac{1}{2} - \theta}{1 - \theta}(v_1(p) + v_2(p)).\tilde{p}_1.$$

We note that $v_1(p) + v_2(p)$ is a linear function of p, hence of $\tilde{p}$, so we may set

$$Q(\tilde{p}) = \frac{\frac{1}{2} - \theta}{1 - \theta}(v_1(p) + v_2(p))$$

with

$$|Q(\tilde{p})| \le C|\tilde{p}|.$$

Hence we get the structure

$$\tilde{L}_1(p) = Q(\tilde{p}).\tilde{p}_1.$$

We may write

$$\tilde{L}_2(p) = Q(\tilde{p}).\tilde{p}_2 + H_2(\tilde{p}),$$

with

$$|H_2(\tilde{p})| \le K(|\tilde{p}_1|^2 + |\tilde{p}_2|^2).$$

With this special structure, we get the desired bounds on $\tilde{u}$.

Turning to our more general situation, where $\sigma \ne \theta$, we infer from the preceding argument that if we can find a real number λ such that the property

$$L_1(p) + \lambda L_2(p) = R(p).(p_1 + \lambda p_2) \tag{3.122}$$

holds with $R(p)$ a linear function, then the special struture detailed above is obtained, setting

$$\tilde{u}_1 = u_1 + \lambda u_2, \tilde{u}_2 = u_2,$$

and related relations as above. When $\sigma = \theta$, the convenient value of λ is -1. So everything amounts to proving that there exists λ such that (3.122) holds. Using the definition of $L_1(p), L_2(p)$, we can write

$$L_1(p) + \lambda L_2(p)$$
$$= \frac{1}{2}(v_1^2 + \lambda v_2^2) + (\theta + \lambda\sigma)v_1.v_2 + (p_1 + \lambda p_2)(v_1 + v_2), \tag{3.123}$$

where we have omitted to write explicitly $v_1(p), v_2(p)$, which are linear functions of p. From (3.114), we can also write

$$-(p_1 + \lambda p_2) = (1 + \lambda\sigma)v_1 + (\lambda + \theta)v_2. \tag{3.124}$$

Considering the forms (3.123), (3.124), we see that finding λ amounts to finding three numbers λ, β, γ such that the following identity holds:

$$\frac{1}{2}(v_1^2 + \lambda v_2^2) + (\theta + \lambda\sigma)v_1.v_2$$

$$= ((1 + \lambda\sigma)v_1 + (\lambda + \theta)v_2)(\beta v_1 + \gamma v_2), \tag{3.125}$$

where in (3.125) v_1, v_2 can take any values. Identifying terms on both sides of (3.125) yields

$$\frac{1}{2} = \beta(1 + \lambda\sigma),$$

$$\frac{\lambda}{2} = (\lambda + \theta)\gamma,$$

$$\theta + \lambda\sigma = (1 + \lambda\sigma)\gamma + (\lambda + \theta)\beta.$$

So we get, obviously,

$$\beta = \frac{1}{2(1 + \lambda\sigma)}, \quad \gamma = \frac{\lambda}{2(\lambda + \theta)}.$$

So the last relation yields

$$\theta + \lambda\sigma = \frac{(1 + \lambda\sigma)\lambda}{2(\lambda + \theta)} + \frac{\lambda + \theta}{2(1 + \lambda\sigma)},$$

which is an equation in λ. Setting

$$T(\lambda) = 2(\lambda + \theta)(1 + \lambda\sigma)(\theta + \lambda\sigma) - \lambda(1 + \lambda\sigma)^2 - (\lambda + \theta)^2,$$

then the problem amounts to finding a real root of

$$T(\lambda) = 0.$$

We express $T(\lambda)$ as follows:

$$T(\lambda) = \lambda^3\sigma^2 + \lambda^2(2\sigma\theta + 2\sigma^2\theta - 1) + \lambda(2\sigma\theta + 2\sigma\theta^2 - 1) + \theta^2.$$

It is obvious that this polynomial has at least one root. Note that whenever $\sigma = \theta$, one has

$$T(\lambda) = \lambda^3\theta^2 + (\lambda^2 + \lambda)(2\theta^2 + 2\theta^3 - 1) + \theta^2,$$

and we obtain that $\lambda = -1$ is a root.

4. Nonlinear Elliptic Systems Arising from Ergodic Control

4.1 Introduction

In Chapter 3 we considered Bellman systems of equations arising from the theory of stochastic games. L^∞ estimates were instrumental in deriving the other types of estimates (in H^1, $W^{1,p}$, C^δ), and the maximum principle was the main tool used to achieve this goal.

In this chapter we are going to consider some situations where it is impossible to obtain L^∞ bounds. This is the standard situation when one considers ergodic instead of ordinary control. More precisely, we are interested in a sequence of functions (depending on a parameter ϵ) each of which is bounded, though there is no global bound.

Consider the system (see (3.37))

$$Au_\epsilon^\nu + \epsilon u_\epsilon^\nu = H^\nu(x, Du_\epsilon), \tag{4.1}$$

where $\nu = 1, \ldots, N$ indexes the equations, and ϵ is a positive parameter that will tend to 0. We write

$$u_\epsilon = (u_\epsilon^1, \ldots, u_\epsilon^\nu, \ldots, u_\epsilon^N).$$

Note already that compared with Chapter 3, the Hamiltonians no longer depend on u. The operator A is a second-order linear differential operator, and the same holds for all equations. However, we are not going to consider Dirichlet conditions, but *periodic* conditions (it is also possible to consider Neumann boundary conditions).

The operator A is given in the divergence form

$$A = -\sum_{ij=1}^{n} \frac{\partial}{\partial x_i} a_{i,j} \frac{\partial}{\partial x_j}, \tag{4.2}$$

where the matrix $a = a_{ij}$ satisfies

$$a_{ij} \in W^{1,\infty}(R^n) \text{ periodic,}$$

$$a(x)\xi.\xi \ge \alpha|\xi|^2, \quad \forall \xi \in R^n, \quad x \in R^n; \quad \alpha > 0. \tag{4.3}$$

Assume also that

$$H(x,p) \text{ is Carathéodory, periodic in } x. \tag{4.4}$$

For convenience, we shall assume that the period is 1 in all components, and we set

$$Y = (0,1)^n.$$

The Hamiltonians will satisfy assumptions such that for positive ϵ the theory developed in Chapter 3 is applicable. In particular, the L^∞ bound is obtained, although not uniformly in ϵ.

The functions u_ϵ^ν will not converge as ϵ tends to 0. However, under appropriate assumptions the functions

$$z_\epsilon^\nu = u_\epsilon^\nu - \bar{u}_\epsilon^\nu,$$

where

$$\bar{u}_\epsilon^\nu = \int_Y u_\epsilon^\nu(x)\,dx,$$

will converge.

The limit problem will look as follows:

$$Az^\nu + \rho^\nu = H^\nu(x, Dz),$$

$$z^\nu \in W^{2,p}(Y), \text{ periodic, } \rho^\nu \text{ constant.} \tag{4.5}$$

In (4.5) the unknown is a pair (z^ν, ρ^ν) consisting of a periodic function and a constant. Of course, each pair is indexed by ν.

4.2 Assumptions and Statement of Results

4.2.1 Assumptions on the Hamiltonians

We first adapt the main assumption (3.43):

Up to a linear transformation the following structure holds:
$$H^\nu(x,p) = Q(x,p).p^\nu + H_0^\nu(x,p),$$
$$|H_0^\nu(x,p)| \le k^\nu + \sum_\mu K_\mu^\nu |p^\mu|^2, \tag{4.6}$$
K_μ^ν satisfies a smallness condition for $\mu > \nu$,
$$|Q(x,p)|^2 \le k + K|p|^2.$$

Remark 4.1. Again, the last Hamiltonian H^N can have general quadratic growth, since there is no smallness condition for $\nu = N$. If there is a function Q such that the structure described in the first part of (4.6) holds for all the other Hamiltonians, then we set

$$H_0^N(x,p) = H^N(x,p) - Q(x,p).p^N.$$

Let us make explicit the smallness condition. It is simpler than (3.44), in the sense that λ^ν does not depend on an argument s; however, $\gamma^\nu(s)$ needs to depend on s, as is apparent in the following inequalities.

Define $\lambda^\nu, \gamma^\nu(s)$ backwards successively by

$$\alpha(\lambda^N)^2 - \lambda^N K_N^N - \frac{K}{4\alpha} > 0, \quad \gamma^N = 1,$$
$$\alpha\lambda^\nu > K_\nu^\nu, \tag{4.7}$$

$$\gamma^\nu(s)\lambda^\nu(\alpha\lambda^\nu - K_\nu^\nu) - \frac{K}{4\alpha} - \sum_{\mu > \nu} \gamma^\mu(s)\lambda^\mu K_\nu^\mu \exp 2s\lambda^\mu > 0,$$

$$\nu = 1, \ldots, N-1.$$

The smallness condition is then

$$\sum_{\mu < \nu} \gamma^\mu(s)\lambda^\mu K_\nu^\mu \exp 2s\lambda^\mu$$
$$< \gamma^\nu(s)\lambda^\nu(\alpha\lambda^\nu - K_\nu^\nu) - \frac{K}{4\alpha} \tag{4.8}$$
$$- \sum_{\mu > \nu} \gamma^\mu(s)\lambda^\mu K_\nu^\mu \exp 2s\lambda^\mu, \quad \forall 2 \le \nu \le N,$$

and again the natural case in which the smallness condition is automatically satisfied is

$$K_\nu^\mu(s) = 0, \quad \forall \mu < \nu.$$

Remark 4.2. When $N = 2$, the assumption again simplifies. It is sufficient to assume that (up to a linear transformation)

$$|H^1(x,p)| \le \bar{k} + \bar{K}|p||p^1|, \tag{4.9}$$

and of course H^2 has general quadratic growth.

The following assumptions are made to derive L^∞ estimates (on z_ϵ^ν). We will mimic the assumptions of Section 3.4.1. We shall consider linear transformations Γ that satisfy the "maximum principle," i.e.,

$$\Gamma u \ge 0 \Rightarrow u \ge 0.$$

We assume that there exists a linear transformation Γ such that

$$\Gamma \text{ or } -\Gamma \text{ satisfies the maximum principle} \tag{4.10}$$

and

$$\sum_\mu \Gamma_{\nu,\mu} H^\mu(x,p) \le \lambda^\nu + \lambda_0^\nu \left| \sum_\mu \Gamma_{\nu,\mu} p^\mu \right|^2. \tag{4.11}$$

Finally, we assume that

There exist strictly positive constants $\alpha_\mu, \lambda, \lambda_0$ such that either Γ satisfies the maximum principle and

$$\sum_\mu \alpha_\mu H^\mu(x, p) \geq -\lambda + \lambda_0 |p|^2 \tag{4.12}$$

or $-\Gamma$ satisfies the maximum principle and

$$\sum_\mu \alpha_\mu H^\mu(x, p) \leq \lambda - \lambda_0 |p|^2.$$

Remark 4.3. These last two assumptions are much more stringent than the corresponding ones in Chapter 3, see (3.63), but we need them to obtain the H^1 estimates. In the bounded case the H^1 estimates are deduced from the L^∞ estimates, but we cannot follow the same route here, and need something else, which is provided by these assumptions.

4.2.2 Statement of Results

We now state the main result.

Theorem 4.4. *We assume (4.3), (4.4), (4.6), (4.10), (4.11), (4.12). Then there exists a solution z, λ of (4.5). Moreover, it is obtained as follows:*

$$u_\epsilon - \bar{u}_\epsilon \to z, \quad \epsilon u_\epsilon \to \rho$$

in $(H^1(Y))^N$ and pointwise.

4.3 Proof of Theorem 4.4

4.3.1 First Estimates

There exists a periodic $W^{2,p}$ solution u_ϵ of (4.1). This follows from Theorem 3.10.

We begin with the following result directly inspired by Theorem 3.9.

Lemma 4.5. *If Γ satisfies the maximum principle , then*

$$\epsilon u_\epsilon^\nu(x) \leq \sum_\mu (\Gamma^{-1})_{\nu,\mu} \lambda^\mu,$$

$$\epsilon u_\epsilon^\nu(x) \geq -\frac{\lambda}{\alpha^\nu} - \sum_{\substack{\mu \neq \nu \\ \varpi}} \frac{\alpha_\mu}{\alpha_\nu} (\Gamma^{-1})_{\mu,\varpi} \lambda^\varpi. \tag{4.13}$$

If $-\Gamma$ satisfies the maximum principle , then

$$\epsilon u_\epsilon^\nu(x) \geq -\sum_\mu (\Gamma^{-1})_{\nu,\mu} \lambda^\mu,$$

$$\epsilon u_\epsilon^\nu(x) \leq \frac{\lambda}{\alpha^\nu} + \sum_{\substack{\mu \neq \nu \\ \varpi}} \frac{\alpha_\mu}{\alpha_\nu} (\Gamma^{-1})_{\mu,\varpi} \lambda^\varpi. \tag{4.14}$$

Proof. If we consider the system obtained after the linear transformation Γ, we have, setting $v_\epsilon = \Gamma u_\epsilon$,

$$Av_\epsilon + \epsilon v_\epsilon = \Gamma H(x, D\Gamma^{-1}v_\epsilon).$$

From the assumption (4.11) it follows that

$$Av_\epsilon^\nu + \epsilon v_\epsilon^\nu \leq \lambda^\nu + \lambda_0^\nu |Dv_\epsilon^\nu|^2, \tag{4.15}$$

and from the weak maximum principle, see Section 2.4, it follows that

$$\epsilon v_\epsilon^\nu \leq \lambda^\nu.$$

Suppose now that Γ satisfies the maximum principle; then by definition we obtain the first estimate in (4.13). Next consider the function

$$\overline{u}_\epsilon(x) = \sum_\mu \alpha_\mu u_\epsilon^\mu(x).$$

Then from the assumption (4.12) we have

$$A\overline{u}_\epsilon + \epsilon\overline{u}_\epsilon \geq -\lambda,$$

and thus again from the weak maximum principle we deduce

$$\epsilon\overline{u}_\epsilon \geq -\lambda. \tag{4.16}$$

The second part of (4.13) follows from (4.16) and the first estimate (4.13). The proof of (4.14) is similar. Hence the proof is complete. $\Diamond$

The next step consists in obtaining H^1 estimates. We have the following lemma.

Lemma 4.6. *The following estimate holds:*

$$\int_Y |Du_\epsilon|^2 \, dx \leq \frac{\lambda + \sum_{\mu,\nu} \alpha^\nu (\Gamma^{-1})_{\nu,\mu} \lambda^\mu}{\lambda_0}. \tag{4.17}$$

Proof. Assume, for instance, that Γ satisfies the maximum principle . We have, thanks to the first part of (4.12),

$$A\sum_\mu \alpha_\mu u_\epsilon^\mu + \epsilon\sum_\mu \alpha_\mu u_\epsilon^\mu \geq -\lambda + \lambda_0 |Du_\epsilon|^2.$$

Integrating over Y and using Lemma 4.5 (see (4.13)) the result follows immediately. The same estimate remains valid in the case where $-\Gamma$ satisfies the maximum principle . $\Diamond$

4.3.2 Estimates on $u_\epsilon^\nu - \bar{u}_\epsilon^\nu$

In the following, the generic term C denotes a constant that does not depend on ϵ.

Lemma 4.7. *Assume that Γ satisfies the maximum principle . Then one has*

$$\sum_\nu \alpha^\nu \left(u_\epsilon^\nu - \bar{u}_\epsilon^\nu \right) \geq -C. \tag{4.18}$$

Next, if $-\Gamma$ satisfies the maximum principle , then one has

$$\sum_\nu \alpha^\nu \left(u_\epsilon^\nu - \bar{u}_\epsilon^\nu \right) \leq C. \tag{4.19}$$

Proof. We prove only the first part. Let x_ϵ be the minimum of $\sum_\nu \alpha_\nu \left(u_\epsilon^\nu - \bar{u}_\epsilon^\nu \right)$. It is a point in Y. Let Q be a large smooth domain such that $\bar{Y} \subset Q$. Consider next the Green function at point x_ϵ, with Dirichlet data on the boundary of Q for the formal adjoint of A. It is the solution of

$$-\sum_{i,j=1}^n \frac{\partial}{\partial x_j} a_{ij} \frac{\partial}{\partial x_i} G^{x_\epsilon} = \delta(x - x_\epsilon),$$

$$G^{x_\epsilon} = 0 \text{ on } \partial Q. \tag{4.20}$$

The preceding equation is purely formal. More precisely, for any ϕ that is $C^1(\bar{Q})$ and equals 0 on the boundary, we have

$$\int_Q a_{ij} \frac{\partial}{\partial x_i} G^{x_\epsilon} \frac{\partial}{\partial x_j} \phi \, dx = \phi(x_\epsilon).$$

Consider also a smooth cutoff function τ that is 1 on $\bar{Y}$ and vanishes on ∂Q. Equations (4.1) yield

$$A \sum_\nu \alpha^\nu \left(u_\epsilon^\nu - \bar{u}_\epsilon^\nu \right) + \epsilon \sum_\nu \alpha_\nu u_\epsilon^\nu = \sum_\nu \alpha_\nu H^\nu(x, Du_\epsilon).$$

We test the above equation with

$$\left(\sum_\nu \alpha_\nu \left(u_\epsilon^\nu - \bar{u}_\epsilon^\nu \right) \right)^- G^{x_\epsilon} \tau,$$

where we recall that

$$a^- = \text{negative part of } a.$$

Integrating over Q and using the definition of G^{x_ϵ}, we easily get the following relation:

$$-\int_Q a_{ij}\frac{\partial}{\partial x_j}\left(\left(\sum_\nu \alpha_\nu(u_\epsilon^\nu-\bar{u}_\epsilon^\nu)\right)^-\right)\frac{\partial}{\partial x_i}\left(\left(\sum_\nu(u_\epsilon^\nu-\bar{u}_\epsilon^\nu)\right)^-\right)G^{x_\epsilon}\tau\,dx$$

$$-\frac{1}{2}\left(\left(\sum_\nu \alpha_\nu(u_\epsilon^\nu-\bar{u}_\epsilon^\nu)\right)^-(x_\epsilon)\right)^2$$

$$+\frac{1}{2}\int_Q\left(\left(\sum_\nu \alpha_\nu(u_\epsilon^\nu-\bar{u}_\epsilon^\nu)\right)^-\right)^2\left(a_{ij}\frac{\partial G^{x_\epsilon}}{\partial x_j}\frac{\partial \tau}{\partial x_i}+a_{ij}\frac{\partial G^{x_\epsilon}}{\partial x_i}\frac{\partial \tau}{\partial x_j}\right.$$
$$\left.+G^{x_\epsilon}a_{ij}\frac{\partial^2\tau}{\partial x_i\partial x_j}+G^{x_\epsilon}\frac{\partial \tau}{\partial x_i}\frac{\partial a_{ij}}{\partial x_j}\right)dx$$

$$+\int_Q \epsilon\sum_\nu \alpha_\nu u_\epsilon^\nu\left(\sum_\nu \alpha_\nu(u_\epsilon^\nu-\bar{u}_\epsilon^\nu)\right)^- G^{x_\epsilon}\tau\,dx$$

$$=\int_Q\left(\sum_\nu \alpha_\nu H^\nu(x,Du_\epsilon)\right)\left(\sum_\nu \alpha_\nu(u_\epsilon^\nu-\bar{u}_\epsilon^\nu)\right)^- G^{x_\epsilon}\tau\,dx.$$

We note that

$$\left(\sum_\nu \alpha_\nu(u_\epsilon^\nu-\bar{u}_\epsilon^\nu)\right)^-(x_\epsilon)=\left\|\left(\sum_\nu \alpha_\nu(u_\epsilon^\nu-\bar{u}_\epsilon^\nu)\right)^-\right\|_\infty,$$

and let us define

$$g^\epsilon(x)=a_{ij}\frac{\partial G^{x_\epsilon}}{\partial x_j}\frac{\partial \tau}{\partial x_i}+a_{ij}\frac{\partial G^{x_\epsilon}}{\partial x_i}\frac{\partial \tau}{\partial x_j}+G^{x_\epsilon}a_{ij}\frac{\partial^2\tau}{\partial x_i\partial x_j}+G^{x_\epsilon}\frac{\partial \tau}{\partial x_i}\frac{\partial a_{ij}}{\partial x_j}.$$

Due to the integrability properties of G^{x_ϵ}, the functions g^ϵ remain in a bounded subset of $L^p(Q)$, for any $p<\frac{n}{n-1}$. Using the assumption (4.12), as well as Lemma 4.5, we can assert that

$$\left\|\left(\sum_\nu \alpha_\nu(u_\epsilon^\nu-\bar{u}_\epsilon^\nu)\right)^-\right\|_\infty^2\le C\left\|\left(\sum_\nu \alpha_\nu(u_\epsilon^\nu-\bar{u}_\epsilon^\nu)\right)^-\right\|_\infty\int_Q G^{x_\epsilon}\,dx$$
$$+\int_Q\left(\left(\sum_\nu \alpha_\nu(u_\epsilon^\nu-\bar{u}_\epsilon^\nu)\right)^-\right)^2 g^\epsilon(x)\,dx.$$

But

$$\int_Y G^{x_\epsilon}\,dx\le C;$$

hence also

$$\left\|\left(\sum_\nu \alpha_\nu(u_\epsilon^\nu - \bar{u}_\epsilon^\nu)\right)^-\right\|_\infty^2 \le C \left\|\left(\sum_\nu \alpha_\nu(u_\epsilon^\nu - \bar{u}_\epsilon^\nu)\right)^-\right\|_\infty^2$$

$$+ \int_Q \left(\left(\sum_\nu \alpha_\nu(u_\epsilon^\nu - \bar{u}_\epsilon^\nu)\right)^-\right)^2 g^\epsilon(x)\, dx \tag{4.21}$$

We may majorize, for any fixed L,

$$\int_Q \left(\left(\sum_\nu \alpha_\nu(u_\epsilon^\nu - \bar{u}_\epsilon^\nu)\right)^-\right)^2 g^\epsilon(x)\, dx \le L^2 \int_Q g^\epsilon(x)\, dx$$

$$+ \left\|\left(\sum_\nu \alpha_\nu(u_\epsilon^\nu - \bar{u}_\epsilon^\nu)\right)^-\right\|_\infty^2 \int_{Q \cap \{(\sum_\nu \alpha_\nu(u_\epsilon^\nu - \bar{u}_\epsilon^\nu))^- \ge L\}} g^\epsilon(x)\, dx.$$

By the Poincaré inequality

$$\|u_\epsilon^\nu - \bar{u}_\epsilon^\nu\|_{L^2} \le C \|Du_\epsilon^\nu\|_{L^2},$$

which together with Lemma 4.6 implies

$$\left|Q \cap \left\{\left(\sum_\nu \alpha_\nu(u_\epsilon^\nu - \bar{u}_\epsilon^\nu)\right)^- \ge L\right\}\right| \le \frac{C}{L^2},$$

and thus, applying Hölder's inequality to the integral $\int g^\epsilon(x)\, dx$,

$$\int_Q \left(\left(\sum_\nu \alpha_\nu(u_\epsilon^\nu - \bar{u}_\epsilon^\nu)\right)^-\right)^2 g^\epsilon(x)\, dx \le CL^2 + C\frac{1}{L^{2/q}} \left\|\left(\sum_\nu \alpha_\nu(u_\epsilon^\nu - \bar{u}_\epsilon^\nu)\right)^-\right\|_\infty^2$$

with

$$q = \frac{p}{p-1}.$$

Using this estimate, with L sufficiently large in (4.21), we obtain the inequality

$$\left\|\left(\sum_\nu \alpha_\nu(u_\epsilon^\nu - \bar{u}_\epsilon^\nu)\right)^-\right\|_\infty^2 \le C \left\|\left(\sum_\nu \alpha_\nu(u_\epsilon^\nu - \bar{u}_\epsilon^\nu)\right)^-\right\|_\infty + C$$

and hence

$$\left\|\left(\sum_\nu \alpha_\nu(u_\epsilon^\nu - \bar{u}_\epsilon^\nu)\right)^-\right\|_\infty \le C,$$

which is equivalent to (4.18), and the proof is complete.   ◇

Lemma 4.8. *If Γ satisfies the maximum principle , then*

$$u_\epsilon^\nu - \bar{u}_\epsilon^\nu \leq C, \ \forall \nu. \tag{4.22}$$

Further, if $-\Gamma$ satisfies the maximum principle , then one has

$$u_\epsilon^\nu - \bar{u}_\epsilon^\nu \geq -C. \tag{4.23}$$

Proof. Let us define

$$E_\epsilon^\nu = \exp \frac{\lambda_0^\nu}{\alpha}(v_\epsilon^\nu - \bar{v}_\epsilon^\nu),$$

where v_ϵ^ν has been defined in Lemma 4.5 and λ_0^ν is the constant entering in (4.11).

Since

$$AE_\epsilon^\nu = \frac{\lambda_0^\nu}{\alpha} E_\epsilon^\nu A u_\epsilon^\nu - \left(\frac{\lambda_0^\nu}{\alpha}\right)^2 E_\epsilon^\nu \sum_{i,j} a_{ij} \frac{\partial v_\epsilon^\nu}{\partial x_j} \frac{\partial v_\epsilon^\nu}{\partial x_i},$$

we make use of inequality (4.15) to deduce

$$AE_\epsilon^\nu + \frac{\lambda_0^\nu}{\alpha} E_\epsilon^\nu \epsilon u_\epsilon^\nu \leq \lambda^\nu \frac{\lambda_0^\nu}{\alpha} E_\epsilon^\nu. \tag{4.24}$$

Let x_ϵ^ν be the maximum of $v_\epsilon^\nu - \bar{v}_\epsilon^\nu$ over Y, and consider the Green function $G^{x_\epsilon^\nu}$ defined as in (4.20) with x_ϵ^ν replaced for x_ϵ. Consider the same cutoff function τ as before.

We test (4.24) with $E_\epsilon^\nu G^{x_\epsilon^\nu} \tau$. Operating with the same argument as in the proof of Lemma 4.7, we derive

$$\|E_\epsilon^\nu\| \leq C,$$

which implies

$$v_\epsilon^\nu - \bar{v}_\epsilon^\nu \leq C, \ \forall \nu.$$

Since Γ satisfies the maximum principle , property (4.22) follows.   ◇

As a consequence of Lemmas 4.7 and 4.8 we get immediately

$$\|u_\epsilon^\nu - \bar{u}_\epsilon^\nu\|_\infty \leq C, \ \forall \nu. \tag{4.25}$$

4.3.3 End of Proof of Theorem 4.4

We can now proceed with the proof of Theorem 4.4. We recall

$$z_\epsilon^\nu = u_\epsilon^\nu - \bar{u}_\epsilon^\nu.$$

Then, using Lemmas 4.5 and 4.6 we deduce

$$\|z_\epsilon\|_{(H^1(Y))^N} \leq C,$$

$$\|\epsilon\, u_\epsilon\|_{(L^\infty(Y))^N} \le C_1.$$

Also, from (4.25) we have

$$\|z_\epsilon\|_{(L^\infty(Y))^N} \le C.$$

Thanks to (4.6) and the regularity theory concomitant with this special structure (see Theorem 3.7) we can assert that

$$\|z_\epsilon\|_{(C^\delta(Y))^N} \le C.$$

Therefore, we can extract a subsequence such that

$$z_\epsilon \to z \text{ in } (H^1(Y))^N \text{ weakly and } (L^\infty(Y))^N \text{ weak-star.}$$

Now, from the equations for z_ϵ and the compactness property that follows from them, we have also

$$z_\epsilon \to z \text{ in } (H^1(Y))^N \text{ and pointwise.}$$

This allows us to pass to the limit in equations (4.1), and we obtain a solution in $H^1 \cap C^\delta$ of equations (4.5). The step from C^δ to $W^{2,p}$ is then standard (cf. Chapter 1). This concludes the proof of Theorem 4.4.

4.4 Verification of the Assumptions

4.4.1 Notation

We consider the Hamiltonians arising from games (see Section 3.6). We recall the formula

$$H^\nu(x,p) = L^\nu(p) + g(x).p^\nu + f^\nu(x), \tag{4.26}$$

where

$$g(x), f^\nu(x) \text{ are periodic and bounded,} \tag{4.27}$$

$$L^\nu = -(p^\nu + \bar{p}^\nu)^*(M + (N-1)\Theta)^{-1}\left(\frac{M}{2} + MBM\right)$$
$$(M + (N-1)\Theta)^{-1}(p^\nu + \bar{p}^\nu) - p^{\nu*}Bp^\nu + 2p^{\nu*}BM \tag{4.28}$$
$$(M + (N-1)\Theta)^{-1}(p^\nu + \bar{p}^\nu),$$

where we have set

$$B = (\Theta - M)^{-1}\left(\Theta - \frac{M}{2}\right)(\Theta - M)^{-1}, \tag{4.29}$$

and we recall that

$$\bar{p}^\nu = \sum_{\mu \neq \nu} p^\mu.$$

Moreover

$$M, \Theta \text{ are } n \times n \text{ symmetric matrices, } M > 0, \tag{4.30}$$

and

$$M - \Theta \text{ and } M + (N-1)\Theta \text{ are invertible.} \tag{4.31}$$

We have seen in Section 3.6.2 that (4.30) and (4.31) guarantee that the assumptions (4.6) and trivially (4.8) are satisfied, thanks to the degree of freedom, which allows use of a linear transformation. We thus need to concentrate on the verification of (4.10), (4.11), (4.12). The main difference, compared with Chapter 3, is the more stringent assumption (4.12).

4.4.2 The Scalar Case

For simplicity, we consider first the scalar case, that is,

$$M = I, \quad \Theta = \theta I, \tag{4.32}$$

where θ is a parameter whose value will separate cases. In order to get (4.31), we must have

$$\theta \neq 1, \quad 1 + (N-1)\theta \neq 0. \tag{4.33}$$

As in Chapter 3, there will be two cases: one for θ "sufficiently" negative, and one for θ "sufficiently" positive.

We first see that we can satisfy (4.10), (4.11) by taking

$$\Gamma = I.$$

Here there is no difference with Chapter 3. First note that

$$H^\nu(x, p) = -\frac{\theta^2}{2(1 + (N-1)\theta)^2(\theta-1)^2} \left| \sum_\mu p^\mu \right|^2$$

$$- \frac{\theta - \frac{1}{2}}{(\theta-1)^2} |p^\nu|^2 + 2\frac{\theta - \frac{1}{2}}{(1 + (N-1)\theta)(\theta-1)^2} p^\nu \cdot \sum_\mu p^\mu$$

$$+ g(x).p^\nu + f^\nu(x).$$

Thus

$$H^\nu(x, s, p) \leq \|f^\nu\| + \|g\| |p^\nu| + \lambda_0 |p^\nu|^2,$$

and (4.10), (4.11) hold.

The main point now is (4.12). Making appropriate assumptions we shall verify (4.12) with

$$\alpha_\mu = 1.$$

We first compute

$$\sum_{\nu} L^{\nu}(p) = -\frac{\theta - \frac{1}{2}}{(\theta - 1)^2} \sum_{\nu} |p^{\nu}|^2$$

$$+ \frac{(\frac{3}{2}N - 2)\theta^2 - (N - 3)\theta - 1}{(1 + (N - 1)\theta)^2(\theta - 1)^2} \left|\sum_{\mu} p^{\mu}\right|^2.$$

Therefore, we also have

$$\sum_{\nu} L^{\nu}(p) = \frac{-(N - 1)^2\theta^3 + \frac{N^2 + 1 - 3N}{2}\theta^2 + \theta - \frac{1}{2}}{(1 + (N - 1)\theta)^2(\theta - 1)^2} \sum_{\nu} |p^{\nu}|^2$$

$$+ 2\frac{(\frac{3}{2}N - 2)\theta^2 - (N - 3)\theta - 1}{(1 + (N - 1)\theta)^2(\theta - 1)^2} \sum_{\mu \neq \nu} p^{\mu}.p\nu. \tag{4.34}$$

From classical considerations on quadratic forms, one can convince oneself that the inequality

$$-(N - 1)^2\theta^3 + \frac{N^2 + 1 - 3N}{2}\theta^2 + \theta - \frac{1}{2}$$

$$-(N - 1)\left|\left(\frac{3}{2}N - 2\right)\theta^2 - (N - 3)\theta - 1\right| > 0 \tag{4.35}$$

is sufficient to obtain

$$\sum_{\nu} L^{\nu}(p) \geq c_0|p|^2,$$

which will clearly imply the property (4.12).

Of course, we could state (4.35) as an assumption on θ to be fulfilled in order to obtain the applicability of Theorem 4.4, but it is possible to check more carefully what it implies about the choice of θ. Let us write

$$F(\theta) = -(N - 1)^2\theta^3 + \frac{N^2 + 1 - 3N}{2}\theta^2 + \theta - \frac{1}{2}$$

$$-(N - 1)\left|\left(\frac{3}{2}N - 2\right)\theta^2 - (N - 3)\theta - 1\right|,$$

and considering the two possibilities for the abolute value, we introduce the two functions

$$F_1(\theta) = -(N - 1)^2\theta^3 + \frac{-2N^2 + 4N - 3}{2}\theta^2 + \theta(N^2 - 4N + 4) + N - \frac{3}{2},$$

$$F_2(\theta) = -(N - 1)^2\theta^3 + \frac{4N^2 - 10N + 5}{2}\theta^2 - \theta(N^2 - 4N + 2) - N + \frac{1}{2}.$$

Consider the two roots of

$$\left(\frac{3}{2}N - 2\right)\theta^2 - (N-3)\theta - 1,$$

namely

$$\theta_0 = \frac{N - 3 - \sqrt{N^2 + 1}}{3N - 4},$$

$$\theta'_0 = \frac{N - 3 + \sqrt{N^2 + 1}}{3N - 4}.$$

Then

$$F(\theta) = F_1(\theta) \text{ if } \theta \le \theta_0 \text{ or } \theta \ge \theta'_0,$$
$$F(\theta) = F_2(\theta) \text{ if } \theta_0 \le \theta \le \theta'_0.$$

Fortunately, $F_2(\theta)$ is quite simple, namely

$$F_2(\theta) = (\theta - 1)^2 \left(-(N-1)^2\theta - N + \frac{1}{2}\right).$$

Let

$$\bar{\theta} = -\frac{N - \frac{1}{2}}{(N-1)^2}.$$

Note that

$$\theta_0 < \bar{\theta} < 0.$$

Since

$$F_2(\theta) > 0 \text{ for } \theta < \bar{\theta},$$

it follows from the definition of $F(\theta)$ that

$$F(\theta) > 0 \text{ for } \theta_0 < \theta < \bar{\theta}. \tag{4.36}$$

The zone

$$\bar{\theta} \le \theta \le \theta'_0$$

must be excluded, since it corresponds to a negative value of $F(\theta)$.

Outside the interval (θ_0, θ'_0) we must discuss the sign of $F_1(\theta)$. Unfortunately, the situation is less easy and depends on the value of N. Consider the two roots of $(F_1)'(\theta)$, called

$$\theta_1 = -\frac{2N^2 - 4N + 3 + \sqrt{16N^4 - 88N^3 + 184N^2 - 168N + 57}}{6(N-1)^2}$$

and

$$\theta'_1 = \frac{-2N^2 + 4N - 3 + \sqrt{16N^4 - 88N^3 + 184N^2 - 168N + 57}}{6(N-1)^2}.$$

The first interesting information is that

$$\theta_1' < \theta_0'.$$

Since $F_1(\theta)$ decreases for $\theta > \theta_1'$, it follows that $F(\theta)$ decreases for $\theta > \theta_0'$. But

$$F(\theta_0') = F_2(\theta_0') < 0.$$

Therefore, we obtain that

$$F(\theta) < 0 \text{ for } \theta > \theta_0'.$$

It follows now that the region

$$\theta \geq \bar{\theta}$$

must be excluded. Taking into consideration (4.36), it remains to explore the region

$$\theta \leq \theta_0.$$

Here is where the situation differs with the value of N. Namely,

$$\theta_1 > \theta_0 \text{ if } N = 2,$$

and

$$\theta_1 < \theta_0 \text{ if } N \geq 3.$$

Note that $\theta_1 = -1$ if $N = 2$, and

$$\theta_1 > -1 \text{ for } N \geq 3.$$

Since $F_1(\theta)$ decreases for $\theta < \theta_1$, we can assert that when $N = 2$, $F(\theta)$ decreases for $\theta < \theta_0$. Noticing that

$$F(\theta_0) = F_2(\theta_0) > 0,$$

we conclude that whenever $N = 2$, the region $\theta \leq \theta_0$ is good (in the sense that $F(\theta) > 0$). Therefore, if $N = 2$, then

$$F(\theta) > 0 \text{ if } \theta < \bar{\theta} = -\frac{3}{2},$$

$$F(\theta) \leq 0 \text{ if } \theta < \bar{\theta} = -\frac{3}{2}.$$

On the other hand, for $N > 2$ not all the region $\theta \leq \theta_0$ is good. Of necessity, $F(\theta_1) = F_1(\theta_1) < 0$, since $F_1(-1) \leq 0$, as was directly checked. An interval around θ_1 must be excluded, namely the interval between the two zeros of $F_1(\theta)$. For instance, for $N = 3$ this interval is

$$\left[-1, -\frac{1 + \sqrt{97}}{16} \right].$$

The number $\theta = -\frac{3}{2}$ belongs to the excluded interval when $N \geq 6$. We summarize:

Theorem 4.9. *We consider the problem (4.5) with Hamiltonians given by (4.26), (4.27), (4.28). We assume (4.3), (4.4), and (4.32). The parameter θ is strictly smaller than*

$$\overline{\theta} = -\frac{N - \frac{1}{2}}{(N-1)^2}$$

and lies outside a particular interval of $(-\infty, \overline{\theta})$ whose endpoints are the two negative roots of $F_1(\theta)$. Then all the assumptions of Theorem 4.4 are satisfied. For $N = 2$, all values of $\theta < -\frac{3}{2}$ are possible. For $N = 3$, all values of $\theta < -\frac{5}{8}$ outside the interval $[-1, -\frac{1+\sqrt{97}}{16}]$ are valid.

Now we check that $\theta > 1$ is also valid. Consider the linear transformation Γ defined by

$$\Gamma_{\nu\,\nu} = -(N-1)\theta, \quad \Gamma_{\nu\,\mu} = 1, \quad \forall \mu \neq \nu.$$

We know from Section 3.6.3 that $-\Gamma$ satisfies the maximum principle , and that (4.10), (4.11) hold. To obtain (4.12) with $\alpha_\mu = 1$, we use again (4.34), and it is necessary and sufficient this time to assume

$$-(N-1)^2\theta^3 + \frac{N^2 + 1 - 3N}{2}\theta^2 + \theta - \frac{1}{2}$$
$$+(N-1)\left|\left(\frac{3}{2}N - 2\right)\theta^2 - (N-3)\theta - 1\right| < 0,$$

which is satisfied with $\theta > \theta_2$, where θ_2 is the positive root of $F_1(\theta)$. One checks easily that

$$\frac{1}{2} \leq \theta_2 < 1,$$

and thus for $\theta > 1$, (4.12) is also valid.

Theorem 4.10. *We consider the problem (4.5) with Hamiltonians given by (4.26), (4.27), (4.28). We assume (4.3), (4.4), and (4.32) with $\theta > 1$. Then all the assumptions of Theorem 4.4 are satisfied.*

4.4.3 The General Case

We do not assume (4.32).

Case $\Theta \leq \frac{M}{2}$

We assume here that

$$\Theta \leq \frac{M}{2}, \quad \Theta \text{ invertible}. \tag{4.37}$$

Note that with this assumption the first condition (4.31) is satisfied.

Given the second assumption (4.31), we have the properties (4.6) and (4.8). We then see that we can satisfy (4.10), (4.11), taking

$$\Gamma = I.$$

This is an easy consequence of formula (4.28). The main point next is to verify (4.12) when

$$\alpha_\mu = 1.$$

We can compute $\sum_\nu L^\nu(p)$ as follows:

$$\sum_\nu L^\nu(p) = \sum_\nu (p^\nu)^*(M + (N-1)\Theta)^{-1}M(\Theta - M)^{-1}$$

$$\left(-(N-1)^2\Theta M^{-1}\Theta M^{-1}\Theta \right.$$

$$+ \frac{N^2 - 3N + 1}{2}\Theta M^{-1}\Theta + \Theta - \frac{M}{2} \bigg)$$

$$(\Theta - M)^{-1}M(M + (N-1)\Theta)^{-1}p^\nu$$

$$+ 2\sum_{\mu \neq \nu}(p^\nu)^*(M + (N-1)\Theta)^{-1}M(\Theta - M)^{-1}$$

$$\times \left(\left(\frac{3N}{2} - 2 \right)\Theta M^{-1}\Theta - (N-3)\Theta - M \right)$$

$$(\Theta - M)^{-1}M(M + (N-1)\Theta)^{-1}.$$

(4.38)

Therefore, if we assume that

$$-(N-1)^2\Theta M^{-1}\Theta M^{-1}\Theta + \frac{N^2 - 3N + 1}{2}\Theta M^{-1}\Theta + \Theta - \frac{M}{2}$$

$$> (N-1)\left\| \left(\frac{3N}{2} - 2 \right)\Theta M^{-1}\Theta - (N-3)\Theta - M \right\| I,$$

(4.39)

then clearly from (4.38), we deduce that

$$\sum_\nu L^\nu(p) \geq c_0|p|^2,$$

which will again imply the property (4.12).

Unfortunately, in contrast to the scalar case, the condition (4.39) is not easy to study. We nevertheless state the following result.

Theorem 4.11. *We consider the problem (4.5) with Hamiltonians given by (4.26), (4.27), (4.28). We assume (4.3), (4.4), (4.30), and the second assumption (4.31). We also assume (4.37) and (4.39). Then all assumptions of Theorem 4.4 are satisfied.*

Case $\Theta \geq \frac{M}{2}$

We assume here that

$$\Theta \geq \frac{M}{2}, \quad \Theta - M \text{ invertible}. \tag{4.40}$$

We follow the presentation of Section 3.6.3. In particular, we make the assumption

$$\left(N - \frac{3}{2}\right) \min_{\rho > \frac{2N-3}{N-1}} \frac{(\rho - 1)(1 + \rho(N - 1))^2}{\rho((N - 1)\rho - (2N - 3))}$$

$$< \inf_{\xi} \frac{\xi^*(N\Theta(\Theta - M)^{-1} - I)(\Theta - \frac{M}{2})(N\Theta(\Theta - M)^{-1} - I)\xi}{\xi^*M(\Theta - M)^{-1}\Theta M^{-1}\Theta(\Theta - M)^{-1}M\xi}.$$

$$(4.41)$$

then we know from Chapter 3 that there exists some linear transformation Γ such that (4.10), (4.11) are satisfied. To verify (4.12), we refer to formula (4.38), and thus we are led to the assumption

$$-(N - 1)^2\Theta M^{-1}\Theta M^{-1}\Theta + \frac{N^2 - 3N + 1}{2}\Theta M^{-1}\Theta + \Theta - \frac{M}{2}$$
$$+ (N - 1)\left\|\left(\frac{3N}{2} - 2\right)\Theta M^{-1}\Theta - (N - 3)\Theta - M\right\|I < 0,$$

$$(4.42)$$

which will imply

$$\sum_\nu L^\nu(p) \leq -c_0|p|^2$$

and thus also the property (4.12). Again (4.42) cannot be solved as easily as in the scalar case.

Theorem 4.12. *We consider the problem (4.5) with Hamiltonians given by (4.26), (4.27), (4.28). We assume (4.3), (4.4), (4.30) and (4.40), (4.41), (4.42). Then all the assumptions of Theorem 4.4 are satisfied.*

4.5 A Variant of Theorem 4.4

4.5.1 Statement of Results

We are going to give a variant of Theorem 4.4 with slightly weaker assumptions, at the expense of restricting to dimension $n = 2$. This is really just a curiosity, since as we shall see, in the important application to differential games, the improvement is not relevant.

We assume (4.3), (4.4). Moreover, we assume

$$H^\nu(x, p) \leq \lambda^\nu + \lambda_0^\nu |p| \|p^\nu\|,$$

$$(4.43)$$

$$\sum_\nu H^\nu(x, p) \geq -\lambda + \lambda_0 |p|^2.$$

$$(4.44)$$

Theorem 4.13. *If we assume (4.3), (4.4), (4.43), (4.44), and $n = 2$, then there exists a solution z, ρ of (4.5).*

Remark 4.14. We note that we do not make the special structure assumption (4.6); also, (4.43) is slightly weaker than (4.11) (applied with $\Gamma = 1$). On the other hand, (4.44) is exactly (4.12) with $\alpha_\mu = 1$.

Remark 4.15. When applied to differential games (see formulas (4.26), (4.28)) with assumptions (4.27), (4.30), (4.31), then (4.43) is automatically satisfied. However, to verify (4.44) we need to have (4.39). But then if we apply to the scalar case, we see that there is no gain.

4.5.2 Proof of Theorem 4.13

Approximate Problem
We begin by proving the existence of a solution of

$$Au_\epsilon^\nu + \epsilon u_\epsilon^\nu = H^\nu(x, Du_\epsilon) \tag{4.45}$$

that is a periodic solution in $H^1(Y) \cap L^\infty(Y)$ and satisfies some a priori estimates.

We cannot exactly apply the results of Theorem 3.10, since we do not have the special structure assumption and because of (4.43), which is weaker than (4.11).

We consider an additional approximation as follows:

$$Au_{\epsilon,\delta}^\nu + \epsilon u_{\epsilon,\delta}^\nu = \frac{H^\nu(x, Du_{\epsilon,\delta})}{1 + \delta |Du_{\epsilon,\delta}|^2}, \tag{4.46}$$

which will have a periodic solution in $W^{2,p}(Y)$.

We first prove that

$$\epsilon u_{\epsilon,\delta}^\nu \leq \lambda^\nu. \tag{4.47}$$

Indeed, set

$$E_{\epsilon,\delta}^\nu = \exp \frac{\beta_\delta^\nu}{\alpha} \left(u_{\epsilon,\delta}^\nu - \frac{\lambda^\nu}{\epsilon} \right)$$

with

$$\beta_\delta^\nu = \frac{(\lambda_0^\nu)^2}{2\delta\lambda^\nu}.$$

An easy calculation shows that by the choice of β_δ^ν and assumption (4.43) we have

$$Au_{\epsilon,\delta}^\nu + \frac{\beta_\delta^\nu}{\alpha} E_{\epsilon,\delta}^\nu (\epsilon u_{\epsilon,\delta}^\nu - \lambda^\nu) \leq 0.$$

Then we can proceed as in Section 2.4.2 to obtain (4.47).

We next prove that

$$\epsilon \sum_\nu u_{\epsilon,\delta}^\nu \geq -\lambda. \tag{4.48}$$

Indeed, we have, thanks to the assumption (4.44),

$$A \sum_{\nu} u^{\nu}_{\epsilon,\delta} + \epsilon \sum_{\nu} u^{\nu}_{\epsilon,\delta} \geq -\lambda,$$

and this easily implies (4.48).

From (4.47) and (4.48) it follows that

$$|\epsilon u^{\nu}_{\epsilon,\delta}| \leq C, \tag{4.49}$$

where C is independent of ϵ, δ.

The next step is to prove an H^1 estimate. Introduce parameters $\theta^{\nu} \in [0,1]$. We note that from (4.44) we have

$$\sum_{\nu} \theta^{\nu} H^{\nu}(x,p) \geq \frac{\lambda_0}{2}|p|^2 - \lambda - \sum_{\nu} \lambda^{\nu},$$

provided that we restrict the θ^{ν} as follows:

$$\sum_{\nu} \lambda^{\nu}_0 (1 - \theta^{\nu}) < \frac{\lambda_0}{2}. \tag{4.50}$$

Since then

$$A \sum_{\nu} \theta^{\nu} u^{\nu}_{\epsilon,\delta} + \epsilon \sum_{\nu} \theta^{\nu} u^{\nu}_{\epsilon,\delta} \geq -\lambda - \sum_{\nu} \lambda^{\nu} \tag{4.51}$$

and

$$-\lambda - \sum_{\nu} \lambda^{\nu} \leq \epsilon \sum_{\nu} \theta^{\nu} u^{\nu}_{\epsilon,\delta} \leq \sum_{\nu} \lambda^{\nu}, \tag{4.52}$$

we obtain easily from these two inequalities

$$\int_Y \left| D \sum_{\nu} \theta^{\nu} u^{\nu}_{\epsilon,\delta} \right|^2 dx \leq C_{\epsilon}.$$

Since the parameters θ^{ν} are restricted only by (4.50), we can take several possible values and prove

$$\|u^{\nu}_{\epsilon,\delta}\|_{H^1} \leq C_{\epsilon}. \tag{4.53}$$

We extract subsequences as $\delta \to 0$ such that

$$u^{\nu}_{\epsilon,\delta} \to u^{\nu}_{\epsilon}, \quad \text{weakly in } H^1(Y) \text{ and pointwise.}$$

To prove strong convergence in $H^1(Y)$, we shall prove successively that as $\delta \to 0$,

$$\left\| \left(\sum_{\nu} \theta^{\nu}(u^{\nu}_{\epsilon,\delta} - u^{\nu}_{\epsilon}) \right)^- \right\|_{H^1} \to 0 \tag{4.54}$$

and

$$\|(u^{\nu}_{\epsilon,\delta} - u^{\nu}_{\epsilon})^+\|_{H^1} \to 0. \tag{4.55}$$

We first prove (4.54). Now, (4.51) and (4.52) yield

$$A \sum_\nu \theta^\nu (u^\nu_{\epsilon,\delta} - u^\nu_\epsilon) \geq \lambda - 2 \sum_\nu \lambda^\nu - A \sum_\nu \theta^\nu u^\nu_\epsilon.$$

We test with $(\sum_\nu \theta^\nu (u^\nu_{\epsilon,\delta} - u^\nu_\epsilon))^-$, which converges to 0 weakly in H^1 and pointwise. It easily follows that

$$\limsup \int_Y \left| D \left(\sum_\nu (u^\nu_{\epsilon,\delta} - u^\nu_\epsilon) \right)^- \right|^2 dx = 0,$$

which is (4.54).

To prove (4.55) we recall that

$$A u^\nu_{\epsilon,\delta} + \epsilon u^\nu_{\epsilon,\delta} \leq \lambda^\nu + \lambda^\nu_0 |Du_{\epsilon,\delta}||Du^\nu_{\epsilon,\delta}|. \tag{4.56}$$

Let β be an arbitrary number; we test (4.56) with

$$(u^\nu_{\epsilon,\delta} - u^\nu_\epsilon)^+ \exp \frac{\lambda^\nu_0}{4\alpha\beta} ((u^\nu_{\epsilon,\delta} - u^\nu_\epsilon)^+)^2,$$

which tends to 0 weakly in H^1 and pointwise.

Set

$$E^\nu_{\epsilon,\delta} = \exp \frac{\lambda^\nu_0}{4\alpha\beta} ((u^\nu_{\epsilon,\delta} - u^\nu_\epsilon)^+)^2.$$

An easy calculation leads to

$$\alpha \int_Y |D(u^\nu_{\epsilon,\delta} - u^\nu_\epsilon)^+|^2 \, dx$$

$$\leq - \int_Y a Du^\nu_\epsilon D(u^\nu_{\epsilon,\delta} - u^\nu_\epsilon)^+ \left(1 + \frac{\lambda^\nu_0}{2\alpha\beta} ((u^\nu_{\epsilon,\delta} - u^\nu_\epsilon)^+)^2 \right) E^\nu_{\epsilon,\delta} \, dx$$

$$+ \int_Y (\lambda^\nu - \epsilon u^\nu_{\epsilon,\delta})(u^\nu_{\epsilon,\delta} - u^\nu_\epsilon)^+ E^\nu_{\epsilon,\delta} \, dx$$

$$+ \lambda^\nu_0 \beta \int_Y |Du_{\epsilon,\delta}|^2 \, dx + \frac{\lambda^\nu_0}{2\beta} \int_Y |Du^\nu_\epsilon|((u^\nu_{\epsilon,\delta} - u^\nu_\epsilon)^+)^2 E^\nu_{\epsilon,\delta} \, dx.$$

We also majorize

$$\int_Y |Du_{\epsilon,\delta}|^2 \, dx \leq C_\epsilon,$$

and we see that we can pass to the limit on the right-hand side of the above inequality. We obtain

$$\limsup_{\delta \to 0} \int_Y |D(u^\nu_{\epsilon,\delta} - u^\nu_\epsilon)^+|^2 \, dx \leq \frac{\lambda^\nu_0 \beta C_\epsilon}{\alpha},$$

and since β is arbitrary, we have obtained (4.55).

From (4.54), (4.55) the strong convergence in H^1 of each $u^\nu_{\epsilon,\delta}$ to u^ν_ϵ is easily obtained.

It is then standard to pass to the limit in (4.46) and obtain (4.45).

Additional Estimates

We now obtain estimates on the solution u^ϵ that has just been found that are independent of ϵ. We first notice that from (4.49) we have

$$|\epsilon u^\nu_\epsilon| \leq C. \tag{4.57}$$

Summing up (4.45) over ν and using (4.44) yields

$$\sum_\nu A u^\nu_\epsilon \geq -\lambda + \lambda_0 |Du_\epsilon|^2.$$

Testing with 1, we obtain

$$\int_Y |Du_\epsilon|^2 \, dx \leq C. \tag{4.58}$$

Next we proceed as in Section 2.3.4, taking account of the similarity between (2.59) and (4.43), (4.44). Consider any point $x_0 \in Y$ and the ball $B_R(x_0)$ with $R < R_0$. The number R_0 is sufficiently large that

$$B_R(x_0) \supset Y, \quad \forall x_0 \in Y.$$

Consider cutoff functions

$$\tau_R \text{ Lipschitz}, \quad \tau = 1 \text{ on } B_R, \quad \text{supp } \tau \subset B_{2R}, \quad |D\tau_R| \leq \frac{1}{R}.$$

Summing up (4.45) over ν and using (4.44) yields

$$\sum_\nu A u^\nu_\epsilon \geq -\lambda + \lambda_0 |Du_\epsilon|^2.$$

Testing with τ_R and using Hölder's inequality, we obtain the "inhomogeneous hole-filling inequality"

$$\int_{B_R} |Du_\epsilon|^2 \, dx \leq K \left(\int_{B_{2R}-B_R} |Du_\epsilon|^2 \, dx \right)^{1/2} + KR^n,$$

where K denotes a generic constant.

We next test (4.45) with

$$(u^\nu_\epsilon - u^\nu_{\epsilon,2R})\tau_R^2,$$

where

$$u^\nu_{\epsilon,2R} = \frac{1}{|B_{2R} - B_R|} \int_{B_{2R}-B_R} u^\nu_\epsilon \, dx.$$

Using the quadratic growth of H^ν and Hölder's inequality, we obtain easily the estimate

$$\int_{B_{2R}} |Du_\epsilon^\nu|^2 \tau_R^2 \, dx \leq K_1 \int_{B_{2R}} |u_\epsilon^\nu - u_{\epsilon,2R}^\nu|^2 |D\tau_R|^2 \, dx$$

$$+ K_2 \int_{B_{2R}} |Du_\epsilon|^2 |u_\epsilon^\nu - u_{\epsilon,2R}^\nu| \tau_R^2 \, dx \qquad (4.59)$$

$$+ K_3 \int_{B_{2R}} |u_\epsilon^\nu - u_{\epsilon,2R}^\nu| \tau_R^2 \, dx.$$

We notice that

$$\int_{B_{2R}} |u_\epsilon^\nu - u_{\epsilon,2R}^\nu|^2 |D\tau_R|^2 \, dx \leq \frac{C}{R^2} \int_{B_{2R}-B_R} |u_\epsilon^\nu - u_{\epsilon,2R}^\nu|^2 \, dx$$

$$\leq C \int_{B_{2R}-B_R} |Du_\epsilon^\nu|^2 \, dx,$$

where we have used the Poincaré inequality.

Now, again using the Poincaré inequality, we have

$$\int_{B_{2R}} |u_\epsilon^\nu - u_{\epsilon,2R}^\nu| \tau_R^2 \, dx \leq CR \int_{B_{2R}} |Du_\epsilon^\nu| \, dx$$

$$\leq \delta \int_{B_{2R}} |Du_\epsilon^\nu|^2 \, dx + \frac{C}{\delta} R^{n+2},$$

where δ is arbitrarily small. We then split the integral over B_{2R} into an integral over B_R and an integral over $B_{2R} - B_R$. Since δ is small, the integral over B_R can be absorbed by the left-hand side of (4.59). Collecting results, we obtain the "Cacciopoli inequality"

$$\int_{B_R} |Du_\epsilon|^2 \, dx \leq K_1 \int_{B_{2R}-B_R} |Du_\epsilon|^2 \, dx$$

$$+ K_2 \int_{B_{2R}} |Du_\epsilon|^2 |u_\epsilon - u_{\epsilon,2R}| \, dx + K_3 R^{n+2}. \qquad (4.60)$$

With a minor adaptation of Theorem 1.24, we obtain from the "inhomogeneous hole-filling inequality," the "Cacciopoli inequality," and the fact that $n = 2$,

$$[u_\epsilon]_\beta \leq C |||u_\epsilon|||_\beta \leq C_0 \qquad (4.61)$$

for a convenient $\beta < 1$. We recall the notation

$$[u_\epsilon]_\beta = \sup \left\{ \frac{|u_\epsilon(x) - u_\epsilon(y)|}{|x-y|^\beta} \mid x, y \in Y \right\},$$

$$|||u_\epsilon|||_\beta^2 = \sup \left\{ \frac{1}{R^{n-2+2\beta}} \int_{B_R(x_0)} |Du_\epsilon|^2 \, dx \mid x_0 \in Y, R < R_0 \right\}.$$

End of Proof

We now conclude the proof of Theorem 4.13. We set

$$z_\epsilon(x) = u_\epsilon(x) - u_\epsilon(x_0).$$

Then, using (4.61) we deduce

$$\|z_\epsilon\|_{(C^\beta(\overline{Y}))^N} \leq C_1,$$

and of course, since we also have (4.58),

$$\|z_\epsilon\|_{(H^1(Y))^N} \leq C_1.$$

Recalling (4.57), we can extract a subsequence such that

$$\epsilon u_\epsilon \to \rho \text{ constant } \in (L^\infty(Y))^N,$$
$$z_\epsilon \to z \text{ in } (L^\infty(Y))^N \text{ strongly },$$

from the compactness of the injection of $(C^\beta(\overline{Y}))^N$ into $(L^\infty(Y))^N$. Also,

$$z_\epsilon \to z \text{ in } (H^1(Y))^N \text{ weakly.}$$

In fact, from the equations, the weak convergence in H^1, and the strong convergence in L^∞ it also follows, by classical arguments, that

$$z_\epsilon \to z \text{ in } (H^1(Y))^N \text{ strongly.}$$

This allows us to pass to the limit in equations (4.1), and we obtain equations (4.5). The passage from C^β to $W^{2,p}$ is then standard. This concludes the proof of Theorem 4.13.

4.6 Ergodic Problems in R^n

4.6.1 Presentation of the Problem

Here we address another type of difficulty arising in the context of ergodic control when we assume that the domain of the space variable x is not bounded, but say is R^n. Because of this new type of difficulty we shall restrict ourselves to scalar equations instead of systems. We rely on the work of the authors [5]. We are interested in the following equation:

$$Az + \rho = H(x, Dz) + f(x), \quad x \in R^n,$$
$$z \in W^{2,s}_{loc}, \quad 2 \leq s < \infty, \quad \rho \text{ scalar} \tag{4.62}$$

The assumptions will be of local nature. We begin with

$$A = - \sum_{i\,j=1}^{n} \frac{\partial}{\partial x_i} a_{i,j} \frac{\partial}{\partial x_j} + \sum_i b_i \frac{\partial}{\partial x_i}, \tag{4.63}$$

where

$$a_{i,j}, b_i \in L^{\infty}_{\text{loc}}(R^n), \tag{4.64}$$

$$a_{i,j} \in W^{1,\infty}_{\text{loc}}(R^n), b_i \in L^{\infty}_{\text{loc}}(R^n), \tag{4.65}$$

$$a(x)\xi.\xi \geq a_0(x)|\xi|^2, \quad \forall \xi \in R^n, \quad \frac{1}{a_0(x)} \in L^{\infty}_{\text{loc}}(R^n). \tag{4.66}$$

The function H satisfies

$$\begin{aligned}
&H(x,p) : R^n \times R^n \to R^n, \\
&H \text{ is measurable and continuous in } p, \\
&-K(x)|p|^2 - \hat{K}(x) \leq H(x,p) \leq -k(x)|p|^2 + \hat{k}(x), \\
&k, \frac{1}{k}, \hat{k}, K, \hat{K} \geq 0, \in L^{\infty}_{\text{loc}}(R^n), \\
&\beta = \sup \frac{K(x) + |b|^2}{a_0(x)} < \infty.
\end{aligned} \tag{4.67}$$

Also, let $f(x)$ be a function such that

$$\begin{aligned}
&f \in L^{\infty}_{\text{loc}}(R^n), \\
&f(x) - \hat{K}(x) \to \infty, \text{ as} |x| \to \infty.
\end{aligned} \tag{4.68}$$

Remark 4.16. One could include f and $b.p$ into the Hamiltonian H for the main part of the theory, but not all. So it proves worthwhile to leave them as they are.

We state our main result.

Theorem 4.17. *If (4.63), (4.64), (4.65), (4.66), (4.67), and (4.68), hold, then there exists a solution (z, ρ) of (4.62), and moreover,*

$$z(x) \geq -K_0, \quad \forall x \in R^n. \tag{4.69}$$

4.6.2 Existence Theorem for an Approximate Solution

We consider, as is natural in the context of (4.62), the following approximation:

$$Au_\epsilon + \epsilon u_\epsilon = H(x, Du_\epsilon) + f(x). \tag{4.70}$$

We recall the notation $B_r(x_0)$ for the ball of center x_0 and radius r. We shall usually work with $x_0 = 0$, in which case the ball is denoted by B_r.

We then state the following result.

Theorem 4.18. *We make the assumptions (4.63), (4.64), (4.65), (4.66), (4.67), (4.68). Then there exists a solution of (4.70)*

$$u_\epsilon \in W^{2,s}_{\text{loc}}(R^n), \quad 1 \leq s < \infty,$$

such that

$$\epsilon u_\epsilon \geq -K_0, \tag{4.71}$$

$$\int_{B_r} |Du_\epsilon|^2 \, dx \leq K_r, \tag{4.72}$$

$$\epsilon \|u_\epsilon\|_{L^\infty(B_r)} \leq K_r, \tag{4.73}$$

$$\text{osc}_{B_r} u_\epsilon \leq K_r, \tag{4.74}$$

where the constants do not depend on ϵ.

Proof.
1. Approximation
We consider the Neumann problem in B_R, with $R \to \infty$. We suppress the explicit ϵ-dependence, except where useful.

$$Au + \epsilon u = H(x, Du) + f(x), \quad \text{a.e. } x \in B_R,$$
$$\tag{4.75}$$
$$\frac{\partial u}{\partial \nu_A}\Big|_{\partial B_R} = 0,$$

where ν denotes the outward unit normal at the boundary of B_R and

$$\frac{\partial u}{\partial \nu_A} = \sum_{i,j} a_{i,j}(x)\nu_i \frac{\partial u}{\partial x_j}.$$

From [66] we can assert that there exists a solution u_R of (4.75) such that

$$u_R \in L^\infty(B_R) \cap H^1(B_R).$$

Of course, the norms depend on R.

Bound from Below
We write u for u_R to simplify the notation. According to the left-hand side of the third assumption (4.67), we have

$$Au + \epsilon u \geq f - \hat{K} - K|Du|^2. \tag{4.76}$$

Write

$$E = \exp[\lambda(u + L)^-],$$

where λ, L are constants to be chosen (possibly depending on R).
    Note that (because we will choose $\lambda > 0$)

$$1 - E \leq 0.$$

Multiplying (4.76) by $1 - E$, integrating over B_R, and taking account of the fact that

$$f(x) - \hat{K}(x) - \frac{1}{4} \geq 0, \text{ for } |x| \geq r_0,$$

we get, for $R \geq r_0$,

$$\epsilon \int_{B_R} u(1 - E) \, dx + \lambda \int_{B_R} a_0 |Du|^2 E \mathbb{1}_{\{u+L\leq 0\}} \, dx$$
$$+ \int_{B_R} (K + |b|^2) |Du|^2 (1 - E) \, dx$$

$$\leq \int_{B_{r_0}} \left(-f + \hat{K} + \frac{1}{4} \right) (E - 1) \, dx. \tag{4.77}$$

Recalling the definition of β (see (4.67)), we deduce from (4.77) that

$$\epsilon L \int_{B_R} (E - 1) \, dx + \int_{B_R} a_0 (\lambda - \beta) |Du|^2 E \mathbb{1}_{\{u+L\leq 0\}}$$

$$\leq \int_{B_{r_0}} \left(-f + \hat{K} + \frac{1}{4} \right) (E - 1) \, dx. \tag{4.78}$$

Choose $\lambda > \beta$ and L such that

$$\epsilon L = - \inf_{\{x \in B_{r_0}\}} \left(f - \hat{K} - \frac{1}{4} \right) (x).$$

With these choices, it follows from (4.78) that

$$E - 1 = 0, \quad \text{a.e. } x \in B_R.$$

Therefore,

$$u \geq -L \quad \text{a.e. } x \in B_R.$$

Hence we have obtained

$$\epsilon u_{\epsilon, R}(x) \geq -K_0, x \in B_R. \tag{4.79}$$

3. Local Bound from Above

Let r be fixed, and consider

$$\tau = 1 \text{ on } B_r, \quad \tau = 0 \text{ outside } B_{2r}, \quad 0 \leq \tau \leq 1; \quad \tau \text{ smooth}.$$

Testing (4.75) with τ^2 we deduce, integrating on B_R,

$$\epsilon \int_{B_R} u\tau^2 \, dx + 2 \int_{B_R} aDu.D\tau\,\tau \, dx + \int_{B_R} b.Du\tau^2 \, dx$$

$$+ \int_{B_R} (k|Du|^2 - \hat{k})\tau^2 \, dx \leq \int_{B_R} f\tau^2 \, dx. \tag{4.80}$$

Since R is large, we may assume without loss of generality that $R \geq 2r$. Since on B_{2r}, $k(x) \geq k_r > 0$, it is easy to deduce from (4.80) that

$$\frac{k_r}{2} \int_{B_R} |Du|^2\tau^2 \, dx + \epsilon \int_{B_R} u(x)\tau^2 \, dx \leq C_r.$$

Then using (4.79), we easily obtain

$$\epsilon \int_{B_r} |u_{\epsilon,R}(x)| \, dx \leq C_r,$$

$$\int_{B_r} |Du_{\epsilon,R}|^2 \, dx \leq C_r, \tag{4.81}$$

where C_r denotes a generic constant depending only on r.

Let us denote by $a^r(x)$ a matrix that coincides with a on B_{2r}, and such that

$$\|a^r(x)\| \leq M_r, \quad \|(a^r)^{-1}(x)\| \leq \frac{1}{a_{0,r}}.$$

To proceed we need to introduce the Green function $G^{x_0}(x)$, where $x_0 \in B_r$, which is the solution of

$$-\frac{\partial}{\partial x_i} a^r_{j,i} \frac{\partial G}{\partial x_j} = \delta(x - x_0), x \in B_{2r},$$

$$G(x) = 0 \text{ on } \partial B_{2r}. \tag{4.82}$$

We recall the estimates on Green functions (see Chapter 1 and (2.34))

$$\|G^{x_0}\|_{W_0^{1,\mu}}(B_{2r}) \leq C_{r,\mu}, \quad \forall \mu \in \left[1, \frac{n}{n-1}\right),$$

$$\|G^{x_0}\|_{L^\nu}(B_{2r}) \leq C_{r,\nu} \quad \forall \nu \in \left[1, \frac{n}{n-2}\right), \tag{4.83}$$

$$k_1^r|x - x_0|^{-(n-2)} \leq G^{x_0}(x) \leq k_2^r|x - x_0|^{-(n-2)}, \quad x \in B_{3r/2},$$

where the constants depend only on $M_r, a_{0,r}$. Of course, $B_{3r/2}$ is chosen in this way for convenience; any neighborhood of x_0 strictly included in B_{2r} would do just as well. We assume here $n \geq 3$; otherwise, appropriate changes must be made.

Now let

$$\tau = 1 \text{ on } B_{3r/2}, \quad \tau = 0 \text{ outside } B_{2r}, \quad 0 \leq \tau \leq 1; \quad \tau \text{ smooth.}$$

We test (4.75) with $\tau^2 G$, and integrate over B_R. We obtain, using (4.82),

$$u(x_0) + \epsilon \int_{B_{2r}} u\tau^2 G\, dx + \int_{B_{2r}} aDu.D\tau^2 G\, dx + \int_{B_{2r}} \text{div}\,(u\,aD\tau^2)\, G\, dx\, ds$$
$$+ \int_{B_{2r}} b.Du\tau^2 G\, dx = \int_{B_{2r}} H(x, Du)\tau^2 G\, dx + \int_{B_{2r}} f\tau^2 G\, dx.$$

Hence we also have, recalling (4.79),

$$u(x_0) + \int_{B_{2r}} aDu.D\tau^2 G\, dx + \int_{B_{2r}} \text{div}\,(u\,aD\tau^2)\, G\, dx$$
$$+ \int_{B_{2r}} b.Du\tau^2 G\, dx \leq \int_{B_{2r}} (f + \hat{k})\tau^2 G\, dx \qquad (4.84)$$
$$\leq c_r \int_{B_{2r}} G\, dx.$$

We note further that for $x \in B_{2r} - B_{3r/2}$ we have $|x - x_0| \geq \frac{r}{2}$; hence from the right inequality (4.83),

$$G^{x_0}(x) \leq k_2^r \left(\frac{r}{2}\right)^{2-n} \leq C_r.$$

Recalling the estimates (4.81), it easily follows from (4.84) that

$$u(x_0) \leq C_r.$$

Thus we have proven that

$$\sup_{x \in B_r} u_{\epsilon,R}(x) \leq C_r, \qquad (4.85)$$

and thus, using (4.79),

$$\epsilon \|u_{\epsilon,R}\|_{L^\infty(B_r)} \leq C_r. \qquad (4.86)$$

4. Proof of

$$\|(u_{\epsilon,R} - c_{\epsilon,r,R})^+\|_{L^\infty(B_r)} \leq C_r \qquad (4.87)$$

Here we denote by $c_{\epsilon,r,R}$ a constant such that

$$|\{x \in B_r | u_{\epsilon,R} - c_{\epsilon,r,R} \geq 0\}| \geq \frac{1}{2}|B_r|,$$
$$\qquad (4.88)$$
$$|\{x \in B_r | u_{\epsilon,R} - c_{\epsilon,r,R} \leq 0\}| \geq \frac{1}{2}|B_r|.$$

Such a number can be found, and

$$|c_{\epsilon,r,R}| \leq \|u_{\epsilon,R}\|_{L^\infty(B_r)}.$$

We shall denote this number simply by c in the following calculations.

We begin by proving that

$$\|(u-c)^+\|_{L^p(B_{\sigma r})} \leq C_{r,p,\sigma}, \quad \forall \sigma \in [1,2), \forall p \in [1,\infty). \tag{4.89}$$

We observe that (4.89) is true with

$$p \leq 2n/n - 2, \quad \sigma \in [1,2].$$

Indeed, from the first estimate (4.81) and the Poincaré inequality, we can write

$$\|(u-c)^+\|_{L^{\frac{2n}{n-2}}(B_{2r})} \leq C_r \|D(u-c)^+\|_{L^2(B_{2r})} \leq C_r.$$

In particular, (4.89) is immediate when $n \leq 2$. Next, consider a number q such that

$$2\left(\frac{n}{n-2}\right)^q > p, \quad q \geq 2,$$

and define successively

$$s_i = 2\left(\frac{n}{n-2}\right)^i, \quad i = 1, \ldots, q.$$

Let θ be a number such that

$$1 < \theta \leq \left(\frac{2}{\sigma}\right)^{1/(q-1)}.$$

We also set

$$r_i = \theta^{q-i}\sigma r.$$

Hence

$$r_q = \sigma r, \quad s_q > p,$$
$$r_{i+1} = \frac{r_i}{\theta},$$
$$s_{i+1} = \frac{n}{n-2}s_i.$$

Let τ_i be a sequence of functions such that

$$\text{supp } \tau_i \subset B_{r_i}, \quad \tau_i = 1 \text{ on } B_{r_{i+1}},$$

$$0 \leq \tau_i \leq 1, \quad \tau_i \text{ smooth}.$$

We test equation (4.75) with the function $((u-c)^+)^{s_i-1}\tau_i^2$ and obtain easily

$$\int_{B_{r_{i+1}}} |Du|^2((u-c)^+)^{s_i-2}\, dx \le K_r \int_{B_{r_i}} ((u-c)^+)^{s_i}\, dx + K_r,$$

where the generic constants K_r depend on σ and q, but not on ϵ, R. Using the Poincaré inequality

$$\left(\int_{B_{r_{i+1}}} ((u-c)^+)^{s_i n/(n-2)}\, dx\right)^{(n-2)/n} \le K_r \int_{B_{r_{i+1}}} |D((u-c)^+)^{s_i/2}|^2\, dx$$

$$\le K_r \int_{B_{r_{i+1}}} |Du|^2((u-c)^+)^{s_i-2}\, dx$$

we obtain

$$\left(\int_{B_{r_{i+1}}} ((u-c)^+)^{s_{i+1}}\, dx\right)^{(n-2)/n} \le K_r + K_r \int_{B_{r_i}} ((u-c)^+)^{s_i}\, dx.$$

Applying this inequality, with i running from 1 to q, and using the fact that

$$\int_{B_{r_1}} ((u-c)^+)^{s_1}\, dx \le C_r,$$

as already shown, we obtain

$$\int_{B_{r_q}} ((u-c)^+)^{s_q}\, dx \le C_r,$$

which implies (4.89).

We now want to proceed and obtain (4.87).

We take

$$\tau = 1 \text{ on } B_{3r/2}, \text{ supp } \tau \subset B_{2r},$$

$$0 \le \tau \le 1, \quad \tau \text{ smooth}.$$

We test the equation (4.75) with

$$(u-c)^+ G^{x_0} \tau^2,$$

where G^{x_0} is again the Green function introduced in (4.82). We obtain

$$\int_{B_{2r}} aDu.D((u-c)^+ G\tau^2)\, dx \le K_r \int_{B_{2r}} |Du|(u-c)^+ G\tau^2\, dx$$

$$+ K_r \int_{B_{2r}} (u-c)^+ G\tau^2\, dx.$$

(4.90)

Consider first

$$I_1 = \int_{B_{2r}} (u - c)^+ G\tau^2 \, dx$$

$$= \int_{B_{r\sigma}} (u - c)^+ G\tau^2 \, dx + \int_{B_{2r-r\sigma}} (u - c)^+ G\tau^2 \, dx.$$

Thanks to (4.89) and the integrability properties of the Green function, the integral over $B_{r\sigma}$ is majorized by K_r. The second one is majorized by

$$K_r \int_{B_{2r-r\sigma}} (u - c)^+ \, dx,$$

since we avoid the singularity of the Green function. But this is bounded by

$$K_r \int_{B_{2r}} |Du|^2 \, dx.$$

Hence we have checked that

$$I_1 \le K_r.$$

Using this in (4.90) and using the definition of the Green function and the estimates already produced, we can assert that

$$((u - c)^+(x_0))^2 + \int_{B_{2r}} |Du|^2 \mathbb{1}_{u-c \ge 0} G\tau^2 \, dx$$

$$\le K_r \int_{B_{2r}} ((u - c)^+)^2 \tau |D\tau| |DG| \, dx + K_r.$$

Introduce $\tilde{\tau}$ such that

$$\tilde{\tau} = \tau \text{ on } B_{2r} - B_{3r/2}, \tilde{\tau} = 0 \text{ on } B_{r+r/4},$$
$$0 \le \tilde{\tau} \le 1, \quad \tilde{\tau} \text{ smooth.}$$

Since obviously

$$\int_{B_{2r}} ((u - c)^+)^2 \tau |D\tau| |DG| \, dx = \int_{B_{2r}} ((u - c)^+)^2 \tilde{\tau} |D\tau| |DG| \, dx,$$

it is sufficient to estimate the quantity

$$J_2 = \int_{B_{2r}} ((u - c)^+)^2 \tilde{\tau}^2 |DG|^2 \, dx.$$

But using the definition of the Green function, and testing (4.82) with $((u - c)^+)^2 \tilde{\tau}^2 G$, we have (since the test function vanishes on the singularity x_0)

$$J_2 \le K_r \left[\int_{B_{2r}} ((u - c)^+)^2 |D\tilde{\tau}|^2 |G|^2 \, dx + \int_{B_{2r}} \tilde{\tau}^2 |G|^2 |Du|^2 \, dx \right].$$

Using the fact that
$$\tilde{r}G \le k'_r, |D\tilde{r}|G \le k'_r,$$
we find that we have already proven the estimates we need, and
$$J_2 \le K_r.$$

Thus (4.87) has been obtained.

5. *Proof of*

$$\int_{B_r} |Du_{\epsilon,R}|^2 \exp 2\gamma(u_{\epsilon,R} - c_{\epsilon,r,R})^- \, dx \le C_r, \text{ for any } \gamma > 0 \qquad (4.91)$$

We may as well assume
$$\gamma \ge \beta,$$
where β has been defined in (4.67). We test (4.75) with the function
$$1 - \exp 2\gamma(u - c)^-.$$

Note that this function does not vanish on ∂B_R, so it is important that we deal with a Neumann approximation. Moreover, notice that it is negative.

By simple calculations we obtain

$$2\gamma \int_{B_R} a_0(x)|Du|^2 \exp 2\gamma(u - c)^- \, \mathbb{1}_{u \le c} \, dx$$
$$+ \epsilon \int_{B_R} (u - c)(1 - \exp 2\gamma(u - c)^-) \, dx$$
$$\le \int_{B_R} (K(x) + |b|^2)|Du|^2 |1 - \exp 2\gamma(u - c)^-| \, dx$$
$$+ \int_{B_R} \left(f(x) - \hat{K}(x) - \epsilon c - \frac{1}{4} \right) (1 - \exp 2\gamma(u - c)^-) \, dx.$$

From the choice of γ we deduce that

$$\gamma \int_{B_R} a_0(x)|Du|^2 \exp 2\gamma(u - c)^- \, \mathbb{1}_{u \le c} \, dx$$
$$\qquad\qquad\qquad (4.92)$$
$$\le \int_{B_R} \left(f(x) - \hat{K}(x) - \epsilon c - \frac{1}{4} \right) (1 - \exp 2\gamma(u - c)^-) \, dx.$$

Denote by J the right-hand side of (4.92). By virtue of (4.86) and the definition of c, we have
$$\epsilon c \le k_r.$$

From the second assumption (4.68) we have

$$f(x) - \hat{K}(x) - k_r - \frac{1}{4} \geq 0, \quad |x| \geq s_r.$$

Therefore, we may majorize

$$J \leq J_1 = \int_{B_{s_r}} \left(f(x) - \hat{K}(x) - \epsilon c - \frac{1}{4} \right) (1 - \exp 2\gamma(u - c)^-)\, dx$$

and thus also

$$J_1 \leq K_r \int_{B_{s_r}} |1 - \exp 2\gamma(u - c)^-|\, dx.$$

We introduce an additional parameter ℓ to be chosen later and set

$$\Gamma_\ell = \{x | (u - c)^- \geq \ell\}.$$

Then

$$\int_{B_{s_r}} |1 - \exp 2\gamma(u - c)^-|\, dx \leq \int_{B_{s_r} \cap \Gamma_\ell} |1 - \exp 2\gamma(u - c)^-|\, dx + K_{r,\ell}.$$

Set

$$\lambda_{r,\ell} = |B_{s_r} \cap \Gamma_\ell|.$$

From Hölder's inequality we have

$$\int_{B_{s_r} \cap \Gamma_\ell} |1 - \exp 2\gamma(u - c)^-|\, dx$$

$$\leq \lambda_{r,\ell}^{1/n} \left(\int_{B_{s_r}} |1 - \exp 2\gamma(u - c)^-|^{n/(n-1)}\, dx \right)^{(n-1)/n}.$$

Since $1 - \exp 2\gamma(u - c)^-$ vanishes on a set of measure greater than or equal to $\frac{1}{2}|B_r|$, we may apply the Poincaré inequality to obtain

$$\left(\int_{B_{s_r}} |1 - \exp 2\gamma(u - c)^-|^{n/(n-1)}\, dx \right)^{(n-1)/n}$$

$$\leq 2C_r \gamma \int_{B_{s_r}} |Du| \exp 2\gamma(u - c)^- \, \mathrm{I\!I}_{u \leq c}\, dx.$$

Now by Hölder's inequality, we can state that

$$2C_r \, \beta \lambda_{r,\ell}^{1/n} \int_{B_{s_r}} |Du| \exp 2\gamma(u - c)^- \, \mathrm{I\!I}_{u \leq c}\, dx$$

$$\leq 2\gamma^2 C_r^2 \lambda_{r,\ell}^{2/n} \int_{B_{s_r}} |Du|^2 \exp 2\gamma(u - c)^- \, \mathrm{I\!I}_{u \leq c}\, dx$$

$$+ \frac{1}{2} \int_{B_{s_r}} \exp 2\gamma(u - c)^- \, \mathrm{I\!I}_{u \leq c}\, dx$$

$$\leq 2\gamma^2 C_r^2 \lambda_{r,\ell}^{2/n} \int_{B_{s_r}} |Du|^2 \exp 2\gamma(u-c)^- \, \mathrm{1\!I}_{u\leq c} \, dx$$

$$+\frac{1}{2} \int_{B_{s_r} \cap \Gamma_\ell} |1 - \exp 2\gamma(u-c)^-| \, dx + K_{r,\ell}.$$

Therefore, collecting results, we have proven that

$$\frac{1}{2} \int_{B_{s_r}} |1 - \exp 2\gamma(u-c)^-| \, dx$$

$$\leq 2\gamma^2 C_r^2 \lambda_{r,\ell}^{2/n} \int_{B_{s_r}} |Du|^2 \exp 2\gamma(u-c)^- \, \mathrm{1\!I}_{u\leq c} \, dx + K_{r,\ell}. \tag{4.93}$$

Again, using the Poincaré inequality, we have

$$\int_{B_{s_r}} (u-c)^- \, dx \leq C_r \int_{B_{s_r}} |Du| \, \mathrm{1\!I}_{u\leq c} \, dx \leq K_r,$$

and thus

$$\lambda_{r,\ell} \leq \frac{K_r}{\ell}.$$

Combining (4.92) and (4.93) yields

$$\int_{B_R} a_0 |Du|^2 \exp 2\gamma(u-c)^- \, \mathrm{1\!I}_{u\leq c} \, dx$$

$$\leq 4\gamma^2 C_r^2 K_r \lambda_{r,\ell}^{2/n} \int_{B_{s_r}} |Du|^2 \exp 2\gamma(u-c)^- \, \mathrm{1\!I}_{u\leq c} \, dx + K'_{r,\ell}, \tag{4.94}$$

and picking $\ell(r)$ sufficiently large, we deduce easily from (4.94) that

$$\int_{B_R} a_0 |Du|^2 \exp 2\gamma(u-c)^- \, \mathrm{1\!I}_{u\leq c} \, dx \leq K_r,$$

which implies (4.91).

6. Proof of

$$\|(u_{\epsilon,R} - c_{\epsilon,r,R})^-\|_{L^\infty(B_r)} \leq C_r \tag{4.95}$$

We test (4.75) with

$$(1 - \exp 2\gamma(u-c)^-)G\tau^2,$$

where G, τ are as in part 4 of this proof. Performing calculations as in part 5, we can state the inequality

$$\gamma \int_{B_{2r}} a_0(x)|Du|^2 \exp 2\gamma(u-c)^- \, G\tau^2 \, \mathbb{I}_{u \le c} \, dx$$

$$+ \int_{B_{2r}} aDu(1 - \exp 2\gamma(u-c)^-)D(G\tau^2) \, dx \qquad (4.96)$$

$$\le K_r \int_{B_{2r}} |1 - \exp 2\gamma(u-c)^-|G\tau^2 \, dx.$$

Consider first the term

$$J_1 = \int_{B_{2r}} |1 - \exp 2\gamma(u-c)^-|G\tau^2 \, dx.$$

By Hölder's inequality and the integrability property of the Green function, we have

$$J_1 \le K_r \left(\int_{B_{2r}} |1 - \exp 2\gamma(u-c)^-|^t \, dx \right)^{1/t}$$

with $t \ge n/2$. But

$$\le \left(\int_{B_{2r}} (\exp 2t\gamma(u-c)^- - 1) \, dx \right)^{1/t}$$

$$\le K_r \left(\int_{B_{2r}} (\exp 2t\gamma(u-c)^- - 1)^2 \, dx \right)^{1/2t},$$

and using the Poincaré inequality,

$$\left(\int_{B_{2r}} |1 - \exp 2\gamma(u-c)^-|^t \, dx \right)^{1/t}$$

$$\le K_r \left(\int_{B_{2r}} |Du|^2 \exp 4t\gamma(u-c)^- \, dx \right)^{1/2t} \le K_r$$

(using part 5 and replacing γ by $4t\gamma$). Thus the right-hand side of (4.96) is majorized by K_r.

Now consider the term

$$J_2 = \int_{B_{2r}} aDu(1 - \exp 2\gamma(u-c)^-)D(G\tau^2) \, dx$$

$$= \int_{B_{2r}} a \, D \left(-(u-c)^- + \frac{\exp 2\gamma(u-c)^-}{2\gamma} - \frac{1}{2\gamma} \right) D(G\tau^2) \, dx.$$

To simplify the notation, let

$$\phi(\xi) = \frac{\exp 2\gamma\xi}{2\gamma} - \xi - \frac{1}{2\gamma}.$$

Then from the definition of the Green function, we can easily deduce that

$$J_2 = \phi((u-c)^-(x_0)) - J_3 + J_4,$$

where we have set

$$J_3 = \int_{B_{2r}} a\,\phi((u-c)^-)D\tau^2 DG\,dx,$$

$$J_4 = \int_{B_{2r}} a\,D\,\phi((u-c)^-)D\tau^2 G\,dx.$$

Since

$$J_4 = 2\gamma \int_{B_{2r}} a(1-\exp 2\gamma(u-c)^-)Du D\tau^2\,G\mathbb{1}_{u\le c}\,dx$$

and since

$$|D\tau^2|G \le c_r \text{ on } B_{2r},$$

we see by previous estimates that

$$|J_4| \le K_r.$$

To estimate J_3 we write

$$J_3 \le K_r \int_{B_{2r}} \phi((u-c)^-)\tau|D\tau||DG|\,dx$$

and

$$J_5 := \int_{B_{2r}} \phi((u-c)^-)\tau|D\tau||DG|\,dx = \int_{B_{2r}} \phi((u-c)^-)\tilde\tau|D\tau||DG|\,dx$$

using the function $\tilde\tau$ introduced in part 4. We are then in a situation similar to that of part 4, with $\phi((u-c)^-)$ replacing $((u-c)^+)^2$. Proceeding along the same lines, one can convince oneself that one can estimate J_5 in terms of the integrals

$$\int_{B_{2r}} \phi((u-c)^-)\,dx, \quad \int_{B_{2r}} \frac{\phi'^2}{\phi}((u-c)^-)|Du|^2\,dx.$$

Using the inequality

$$\frac{\phi'^2}{\phi} \le 4\gamma(\phi'+2),$$

the second integral can be estimated by part 5 and (4.81). Finally,

$$\int_{B_{2r}} \phi((u-c)^-)\,dx \le \int_{B_{2r}} (u-c)^-\,dx + \frac{1}{2\gamma}\int_{B_{2r}} (\exp 2\gamma(u-c)^- - 1)\,dx,$$

which can be estimated with the Poincaré inequality and previous techniques. Collecting results, we have shown that

$$|J_3| \le K_r.$$

Going back to J_2, we deduce that

$$J_2 \geq \phi((u-c)^-(x_0)) - K_r \geq \gamma((u-c)^-(x_0)^2 - K_r,$$

and in (4.96) one obtains

$$(u-c)^-(x_0)^2 \leq K_r,$$

which is property (4.95).

7. End of Proof

It follows from (4.87) and (4.95) that

$$\operatorname{osc}_{B_r} u_{c,R} \leq K_r, \tag{4.97}$$

where we recall that "osc" denotes the oscillation function. We can now let R tend to ∞ in equation (4.75). Taking a sequence converging in $H^1_{\text{loc}}(R^n)$ weakly and in $L^\infty_{\text{loc}}(R^n)$ weak-star (ϵ is fixed here), it is a standard matter to prove the strong convergence in $H^1_{\text{loc}}(R^n)$. Then it is easy to pass to the limit. Once one has a solution in $H^1_{\text{loc}}(R^n) \cap L^\infty_{\text{loc}}(R^n)$, the regularity is again standard; see Chapter 1.

4.6.3 Proof of Theorem 4.17

We now have a regular solution u_ϵ of (4.70), and the estimates (4.71), (4.72), (4.73), (4.74) hold. In fact, as we shall see, we even have an additional estimate, which we state now. Let us set

$$u_{\epsilon,r} = \fint_{B_r} u_\epsilon(x)\, dx.$$

Then

$$u_\epsilon(x) - u_{\epsilon,r} \geq -K_r, \quad \forall x \in R^n. \tag{4.98}$$

Remark 4.19. Of course, from (4.74) we know that (4.98) holds in B_r, but in fact, it holds on all of R^n.

The result (4.98) will follow from the following lemma.

Lemma 4.20. *The solution $u_{\epsilon,R}(x)$ of (4.75) satisfies*

$$u_{\epsilon,R}(x) - u_{\epsilon,R,r} \geq -K_r, \quad \forall x \in B_R. \tag{4.99}$$

Proof. Of course, we have set

$$u_{\epsilon,R,r} = \fint_{B_r} u_{\epsilon,R}(x)\, dx.$$

Moreover, it is clear that (4.99) will imply (4.98).

Let us prove (4.98). We first notice that from (4.86) we can assert that

$$|\epsilon u_{\epsilon,R,r}| \leq C_r.$$

Let $s(r)$ be such that

$$f(x) - \hat{K}(x) - \frac{1}{4} \quad C_r \geq 0, \quad \forall x \text{ with } |x| \geq s(r).$$

Without loss of generality we can assume $s(r) \geq r$; hence from (4.74) it follows that

$$|u_{\epsilon,R}(x) - u_{\epsilon,R,r}| \leq K_r, \forall x \in B_{s(r)}.$$

We test (4.75) with the function

$$1 - \exp 2\beta(u_{\epsilon,R}(x) - u_{\epsilon,R,r} + K_r)^-,$$

which by construction vanishes on $B_{s(r)}$.

We now proceed as in the proof of Theorem (4.18), part 5, and obtain

$$2\beta \int_{B_R} a_0(x)|Du|^2 \exp 2\beta(u-c)^- \, \mathbb{1}_{u \leq c} \, dx$$

$$+\epsilon \int_{B_R} (u-c)(1 - \exp 2\beta(u-c)^-) \, dx$$

$$\leq \int_{B_R} (K(x) + |b|^2)|Du|^2|1 - \exp 2\beta(u-c)^-| \, dx$$

$$+ \int_{B_R} \left(f(x) - \hat{K}(x) - \epsilon c - \frac{1}{4} \right) (1 - \exp 2\beta(u-c)^-) \, dx.$$

(4.100)

Of course, here c represents a different constant, namely

$$c = u_{\epsilon,R,r} - K_r.$$

Now

$$f(x) - \hat{K}(x) - \epsilon c - \frac{1}{4} = f(x) - \hat{K}(x) - \epsilon u_{\epsilon,R,r} + \epsilon K_r - \frac{1}{4}$$

$$\geq f(x) - \hat{K}(x) - \epsilon C_r + \epsilon K_r - \frac{1}{4} \geq 0, \quad x \in B_R - B_{s(r)}.$$

Since in all integrals in (4.100) the domain of integration reduces to $B_R - B_{s(r)}$, the second integral on the right-hand side of (4.100) is negative.

From the definition of β, one deduces from (4.100) that

$$\beta \int_{B_R} a_0(x)|Du|^2 \exp 2\beta(u-c)^- \, \mathbb{1}_{u \leq c} \, dx \leq 0,$$

and thus (4.99) follows.   ◇

We proceed with the proof of Theorem 4.17. We define

$$z_\epsilon = u_\epsilon - u_{\epsilon,1}.$$

Then we have

$$z_\epsilon \geq K_0,$$

where $K_0 = K_1(K_r$ with $r = 1)$. Moreover, z_ϵ remains bounded in $H^1_{\text{loc}}(R^n) \cap L^\infty_{\text{loc}}(R^n)$. We extract a subsequence such that

$$z_\epsilon \to z \text{ in } H^1_{\text{loc}}(R^n) \text{ weakly, and pointwise,}$$

$$\epsilon u_{\epsilon,1} \to \rho \text{ pointwise.}$$

Note that (4.70) becomes

$$Az_\epsilon + \epsilon z_\epsilon + \epsilon u_{\epsilon,1} = H(x, Dz_\epsilon) + f. \tag{4.101}$$

The strong convergence in $H^1_{\text{loc}}(R^n)$ is next checked by testing (4.101) with $\tau^2(z_\epsilon - z) \exp 2\beta(z_\epsilon - z)^2$. This is standard. We can then pass to the limit in (4.101) and obtain a solution of (4.62). Moreover, (4.69) has already been proven. The regularity follows in a standard way. The proof is complete.

4.6.4 Growth at Infinity

Our objective in this section is to prove the following result.

Theorem 4.21. *We make the assumptions of Theorem 4.17 and that*

$$r^2 \inf_{|x| \geq r} \frac{f(x) - \hat{K}(x)}{K(x)} \to \infty, \text{ as } r \to \infty,$$

$$r^2 \inf_{|x| \geq r} \frac{f(x) - \hat{K}(x)}{\text{tr } a(x)} \to \infty, \text{ as } r \to \infty, \tag{4.102}$$

$$r \inf_{|x| \geq r} \frac{f(x) - \hat{K}(x)}{(\sum_j (\sum_i \frac{\partial a_{i,j}}{\partial x_i} - b_j)^2)^{1/2}} \to \infty, \text{ as } r \to \infty.$$

Then the solutions of (4.62) such that $z(x) \geq -K_0$ satisfy

$$\inf_{|x| \geq r} z(x) \to \infty, \text{ as } r \to \infty. \tag{4.103}$$

Proof. We first choose a sequence $\lambda_r \to \infty$ such that

$$\frac{f(x) - \hat{K}(x)}{K(x)} \geq 96 \frac{\lambda_r^2}{r^2}, \quad |x| \geq \frac{r}{2}, \tag{4.104}$$

$$\frac{f(x) - \hat{K}(x)}{\text{tr } a(x)} \geq 48 \frac{\lambda_r}{r^2}, \quad |x| \geq \frac{r}{2}, \tag{4.105}$$

$$\frac{f(x) - \hat{K}(x)}{(\sum_j (\sum_i \frac{\partial a_{i,j}}{\partial x_i} - b_j)^2)^{1/2}} \geq 24 \frac{\lambda_r}{r}, \quad |x| \geq \frac{r}{2}. \tag{4.106}$$

Consider the function

$$\chi_r(x) = \lambda_r \left(1 - 4 \frac{|x - x_r|^2}{r^2} \right)$$

with

$$|x_r| \geq r$$

and λ_r satisfying the conditions above. Set

$$\zeta(x) = z(x) - \chi_r(x) \text{ on } B_{r/2}(x_r).$$

Note that

$$\chi_r(x) = 0 \text{ if } x \in \partial B_{r/2}(x_r).$$

Hence we can assert that

$$\zeta(x) \geq -K_0 \text{ if } x \in \partial B_{r/2}(x_r). \tag{4.107}$$

We prepare ourselves to apply the maximum principle on $B_{r/2}(x_r)$. To this end, we compute (skipping easy steps)

$$A\zeta + K(x)D\zeta.(D\chi_r + Dz) \geq h(x)$$

with

$$h(x) = f(x) - \rho - \hat{K}(x) - K(x)|D\chi_r|^2 - A\chi_r.$$

Since the solution z is in $W_{loc}^{2,p}$, the function Dz is bounded on $B_{r/2}(x_r)$. From the conditions (4.104), (4.105), (4.106), it follows that

$$h(x) \geq 1/2(f(x) - \hat{K}(x)) \geq 0, \quad x \in B_{r/2}(x_r).$$

Then, from (4.107),

$$\zeta(x) \geq -K_0, \text{ on } B_{r/2}(x_r).$$

In particular,

$$z(x_r) \geq \lambda_r - K_0,$$

which means that in fact, since x_r is arbitrary with norm greater than or equal to r,

$$z(x) \geq \lambda_r - K_0, \quad |x| \geq r,$$

and the proof is complete.   ◇

4.6.5 Uniqueness

We can give a uniqueness theorem.

Theorem 4.22. *We make assumptions (4.63), (4.64), (4.65), (4.66). We assume that $H(x,p)$ satisfies*

$$H(x,p) : R^n \times R^n \to R^n,$$
H *is measurable and continuous in p, locally Lipschitz in p,*
$$H(x,p) \le -k_0\, a(x)p.p, k_0 > 0,$$ (4.108)
$$H(x,\lambda p) \le \lambda^2 H(x,p), \; \forall x, \; \forall \lambda \ge 1.$$

Next, the function f satisfies

$$f \in L_{loc}^\infty(R^n),$$
$$\inf_{|x| \ge r} f(x) \to \infty \text{ as } r \to \infty.$$ (4.109)

Then there exists at most one solution z, ρ of (4.62) (up to a constant in z) such that

$$z \in W_{loc}^{1,\infty}(R^n), \; \rho \text{ constant},$$
$$\inf_{|x| \ge r} z(x) \to \infty \text{ as } r \to \infty.$$ (4.110)

Proof. Let $(z_1, \rho_1), (z_2, \rho_2)$ be two solutions of (4.62). We may assume that $\rho_1 \ge \rho_2$. We are going to prove that for r sufficiently large we have

$$\max_{x \in \overline{B}_r}(z_1(x) - z_2(x)) = \max_{x \in R^n}(z_1(x) - z_2(x)).$$ (4.111)

First suppose (4.111) has been proven.

Let x_r be this maximum. Then define

$$\psi = z_1(x_r) - z_2(x_r) - z_1(x) + z_2(x) \ge 0.$$

We can write

$$A\psi - g(x).D\psi = \rho_1 - \rho_2 \ge 0,$$

where

$$g(x) = \int_0^1 \frac{\partial H}{\partial p}(x, Dz_1 + \lambda D\psi)\, d\lambda,$$

and from the assumptions $g \in L_{loc}^\infty(R^n)$. We shall apply Harnack's inequality (see D. Gilbarg, N.S. Trudinger [46]). We can assert that for any $R > 0$,

$$\int_{B_{2R}(x_r)} \psi(x)\, dx \le c_R \min_{x \in B_R(x_r)} \psi(x).$$

Since $\psi(x_r) = 0$, it follows that

$$\psi = 0, \quad \text{on } B_{2R}(x_r).$$

Since R is arbitary, necessarily $\psi = 0$; hence $z_1 - z_2$ is a constant. It follows also that $\rho_1 = \rho_2$.

Let us prove (4.111). We introduce for each r a constant

$$\sigma = \sigma_r$$

such that

$$z_1(x) + \sigma > z_2(x), \quad \forall x \in B_r.$$

Next introduce a constant

$$\gamma > 0, \ \gamma < k_0$$

(the constant entering in (4.108)) and set

$$w_\gamma(x) = \exp[-\gamma(z_1(x) + \sigma)] - \exp[-\gamma z_2(x)].$$

We shall prove that

$$\exists r \text{ sufficiently large that}$$
$$\min_{x \in \bar{B}_r} w_\gamma(x) = \min_{x \in R^n} w_\gamma(x), \ \forall \gamma. \tag{4.112}$$

Then (4.111) will result from (4.112). Indeed, let $x_{r,\gamma}$ be this minimum. We write

$$\frac{\exp[-\gamma(z_1(x) + \sigma)] - \exp[-\gamma z_2(x)]}{\gamma}$$
$$\geq \frac{\exp[-\gamma(z_1(x_{r,\gamma}) + \sigma)] - \exp[-\gamma z_2(x_{r,\gamma})]}{\gamma}.$$

Letting γ tend to 0, and recalling that $x_{r,\gamma} \in B_r$, we can extract subsequences converging to x_r, which will be a maximum of $z_1 - z_2$; hence (4.111) follows.

It remains to prove (4.112). We write w for w_γ to simplify the notation. We are going to prove

$$\exists r \text{ large enough, independent of } \gamma, \text{ such that } \left[w(x) - \min_{B_r} w \right]^- = 0, \tag{4.113}$$

which of course implies (4.112). By the choice of σ,

$$\min_{B_r} w(x) < 0.$$

Since clearly

$$\inf_{|x| \geq R} w(x) \to 0 \text{ as } R \to \infty,$$

we have

$$\left[w(x) - \min_{B_r} w \right]^- = 0 \text{ for } x \in B_r \text{ or } |x| \geq \bar{R}_r.$$

Define

$$H_\gamma(x, p) = H(x, p) + \gamma a(x) p.p \leq 0,$$

where the inequality follows from the choice of γ and the third assumption (4.108).

We can perform a calculation to yield

$$Aw + \gamma f\, w + \gamma \exp -\gamma(z_1 + \sigma)\, H_\gamma(Dz_1) - \gamma \exp -\gamma z_2\, H_\gamma(Dz_2) = 0. \quad (4.114)$$

Take points x such that $w(x) \le 0$. Then

$$\frac{\exp -\gamma z_2(x)}{\exp -\gamma(z_1(x) + \sigma)} \ge 1.$$

On those points, using the fourth assumption (4.108), we obtain

$$Aw + \gamma f\, w - h(x).Dw \ge 0, \quad\quad\quad (4.115)$$

where

$$h(x) = \int_0^1 \frac{\partial H_\gamma}{\partial p}\left(Dz_1 + \tau\left(\frac{\exp -\gamma z_2(x)}{\exp -\gamma(z_1(x) + \sigma)}Dz_2 - Dz_1\right)\right) d\tau.$$

From the second assumption (4.108) and the regularity properties of z_1 and z_2, h is in $L^\infty_{\text{loc}}(R^n)$.

From the second assumption (4.109), we can assume without loss of generality that

$$f(x) \ge 0, \quad |x| \ge r_0.$$

We then pick $r \ge r_0$ and test (4.114) with $-([w(x) - \min_{B_r} w]^-)^m$, $m \ge 2$. On the support of this function, (4.115) holds. Hence we obtain

$$m \int_{B_{\bar{R}_r} - B_r} a_0 |Dw|^2 \left(\left[w(x) - \min_{B_r} w\right]^-\right)^{m-1} dx$$

$$- \int_{B_{\bar{R}_r} - B_r} (b + h).Dw \left(\left[w(x) - \min_{B_r} w\right]^-\right)^m dx \quad (4.116)$$

$$+ \gamma \int_{B_{\bar{R}_r} - B_r} f \left(\left[w(x) - \min_{B_r} w\right]^-\right)^{m+1} dx \le 0.$$

Choosing m such that

$$4ma_0(x) - \frac{1}{\gamma f(x)}|(h + b)(x)|^2 > 0, \quad r \le |x| \le \bar{R}_r,$$

it readily follows from (4.116) that (4.113) holds.

The proof is complete. $\diamond$

5. Harmonic Mappings

5.1 Introduction

We treat here only a small aspect of this large field. In particular, we do not consider harmonic mappings between manifolds, but only from a domain Ω of R^n into the unit sphere of R^N. This is, however, sufficient to cover many of the analytical difficulties. There is a large litterature, starting with the famous paper of J. Eells, J.H. Sampson [22]. Let us introduce the problem: Let Ω be a bounded open subset of R^n, $n \geq 2$, and

$$g : \overline{\Omega} \to R^N, \text{ Lipschitz, } |g(x)| = 1 \ \forall x. \tag{5.1}$$

Find u such that

$$u \in H^1(\Omega; R^N) \ |u| = 1, \ u|_{\partial\Omega} = g,$$
$$u \text{ minimizes } \int_\Omega |Du|^2 \, dx. \tag{5.2}$$

It is clear that a minimum exists but may be not unique. Note that although in the formulation of the problem only the values of g on $\partial\Omega$ play a role, it is important that an extension exist on Ω that is sufficiently regular. For instance, if $n = 2$ and Ω is the unit ball B_1 of center 0, and if

$$g(x) = x \quad \text{on } \partial B_1,$$

then there is no function g satisfying (5.1). There is not even a function u satisfying (5.2). In other words, the set of admissible functions in the optimization problem (5.2) is empty. This is consistent with the topological result that in R^n, in general, there is no continuous mapping from the unit ball B_1 into ∂B_1, since ∂B_1 is not a retract of B_1. Since our objective in this chapter is to obtain (in the case $n = 2$ and for any $p < \infty$) the $W^{2,p}_{\text{loc}}(\Omega)$ regularity not only for minima, but for extremals in general, this regularity result could look at first glance like a contradiction with the topological result.

5.2 Extremals

Theorem 5.1. *A minimum u of the problem (5.2) satisfies the relation*

$$\int_\Omega Du.D\phi \, dx - \int_\Omega |Du|^2 u.\phi \, dx = 0, \quad \forall \phi \in H_0^1 \cap L^\infty(\Omega; R^N). \tag{5.3}$$

Proof. Let $\phi \in C_0^\infty(\Omega; R^N)$. For $|t|$ small, the function

$$\frac{u + t\phi}{|u + t\phi|}$$

belongs to the set on which u is a minimum. It then follows easily that

$$\frac{d}{dt} \int_\Omega \left| D\left(\frac{u + t\phi}{|u + t\phi|}\right) \right|^2 dx \, |_{t=0} = 0.$$

By a simple calculation, we have

$$\left(Du, D\left(\frac{\phi}{|u + t\phi|} - (u + t\phi)\frac{(u + t\phi).\phi}{|u + t\phi|^{3/2}}\right)\right)|_{t=0} = 0.$$

Hence, recalling that $|u| = 1$, we obtain

$$(Du, D\phi) - (Du, D(u\,u.\phi)) = 0.$$

We have

$$\sum_{j,k=1}^N Du^k.D(u^k \, u^j \phi^j) = |Du|^2 u.\phi,$$

since $|u| = 1$, as is easily checked. Therefore, we have (5.3) for ϕ as above. By regularization we produce (5.3) itself.   ◇

One can consider a large class of problems of this type. We refer to J. Jost [60] for a presentation of harmonic mappings. Define symmetric matrices

$$\gamma_{\alpha\beta}(x), \quad \alpha, \beta = 1, \ldots, n,$$

$$g_{ij}(u), \quad i, j = 1, \ldots, N,$$

with inverses denoted by

$$\gamma^{\alpha\beta}(x), g^{ij}(u), \quad i, j = 1, \ldots, N.$$

Let

$$\gamma = \det(\gamma_{\alpha\beta}).$$

The energy density is defined by the formula

$$e(u)(x) = \frac{1}{2}\gamma^{\alpha\beta}(x)g_{ij}(u(x))D_\alpha u^i D_\beta u^j,$$

and the global energy is given by

$$E(u) = \int_\Omega e(u)\sqrt{\gamma}\,dx.$$

The vector function u is harmonic whenever it minimizes the energy $E(u)$.

A proof analogous to that of Theorem 5.1 yields the corresponding Euler–Lagrange equations

$$\frac{1}{\sqrt{\gamma}}D_\alpha(\sqrt{\gamma}\gamma^{\alpha\beta}D_\beta u^i) + \gamma^{\alpha\beta}\Gamma^i_{jk}D_\alpha u^j D_\beta u^k = 0,$$

where

$$\Gamma^i_{jk} = \frac{1}{2}g^{il}(g_{l\,k,j} + g_{l\,j,k} - g_{j\,k,l})$$

are the Christoffel symbols. We notice that we get a general type of quadratic term in the first derivatives.

We go back to (5.3) and obtain the following identity:

Lemma 5.2. *Define*

$$f^{kj} = u^k Du^j - u^j Du^k. \tag{5.4}$$

Then one has the properties

$$u^k|Du|^2 = f^{kj}.Du^j, \tag{5.5}$$

$$\operatorname{div} f^{kj} = 0. \tag{5.6}$$

Proof. Note that

$$f^{kj}.Du^j = u^k Du^j.Du^j,$$

since

$$u^j Du^k.Du^j = Du^k.D\left(\frac{1}{2}|u|^2\right) = 0.$$

Hence we obtain (5.5). Next

$$\operatorname{div} f^{kj} = \sum_{l=1}^n D_l(u^k D_l u^j - u^j D_l u^k)$$

$$= \sum_{l=1}^n (D_l u^k D_l u^j + u^k \Delta u^j - D_l u^j D_l u^k - u^j \Delta u^k)$$

$$= u^k \Delta u^j - u^j \Delta u^k,$$

and taking account of (5.3),

$$= u^k u^j |Du|^2 - u^j u^k |Du|^2 = 0.$$

$\diamondsuit$

5.3 Regularity

Our objective is to prove that a solution u of (5.3) locally has the $W^{2,p}$ regularity, for $1 \leq p < \infty$. Note that we have

$$-\Delta u = u|Du|^2 \tag{5.7}$$

and

$$u \in H^1(\Omega; R^N), \quad |u| = 1. \tag{5.8}$$

Therefore, we are in the same situation as Theorem 2.23 except that we do not have the main assumption (2.39). In dimension $n = 2$, the result has been proven by F. Helein [56]. Note that

$$\frac{x}{|x|}$$

is an extremal for $n \geq 3$ and is not smooth. To obtain smoothness in dimension 2, the idea is to prove that

$$u \in W^{2,1}_{\text{loc}}(\Omega; R^N). \tag{5.9}$$

Indeed, in that case, from (1.2) we obtain

$$u \in C^0(\Omega; R^N), \tag{5.10}$$

and then we can refer to Theorem 2.24 to derive the result.

If we look at the right-hand side of (5.7), we note that it belongs to $L^1(\Omega; R^N)$, and thus (5.9) does not follow in a natural manner from the regularity of the solution of the Laplace equation. Fortunately, an additional property is available. We have, from Lemma 5.2,

$$u^k|Du|^2 = \sum_j f^{kj}.Du^j, \tag{5.11}$$

where each vector f^{kj} satisfies (5.6). We state the following result.

Theorem 5.3. *Any solution of (5.7), (5.8) with $n = 2$ belongs to*

$$W^{2,p}_{\text{loc}}(\Omega; R^N).$$

Proof. One observes that

$$u^k|Du|^2 \in \mathcal{H}^1_{\text{loc}}(\Omega), \tag{5.12}$$

the Hardy space, whose definition will be made precise later. Then from (5.7),

$$\Delta u^k \in \mathcal{H}^1_{\text{loc}}(\Omega),$$

which implies

$$D^2 u^k \in \mathcal{H}^1_{\text{loc}}(\Omega) \subset L^1_{\text{loc}}(\Omega).$$

This result is true for any value of n: When $n = 2$, we then have (5.9) and (5.10), and from Theorem 2.24 the result follows.   ◇

5.4 Hardy Spaces

5.4.1 Basic Properties

The main motivation for introducing these spaces is already contained in the statement of Theorem 5.3. If we consider the Dirichlet problem

$$-\Delta u = f$$

with $f \in L^1$, then the solution u may not have second derivatives in L^1. However, this is the case whenever $f \in \mathcal{H}^1$. We shall give in this section some basic notions about Hardy spaces. We begin with the definition (we follow the presentation of S. Semmes [92]).

Let

$$\mathcal{F} = \{\phi \in C^\infty(R^n) : \text{supp } \phi \subset B(0,1), \ ||D\phi||_\infty \leq 1\}. \tag{5.13}$$

Note that

$$||\phi||_\infty \leq 1.$$

For $t > 0$, we write

$$\phi_t(x) = \frac{\phi(\frac{x}{t})}{t^n}.$$

If $f \in L^1_{\text{loc}}(R^n)$, we write

$$f^*(x) = \sup_{t>0} \sup_{\phi \in \mathcal{F}} |\phi_t * f(x)|, \tag{5.14}$$

where $\phi_t * f$ denotes a convolution. The function f^* is called the "grand maximal function" of f.

Definition 5.4. We say that f lies in the Hardy space $\mathcal{H}^1(R^n)$ if

$$f^* \in L^1(R^n),$$

and the Hardy space norm is defined by

$$||f||_{\mathcal{H}^1} = ||f^*||_{L^1}.$$

It is easy to check that

$$|f(x)| \leq f^*, \ \text{a.e.}$$

Hence we have the inclusion

$$\mathcal{H}^1 \subset L^1$$

with continuous injection.

The grand maximal function is an alternative to the maximal function defined by

$$Mf(x) = \sup_{t>0} \frac{1}{|B_t(x)|} \int_{B_t(x)} |f(y)|dy = \sup_{t>0} \fint_{B_t(x)} |f(y)|dy, \qquad (5.15)$$

where $B_t(x)$ denotes as usual the ball of center x and radius t. Clearly,

$$f^*(x) \le CMf(x). \qquad (5.16)$$

An important result is

$$\|Mf\|_p \le C_p\|f\|_p, \quad 1 < p < \infty, \qquad (5.17)$$

but this result fails when $p = 1$ [95].

Note that

$$Mf \in L^1(R^n) \Rightarrow f = 0,$$

since whenever

$$|f| > 0 \text{ on } \Omega_0, \text{ with } |\Omega_0| > 0,$$

then, as is easily seen,

$$Mf(x) \ge \frac{C}{1 + |x|^n}.$$

This explains the intermediary role played by the grand maximal function. Also from (5.16), (5.17), the space $\mathcal{H}^p(R^n)$ coincides with $L^p(R^n)$ when $1 < p < \infty$.

The property (5.12) will follow from a result that we write in full generality, following R. Coifman, P.L. Lions, Y. Meyer, S. Semmes [11]. We consider

$$E(.) \in (L^p(R^n))^n, \text{ div } E = 0, \qquad (5.18)$$

$$B(.) \in (L^{p\prime}(R^n))^n, \quad B = D\pi, \ \pi \in (H^1)^{p\prime}(R^n), \qquad (5.19)$$

with

$$1 < p < \infty, \quad 1 < p\prime < \infty, \quad \frac{1}{p} + \frac{1}{p\prime} = 1. \qquad (5.20)$$

Then we have the following property.

Theorem 5.5. *If (5.18), (5.19), (5.20) hold, then*

$$E.B \in \mathcal{H}^1(R^n). \qquad (5.21)$$

Proof. Clearly, we have

$$E.B = \text{ div } (E\pi).$$

Hence

$$\phi_t * (E.B)(x) = \int \phi\left(\frac{x-y}{t}\right) \frac{1}{t^n} \text{ div } (E\pi)(y)\,dy$$

$$= \int D\phi\left(\frac{x-y}{t}\right) \frac{1}{t^{n+1}} E(y)\pi(y)\,dy.$$

Also, we can write

$$\phi_t * (E.B)(x) = \int D\phi\left(\frac{x-y}{t}\right)\frac{1}{t^{n+1}}E(y)\left(\pi(y) - \fint_{B_t(x)}\pi\right)dy,$$

making use of property (5.18).

We pick α, β such that

$$1 < \frac{1}{\alpha} + \frac{1}{\beta} = 1 + \frac{1}{n}$$

and

$$1 < \alpha < p', \quad 1 < \beta < p.$$

This is possible. Using the properties of ϕ and Hölder's inequality, we easily obtain

$$|\phi_t * (E.B)(x)| \le C\left(\fint_{B_t(x)}|E|^\beta\right)^{1/\beta}\left(\fint_{B_t(x)}\left(\frac{|\pi - \fint_{B_t(x)}\pi|}{t}\right)^{\beta'}\right)^{1/\beta'}.$$

Making use of the Poincaré inequality, we have

$$\left(\fint_{B_t(x)}\left(\frac{|\pi - \fint_{B_t(x)}\pi|}{t}\right)^{\beta'}\right)^{1/\beta'} \le C\left(\fint_{B_t(x)}|D\pi|^\alpha\right)^{1/\alpha},$$

since

$$n\left(\frac{1}{\beta'} - \frac{1}{\alpha}\right) + 1 = 0.$$

Therefore, we can assert

$$|\phi_t * (E.B)(x)| \le C\left(\fint_{B_t(x)}|E|^\beta\right)^{1/\beta}\left(\fint_{B_t(x)}|B|^\alpha\right)^{1/\alpha}.$$

Recalling the definition of the maximal function, we get

$$\sup_{t>0}\sup_{\phi\in\mathcal{F}}|\phi_t * (E.B)(x)| \le C(M(|E|^\beta))^{1/\beta}(M(|B|^\alpha))^{1/\alpha}. \tag{5.22}$$

Using the properties (5.17) of maximal functions we can write

$$\int (M(|E|^\beta))^{p/\beta}\, dx \leq C \int |E|^p\, dx,$$

$$\int (M(|B|^\alpha))^{p'/\alpha}\, dx \leq C \int |B|^{p'}\, dx. \tag{5.23}$$

Making use of (5.23), after using Hölder's inequality in integrating (5.22), yields

$$\|E.B\|_{\mathcal{H}^1} \leq C\|E\|_{L^p}\|B\|_{L^{p'}},$$

and the proof is complete.

$\diamond$

To proceed, we need the definition of the local Hardy space $\mathcal{H}^1_{\text{loc}}(\Omega)$. Let

$$K \text{ any compact subset of } \Omega, \ \epsilon_K = \text{ dist } (K, R^n - \Omega).$$

Definition 5.6. We say that f lies in the local Hardy space $\mathcal{H}^1_{\text{loc}}(\Omega)$ if for any K as above

$$\int_K \left(\sup_{0<t<\epsilon_K} \sup_{\phi \in \mathcal{F}} |\phi_t * f(x)| \right) dx < +\infty \tag{5.24}$$

It is clear that in (5.24), f needs to be defined only on Ω.

Then we can state the local version of Theorem 5.5:

Theorem 5.7. *We assume*

$$E(.) \in (L^p(\Omega))^n, \text{ div } E = 0, \tag{5.25}$$

$$B(.) \in (L^{p'}(\Omega))^n, \ B = D\pi, \tag{5.26}$$

with

$$1 < p < \infty, \quad 1 < p\prime < \infty, \quad \frac{1}{p} + \frac{1}{p\prime} = 1. \tag{5.27}$$

then

$$E.B \in \mathcal{H}^1_{\text{loc}}(\Omega). \tag{5.28}$$

Therefore, we see that each term $f^{kj}.Du^j$ in (5.11) is of the form $E.B$, with all assumptions of Theorem 5.7 satisfied ($p = p\prime = 2$). Hence the property (5.12) is established.

5.4.2 Main Regularity Result in the Hardy Space

Thanks to the developments of the previous section, to complete the proof of Theorem 5.3 we are reduced to the following scalar problem. Consider

$$-\Delta v = f, \text{ in } \Omega, v \in H^1(\Omega), f \in \mathcal{H}^1_{\text{loc}}(\Omega). \tag{5.29}$$

Then we have the following theorem.

Theorem 5.8. *Assume (5.29) and* $n = 2$. *Then*

$$v \in C^0(\Omega), \tag{5.30}$$

$$D^2 v \in \mathcal{H}^1_{\text{loc}}(\Omega) \subset L^1_{\text{loc}}(\Omega). \tag{5.31}$$

Remark 5.9. In fact, in view of the special right-hand side, (5.30) will follow from (5.31), and for our original problem, (5.7) is enough for us. In the setup of Theorem 5.8, these properties are proved independently.

Remark 5.10. The assumption $v \in H^1(\Omega)$ can be removed. It facilitates the proof, but by the Sobolev embedding, it is not sufficient to imply (5.30).

Before proving Theorem 5.8 we shall need a few basic properties, and will reduce the problem to R^n.

Proposition 5.11. *If* $f \in \mathcal{H}^1(R^n)$, *then*

$$\int_{R^n} f(x)\,dx = 0. \tag{5.32}$$

Proof. Pick

$$\theta \in C^\infty(R^n), \quad \text{supp } \theta \subset B_1(0), \quad \theta(0) = 1.$$

For fixed x and $s \geq |x|$, we can write

$$\frac{1}{s^n} \int_{R^n} \theta\left(\frac{y}{s}\right) f(y)\,dy = \frac{C(\theta)}{s^n} \int_{R^n} \phi\left(\frac{x-y}{2s}\right) f(y)\,dy,$$

where

$$\phi(\xi) = \frac{\theta\left(\frac{x}{s} - 2\xi\right)}{C(\theta)}.$$

Of course, ϕ depends on x, s, which are fixed. Clearly, from the condition on s, the support of ϕ is in $B_1(0)$. We can always fix the constant $C(\theta)$ in order to get

$$\|D\phi\|_\infty \leq 1.$$

Therefore, we deduce

$$\left| \frac{1}{s^n} \int_{R^n} \theta\left(\frac{y}{s}\right) f(y)\,dy \right| \leq 2^n C(\theta) f^*(x)$$

for $s \geq |x|$. In particular,

$$\left| \int_{R^n} \theta\left(\frac{y}{|x|}\right) f(y)\,dy \right| \leq 2^n C(\theta) f^*(x)|x|^n.$$

Letting $|x| \to \infty$ yields

$$\left| \int_{R^n} f(y) \, dy \right| \leq 2^n C(\theta) \liminf_{|x| \to \infty} f^*(x)|x|^n = 0,$$

since $f^* \in L^1$.   ◇

We next associate to $f \in \mathcal{H}^1_{\text{loc}}(\Omega)$ a function in $\mathcal{H}^1(R^n)$. We extend f by 0 outside Ω; this extension is not in general in $\mathcal{H}^1(R^n)$. If we take a C^∞ function θ with compact support in Ω, the function θf is not in $\mathcal{H}^1(R^n)$ because condition (5.32) is not satisfied.

Proposition 5.12. *If $f \in \mathcal{H}^1_{\text{loc}}(\Omega)$ for any C^∞ function θ with compact support in Ω such that*

$$\int \theta \neq 0,$$

then

$$\theta(f - \lambda) \in \mathcal{H}^1(R^n) \tag{5.33}$$

with λ a constant such that

$$\int \theta(f - \lambda) \, dx = 0. \tag{5.34}$$

Proof. Set

$$g = \theta(f - \lambda).$$

Take any compact set K in Ω that is strictly larger than supp θ. We recall the definition

$$\epsilon_K = \text{dist } (K, R^n - \Omega).$$

We shall estimate separately

$$I_K = \int_K \sup_{t > 0} \sup_{\phi \in \mathcal{F}} |\phi_t * g(x)| dx$$

and

$$I_{R^n - K} = \int_{R^n - K} \sup_{t > 0} \sup_{\phi \in \mathcal{F}} |\phi_t * g(x)| dx.$$

Since g vanishes on the support of θ, we have

$$I_{R^n - K} = \int_{R^n - K} \sup_{t > \text{dist}(x, \, \text{supp } \theta)} \sup_{\phi \in \mathcal{F}} |\phi_t * g(x)| dx.$$

Since $\int g = 0$, we can assert that

$$|\phi_t * g(x)| \leq \|D\phi_t\|_\infty \|g\|_1 \leq \frac{1}{t^{n+1}} \|g\|_1.$$

Therefore, we deduce that

$$I_{R^n - K} \leq \|g\|_1 \int_{R^n - K} \frac{dx}{(\operatorname{dist}(x, \ \operatorname{supp} \ \theta))^{n+1}}.$$

It is easy to check that this integral is finite.

Next we write

$$I_K \leq I_K^1 + I_K^2,$$

where

$$I_K^1 = \int_K \sup_{0 < t < \epsilon_K} \sup_{\phi \in \mathcal{F}} |\phi_t * g(x)| dx$$

and

$$I_K^2 = \int_K \sup_{t \geq \epsilon_K} \sup_{\phi \in \mathcal{F}} |\phi_t * g(x)| dx.$$

Then

$$I_K^2 \leq \frac{\|g\|_1 |K|}{\epsilon_K^n}.$$

Thus I_K^2 is also finite. It remains to estimate I_K^1.

We first write

$$\phi_t * g(x) = -\lambda \phi_t * \theta(x) + \psi_t * f(x)$$

with

$$\psi(y) = \phi(y)\theta(x - ty),$$

and take care to remember that x, t enter in the definition of the function ψ as parameters. Computing

$$D\psi(y) = D\phi(y)\theta(x - ty) - t\phi(y)D\theta(x - ty),$$

we see that if $t < \epsilon_K$, we have

$$\|D\psi(y)\|_\infty \leq C(K, \theta).$$

One can then check that

$$\frac{\psi}{C(K, \theta)} \in \mathcal{F}.$$

Therefore, again for $t < \epsilon_K$, we have

$$\sup_{\phi \in \mathcal{F}} |\phi_t * g(x)| \leq C'(K, \theta) \left(1 + \sup_{\phi \in \mathcal{F}} |\phi_t * f(x)| \right).$$

Thus integrating over K and making use of property (5.24), we obtain that I_K^1 is also finite.

Hence the proof is complete.   ◇

We shall now state the following fundamental results, whose proof is delayed until the next section.

Theorem 5.13. *If $f \in \mathcal{H}^1(R^n)$, $n \geq 2$, then*

$$(-\Delta)^{-1}f \in L^{n/(n-2)}(R^n), \quad \text{if } n > 2, \tag{5.35}$$

and it is continuous when $n = 2$; Moreover

$$(-\Delta)^{-1}f \in W^{1,n/(n-1)}(R^n), \tag{5.36}$$

$$(-\Delta)^{-1}f \in W^{2,1}(R^n). \tag{5.37}$$

We can now proceed with the

Proof of Theorem 5.8. Let K be a compact subset of Ω, and let θ be a positive C^∞ function with compact support in Ω, and $\theta - 1$ on a neighborhood of K. Recalling (5.29) and using Proposition 5.12, we have

$$\theta(f - \lambda) \in \mathcal{H}^1(R^2).$$

If we set

$$w = (-\Delta)^{-1}(\theta(f - \lambda)),$$

then from Theorem 5.13, we have

$$w \in C^0(R^2), \; H^1(R^2), \; W^{2,1}(R^2).$$

Define

$$z = w - v - \frac{\lambda}{2n}|x|^2.$$

Then we have

$$-\Delta z = \theta(f - \lambda) - f + \lambda, \quad \text{in } \Omega, \; z \in H^1(\Omega).$$

In particular, from the definition of θ,

$$-\Delta z = 0$$

on a neighborhood of K. This is sufficient to imply that z is $W^{2,p}$ (for any $1 \leq p < \infty$) in a neighborhood of K contained in $\theta^{-1}(1)$. The regularity of z and w implies that of v, (5.30) and (5.31).

So the proof of Theorem 5.8 is complete.   ◇

5.5 Proof of Theorem 5.13

5.5.1 Continuity when $n = 2$

When $n = 2$, we have the explicit formula

$$(-\Delta)^{-1}f(x) = -\frac{1}{2\pi}\int_{R^n} \log|x - y| f(y)\,dy. \tag{5.38}$$

We shall need a few intermediate results that are important in their own right.

Proposition 5.14. *We have the estimate*

$$\left|\int_{R^n} f(x) \log |x| \, dx\right| \leq C\|f\|_{\mathcal{H}^1}. \tag{5.39}$$

Proof. Let τ be C^∞, $0 \leq \tau \leq 1, \tau = 1$ on $B_1(0), \tau = 0$ outside $B_2(0)$. Consider the function

$$l(x) = -\sum_1^\infty \tau(2^j x) + \sum_1^\infty (1 - \tau(2^{-j} x)).$$

Clearly, we have

$$\begin{aligned}
l(x) &= -n - \tau(2^{n+1} x), \quad 2^{-n-1} \leq |x| < 2^{-n}, \quad n \geq 0, \\
l(x) &= 0, \quad 1 \leq |x| < 2, \\
l(x) &= n - \tau(2^{-n} x), \quad 2^n \leq |x| < 2^{n+1}, \quad n \geq 1.
\end{aligned}$$

The function $l(x)$ has the following important property:

$$|\log |x| - \log 2l(x)| \leq (1 + \log 2).$$

Therefore, in order to prove (5.39), it is enough to prove

$$\left|\int_{R^n} f(x) l(x) \, dx\right| \leq C\|f\|_{\mathcal{H}^1}. \tag{5.40}$$

We shall prove

$$\left|\int_{R^n} f(x)\tau(2^j x) \, dx\right| \leq C 2^{-jn} \inf_{z \in B_{2^{-j}}} f^*(z), \quad j \geq 1, \tag{5.41}$$

$$\left|\int_{R^n} f(x)\tau(2^{-j} x) \, dx\right| \leq C 2^{jn} \inf_{z \in B_{2^j}} f^*(z), \quad j \geq 1. \tag{5.42}$$

We prove only (5.41). The proof of (5.42) is similar. Pick $z \in B_{2^{-j}}$. Define

$$\phi(y) = \frac{\tau(2^j z - 4y)}{4\|D\tau\|_\infty}.$$

Then $\phi \in \mathcal{F}$, and

$$2^{jn} \int_{R^n} f(x)\tau(2^j x) \, dx = C\phi_{2^{-j+2}} * f(z).$$

Therefore, since z is arbitrary in the ball B_{2^j}, we obtain (5.41).

It follows from (5.41), (5.42) that

$$\left| \int_{R^n} f(x) l(x) \, dx \right| \leq C \sum_{j=1}^{\infty} 2^{-jn} \inf_{z \in B_{2^{-j}}} f^*(z) + C \sum_{j=1}^{\infty} 2^{jn} \inf_{z \in B_{2^j}} f^*(z)$$

$$\leq \frac{C}{1 - 2^{-n}} \left(\sum_{j=1}^{\infty} (2^{-jn} - 2^{-jn-n}) \inf_{z \in B_{2^{-j}} - B_{2^{-j-1}}} f^*(z) \right.$$

$$\left. + \sum_{j=1}^{\infty} (2^{jn} - 2^{jn-n}) \inf_{z \in B_{2^j} - B_{2^{j-1}}} f^*(z) \right).$$

Therefore, we also have

$$\left| \int_{R^n} f(x) l(x) \, dx \right| \leq C \sum_{j=1}^{\infty} \int_{B_{2^{-j}} - B_{2^{-j-1}}} f^*(z) \, dz + \sum_{j=1}^{\infty} \int_{B_{2^j} - B_{2^{j-1}}} f^*(z) \, dz$$

$$\leq C \int_{R^n - B_1} f^*(z) \, dz + C \int_{B_{1/2}} f^*(z) \, dz \leq C \|f\|_{\mathcal{H}^1}.$$

Thus we obtain (5.40).   ◊

We give an easy consequence of Proposition 5.14.

Proposition 5.15. *We have the estimate*

$$\left| \int_{R^n} f(y) \log |x - y| \, dy \right| \leq C \|f\|_{\mathcal{H}^1}, \quad \forall x. \tag{5.43}$$

Proof. Set

$$f_x(y) = f(x - y).$$

Clearly,

$$\int_{R^n} f(y) \log |x - y| \, dy = \int_{R^n} f_x(y) \log |y| \, dy,$$

and from Proposition 5.14, it follows that

$$\left| \int_{R^n} f_x(y) \log |y| \, dy \right| \leq C \|f_x\|_{\mathcal{H}^1}.$$

It suffices to observe that

$$f_x^*(z) = f^*(x - z)$$

to conclude immediately.   ◊

The next step to consider is the possibility of approximating $\mathcal{H}^1$ functions by smooth functions with compact support. Namely, introduce

$$\mathcal{D}_0 = \text{space of } C^\infty \text{ functions with compact support and integral } 0.$$

Proposition 5.16. $\mathcal{D}_0$ *is a dense subset of* $\mathcal{H}^1$.

We shall prove Proposition 5.16 by cutting off at infinity and regularizing.
    Introduce, for R large, $\eta_R(x)$, a smooth function satisfying

$$\eta_R = 1 \text{ on } B_{2R}, \ \eta_R = 0 \text{ outside } B_{3R},$$

$$0 \leq \eta_R \leq 1, \quad \|D\eta_R\|_\infty \leq \frac{C_0}{R}.$$

Define

$$P_R(f) = \eta_R f - \eta_R \frac{\int \eta_R f}{\int \eta_R}.$$

Our objective is to prove the following lemma.

Lemma 5.17.

$$\int_{R^n} (f - P_R f)^* \, dx \leq C \int_{R^n - B_R} f^* \, dx; \tag{5.44}$$

hence

$$P_R f \to f \text{ in } \mathcal{H}^1, \text{ as } R \to \infty. \tag{5.45}$$

Proof. We first note the inequality

$$\inf_{B_{4R}} f^*(z) \leq \frac{C}{R^n} \int_{R^n - B_R} f^*(y) \, dy. \tag{5.46}$$

This follows from the trivial remark that

$$\int_{B_{4R} - B_R} f^*(z) \, dz \leq \int_{R^n - B_R} f^*(z) \, dz.$$

A second remark is

$$\frac{|\int_{R^n} \eta_R f \, dx|}{\int_{R^n} \eta_R \, dx} \leq C \inf_{B_{4R}} f^*(z). \tag{5.47}$$

This is obtained by noting that

$$\int_{R^n} \eta_R \, dx \cong R^n$$

and

$$R^{-n} \int_{R^n} \eta_R f \, dx = C\phi_{7R} * f(z)$$

with

$$\phi(y) = C'\eta_R(z - 7Ry).$$

We next establish the inequality

$$(f - P_R f)^*(x) \leq Cf^*(x) + C\frac{R^{n+1}}{(|x| + R)^{n+1}} \inf_{B_{4R}} f^*(z).$$

Since

$$(f - P_R f)^*(x) \le f^*(x) + (P_R f)^*(x),$$

it is sufficient to prove that

$$(P_R f)^*(x) \le C f^*(x) + C \frac{R^{n+1}}{(|x| + R)^{n+1}} \inf_{B_{4R}} f^*(z). \tag{5.48}$$

We shall prove successively that for $\phi \in \mathcal{F}$,

$$|\phi_t * (\eta_R f)(x)| \le C f^*(x), \ \ 0 < t \le R, \tag{5.49}$$

$$\frac{\int \eta_R f}{\int \eta_R} |\phi_t * \eta_R(x)| \le C \frac{R^{n+1}}{(|x| + R)^{n+1}} \inf_{B_{4R}} f^*(z), \ \ 0 < t \le R, \tag{5.50}$$

$$|\phi_t * (P_R f)(x)| \le C \frac{R^{n+1}}{(|x| + R)^{n+1}} \inf_{B_{4R}} f^*(z), \ \ t \ge R. \tag{5.51}$$

Of course, (5.49), (5.50), and (5.51) imply (5.48).

The proof of (5.49) comes from writing

$$\phi_t * (\eta_R f)(x) = \psi_t * f(x)$$

with

$$\psi(y) = \phi(y) \eta_R(x - ty)$$

and observing that, thanks to the restriction on t, $D\psi$ is bounded by a constant, and hence ψ divided by a constant belongs to $\mathcal{F}$.

To prove (5.50), we first observe that the left-hand side vanishes when $|x| > 4R$. When $|x| \le 4R$, we just notice that

$$\|\phi_t * \eta_R\|_\infty \le C,$$

$$\frac{R}{|x| + R} \ge \frac{1}{5},$$

and we just make use of (5.47).

To prove (5.51), we write

$$\phi_t * (P_R f)(x) = \psi_{7R} * f(z)$$

for a convenient ψ (which we do not make explicit), with the important observation that

$$C \left(\frac{t}{R} \right)^{n+1} \in \mathcal{F}.$$

Hence we get

$$|\phi_t * (P_R f)(x)| \le C \frac{R^{n+1}}{t^{n+1}} \inf_{B_{4R}} f^*(z), \ \ t \ge R.$$

We notice that the left-hand side vanishes when $|x| \geq t + 3R$, and

$$\frac{1}{t} \leq \frac{5}{|x| + R} \quad \text{if } |x| \leq t + 3R.$$

Thus we have proven (5.48).

We then deduce

$$\int_{R^n - B_R} (f - P_R f)^*(x) \, dx \leq \int_{R^n - B_R} f^*(x) dx + C R^n \inf_{B_{4R}} f^*(z),$$

and making use of (5.46) we obtain

$$\int_{R^n - B_R} (f - P_R f)^*(x) \, dx \leq C \int_{R^n - B_R} f^*(x) dx. \qquad (5.52)$$

Finally, we notice that $\eta_R = 1$ when $|x| \leq R$; hence

$$f - P_R(f) = \frac{\int_{R^n} \eta_R f \, dx}{\int_{R^n} \eta_R \, dx}.$$

Thus from (5.47) it follows that

$$|f - P_R(f)(x)| \leq \inf_{B_{4R}} f^*(z) \quad \text{when } |x| \leq R.$$

Therefore, making use of (5.46), we also have

$$\int_{B_R} (f - P_R f)^*(x) \, dx \leq C \int_{R^n - B_R} f^*(x) dx,$$

which combined with (5.52) completes the proof. $\qquad \qquad \diamondsuit$

The next lemma gives the corresponding regularization result.

Lemma 5.18. Let θ be a C^∞ function with support in $B_1(0)$ and $\int \theta = 1$. If $f \in \mathcal{H}^1$, then

$$\theta_\epsilon * f \to f \text{ in } \mathcal{H}^1. \qquad (5.53)$$

Proof. Let us first check that

$$(\theta_\epsilon * f)^*(x) \leq C(\theta) f^*(x), \quad \forall x, \forall \epsilon. \qquad (5.54)$$

Indeed,

$$(\theta_\epsilon * f)^*(x) = \sup_{t>0} \sup_{\phi \in \mathcal{F}} |\phi_t * (\theta_\epsilon * f)(x)|$$

$$= \sup_{t>0} \sup_{\phi \in \mathcal{F}} |(\phi_t * \theta_\epsilon) * f(x)|.$$

Then we notice the easy identities

$$t < \epsilon \Rightarrow \phi_t * \theta_\epsilon(x) = 2^{n+1} \|D\theta\|_\infty \tilde{\theta}_{2\epsilon}(x),$$

where

$$\tilde{\theta}(x) = \frac{1}{2\|D\theta\|_\infty} \int \theta \left(2x - \frac{t}{\epsilon} y \right) \phi(y) \, dy$$

and by construction

$$\tilde{\theta} \in \mathcal{F}.$$

Therefore,

$$t < \epsilon \Rightarrow |(\phi_t * \theta_\epsilon) * f(x)| \leq 2^{n+1} \|D\theta\|_\infty f^*(x).$$

Suppose next $t \geq \epsilon$; then we write

$$\phi_t * \theta_\epsilon(x) = 2^{n+1} \tilde{\phi}_{2t}(x),$$

where

$$\tilde{\phi}(x) = \frac{1}{2} \int \phi \left(2x - \frac{\epsilon}{t} y \right) \theta(y) \, dy$$

and by construction

$$\tilde{\phi} \in \mathcal{F}.$$

Therefore, again

$$t \geq \epsilon \Rightarrow |(\phi_t * \theta_\epsilon) * f(x)| \leq 2^{n+1} f^*(x),$$

and then (5.54) has been proven.

Thanks to (5.54) and Lebesgue's theorem, we need only show

$$(\theta_\epsilon * f)^*(x) \to f^*(x), \text{ a.e. } x. \tag{5.55}$$

It is enough to show

$$\limsup_{\epsilon \to 0} \sup_{t>0} \sup_{\phi \in \mathcal{F}} |(\phi_t * \theta_\epsilon) * f(x) - \phi_t * f(x)| = 0, \text{ a.e. } x. \tag{5.56}$$

We first note that

$$\sup_{t > \epsilon^{1/2(n+1)}} \sup_{\phi \in \mathcal{F}} |(\phi_t * \theta_\epsilon) * f(x) - \phi_t * f(x)| \leq C\sqrt{\epsilon},$$

which follows from the fact that

$$\|\phi_t * \theta_\epsilon - \phi_t\|_\infty \leq C \frac{\epsilon}{t^{n+1}}.$$

It is then sufficient to prove

$$\lim_{\epsilon \to 0} \sup_{t \leq \epsilon^{1/2(n+1)}} \sup_{\phi \in \mathcal{F}} |(\phi_t * \theta_\epsilon) * f(x) - \phi_t * f(x)| = 0, \text{ a.e. } x.$$

We pick x a Lebesgue point of f. We are going to show that

$$\lim_{\epsilon \to 0} \sup_{t \le \epsilon^{1/2(n+1)}} \sup_{\phi \in \mathcal{F}} |(\phi_t * \theta_\epsilon) * f(x) - f(x)| = 0, \tag{5.57}$$

and similarly

$$\lim_{\epsilon \to 0} \sup_{t \le \epsilon^{1/2(n+1)}} \sup_{\phi \in \mathcal{F}} |\phi_t * f(x) - f(x)| = 0.$$

We concentrate on (5.57), which is the hardest part. We shall use the two following equivalent formulas:

$$(\phi_t * \theta_\epsilon) * f(x) - f(x) = \frac{1}{t^n \epsilon^n} \int \int \phi\left(\frac{x - y - z}{t}\right) \theta\left(\frac{y}{\epsilon}\right) (f(z) - f(x)) \, dy \, dz$$

$$= \frac{1}{t^n \epsilon^n} \int \int \theta\left(\frac{x - y - z}{\epsilon}\right) \phi\left(\frac{y}{t}\right) (f(z) - f(x)) \, dy \, dz.$$

Assume $t < \epsilon$. Then from the second formula,

$$|(\phi_t * \theta_\epsilon) * f(x) - f(x)| \le C \frac{1}{t^n \epsilon^n} \int \int_{z + y \in B_\epsilon(x), y \in B_t(0)} |f(z) - f(x)| \, dy \, dz$$

$$\le C \frac{1}{\epsilon^n} \int_{B_{2\epsilon}(x)} |f(z) - f(x)| \, dz,$$

and thus clearly,

$$\sup_{t \le \epsilon} \sup_{\phi \in \mathcal{F}} |(\phi_t * \theta_\epsilon) * f(x) - f(x)| \le C \frac{1}{\epsilon^n} \int_{B_{2\epsilon}(x)} |f(z) - f(x)| \, dz. \tag{5.58}$$

Similarly, using the first formula, for $\epsilon \le t \le \epsilon^{1/2(n+1)}$, we have

$$|(\phi_t * \theta_\epsilon) * f(x) - f(x)| \le C \frac{1}{t^n \epsilon^n} \int \int_{z + y \in B_t(x), y \in B_\epsilon(0)} |f(z) - f(x)| \, dy \, dz$$

$$\le C \sup_{t < \epsilon^{1/2(n+1)}} \frac{1}{t^n} \int_{B_{2t}(x)} |f(z) - f(x)| \, dz.$$

Thus

$$\sup_{\epsilon \le t \le \epsilon^{1/2(n+1)}} \sup_{\phi \in \mathcal{F}} |(\phi_t * \theta_\epsilon) * f(x) - f(x)| \le C \sup_{t < \epsilon^{1/2(n+1)}} \frac{1}{t^n} \int_{B_{2t}(x)} |f(z) - f(x)| \, dz,$$
$$\tag{5.59}$$

and from (5.58), (5.59), since x is a Lebesgue point of f, the result (5.57) follows. The proof of (5.53) is now complete.

From Lemmas 5.17, 5.18 the proof of Proposition 5.16 is now complete. ◇

Proof of continuity of

$$(-\Delta)^{-1} f, \text{ when } f \in \mathcal{H}^1.$$

Let us write

$$Tf = (-\Delta)^{-1}f, \qquad (5.60)$$

defining a linear operator on $\mathcal{H}^1$. From formula (5.38) and estimate (5.43), T maps $\mathcal{H}^1$ to L^∞ and

$$\|Tf\|_{L^\infty} \leq C\|f\|_{\mathcal{H}^1}.$$

Now, Tf is clearly continuous for $f \in \mathcal{D}_0$. Using Proposition 5.16, the result follows.   ◇

5.5.2 Proof of (5.35) and (5.36)

Here $n \geq 2$. For $n > 2$, we have the formula

$$Tf = (-\Delta)^{-1}f = \frac{1}{(n-2)|S_{n-1}|} \int_{R^n} \frac{f(y)}{|x-y|^{n-2}}\, dy. \qquad (5.61)$$

Let us introduce the family of operators Af defined by

$$Af(x) = \int_{R^n} a(x-y)f(y)\, dy, \qquad (5.62)$$

where $a(x)$ is a C^∞ function on $R^n/0$ satisfying

$$|a(x)| \leq C_0|x|^{\alpha-n},$$
$$|Da(x)| \leq C_0|x|^{\alpha-n-1} \qquad (5.63)$$

with $0 < \alpha < n$. Clearly, from (5.61), Tf is of the type Af for $n > 2$, with

$$a(x) = \frac{1}{(n-2)|S_{n-1}|}|x|^{2-n},$$

and hence $\alpha = 2$.

Similarly, for any $n \geq 2$, D_iTf is of the type Af, with

$$a(x) = -\frac{1}{|S_{n-1}|}|x|^{1-n}\frac{x_i}{|x|}.$$

Hence $\alpha = 1$. Therefore, (5.35) and (5.36) will be a consequence of the following general result.

Proposition 5.19. *Assume (5.63). Then we have*

$$\|Af\|_q \leq C\|f\|_{\mathcal{H}^1}, \quad \frac{1}{q} = 1 - \frac{\alpha}{n}, \qquad (5.64)$$

where C depends only on α, n, and C_0.

To prove the proposition we rely on a classical result on Riesz potentials (see [95]). Writing

$$I_\alpha f(x) = \int_{R^n} |x - y|^{-n+\alpha} f(y) \, dy$$

with $0 < \alpha < n$, one then has

$$\|I_\alpha f\|_q \le C(p,q)\|f\|_p, \quad 1 < p < \frac{n}{\alpha}, \quad \frac{1}{q} = \frac{1}{p} - \frac{\alpha}{n}. \tag{5.65}$$

The result is false whenever $p = 1$, and to some extent (5.63) provides a substitute for (5.65).

Proof of Proposition (5.19). We shall check the pointwise property

$$|Af| \le C(s) \left(I_{\alpha s}((f^*)^s) \right)^{1/s}, \quad \forall 0 < s \le 1. \tag{5.66}$$

Suppose (5.66) is proven; then picking $s < 1$, and noting that

$$(f^*)^s \in L^{1/s},$$

we make use of (5.65) to assert that

$$\|I_{\alpha s}(f^{*s})\|_{L^{q/s}} \le C(\alpha, s)\|f^{*s}\|_{L^{1/s}}$$

with

$$\frac{1}{q} = 1 - \frac{\alpha}{n}.$$

Then (5.64) follows easily. It remains to prove (5.66).

Let us introduce a function $\theta \in C^\infty$, $\theta \ge 0$, with

$$\theta(x) = 0, \quad \text{if } |x| \le \frac{1}{2} \text{ or } |x| \ge 3,$$

$$\theta(x) > 0, \quad \text{when } 1 \le |x| \le 2.$$

We may without loss of generality assume

$$\sum_{k \in Z} \theta(2^{-k}x) = 1, \quad \forall x \in R^n - \{0\}.$$

Indeed, if not, we replace $\theta(x)$ by

$$\frac{\theta(x)}{\sum_{k \in Z} \theta(2^{-k}x)},$$

and we check that the properties of θ are preserved and the property pertaining to adding is achieved. Define for, $j \in Z$,

$$\eta^j(x) = 2^{-j\alpha}\theta(2^{-jx})a(x).$$

By construction, we have

$$a(x) = \sum_{j\in Z} 2^{j\alpha}\eta^j(x).$$

Hence

$$Af(x) = \sum_{j\in Z} 2^{j\alpha}\eta^j * f(x).$$

Let $z \in B_{2^j}(x)$ be fixed. Define

$$\phi^j(\eta) = 2^{j(n-\alpha)+2n}\theta(2^{-j}(x-z)+2^2\eta)a(x-z+\eta 2^{j+2}).$$

Then by construction

$$\eta^j * f(x) = \phi^j_{2^{j+2}} * f(z).$$

Moreover, for a convenient constant K depending only on θ,

$$\frac{\phi^j}{K} \in \mathcal{F},$$

Hence we can assert that

$$|\eta^j * f(x)| \le K \inf_{z\in B_{2^j}(x)} f^*(z).$$

Therefore,

$$|Af(x)|^s \le \sum_Z 2^{j\alpha s}|\eta^j * f(x)|^s$$

$$\le K^s \sum_Z 2^{j\alpha s} \inf_{z\in B_{2^j}(x)} (f^*(z))^s$$

$$\le C_s \sum_Z 2^{j(\alpha s-n)} \int_{B_{2^j}(x)-B_{2^{j-1}}(x)} (f^*(z))^s\, dz$$

$$\le C_s \sum_Z \int_{B_{2^j}(x)-B_{2^{j-1}}(x)} \frac{(f^*(z))^s}{|z-x|^{n-\alpha s}}\, dz.$$

Therefore, (5.66) has been obtained.   ◇

5.5.3 Proof of (5.37)

We have

$$D_i D_j Tf(x) = -\frac{1}{|S_{n-1}|}\int_{R^n}\left(\frac{\delta_{ij}}{|x-y|^n} - n\frac{(x_i-y_i)(x_j-y_j)}{|x-y|^{n+2}}\right)f(y)\, dy.$$

Write

$$\Gamma_{ij}(x) = -\frac{1}{|S_{n-1}|}\left(\frac{\delta_{ij}}{|x|^n} - n\frac{x_i x_j}{|x|^{n+2}}\right).$$

Hence

$$D_i D_j T f(x) = \Gamma_{ij} * f(x).$$

We are going to emphasize the properties of Γ_{ij} that are relevant for our purpose. We begin with a lemma.

Lemma 5.20. *The Fourier transform satisfies*

$$\widehat{\Gamma_{ij}}(\xi) = \int_{R^n} \exp(-2\pi\, i\xi.x)\, \Gamma_{ij}(x)\, dx = -\frac{\xi_i \xi_j}{|\xi|^2}.$$

Proof. Set

$$g_\xi(x) = \exp(-2\pi i\xi.x).$$

By the symmetry of Γ_{ij}, we have

$$\widehat{\Gamma_{ij}}(\xi) = \Gamma_{ij} * g_\xi(0).$$

But

$$\Gamma_{ij} * g_\xi(x) = D_i D_j (-\Delta)^{-1} g_\xi(x),$$

and we notice that

$$(-\Delta)^{-1} g_\xi(x) = \frac{\exp(-2\pi\xi.x)}{4\pi^2 |\xi|^2}.$$

The result follows easily.   ◇

Corollary 5.21.

$$\|\Gamma_{ij} * f\|_{L^2} \le \|f\|_{L^2}. \tag{5.67}$$

Proof. This is a consequence of the Plancherel theorem and of the boundedness of $\widehat{\Gamma_{ij}}(\xi)$.   ◇

Lemma 5.22. *We have the property*

$$\int_{|x|\ge 2|y|} |\Gamma_{ij}(x-y) - \Gamma_{ij}(x)|\, dx \le C, \quad \forall y \ne 0. \tag{5.68}$$

Proof. This follows easily from the fact that

$$|D\Gamma_{ij}(x)| \le \frac{C}{|x|^{n+1}}.$$

◇

Proof of (5.37). We can now proceed with the proof of (5.37). Omitting the indices i, j we shall prove that

$$\|\Gamma * f\|_{L^1} \le C\|f\|_{\mathcal{H}^1}.$$

In fact, this follows from an important property of the Hardy space $\mathcal{H}^1$, which is the atomic decomposition (see [96]). We shall present it in the next section.

Definition 5.23. An atom is a function a such that

$$a \text{ is supported in a ball } B,$$
$$|a| \leq |B|, \text{a.e.,}$$
$$\int a(x)\, dx = 0.$$

The point is that any function $f \in \mathcal{H}^1$ can be written as

$$f = \sum_k \lambda_k a_k \tag{5.69}$$

where the (possibly infinite) family a_k is a family of atoms and λ_k are complex numbers such that

$$\sum_k |\lambda_k| \leq c\|f\|_{\mathcal{H}^1}$$

and the sum (5.69) converges in $\mathcal{H}^1$.

Noting that for a generic atom a,

$$a_h(x) = a(x + h),$$

it is clear that

$$\Gamma * a(x + h) = \Gamma * a_h(x)$$

and thus

$$\|\Gamma * a\|_{L^1} = \|\Gamma * a_h\|_{L^1}.$$

Therefore, in computing $\|\Gamma * a\|_{L^1}$ we can, without loss of generality, make any translation we like on a. In this context we can then always assume that the ball B that contains the support of a is centered at the origin. So

$$B = B_r(0) = B_r,$$

where r denotes the radius of B. Using (5.67) we have

$$\|\Gamma * a\|_{L^2} \leq \|a\|_{L^2} \leq \frac{1}{|B_r|}.$$

As a consequence, we get by the Schwarz inequality

$$\int_{B_{2r}} |\Gamma * a(x)|\, dx \leq \left(\frac{|B_{2r}|}{|B_r|}\right)^{1/2} = 2^{n/2}. \tag{5.70}$$

Next, for $x \notin B_{2r}$, we write

$$\Gamma * a(x) = \int_{R^n} (\Gamma(x - y) - \Gamma(x))a(y)\, dy$$

thanks to the last property of atoms. Using property (5.68) we then have

$$\int_{R^n - B_{2r}} |\Gamma * a(x)| \, dx \leq \int_{|y| \leq r} |a(y)| \int_{|x| \geq 2|y|} |\Gamma(x - y) - \Gamma(x)| \, dx \, dy \leq C.$$

Combining this estimate with (5.70) yields

$$\|\Gamma * a\|_{L^1} \leq C_n,$$

where the constant C_n does not depend on the specific atom. In view of this and the atomic decomposition, we have

$$\|\Gamma * f\|_{L^1} \leq C_n \sum_k |\lambda_k| \leq C \|f\|_{\mathcal{H}^1},$$

and the proof is complete. ◇

5.5.4 Atomic decomposition

Our purpose here is to establish the important property (5.69). Let us first check that an atom is an element of $\mathcal{H}^1$. Let us denote by x_0, R_0 the center and the radius of B. We also set

$$B^* = B(x_0, 2R_0).$$

First, it is clear that

$$|a^*(x)| \leq \frac{1}{|B|}.$$

Assume now $x \notin B^*$. Considering any test function $\phi \in \mathcal{F}$ and the expression $\phi_t * a(x)$, we first write

$$\phi_t * a(x) = \int a(y)[\phi_t(x - y) - \phi_t(x - x - 0)] dy$$

thanks to the last property of atoms. Next, $y \in B, x \notin B^*$ implies

$$|x - y| \geq \frac{1}{2}|x - x_0|.$$

Therefore, we deduce that

$$t < \frac{1}{2}|x - x_0| \Rightarrow \phi_t * a(x) = 0.$$

So, we assume

$$t \geq \frac{1}{2}|x - x_0|.$$

Note that

$$|\phi_t * a(x)| \leq \int_B |a(y)| \frac{|y - x_0|}{t^{n+1}} dy.$$

Thus

$$|\phi_t * a(x)| \leq \frac{R_0}{2^{n+1}|x - x_0|^{n+1}}, \quad \forall x \notin B^*.$$

These estimates imply

$$\|a^*\|_{L^1} \leq C$$

with a constant not depending on x_0, R_0.

We now proceed in constructing the atomic decomposition of a function $f \in \mathcal{H}^1$. We consider the Whitney decomposition associated with the open set

$$\mathcal{O} = \{f^* > \alpha\}.$$

The fact that this set is open is a consequence of the lower semicontinuity of f^*. So we have cubes

$$Q_k, \quad \tilde{Q}_k = aQ_k, \quad Q_k^* = bQ_k,$$

where we set (see Section 1.1.4)

$$a = \left(1 + \frac{\epsilon}{2}\right), \quad b = 1 + \epsilon.$$

The number ϵ is arbitrary, provided that $0 < \epsilon < \frac{1}{4}$. The cubes Q_k have sides parallel to the axes and have side-lengths that are powers of 2 (positive or negative). Moreover,

$$\cup_k Q_k = \mathcal{O}, \text{ the } Q_k \text{ have disjoint interiors,}$$
$$\text{diam } Q_k \leq \text{ dist } (Q_k, F) \leq 4 \text{ diam } Q_k,$$
$$\cup_k Q_k^* = \mathcal{O}$$

(where F denotes the complement of $\mathcal{O}$), and the Q_k^* have the bounded intersection property (each point of $\mathcal{O}$ is contained in as most $N = 12^n$ cubes Q_k^*).

Let x_k, ρ_k denote the center and the side-length of the cube Q_k. Note that the ball $B(x_k, \beta\rho_k)$, with $\beta > \frac{9}{2}\sqrt{n}$, contains Q_k^* and intersects F.

There is a partition of unity of $\mathcal{O}$ associated with the Whitney decomposition, namely

$$\sum_k \eta_k = \mathbb{1}_{\mathcal{O}},$$

where $\eta_k = 1$ on Q_k, its support is in $\tilde{Q}_k$, and

$$0 \leq \eta_k \leq 1, \quad |D\eta_k| \leq \frac{c_0}{\rho_k},$$

where c_0 does not depend on $\mathcal{O}$.

Let

$$c_k = \frac{\int f\eta_k \, dx}{\int \eta_k \, dx}.$$

Then we have the following result.

Lemma 5.24. *Given the above definitions,*

$$|c_k| \leq C\alpha. \tag{5.71}$$

Moreover,

$$|c_k| \leq Cf^*(x), \quad \forall x \notin Q_k^*, \tag{5.72}$$

where the generic constants C do not depend on $\mathcal{O}$.

Proof. Let $\bar{x}$ be a point of F in the ball $B(x_k, \beta\rho_k)$, which is possible because of the choice of β. Consider the function

$$\phi(\xi) = \frac{\rho_k^n}{c_0(\frac{a}{2}+\beta)\sqrt{n}} \frac{\eta_k(\bar{x} - \xi(\frac{a}{2}+\beta)\rho_k\sqrt{n})}{\int \eta_k}.$$

It is easy to check that it belongs to the set of test functions $\mathcal{F}$ and that

$$c_k = c_0 \left(\left(\frac{a}{2}+\beta\right)\sqrt{n}\right)^{n-1} \phi_{(\frac{a}{2}+\beta)\rho_k\sqrt{n}} * f(\bar{x}),$$

and thus

$$|c_k| \leq c_0 \left(\left(\frac{a}{2}+\beta\right)\sqrt{n}\right)^{n-1} f^*(\bar{x}).$$

Noting that $f^*(\bar{x}) \leq \alpha$, the result (5.71) follows.

We prove (5.72) similarly, with $b/2$ replacing β. ◇

Lemma 5.25. *Define the function*

$$b_k = (f - c_k)\eta_k. \tag{5.73}$$

Then we have the properties

$$\begin{aligned}
b_k^*(x) &\leq Cf^*(x), \quad \forall x \in Q_k^*, \\
b_k^*(x) &\leq C\alpha\frac{\rho_k^{n+1}}{|x - x_k|^{n+1}}, \quad \forall x \notin Q_k^*.
\end{aligned} \tag{5.74}$$

Proof.
A. *Case $x \in Q_k^*$*
Thanks to (5.72) it is sufficient to prove

$$(f\eta_k)^*(x) \leq Cf^*(x), \quad \forall x \in Q_k^*. \tag{5.75}$$

Pick any $\phi \in \mathcal{F}$ and consider $\phi_t * (f\eta_k)(x)$, for a given t. Two cases are possible:

Case 1: $t \leq \rho_k$.
We set

$$\psi(\xi) = \frac{1}{C}\phi(\xi)\eta_k(x - t\xi)$$

with $C = 1 + c_0$. Then one has

$$\phi_t * (f\eta_k)(x) = C\psi_t * f(x),$$

and ψ belongs to $\mathcal{F}$. Therefore, (5.75) holds.

Case 2: $t > \rho_k$
This time we set

$$\psi(\xi) = \frac{1}{C} \left(\frac{\left(\frac{a+b}{2}\right)\rho_k\sqrt{n}}{t} \right)^n \phi\left(\xi \frac{\left(\frac{a+b}{2}\right)\rho_k\sqrt{n}}{t} \right) \eta_k\left(x - \xi\left(\frac{a+b}{2}\right)\rho_k\sqrt{n} \right)$$

with

$$C = \left(\left(\frac{a+b}{2}\right)\sqrt{n} \right)^{n+1} (1 + c_0).$$

Then one checks that ψ belongs to $\mathcal{F}$, thanks to the fact that $x \in Q_k^*$ and

$$\phi_t * (f\eta_k)(x) = C\psi_{((a+b)/2)\rho_k\sqrt{n}} * f(x).$$

Therefore, (5.75) holds.

B. Case $x \notin Q_k^*$
We now write

$$\phi_t * b_k(x) = \int b_k(y)[\phi_t(x - y) - \phi_t(x - x_k)]\, dy.$$

Recalling the definition of b_k and in view of the fact that $x \notin Q_k^*$, we can see that

$$\phi_t * b_k(x) = 0 \quad \text{if } t \le \frac{b-a}{b\sqrt{n}}|x - x_k|.$$

So we may assume

$$t > \frac{b-a}{b\sqrt{n}}|x - x_k|.$$

We write

$$\phi_t * b_k(x) = I_1 - I_2$$

with

$$I_1 = \int f(y)\eta_k(y)[\phi_t(x - y) - \phi_t(x - x_k)]\, dy$$

and

$$I_2 = c_k \int \eta_k(y)[\phi_t(x - y) - \phi_t(x - x_k)]\, dy.$$

From the assumption on t, we first have

$$|I_1| \le \left(\frac{b\sqrt{n}}{b-a} \right)^n \frac{1}{|x - x_k|} I_1'$$

with

$$I_1' = \int f(y)\eta_k(y) \left[\phi\left(\frac{x-y}{t}\right) - \phi\left(\frac{x-x_k}{t}\right) \right] dy.$$

As above, consider $\bar{x}$ a point of F in the ball $B(x_k, \beta\rho_k)$. Then we may define

$$\psi(\xi) = \frac{|x-x_k|}{C\rho_k} \left(\left(\frac{a}{2}+\beta\right)\sqrt{n}\right)^n \eta_k \left(\bar{x}-\xi\left(\frac{a}{2}+\beta\right)\rho_k\sqrt{n}\right)$$

$$\times \left[\phi\left(\frac{x-\bar{x}+\xi\left(\frac{a}{2}+\beta\right)\rho_k\sqrt{n}}{t}\right) - \phi\left(\frac{x-x_k}{t}\right) \right]$$

with

$$C = \frac{\left(\left(\frac{a}{2}+\beta\right)\sqrt{n}\right)^{n+1} b\sqrt{n}}{b-a} \left(1 + c_0\left(\beta + \left(\frac{a}{2}+\beta\right)\sqrt{n}\right)\right).$$

The function ψ belongs to $\mathcal{F}$, and

$$I_1' = \frac{C\rho_k^{n+1}}{|x-x_k|} \psi_{\left(\frac{a}{2}+\beta\right)\rho_k\sqrt{n}} * f(\bar{x}).$$

Hence

$$|I_1'| \leq \frac{C\rho_k^{n+1}}{|x-x_k|} \alpha,$$

and collecting results we get

$$|I_1| \leq C\alpha \frac{\rho_k^{n+1}}{|x-x_k|^{n+1}}$$

with

$$C = \left(\frac{\left(\frac{a}{2}+\beta\right)nb}{b-a}\right)^{n+1} \left(1 + c_0\left(\beta + \left(\frac{a}{2}+\beta\right)\sqrt{n}\right)\right).$$

We majorize $|I_2|$ by

$$|I_2| \leq \frac{|c_k|}{t^{n+1}} \int |y-x_k|\eta_k \, dy.$$

Using the condition on t, the information on the support of η_k, and (5.71) we conclude that

$$|I_2| \leq C\alpha \frac{\rho_k^{n+1}}{|x-x_k|^{n+1}},$$

where

$$C = \left(\frac{a}{2}+\beta\right)^{n-1} \frac{c_0}{2}(ab)^{n+1}(\sqrt{n})^{2n+1}.$$

Hence the second part of the assertion (5.74) has been proven.   ◇

We state without proof an easy consequence of Lemma 5.25.

Corollary 5.26. *We have the estimate*

$$\int_{R^n} b_k^*(x)\,dx \le C \int_{Q_k^*} f^*(x)\,dx,$$

where C depends only on n, a, b, c_0.

We proceed now with the definition of the atomic decomposition. We consider the sequence of values of $\alpha = 2^j$, with j a positive or negative integer. We denote the corresponding sequence of open sets $\mathcal{O}$ by $\mathcal{O}^j$. Of course, we have the inclusion

$$\mathcal{O}^{j+1} \subset \mathcal{O}^j.$$

The Whitney decomposition associated with $\mathcal{O}^j$ is denoted by Q_k^j. So we have now two Whitney decompositions, Q_k^j and Q_l^{j+1}. We have also two partitions of unity η_k^j and η_l^{j+1}. It is important to recall the result of Lemma 1.9 that

$$(Q^j)_k^* \cap (Q^{j+1})_l^* \ne \emptyset \Rightarrow \operatorname{diam}(Q^{j+1})_l^* \le 4 \operatorname{diam}(Q^j)_k^*. \tag{5.76}$$

We define the averages c_k^j, c_l^{j+1} and introduce the double average

$$c_{kl}^j = \frac{\int (f - c_l^{j+1})\eta_l^{j+1}\eta_k^j\,dx}{\int \eta_l^{j+1}\,dx}. \tag{5.77}$$

Thanks to (5.76) and reasoning as for (5.71) one also obtains

$$|c_{kl}^j| \le C2^j.$$

Indeed, one picks a point $\bar{x}$ in $B\left(x_l^{j+1}, \beta\rho_l^{j+1}\right)$ as in the case of estimating c_l^{j+1}. The only trouble comes when one computes the gradient of the candidate as a test function, since it also involves η_k^j. Thanks to (5.76), one can push the calculation through nevertheless. The details are easy.

We define the functions

$$A_k^j = (f - c_k^j)\eta_k^j - \sum_l (f - c_l^{j+1})\eta_l^{j+1}\eta_k^j + \sum_l c_{kl}^j \eta_l^{j+1}$$

and state the following properties.

1. A_k^j is supported in the ball $B\left(x_k^j, 5\sqrt{n}\rho_k^j\right) = B_k^j$. This is again a consequence of (5.76). This implies

$$|B_k^j| = C|Q_k^j|.$$

2.
$$|A_k^j| \le C\,2^j.$$

Indeed, we can also write

$$A_k^j = f\,\mathbb{1}_{F^{j+1}}\eta_k^j - c_k^j\eta_k^j + \sum_l c_l^{j+1}\eta_l^{j+1}\eta_k^j + \sum_l c_{kl}^j\eta_l^{j+1},$$

and in this form the property is clear.

3.

$$\int A_k^j \, dx = 0$$

by construction.

By summing over k, we have

$$\sum_k A_k^j = \sum_k \left(f - c_k^j \right) \eta_k^j - \sum_l \left(f - c_l^{j+1} \right) \eta_l^{j+1}.$$

Hence if we write

$$b^j = \sum_k b_k^j$$

with

$$b_k^j = \left(f - c_k^j \right) \eta_k^j,$$

we can write

$$\sum_k A_k^j = b^j - b^{j+1}.$$

From Corollary 5.26, we have

$$\int (b_k^j)^* \, dx \le C \int_{(Q_k^j)^*} f^* \, dx.$$

Hence

$$\| b^j \|_{\mathcal{H}^1} \le \sum_k \int_{(Q_k^j)^*} f^* \, dx.$$

From the bounded intersection property we have

$$\sum_k \mathbb{1}_{(Q_k^j)^*} \le N \mathbb{1}_{\mathcal{O}^j}$$

and thus

$$\| b^j \|_{\mathcal{H}^1} \le N \int_{\mathcal{O}^j} f^* \, dx.$$

Therefore,

$$\| b^j \|_{\mathcal{H}^1} \to 0, \text{ as } j \to +\infty.$$

Next, considering

$$f - b^{-j} = f \mathbb{1}_{F^{-j}} + \sum_k c_k^{-j} \eta_k^{-j},$$

clearly

$$\| f - b^{-j} \|_{\mathcal{H}^1} \le 2^{-j+1}.$$

From these considerations we may assert that

$$f = \sum_{jk} A_k^j$$

in the sense of convergence of $\mathcal{H}^1$.

To conclude, it remains to write

$$a_k^j = \frac{A_k^j}{\|A_k^j\|_{L^\infty}|B_k^j|},$$
$$\lambda_k^j = \|A_k^j\|_{L^\infty}|B_k^j|,$$

and we note that

$$\sum_{kj} |\lambda_k^j| \leq C \sum_{kj} 2^j |Q_k^j| \leq C \sum_j 2^j |\mathcal{O}^j|.$$

However, we have

$$
\begin{aligned}
\int f^*(x)\, dx &= \sum_j \int_{2^{j+1} \geq f^* > 2^j} f^*(x)\, dx \\
&\geq \sum_j 2^j \operatorname{Meas} \{2^{j+1} \geq f^* > 2^j\} \\
&= \sum_j 2^j \left(\operatorname{Meas} \{f^* > 2^j\} - \operatorname{Meas} \{f^* > 2^{j+1}\}\right) \\
&= \frac{1}{2} \sum_j 2^j \operatorname{Meas} \{f^* > 2^j\}.
\end{aligned}
$$

Collecting results, we see that

$$\sum_j 2^j |\mathcal{O}^j| \leq 2 \int f^*(x)\, dx.$$

Since the a_k^j are atoms, the proof of the atomic decomposition is complete.

6. Nonlinear Elliptic Systems Arising from the Theory of Semiconductors

6.1 Physical Background

We refer for a full justification of the physical setting to the articles [4], [35], [50], [51], [88], [90], [100]. We follow here the presentation of [83].

We begin with the basic equations of carrier transport. We consider a semiconductor device represented by a bounded domain Ω. The carriers are electrons and holes whose respective densities are denoted by n and p. The following continuity equations for electrons and holes have to be satisfied:

$$\frac{\partial n}{\partial t} - \operatorname{div} J_n = -R,$$
$$\frac{\partial p}{\partial t} + \operatorname{div} J_p = -R, \tag{6.1}$$

where

$$
\begin{aligned}
J_n &= \text{conduction current density of electrons,} \\
J_p &= \text{conduction current density of holes,} \\
R &= \text{generation of carriers.}
\end{aligned}
$$

With these continuity equations is associated a potential equation written as follows:

$$- \operatorname{div}(aD\psi) = p - n + f, \tag{6.2}$$

where ψ is the electrostatic potential, a is the permittivity matrix, and f the net impurity (doping profile).

The next step is to express the current densities in terms of the quasi-Fermi potentials, namely

$$J_n = -\mu_n\, nD\phi_n; \quad J_p = -\mu_p\, pD\phi_p, \tag{6.3}$$

where μ_n, μ_p represent the mobility of electrons and holes, and ϕ_n, ϕ_p represent the quasi-Fermi potentials of electrons and holes. The mobilities are not necessarily constants and may depend in particular on $D\phi_n, D\phi_p$. The last step is to express the densities n, p in terms of the quasi-Fermi potentials and the electrostatic potential. In the present setup, we take

$$n = F_n(\psi - \phi_n); \quad p = F_p(\phi_p - \psi), \tag{6.4}$$

where F_n, F_p are positive functions. To simplify a little bit further we shall take

$$F_n(s) = F_p(s) = F(s).$$

Examples.

Boltzmann distribution

$$F(s) = e^s.$$

Fermi Dirac distribution

$$F(s) = \frac{2}{\sqrt{\pi}} \int_0^{+\infty} \frac{\sqrt{\tau}}{1 + \exp \tau - s} \, d\tau.$$

It remains to express the mode of generation of carriers. Two main cases will be considered, the case of recombination–generation and the case of generation by impact ionization. A superposition of both factors is possible and will be considered. We ignore for the time being impact ionization, and consider recombination–generation. In this case the term R must satisfy the following sign conditions:

$$\phi_n > \phi_p \Rightarrow R < 0 \text{ (generation)},$$
$$\phi_n = \phi_p \Rightarrow R = 0 \text{ (thermodynamic equilibrium)},$$
$$\phi_n < \phi_p \Rightarrow R > 0 \text{ (recombination)},$$

which motivates the model

$$R = r(\phi_n, \phi_p, \psi)(e^{\phi_p - \phi_n} - 1),$$

where $r > 0$.

6.2 Stationary Case Without Impact Ionization

6.2.1 Mathematical Setting

Referring to Section 6.1 we consider as independent unknowns

$$u = -\phi_n, \quad v = \phi_p.$$

We next assume

$$\mu_n = \mu(Du); \quad \mu_p = \nu(Dv).$$

We arrive at the equations

$$-\operatorname{div}(\mu(Du)F(u + \psi)Du) = r(u, v, \psi)(1 - e^{u+v}),$$
$$-\operatorname{div}(\nu(Dv)F(v - \psi)Dv) = r(u, v, \psi)(1 - e^{u+v}), \tag{6.5}$$
$$-\operatorname{div}(a\, D\psi) = F(v - \psi) - F(u + \psi) + f.$$

We denote by N the dimension (to avoid confusion with the density of electrons n). In practice, $N = 2$ or $N = 3$. Hence Ω is a bounded domain of R^N, which is a Lipschitz domain whose boundary is divided into a Dirichlet part Γ_D and a Neumann part Γ_N, where the Dirichlet part is made up of a finite number of simply connected pieces, and the sphere condition is satisfied; more precisely, we assume (1.22), (2.51), (2.52).

We assume further that

$$a(x) \text{ measurable,}$$
$$\alpha|\xi|^2 \le a(x)\xi.\xi \le \alpha_1|\xi|^2, \ \alpha > 0, \ \forall \xi \in R^N, \tag{6.6}$$

$$F(s) \in C^0, \quad F(s) > 0, \quad \forall s \in R,$$
$$F(s_1) \le F(s_2), \quad \text{if } s_1 \le s_2, \tag{6.7}$$

$$r(s_1, s_2, s_3) \ge 0, \quad r, \mu, \nu \text{ continuous,} \tag{6.8}$$

$$(\mu(\xi)\xi - \mu(\eta)\eta).(\xi - \eta) \ge 0, \quad \forall \xi, \eta \in R^N,$$
$$\mu_1|\xi|^p \le \mu(\xi)|\xi|^2 \le \mu_2|\xi|^p + \epsilon_0|\xi|,$$
$$\mu_1 > 0, \quad 1 < p < \infty, \quad \epsilon_0 \ge 0 \ \forall \xi \in R^N, \tag{6.9}$$

$$(\nu(\xi)\xi - \nu(\eta)\eta).(\xi - \eta) \ge 0, \quad \forall \xi, \eta \in R^N,$$
$$\nu_1|\xi|^p \le \nu(\xi)|\xi|^2 \le \nu_2|\xi|^p + \epsilon_1|\xi|,$$
$$\nu_1 > 0, \quad 1 < p < \infty, \quad \epsilon_1 \ge 0 \ \forall \xi \in R^N, \tag{6.10}$$

$$f \in L^q(\Omega), \quad q > \frac{N}{2}. \tag{6.11}$$

We complete equations (6.5) with the following boundary conditions:

$$u = u_0, \quad v = v_0, \quad \psi = \psi_0 \text{ on } \Gamma_D,$$
$$\frac{\partial u}{\partial n} = \frac{\partial v}{\partial n} = \frac{\partial \psi}{\partial n_a} = 0 \text{ on } \Gamma_N, \tag{6.12}$$

where

$$u_0, v_0 \in L^\infty \cap W^{1,\max(2,p)}(\Omega),$$
$$\psi_0 \in L^\infty \cap H^1(\Omega), \tag{6.13}$$

and where we have used the notation

$$\frac{\partial \psi}{\partial n_a} = \sum_{ij} n_i a_{ij} \frac{\partial \psi}{\partial x_j} \quad \text{on } \Gamma,$$

in which n represents the outward unit normal. Moreover,

$$\frac{\partial u}{\partial n} = \frac{\partial u}{\partial n_I}$$

is the normal derivative.

Define the function $\bar{\psi}$ that is the solution of

$$
\begin{aligned}
-\operatorname{div}(aD\bar{\psi}) &= f, \quad \text{on } \Omega, \\
\bar{\psi} &= 0 \text{ on } \Gamma_D, \\
\frac{\partial \bar{\psi}}{\partial n_a} &- 0 \text{ on } \Gamma_N.
\end{aligned}
\tag{6.14}
$$

Since $2q > N$, we know that ψ is Hölder on $\overline{\Omega}$ (see Theorem 2.14), thanks to the assumptions (1.22), (2.51), (2.52).

Introduce the following constants:

$$
\begin{aligned}
\alpha_0 &= \max\{\operatorname{ess\,sup}_{\Gamma_D}|u_0|, \operatorname{ess\,sup}_{\Gamma_D}|v_0|\}, \\
\beta_0 &= \operatorname{ess\,sup}_{\Gamma_D}|\psi_0|, \\
\gamma_0 &= \operatorname{ess\,sup}_{\overline{\Omega}}|\bar{\psi}|,
\end{aligned}
\tag{6.15}
$$

and further,

$$
\begin{aligned}
\delta_0 &= F(-2\alpha_0 - 2\gamma_0 - \beta_0), \\
\delta_1 &= F(2\alpha_0 + 2\gamma_0 + \beta_0).
\end{aligned}
\tag{6.16}
$$

Our objective is to prove the following main result.

Theorem 6.1. *We assume (1.22), (2.51), (2.52), (6.6), (6.7), (6.8), (6.9), (6.10), (6.11), (6.13). Then there exists one solution*

$$u, v \in L^\infty \cap W^{1,p}(\Omega),$$

$$\psi \in L^\infty \cap H^1(\Omega),$$

of the system (6.5), (6.12).

Note that thanks to the second parts of the assumptions (6.9), (6.10), we can assert that

$$\mu(Du)Du, \ \nu(Dv)Dv \in W^{1,p'}(\Omega).$$

Then as usual, (6.5) has to be interpreted in the weak sense; namely, pick any test function

$$
\begin{aligned}
\phi_1, \phi_2 &\in L^\infty \cap W^{1,p}(\Omega), \\
\phi_1, \phi_2 &= 0 \text{ on } \Gamma_D.
\end{aligned}
$$

Then one has

$$\int_\Omega F(u+\psi)\,\mu(Du)Du.D\phi_1\,dx = \int_\Omega r(u,v,\psi)(1-e^{u+v})\phi_1\,dx,$$

$$(6.17)$$

$$\int_\Omega F(v-\psi)\,\nu(Dv)Dv.D\phi_2\,dx = \int_\Omega r(u,v,\psi)(1-e^{u+v})\phi_2\,dx.$$

Next, for any test function

$$\phi \in L^\infty \cap H^1(\Omega), \quad \phi = 0 \text{ on } \Gamma_D,$$

one has

$$\int_\Omega a(x)D\psi.D\phi\,dx = \int_\Omega (F(v-\psi) - F(u+\psi) + f)\phi\,dx. \tag{6.18}$$

6.2.2 Proof of Theorem 6.1

• A Priori Estimates
Suppose we have a solution u, v, ψ of (6.5), (6.12). Then one has the estimates

$$\text{ess sup}_{\overline{\Omega}}|u|, \ \text{ess sup}_{\overline{\Omega}}|v| \le \alpha_0,$$

$$\text{ess sup}_{\overline{\Omega}}|\psi| \le 2\gamma_0 + \alpha_0 + \beta_0. \tag{6.19}$$

Indeed, let us check that

$$\|u\|_\infty \le \max\{\text{ess sup}_{\Gamma_D}|u_0|, \|v\|_\infty\},$$

$$\|v\|_\infty \le \max\{\text{ess sup}_{\Gamma_D}|v_0|, \|u\|_\infty\}. \tag{6.20}$$

To check (6.20), set

$$k = \max\{\text{ess sup}_{\Gamma_D}|u_0|, \|v\|_\infty\}.$$

We can test the first equation (6.5) with

$$\phi_1 = (u-k)^+$$

or take it directly in the variational form (6.17). Since

$$u = u_0 \text{ on } \Gamma_D,$$
$$(u-k)^+ = 0 \text{ on } \Gamma_D,$$

it follows that $(u-k)^+$ is a test function admissible in (6.17). Moreover, clearly

$$u(x) > k \Rightarrow u(x) + v(x) > 0;$$

hence

$$(u - k)^+(1 - e^{u+v}) \leq 0.$$

Since the left-hand side of (6.17) can be written

$$\int_\Omega F(u + \psi)\,\mu(Du)|D(u - k)^+|^2\,dx,$$

necessarily

$$F(u + \psi)\,\mu(Du)|D(u - k)^+|^2 = 0 \text{ a.e.}$$

Hence also from assumption (6.7) on F and assumption (6.9) on μ we deduce

$$D(u - k)^+ = 0 \text{ a.e.},$$

which, combined with the value on Γ_D, implies

$$(u - k)^+ = 0 \text{ on } \overline{\Omega}.$$

By similar arguments we complete the proof of (6.20). Now if $\|v\|_\infty \neq \|u\|_\infty$, from (6.20) it is an easy exercise to deduce the first a priori estimate (6.19). If $\|v\|_\infty = \|u\|_\infty$, then consider $u_\epsilon = u + \epsilon$, $v_\epsilon = v - \epsilon$. The pair u_ϵ, v_ϵ satisfies

$$\int_\Omega F(u + \psi)\,\mu(Du)Du_\epsilon.D\phi_1\,dx = \int_\Omega r(u, v, \psi)(1 - e^{u_\epsilon + v_\epsilon})\phi_1\,dx,$$

$$\int_\Omega F(v - \psi)\,\nu(Dv)Dv_\epsilon.D\phi_2\,dx = \int_\Omega r(u, v, \psi)(1 - e^{u_\epsilon + v_\epsilon})\phi_2\,dx,$$

for any test function such as in (6.17), with the boundary conditions

$$u_\epsilon = u_0 + \epsilon, \quad v_\epsilon = v_0 - \epsilon \text{ on } \Gamma_D,$$
$$\frac{\partial u_\epsilon}{\partial n} = \frac{\partial v_\epsilon}{\partial n} = 0 \text{ on } \Gamma_N.$$

Using the same approach, we can assert (noting that $\|u_\epsilon\|_\infty > \|v_\epsilon\|_\infty$) that

$$\|u_\epsilon\|_\infty \leq \text{ess sup}_{\Gamma_D}|u_0| + \epsilon,$$

and thus we easily complete the proof.

Let us turn to the L^∞ estimate on ψ. First define

$$\tilde{\psi} = \psi - \bar{\psi}.$$

Then from (6.18) one can write

$$\int_\Omega a(x)D\tilde{\psi}.D\phi\,dx = \int_\Omega (F(v - \tilde{\psi} - \bar{\psi}) - F(u + \tilde{\psi} + \bar{\psi}))\phi\,dx \qquad (6.21)$$

for any ϕ such that

$$\phi \in L^\infty \cap H^1(\Omega), \quad \phi = 0 \text{ on } \Gamma_D,$$

and
$$\tilde{\psi} = \psi_0 \text{ on } \Gamma_D.$$

Let
$$k = \alpha_0 + \beta_0 + \gamma_0$$

and take
$$\phi = (\tilde{\psi} - k)^+$$

in (6.21), which is an admissible test function. Note that
$$\tilde{\psi}(x) > k \Leftarrow \tilde{\psi}(x) + \bar{\psi}(x) \geq \alpha_0.$$

Hence
$$2(\tilde{\psi}(x) + \bar{\psi}(x)) \geq v(x) - u(x)$$

and also
$$\tilde{\psi}(x) + \bar{\psi}(x) + u(x) \geq v(x) - \tilde{\psi}(x) - \bar{\psi}(x),$$

which implies that
$$F(v(x) - \tilde{\psi}(x) - \bar{\psi}(x)) \leq F(\tilde{\psi}(x) + \bar{\psi}(x) + u(x)).$$

Therefore, from (6.21) it follows that
$$(\tilde{\psi} - k)^+ = 0 \text{ a.e.},$$

and we can easily complete the proof of the second estimate (6.19). $\diamondsuit$

• **Replacing** μ, ν **by** $\mu + \epsilon, \nu + \epsilon$
In this paragraph we check that without loss of generality, we can make stronger assumptions than (6.9) and (6.10), namely

$$(\mu(\xi)\xi - \mu(\eta)\eta).(\xi - \eta) \geq \mu_0|\xi - \eta|^2, \quad \forall \xi, \eta \in R^N, \ \mu_0 > 0, \qquad (6.22)$$

$$(\nu(\xi)\xi - \nu(\eta)\eta).(\xi - \eta) \geq \nu_0|\xi - \eta|^2, \quad \forall \xi, \eta \in R^N, \ \nu_0 > 0, \qquad (6.23)$$

and
$$p \geq 2. \qquad (6.24)$$

Indeed, let us suppose that the theorem has been proven with assumptions (6.22), (6.23), (6.24), instead of the corresponding parts of (6.9), (6.10) and $p > 1$. Define
$$\mu_\epsilon = \mu + \epsilon, \quad \nu_\epsilon = \nu + \epsilon.$$

Everything remains unchanged except that (6.22), (6.23) are now satisfied. Moreover, if in the original assumptions (6.9), (6.10) p was less than 2, we can assert that (6.9), (6.10) hold now with $p = 2$ and

$$\mu_1 = \nu_1 = \mu_2 = \nu_2 = \epsilon,$$
$$\epsilon_0 = \mu_2 + \epsilon_0,$$
$$\epsilon_1 = \nu_2 + \epsilon_1.$$

On the other hand, if in the original assumptions (6.9), (6.10) p was greater than or equal to 2, we claim that the assumptions (6.9), (6.10) hold now with the same p, the same μ_1, ν_1, and values of $\mu_2, \nu_2, \epsilon_0, \epsilon_1$ modified as follows:

$$\mu_2 \to \mu_2 + \frac{\epsilon}{p-1},$$

$$\epsilon_0 \to \epsilon_0 + \epsilon\frac{p-2}{p-1},$$

$$\nu_2 \to \nu_2 + \frac{\epsilon}{p-1},$$

$$\epsilon_1 \to \epsilon_1 + \epsilon\frac{p-2}{p-1}.$$

At any rate, (6.24) is now satisfied.

Therefore, we can state that there exists a triple $u_\epsilon, v_\epsilon, \psi_\epsilon$ satisfying

$$\int_\Omega F(u_\epsilon + \psi_\epsilon)\,\mu_\epsilon(Du_\epsilon)Du_\epsilon.D\phi_1\,dx = \int_\Omega r(u_\epsilon, v_\epsilon, \psi_\epsilon)(1 - e^{u_\epsilon + v_\epsilon})\phi_1\,dx,$$

$$\int_\Omega F(v_\epsilon - \psi_\epsilon)\,\nu_\epsilon(Dv_\epsilon)Dv_\epsilon.D\phi_2\,dx = \int_\Omega r(u_\epsilon, v_\epsilon, \psi_\epsilon)(1 - e^{u_\epsilon + v_\epsilon})\phi_2\,dx,$$

$$\tag{6.25}$$

for any test functions

$$\phi_1, \phi_2 \in L^\infty \cap W^{1,\max(p,2)}(\Omega),$$
$$\phi_1, \phi_2 = 0 \text{ on } \Gamma_D,$$

and furthermore,

$$u_\epsilon = u_0, \quad v_\epsilon = v_0 \text{ on } \Gamma_D,$$
$$\frac{\partial u_\epsilon}{\partial n} = \frac{\partial v_\epsilon}{\partial n} = 0 \text{ on } \Gamma_N.$$
$$u_\epsilon, v_\epsilon \in L^\infty \cap W^{1,\max(p,2)}(\Omega).$$

Next, one has

$$\int_\Omega a(x)D\psi_\epsilon.D\phi\,dx = \int_\Omega (F(v_\epsilon - \psi_\epsilon) - F(u_\epsilon + \psi_\epsilon) + f)\phi\,dx$$

for any test function

$$\phi \in L^\infty \cap H^1(\Omega), \quad \phi = 0 \text{ on } \gamma_D,$$

and

$$\psi_\epsilon \in L^\infty \cap H^1(\Omega),$$
$$\psi_\epsilon = \psi_0 \text{ on } \Gamma_D, \quad \frac{\partial \psi_\epsilon}{\partial n_a} = 0 \text{ on } \Gamma_N.$$

We then derive estimates independent of ϵ.

Looking first at the a priori estimates established above, we can immediately state the following estimates:

$$\|u_\epsilon\|_\infty, \|v_\epsilon\|_\infty \leq \alpha_0,$$
$$\|\psi_\epsilon\|_\infty \leq 2\gamma_0 + \alpha_0 + \beta_0. \tag{6.26}$$

We next establish estimates in Sobolev spaces. Note that

$$\delta_0 \leq F(u_\epsilon + \psi_\epsilon) \leq \delta_1$$

and similarly

$$\delta_0 \leq F(v_\epsilon - \psi_\epsilon) \leq \delta_1.$$

Let

$$r_0 = \sup_{|s_1|,|s_2|\leq\alpha_0, |s_3|\leq\alpha_0+\beta_0+2\gamma_0} r(s_1, s_2, s_3).$$

We deduce from (6.25) that

$$\int_\Omega F(u_\epsilon + \psi_\epsilon) \mu_\epsilon(|Du_\epsilon|^2 - Du_\epsilon . Du_0)\, dx$$
$$= \int_\Omega r(u_\epsilon, v_\epsilon, \psi_\epsilon)(1 - e^{u_\epsilon + v_\epsilon})(u_\epsilon - u_0)\, dx,$$
$$\int_\Omega F(v_\epsilon - \psi_\epsilon) \nu_\epsilon(|Dv_\epsilon|^2 - Dv_\epsilon . Dv_0)\, dx$$
$$= \int_\Omega r(u_\epsilon, v_\epsilon, \psi_\epsilon)(1 - e^{u_\epsilon + v_\epsilon})(v_\epsilon - v_0)\, dx. \tag{6.27}$$

Indeed, $u_\epsilon - u_0$, $v_\epsilon - v_0$ are valid test functions in (6.25). We deduce from the first relation (6.27) that

$$\delta_0\mu_1|Du_\epsilon|_{L^p}^p \leq \delta_1\mu_2|Du_\epsilon|_{L^p}^{p-1}|Du_0|_{L^p} + \delta_1\epsilon_0|Du_0|_{L^1} + \epsilon\frac{\delta_1^2}{4\delta_0}|Du_0|_{L^2}^2 + 2r_0\alpha_0e^{2\alpha_0},$$

and thus, assuming $\epsilon \leq 1$,

$$\delta_0\mu_1|Du_\epsilon|_{L^p}^p \leq \delta_1\mu_2|Du_\epsilon|_{L^p}^{p-1}|Du_0|_{L^p} + \delta_1\epsilon_0|Du_0|_{L^1} + \frac{\delta_1^2}{4\delta_0}|Du_0|_{L^2}^2 + 2r_0\alpha_0e^{2\alpha_0}.$$

Hence we get

$$|Du_\epsilon|_{L^p} \leq M_1, \tag{6.28}$$

where M_1 is the positive root of the equation

$$\delta_0\mu_1 x^p - \delta_1\mu_2|Du_0|_{L^p} x^{p-1} - \delta_1\epsilon_0|Du_0|_{L^1} - \frac{\delta_1^2}{4\delta_0}|Du_0|_{L^2}^2 - 2r_0\alpha_0e^{2\alpha_0} = 0.$$

Similarly, we get

$$|Dv_\epsilon|_{L^p} \leq M_2, \tag{6.29}$$

where M_2 is defined as M_1 with μ_1, μ_2 replaced by ν_1, ν_2, with u_0 replaced by v_0, and with ϵ_0 replaced by ϵ_1.

In the same way, we have

$$\int_\Omega a(x)D\psi_\epsilon.D(\psi_\epsilon - \psi_0)\,dx = \int_\Omega (F(v_\epsilon - \psi_\epsilon) - F(u_\epsilon + \psi_\epsilon) + f)(\psi_\epsilon - \psi_0)\,dx.$$

Hence

$$\int_\Omega a(x)D\psi_\epsilon.D(\psi_\epsilon - \psi_0)\,dx \le \int_\Omega (\delta_1 - \delta_0 + |f|)(2\gamma_0 + \alpha_0 + \beta_0 + \|\psi_0\|_\infty)\,dx,$$

and therefore

$$|D\psi_\epsilon|_{L^2} \le M_3, \tag{6.30}$$

where M_3 is the positive root of the equation

$$\alpha\,x^2 - \|a\|_\infty|D\psi_0|x - (2\gamma_0 + \alpha_0 + \beta_0 + \|\psi_0\|_\infty)((\delta_1 - \delta_0)|\Omega| + |f|_{L^1}) = 0.$$

From the estimates $(6.26),(6.28),(6.29),(6.30)$ we can see that

$$u_\epsilon, v_\epsilon \text{ remains in a bounded subset of } L^\infty \cap W^{1,p}(\Omega)$$

and

$$\psi_\epsilon \text{ remains in a bounded subset of } L^\infty \cap H^1(\Omega).$$

We extract subsequences such that

$$u_\epsilon \rightharpoonup u \text{ in } W^{1,p} \text{ weakly, } L^\infty \text{ weak-star, a.e.,}$$
$$v_\epsilon \rightharpoonup v \text{ in } W^{1,p} \text{ weakly, } L^\infty \text{ weak-star, a.e.,}$$
$$\psi_\epsilon \rightharpoonup \psi \text{ in } H^1 \text{ weakly, } L^\infty \text{ weak-star, a.e.}$$

Letting $\phi_1 \in L^\infty \cap W^{1,p}(\Omega)$, we replace ϕ_1 by $u_\epsilon - u - \theta\phi_1$ in the first equation (6.25), where $\theta > 0$.

We obtain

$$\int_\Omega F(u_\epsilon + \psi_\epsilon)\,\mu_\epsilon(Du_\epsilon)Du_\epsilon.D(u_\epsilon - u - \theta\phi_1)\,dx$$
$$= \int_\Omega r(u_\epsilon, v_\epsilon, \psi_\epsilon)(1 - e^{u_\epsilon + v_\epsilon})(u_\epsilon - u - \theta\phi_1)\,dx.$$

From assumption (6.9) we deduce

$$\int_\Omega F(u_\epsilon + \psi_\epsilon)\,\mu_\epsilon(D(u + \theta\phi_1))D(u + \theta\phi_1).D(u_\epsilon - u - \theta\phi_1)\,dx$$
$$\le \int_\Omega r(u_\epsilon, v_\epsilon, \psi_\epsilon)(1 - e^{u_\epsilon + v_\epsilon})(u_\epsilon - u - \theta\phi_1)\,dx.$$

Letting $\epsilon \to 0$, we obtain

$$\int_\Omega F(u+\psi)\,\mu(D(u+\theta\phi_1))D(u+\theta\phi_1).D\phi_1\,dx \geq \int_\Omega r(u,v,\psi)(1-e^{u+v})\phi_1\,dx.$$

Replacing ϕ_1 with $-\phi_1$, we in fact get equality. Hence the first equation in (6.17) is satisfied. Similar reasoning is used to obtain the second relation (6.17) and (6.18). $\diamond$

• Fixed Point Argument

From now on, we assume that (6.22), (6.23), (6.24) hold. Let us define

$$\mathcal{K} = \left\{ u,v \in (L^p)^2 \left| \begin{array}{l} \|u\|_\infty, \|v\|_\infty \leq \alpha_0 \\ |Du|_{L^p} \leq M_1, |Dv|_{L^p} \leq M_2 \\ u = u_0, v = v_0 \text{ on } \Gamma_0 \end{array} \right. \right\},$$

which is a convex compact subset of $(L^p)^2$. For any pair $u,v \in \mathcal{K}$, define u^*, v^*, ψ to be the solution of the equations

$$\begin{aligned}
-\,\text{div}\,(\mu(Du^*)F(u+\psi)Du^*) &= r(u,v,\psi)(1-e^{u^*+v}), \\
-\,\text{div}\,(\nu(Dv^*)F(v-\psi)Dv^*) &= r(u,v,\psi)(1-e^{u+v^*}), \\
-\,\text{div}\,(a\,D\psi) &= F(v-\psi) - F(u+\psi) + f,
\end{aligned} \tag{6.31}$$

with boundary conditions

$$\begin{aligned}
u^* = u_0, \quad v^* = v_0, \quad \psi = \psi_0 \text{ on } \Gamma_D, \\
\frac{\partial u^*}{\partial n} = \frac{\partial v^*}{\partial n} = \frac{\partial \psi}{\partial n_a} = 0 \text{ on } \Gamma_N.
\end{aligned} \tag{6.32}$$

To solve (6.31), (6.32), we begin by solving the equation for ψ in the space

$$\psi \in L^\infty \cap H^1(\Omega).$$

Given the a priori estimate

$$\|\psi\|_\infty \leq \alpha_0 + \beta_0 + 2\gamma_0,$$

it is easy to prove the existence and uniqueness of ψ (for instance, use an elementary fixed point argument). Note, then, that in (6.31) we can assert that

$$\delta_0 \leq F(u+\psi), \quad F(v-\psi) \leq \delta_1,$$

and

$$r(u,v,\psi) \leq r_0.$$

We note that from the assumptions,

$$\begin{aligned}
\int_\Omega F(u+\psi)(\mu(Du_1)Du_1 - \mu(Du_2)Du_2)D(u_1-u_2)\,dx \\
+ \int_\Omega r(u,v,\psi)e^v(e^{u_1}-e^{u_2})(u_1-u_2)\,dx \geq \delta_0\mu_0 \int_\Omega |D(u_1-u_2)|^2\,dx.
\end{aligned}$$

Thus the equations giving u^*, v^* are classical monotone operator equations, having well-defined unique solutions. It is easy to check that

$$u^*, v^* \in \mathcal{K},$$

which follows from arguments already described in obtaining the a priori estimates. Thus we have defined a map

$$u, v \rightarrow u^*, v^*$$

from $\mathcal{K}$ into itself. Let us check that this map is continuous for the topology of L^p. Indeed, let

$$u_n, v_n \rightarrow u, v \in (L^p)^2$$

and u_n^*, v_n^*, ψ_n be the corresponding solutions of (6.31). Since $(u_n^*, v_n^*) \in \mathcal{K}$, we can assert that

$$u_n^*, v_n^* \rightharpoonup u^*, v^* \in (W^{1,p})^2 \text{ weakly and a.e.,}$$

and also that

$$\psi_n \rightharpoonup \psi \in H^1 \text{ weakly and a.e.,}$$

where ψ is the solution to the third equation (6.31). By monotone operator arguments, similar to those already used in the second part of this proof, we obtain easily that the pair u^*, v^* is the image of u, v under the map, which is thus continuous. From the Leray–Schauder theorem, we can assert the existence of a fixed point. The desired results have been proven.   ◊

6.2.3 A Uniqueness Result

We need more stringent assumptions for uniqueness, namely,

$$(\mu(\xi)\xi - \mu(\eta)\eta).(\xi - \eta) \geq \mu_0|\xi - \eta|^2, \quad \mu_0 > 0, \quad \forall \xi, \eta \in R^N,$$
$$\mu(\xi) \leq \mu_2, \tag{6.33}$$

$$(\nu(\xi)\xi - \nu(\eta)\eta).(\xi - \eta) \geq \nu_0|\xi - \eta|^2, \quad \mu_0 > 0, \quad \forall \xi, \eta \in R^N,$$
$$\nu(\xi) \leq \nu_2, \tag{6.34}$$

$$|F(s_1) - F(s_2)| \leq K(\rho)|s_1 - s_2|, \forall s_i, |s_i| \leq \rho, i = 1, 2,$$
$$F(s_1) \leq F(s_2), \text{ if }, s_1 \leq s_2; \quad F(s) > 0, \tag{6.35}$$

$$|r(s_1, s_1', s_1'') - r(s_2, s_2', s_2'')| \leq M(\rho)(|s_1 - s_2| + |s_1' - s_2'| + |s_1'' - s_2''|),$$
$$\forall s_i, s_i', s_i'', |s_i|, |s_i'|, |s_i''| \leq \rho, i = 1, 2,$$
$$r(s, s', s'') \geq 0. \tag{6.36}$$

We also assume that

$$N = 2, \tag{6.37}$$

$$u_0, v_0 \in L^\infty \cap W^{1,s_0}(\Omega), s_0 > 2,$$
$$\psi_0 \in L^\infty \cap H^1(\Omega). \tag{6.38}$$

We shall set

$$\rho_0 = 2\alpha_0 + 2\gamma_0 + \beta_0$$

Our objective is to prove the following result.

Theorem 6.2. *We assume (1.22), (2.51), (2.52), (6.6), (6.11), (6.33), (6.34), (6.35), (6.36), (6.37), (6.38). Then there exists a positive number ϵ depending only on Ω, α, α_1, μ_0, μ_2, ν_0, ν_2, ρ_0, δ_0, δ_1, $r(0)$, $K(\rho_0)$, $M(\rho_0)$, s_0, $\|u_0 + v_0\|_\infty$, such that if*

$$|1 - e^{u_0+v_0}|_{L^2} + \|Du_0\|_{L^{s_0}} + \|Dv_0\|_{L^{s_0}} \le \epsilon, \tag{6.39}$$

then the solution of the system (6.5), (6.12) is unique.

We shall denote by C a generic constant, depending only on the list of parameters given in the statement of the theorem (henceforth referred to as the list of parameters). We begin by proving an intermediate result.

Lemma 6.3. *There exists $2 < s \le s_0$, depending only on the list of parameters, such that if u, v, ψ is a solution of (6.5), (6.12),*

$$\|Du\|_{L^s} + \|Dv\|_{L^s} \le |1 - e^{u_0+v_0}|_{L^2} + \|Du_0\|_{L^{s_0}} + \|Dv_0\|_{L^{s_0}}. \tag{6.40}$$

Proof. Note that from (6.17), $u - u_0$ satisfies

$$\int_\Omega F(u + \psi)\, \mu(Du) D(u - u_0).D\phi\, dx = \int_\Omega r(u, v, \psi)(1 - e^{u+v})\phi\, dx$$
$$- \int_\Omega F(u + \psi)\, \mu(Du) Du_0.D\phi\, dx, \tag{6.41}$$

$$\forall \phi \in H^1_{\Gamma_D}, \quad u - u_0 \in H^1_{\Gamma_D},$$

and a similar relation holds for $v - v_0$. But we may consider $F(u + \psi)\, \mu(Du)$ as a given function that satisfies

$$\mu_0 \delta_0 \le F(u + \psi)\, \mu(Du) \le \mu_2 \delta_1,$$

and we may also consider $r(u, v, \psi)(1 - e^{u+v})$ and $-F(u + \psi)\, \mu(Du)Du_0$ as right-hand sides. From these considerations (6.41) appears as a linear mixed boundary value problem. Therefore, we apply Theorem 2.2, to assert that there exists an ϵ depending only on $\Omega, \mu_0\delta_0, \mu_2\delta_1$, such that for all

$$2 \le s < \min(2 + \epsilon, s_0)$$

we have

$$\|D(u - u_0)\|_{L^s} \leq C_s(|r(u, v, \psi)(1 - e^{u+v})|_{L^2} + \|F(u + \psi)\,\mu(Du)Du_0\|_{L^{s_0}}).$$

From now on, we fix such an $s > 2$. But as is easily checked,

$$|1 - e^{u+v}|_{L^2} \leq C[|1 - e^{u_0+v_0}|_{L^2} + |u - u_0|_{L^2} + |v - v_0|_{L^2}].$$

Hence we have

$$\|D(u-u_0)\|_{L^s} \leq C[|1-e^{u_0+v_0}|_{L^2}+|u-u_0|_{L^2}+|v-v_0|_{L^2}+\|Du_0\|_{L^{s_0}}]. \quad (6.42)$$

Now in (6.41) we take $\phi = u - u_0$, which yields

$$\int_\Omega F(u + \psi)\,\mu(Du)|D(u - u_0)|^2\,dx = \int_\Omega r(u, v, \psi)(1 - e^{u+v})(u - u_0)\,dx$$
$$- \int_\Omega F(u + \psi)\,\mu(Du)Du_0.D(u - u_0)\,dx.$$

We write a similar relation for v, and summing up we obtain

$$\int_\Omega (F(u + \psi)\,\mu(Du)|D(u - u_0)|^2 + F(v - \psi)\,\nu(Dv)|D(v - v_0)|^2)\,dx$$
$$= \int_\Omega r(u, v, \psi)(1 - e^{u+v})(u - u_0 + v - v_0)\,dx$$
$$- \int_\Omega (F(u + \psi)\,\mu(Du)Du_0.D(u - u_0)$$
$$+ F(v - \psi)\,\nu(Dv)Dv_0.D(v - v_0))\,dx.$$

$$(6.43)$$

Using the algebraic relation

$$\left(e^{x+y} - e^{x'+y'}\right)(x - x' + y - y') \geq 0 \quad (6.44)$$

we deduce from (6.43) that

$$\int_\Omega (|D(u - u_0)|^2 + |D(v - v_0)|^2)\,dx$$
$$\leq C[|1 - e^{u_0+v_0}|_{L^2}|u - u_0 + v - v_0|_{L^2} + |Du_0|^2 + |Dv_0|^2].$$

Using the Poincaré inequality, since $u - u_0 + v - v_0$ vanishes on Γ_D, we obtain

$$\int_\Omega (|D(u - u_0)|^2 + |D(v - v_0)|^2)\,dx \leq C\left[|1 - e^{u_0+v_0}|_{L^2}^2 + |Du_0|^2 + |Dv_0|^2\right].$$

We can thus estimate, on the right-hand side of (6.42), the quantity $|u - u_0|_{L^2} + |v - v_0|_{L^2}$ by

$$C\left[|1 - e^{u_0+v_0}|_{L^2} + |Du_0|_{L^2} + |Dv_0|_{L^2}\right],$$

and since $2 < s_0$, we easily obtain the expected result. $\qquad \diamond$

Proof of Theorem 6.2.
Let (u_1, v_1, ψ_1), (u_2, v_2, ψ_2) be two solutions of (6.5). We first derive from the third relation in (6.5)

$$\int_\Omega aD(\psi_1 - \psi_2).D(\psi_1 - \psi_2)\,dx$$
$$= \int_\Omega [F(v_1 - \psi_1) - F(u_1 + \psi_1) - F(v_2 - \psi_2) + F(u_2 + \psi_2)](\psi_1 - \psi_2)\,dx,$$

and since F is monotone nondecreasing, it follows that the right-hand side of the above equation is less than

$$\int_\Omega [F(v_1 - \psi_2) - F(u_1 + \psi_2) - F(v_2 - \psi_2) + F(u_2 + \psi_2)](\psi_1 - \psi_2)\,dx.$$

Thus, using (6.35), we have

$$\alpha |D(\psi_1 - \psi_2)|_{L^2}^2 \le K(\rho)\int_\Omega (|v_1 - v_2| + |u_1 - u_2|)|\psi_1 - \psi_2|\,dx. \qquad (6.45)$$

Again making use of the Poincaré inequality in (6.45), we obtain

$$|D(\psi_1 - \psi_2)|_{L^2} \le C(|D(v_1 - v_2)|_{L^2} + |D(u_1 - u_2)|_{L^2}). \qquad (6.46)$$

We test the first relation in (6.5) with $u_1 - u_2$, giving us

$$\int_\Omega [(F(u_1 + \psi_1)\mu(Du_1)Du_1 - F(u_2 + \psi_2)\mu(Du_2)Du_2).D(u_1 - u_2)]dx$$
$$= \int_\Omega [r(u_1, v_1, \psi_1)(1 - e^{u_1 + v_1}) - r(u_2, v_2, \psi_2)(1 - e^{u_2 + v_2})](u_1 - u_2)\,dx.$$

Hence from the strong monotonicity (6.33), we deduce

$$\delta_0\mu_0 \int_\Omega |D(u_1 - u_2)|^2\,dx$$
$$\le \mu_2 \int_\Omega |Du_2||F(u_1 + \psi_1) - F(u_2 + \psi_2)||D(u_1 - u_2)|\,dx$$
$$+ \int_\Omega [r(u_1, v_1, \psi_1)(1 - e^{u_1 + v_1}) - r(u_2, v_2, \psi_2)(1 - e^{u_2 + v_2})](u_1 - u_2)\,dx.$$

Writing the corresponding relation for $v_1 - v_2$, and again using (6.44), we arrive at

$$\delta_0 \min(\mu_0, \nu_0)\int_\Omega (|D(u_1 - u_2)|^2 + |D(v_1 - v_2)|^2)\,dx$$
$$\le \mu_2 \int_\Omega |Du_2||F(u_1 + \psi_1) - F(u_2 + \psi_2)||D(u_1 - u_2)|\,dx$$
$$+ \nu_2 \int_\Omega |Dv_2||F(v_1 - \psi_1) - F(v_2 - \psi_2)||D(v_1 - v_2)|\,dx$$
$$+ \int_\Omega [r(u_1, v_1, \psi_1) - r(u_2, v_2, \psi_2)](1 - e^{u_1 + v_1})(u_1 - u_2 + v_1 - v_2)\,dx.$$
$$(6.47)$$

We shall use the following notation:

$$I_1 = \int_\Omega |Du_2| \|F(u_1 + \psi_1) - F(u_2 + \psi_2)\| |D(u_1 - u_2)| \, dx,$$

$$I_2 = \int_\Omega |Dv_2| \|F(v_1 - \psi_1) - F(v_2 - \psi_2)\| |D(v_1 - v_2)| \, dx,$$

$$I_3 = \int_\Omega [r(u_1, v_1, \psi_1) - r(u_2, v_2, \psi_2)](1 - e^{u_1 + v_1})(u_1 - u_2 + v_1 - v_2) \, dx,$$

and also

$$\zeta_0 = |1 - e^{u_0 + v_0}|_{L^2} + \|Du_0\|_{L^{s_0}} + \|Dv_0\|_{L^{s_0}}.$$

From the assumption (6.36), we can estimate

$$I_3 \leq M(\rho) \int_\Omega (|u_1 - u_2| + |v_1 - v_2| + |\psi_1 - \psi_2|)$$
$$|1 - e^{u_1 + v_1}|(|u_1 - u_2| + |v_1 - v_2|) \, dx$$
$$\leq M(\rho)|1 - e^{u_1 + v_1}|_{L^2}[\|u_1 - u_2\|_{L^4}^2 + \|v_1 - v_2\|_{L^4}^2 + \|\psi_1 - \psi_2\|_{L^4}^2].$$

Since $N = 2 \leq 4$, we can use the Poincaré inequality to write

$$\|u_1 - u_2\|_{L^4}^2 \leq C|D(u_1 - u_2)|_{L^2}^2$$

and similar relations for $\|v_1 - v_2\|_{L^4}^2$ and $\|\psi_1 - \psi_2\|_{L^4}^2$. Therefore,

$$I_3 \leq C|1 - e^{u_1 + v_1}|_{L^2}[|D(u_1 - u_2)|_{L^2}^2 + |D(v_1 - v_2)|_{L^2}^2 + |D(\psi_1 - \psi_2)|_{L^2}^2].$$

Now

$$|1 - e^{u_1 + v_1}|_{L^2} \leq C|u_1 + v_1|_{L^2},$$

and using Lemma 6.3 as well as (6.46) we can assert that

$$I_3 \leq C\zeta_0(|D(u_1 - u_2)|_{L^2}^2 + |D(v_1 - v_2)|_{L^2}^2). \tag{6.48}$$

Next, using assumption (6.35), we have

$$I_1 \leq K(\rho_0)|D(u_1 - u_2)|_{L^2}\left[\left(\int_\Omega |u_1 - u_2|^2 |Du_2|^2 \, dx\right)^{1/2} + \left(\int_\Omega |\psi_1 - \psi_2|^2 |Du_2|^2 \, dx\right)^{1/2}\right].$$

Now from Hölder's inequality

$$\leq \|Du_2\|_{L^s} \|u_1 - u_2\|_{2s/s-2},$$

and if $s \geq N$ (this is where the restriction $N = 2$ is necessary),

$$\left(\int_\Omega |u_1 - u_2|^2 |Du_2|^2 \, dx\right)^{1/2} \leq C\|Du_2\|_{L^s} |D(u_1 - u_2)|_{L^2}.$$

Similarly,

$$\left(\int_\Omega |\psi_1 - \psi_2|^2 |Du_2|^2 \, dx\right)^{1/2} \leq C\|Du_2\|_{L^s} |D(\psi_1 - \psi_2)|_{L^2}$$
$$\leq C\|Du_2\|_{L^s} (|D(u_1 - u_2)|_{L^2} + |D(v_1 - v_2)|_{L^2}),$$

and thus (6.48) implies

$$I_1 \leq C\|Du_2\|_{L^s} (|D(u_1 - u_2)|_{L^2}^2 + |D(v_1 - v_2)|_{L^2}^2).$$

Using Lemma 6.3, it follows that

$$I_1 \leq C\zeta_0 (|D(u_1 - u_2)|_{L^2}^2 + |D(v_1 - v_2)|_{L^2}^2), \tag{6.49}$$

and in the same way,

$$I_2 \leq C\zeta_0 (|D(u_1 - u_2)|_{L^2}^2 + |D(v_1 - v_2)|_{L^2}^2). \tag{6.50}$$

It then follows from (6.48), (6.49), (6.50) in (6.47) that

$$|D(u_1 - u_2)|_{L^2}^2 + |D(v_1 - v_2)|_{L^2}^2 \leq C\zeta_0 \left(|D(u_1 - u_2)|_{L^2}^2 + |D(v_1 - v_2)|_{L^2}^2\right),$$

and if ζ_0 sufficiently small, we have

$$|D(u_1 - u_2)|_{L^2}^2 + |D(v_1 - v_2)|_{L^2}^2 = 0,$$

and hence the uniqueness.   ◇

6.2.4 Local Regularity

Our objective is to prove the following result.

Theorem 6.4. *We assume (6.6), (6.8), (6.11), (6.33), (6.34), (6.35) and also that*

$$|\mu(\xi)\xi - \mu(\eta)\eta| \leq K|\xi - \eta|, \quad a(x) \text{ Lipschitz.}$$

Then a solution $u, v, \psi \in H^1 \cap L^\infty$ *of (6.5) satisfies*

$$u, v, \psi \in H^2_{\text{loc}}(\Omega).$$

Proof. For ψ this is just a consequence of the theory of linear elliptic equations. In fact, $\psi \in W^{1,4}_{\text{loc}}$ (which will be needed). Moreover, we know that

$$u, v \in W^{1,p_0}, \quad p_0 > 2, \text{ and } \in C^{0,\delta}$$

locally, which is sufficient to apply Theorem 2.20. We can check that all the assumptions are satisfied. The result follows.   ◇

6.3 Stationary Case with Impact Ionization

6.3.1 Setting of the Model

In the following model[1] it will be preferable to work with the densities n, p and not with the corresponding quasi-Fermi potentials. We use the Boltzmann distribution $F(s) = e^s$. Hence we have (see Section 6.1)

$$J_n = \mu_n(Dn - nD\psi), \quad J_p = -\mu_p(Dp + pD\psi).$$

We take μ_n, μ_p to be positive constants (mobilities). We recall that

$$-\operatorname{div} J_n = -R,$$
$$-\operatorname{div} J_p = R,$$

where R is the term describing the generation of carriers. We shall add to the recombination–generation term considered in Section 6.2, a term describing impact ionization. More precisely, we take

$$-R = r(n,p)(1 - np) + g(n,p,Dn,Dp,D\psi),$$

where ψ is the electrostatic potential. We also take the permittivity matrix to be the identity. We then write the equations

$$
\begin{aligned}
-\mu_n \operatorname{div}(Dn - nD\psi) &= r(n,p)(1 - np) + g(n,p,Dn,Dp,D\psi),\\
-\mu_p \operatorname{div}(Dp + pD\psi) &= r(n,p)(1 - np) + g(n,p,Dn,Dp,D\psi),\\
-\Delta\psi &= p - n + f,
\end{aligned}
\tag{6.51}
$$

to which we add the boundary conditions

$$
\begin{aligned}
n = n_0, \quad p = p_0, \quad \psi &= \psi_0 \text{ on } \Gamma_D,\\
\frac{\partial n}{\partial \nu} = \frac{\partial p}{\partial \nu} = \frac{\partial \psi}{\partial \nu} &= 0 \text{ on } \Gamma_N,
\end{aligned}
\tag{6.52}
$$

in which ν denotes the outward unit normal (instead of n to avoid confusion). Note that this is equivalent to the variational formulation

$$
\begin{aligned}
\int_\Omega (Dn - nD\psi).Dz\, dx &= \int_\Omega (r(1 - np) + g)z\, dx,\\
\int_\Omega (Dp + pD\psi).Dz\, dx &= \int_\Omega (r(1 - np) + g)z\, dx,\\
\int_\Omega D\psi.Dz\, dx &= \int_\Omega (p - n + f)z\, dx,\\
\forall z \in H^1(\Omega), \quad z &= 0 \text{ on } \Gamma_D,\\
n = n_0, \quad p = p_0, \quad \psi &= \psi_0 \text{ on } \Gamma_D.
\end{aligned}
\tag{6.53}
$$

---

[1] We consider this model, since it is supposed to model the avalanche effect in the time-delay case.

We now make precise our assumptions:

$$\Omega \text{ is a Lipschitz domain,} \tag{6.54}$$
$$\Gamma = \Gamma_D \cup \Gamma_N, \quad \Gamma_D \cap \Gamma_N = \emptyset, \quad |\Gamma_D| > 0,$$
$$N = 2, 3, \tag{6.55}$$
$$f \in L^\ell(\Omega), \ell > N, \tag{6.56}$$
$$n_0, p_0 \in H^{1,3}(\Omega), n_0, p_0 \geq 0, \tag{6.57}$$
$$\text{ess sup}_{\Gamma_D}(|n_0|, |p_0|) < \infty,$$
$$\psi_0 \in H^{1,3}(\Omega) \tag{6.58}$$

$$r \text{ is continuous on } R^+ \times R^+,$$
$$\rho_0(s + t) \leq r(s, t) \leq \rho_1(1 + s + t), \quad \forall s, t \geq 0, \tag{6.59}$$
$$\rho_0, \rho_1, \text{positive constants.}$$

This model includes the cases

$$r(s, t) = \beta_1 s + \beta_2 t$$

(which corresponds to the Auger recombination–generation term) and

$$r(s, t) = \beta_1 s + \beta_2 t + \frac{1}{\gamma_0 + \gamma_1 s + \gamma_2 t}$$

(in which the Shockley–Read–Hall recombination–generation term is added). See [73], [90], [100] for details. Note that the Auger term has a stabilizing effect that might prevent the blowup of solutions. However, the constants β_1, β_2 are in practice very small, so the a priori estimate does not mean much for applications. Nevertheless, this provides ideas for further analytic sudies. We proceed with the assumptions

$$g(s, t, \xi, \eta, \zeta) \text{ is continuous on } R^+ \times R^+ \times R^N \times R^N \times R^N,$$
$$0 \leq g \leq g_0(|\xi| + |\eta| + (s + t)|\zeta|), \quad g_0 > \text{constant.} \tag{6.60}$$

In practice, we shall consider the model

$$g(n, p, Dn, Dp, D\psi) = \alpha_n(D\psi)|J_n| + \alpha_p(D\psi)|J_p|,$$

where α_n, α_p are bounded continuous functions. This models the effect of impact ionization (avalanche generation) in a semiconductor. The coefficients α_n, α_p represent the the ionization rates for electrons and holes, and possible functions are

$$\alpha_n(\zeta) = \begin{cases} \alpha_{0n} \exp\left(-\frac{\beta_{0n}}{|\zeta|}\right), & \text{if } \zeta \neq 0, \\ 0, & \text{if } \zeta = 0. \end{cases}$$

Our objective is to prove the following result.

Theorem 6.5. *If we assume (6.54) to (6.60), then there exists a solution of (6.51), (6.52), such that*

$$n, p, \psi \in H^1(\Omega),$$
$$n, p \geq 0; \quad n, p \in L^\infty(\Omega).$$

Remark 6.6. The presentation is based on [27].

6.3.2 Proof of Theorem 6.5

We must begin by proving an a priori estimate in H^1, and then we prove only the L^∞ estimate. The latter does not rely on the maximum principle as in the case $g = 0$ (see (6.1)). Moreover, the H^1 estimate requires less-stringent conditions on f, n_0, p_0, ψ_0. We need to assume

$$n_0, p_0 \in H^{1,3} \cap L^6(\Omega), \quad \psi_0 \in H^{1,3}(\Omega), \quad f \in L^3(\Omega),$$
N arbitrary.

A Priori H^1 Estimate

We assume that we have a solution $n, p, \psi \in (H^1(\Omega))^3$ such that

$$n, p \geq 0$$

and

$$n^2, p^2 \in H^1.$$

Then

$$\|n\|_{H^1 \cap L^3}, \|p\|_{H^1 \cap L^3}, \|\psi\|_{H^1}, \|(n+p)|D\psi|^2\|_{L^1}$$
$$\leq F(g_0, \rho_0, \rho_1, \mu_n, \mu_p, \|f\|_{L^3}, \|n_0\|_{W^{1,3} \cap L^6}, \|p_0\|_{W^{1,3} \cap L^6}, \|\psi_0\|_{W^{1,3}}),$$

where F is bounded on bounded arguments.

We test the first two equations in (6.51), by $n - n_0$, $p - p_0$. Adding up, we get

$$\int_\Omega [(Dn - nD\psi).D(n - n_0) + (Dp + pD\psi).D(p - p_0)] \, dx$$
$$= \int_\Omega (r(1 - np) + g) \left[\frac{1}{\mu_n}(n - n_0) + \frac{1}{\mu_p}(p - p_0) \right] dx. \tag{6.61}$$

On the other hand, testing the third equation in (6.51) with $\frac{1}{2}(p^2 - n^2 - p_0^2 + n_0^2)$ yields

$$\int_\Omega D\psi.(pDp - nDn) \, dx$$
$$= \frac{1}{2} \int_\Omega (p - n + f)(p^2 - n^2 - p_0^2 + n_0^2) \, dx + \frac{1}{2} \int_\Omega D\psi.D(p_0^2 - n_0^2) \, dx,$$

and combined with (6.61) we obtain

$$\int_\Omega (|Dn|^2 + |Dp|^2)\, dx + \int_\Omega \left[\frac{1}{2}(p-n)(p^2-n^2) + rnp\left(\frac{n}{\mu_n} + \frac{p}{\mu_p}\right)\right] dx$$

$$= -\frac{1}{2}\int_\Omega D\psi.D(p_0^2 - n_0^2)\, dx - \frac{1}{2}\int_\Omega f(p^2 - n^2 - p_0^2 + n_0^2)\, dx$$

$$+ \frac{1}{2}\int_\Omega (p-n)(p_0^2 - n_0^2)\, dx$$

$$+ \int_\Omega [Dn_0.(Dn - nD\psi) + Dp_0.(Dp + pD\psi)]\, dx$$

$$- \int_\Omega (r(1-np) + g)\left(\frac{n_0}{\mu_n} + \frac{p_0}{\mu_p}\right) dx + \int_\Omega (r+g)\left(\frac{n}{\mu_n} + \frac{p}{\mu_p}\right) dx.$$

We use assumptions (6.59), (6.60) to obtain

$$\int_\Omega (|Dn|^2 + |Dp|^2)\, dx + \int_\Omega \left[\frac{1}{2}(p-n)(p^2 - n^2)\right.$$

$$\left. + \frac{\rho_0}{\max(\mu_n, \mu_p)} np(n+p)^2\right] dx \le -\frac{1}{2}\int_\Omega D\psi.D(p_0^2 - n_0^2)\, dx$$

$$- \frac{1}{2}\int_\Omega f(p^2 - n^2 - p_0^2 + n_0^2)\, dx + \frac{1}{2}\int_\Omega (p-n)(p_0^2 - n_0^2)\, dx$$

$$+ \int_\Omega [Dn_0.(Dn - nD\psi) + Dp_0.(Dp + pD\psi)]\, dx \qquad\qquad (6.62)$$

$$+ \frac{g_0}{\min(\mu_n, \mu_p)}\int_\Omega [(n+p)(|Dn| + |Dp|) + |D\psi|(n+p)^2]\, dx$$

$$+ \frac{\rho_1}{\min(\mu_n, \mu_p)}\int_\Omega [n + p + (n+p)^2 + np(n_0 + p_0)$$

$$+ np(n+p)(n_0 + p_0)]\, dx.$$

With Young's inequality we absorb the term in $np(n+p)$ at the very end of equation (6.62) into the term on the left-hand side involving $np(n+p)^2$. Let

$$\beta_0 = \frac{1}{2}\min\left(1, \frac{\rho}{\max(\mu_n, \mu_p)}\right)$$

and

$$\beta_1 = \frac{\max(\mu_n, \mu_p)\rho_1^2}{\rho_0(\min(\mu_n, \mu_p))^2}.$$

We also notice the algebraic relation

$$np(n+p)^2 + (n-p)^2(n+p) \ge (n+p)^3 - \frac{27}{4},$$

and then we can derive from (6.62) the inequality

$$\int_\Omega (|Dn|^2 + |Dp|^2)\, dx + \beta_0 \int_\Omega (n+p)^3\, dx$$

$$\leq \frac{27}{4}\beta_0|\Omega| - \frac{1}{2}\int_\Omega D\psi.D(p_0^2 - n_0^2)\, dx$$

$$-\frac{1}{2}\int_\Omega f(p^2 - n^2 - p_0^2 + n_0^2)\, dx + \frac{1}{2}\int_\Omega (p-n)(p_0^2 - n_0^2)\, dx$$

$$+ \int_\Omega [Dn_0.(Dn - nD\psi) + Dp_0.(Dp + pD\psi)]\, dx$$

$$+ \frac{g_0}{\min(\mu_n, \mu_p)}\int_\Omega [(n+p)(|Dn| + |Dp|) + |D\psi|(n+p)^2]\, dx$$

$$+ \beta_1 \int_\Omega [n + p + (n+p)^2 + np(n_0 + p_0) + np(n_0 + p_0)^2]\, dx.$$

We then use Young's inequality in various places to deduce

$$\int_\Omega (|Dn|^2 + |Dp|^2 + (n+p)^3)\, dx \leq C \int_\Omega |D\psi|^2(1 + n + p)\, dx + C, \quad (6.63)$$

where the constant C depends only on the parameters and on the norms

$$\|f\|_{L^3}, \|n_0\|_{H^{1,3} \cap L^6}, \|p_0\|_{H^{1,3} \cap L^6}$$

and is bounded if these arguments are bounded.

Now test the two first equations (6.51) by $\psi - \psi_0$ and subtract. We get

$$\int_\Omega |D\psi|^2(\mu_n n + \mu_p p)\, dx$$

$$= \int_\Omega D(\psi - \psi_0).(\mu_n Dn - \mu_p Dp)\, dx + \int_\Omega D\psi.D\psi_0(\mu_n n + \mu_p p)\, dx$$

Combining this relation with (6.63), to estimate the integral $\int_\Omega |D\psi|^2(n+p)dx$, we easily deduce

$$\int_\Omega (|Dn|^2 + |Dp|^2 + (n+p)^3)\, dx \leq c_0 \int_\Omega |D\psi|^2\, dx + C, \quad (6.64)$$

where the new constant C now also depends on $\|\psi\|_{H^{1,3}}$, and c_0 depends only on the parameters. But then we can use the third equation (6.51) tested by $\psi - \psi_0$ to derive

$$\int_\Omega |D\psi|^2\, dx = \int_\Omega D\psi.D\psi_0\, dx + \int_\Omega (n - p + f)(\psi - \psi_0)\, dx.$$

Using the Poincaré inequality

$$\int_\Omega (\psi - \psi_0)^2\, dx \leq k_0 \int_\Omega |D\psi - D\psi_0|^2\, dx$$

we obtain

$$\int_\Omega |D\psi|^2\, dx \le c_1 \int_\Omega (p+n)^2\, dx + C,$$

where C depends only on the L^2 norms of f and $D\psi_0$, and c_1 is a fixed constant. We can combine this last inequality with (6.64) to immediately obtain

$$\int_\Omega (|Dn|^2 + |Dp|^2 + (n+p)^3)\, dx \le C.$$

Remembering that we can also estimate $\int_\Omega |D\psi|^2(n+p)\, dx$ by this same quantity, the a priori estimate has been proven.

• **L^∞ a Priori Estimate**

We proceed now with the L^∞ estimate, using the H^1 estimate and the full weight of our assumptions. Note that since $N \le 3$, it follows from the Sobolev embedding theorems that

$$n, p \in L^6(\Omega).$$

Let

$$k_0 = \max(\text{ess sup}_{\Gamma_D} n_0, \text{ess sup}_{\Gamma_D} p_0)$$

and consider $k > k_0$. We test the first two equations in (6.51) with $(n-k)^+$, $(p-k)^+$. We obtain

$$\int_\Omega (Dn - nD\psi).D((n-k)^+)\, dx = \frac{1}{\mu_n}\int_\Omega (r(1-np)+g)((n-k)^+)\, dx,$$

$$(6.65)$$

$$\int_\Omega (Dp + pD\psi).D((p-k)^+)\, dx = \frac{1}{\mu_p}\int_\Omega (r(1-np)+g)((p-k)^+)\, dx.$$

Now we notice that

$$\frac{1}{2}D((p^2 - k^2)^+) = p\,D((p-k)^+).$$

We then test the third equation (6.51) with

$$\frac{1}{2}\left[(p^2 - k^2)^+ - (n^2 - k^2)^+\right]$$

and combine the result with (6.65) to obtain

$$\int_\Omega (|D((n-k)^+)|^2 + |D((p-k)^+)|^2)\, dx$$
$$+ \frac{1}{2}\int_\Omega (p - n + f)[(p^2 - k^2)^+ - (n^2 - k^2)^+]\, dx \qquad (6.66)$$
$$= \int_\Omega (r(1-np)+g)\left(\frac{1}{\mu_n}(n-k)^+ + \frac{1}{\mu_p}(p-k)^+\right)\, dx.$$

Using the assumptions (6.59), (6.60), we can majorize (6.66) as follows:

$$\int_\Omega (|D((n-k)^+)|^2 + |D((p-k)^+)|^2)\, dx$$

$$\leq \frac{1}{2} \int_\Omega |f|[(p^2-k^2)^+ + (n^2-k^2)^+]\, dx$$

$$+ \int_\Omega (\phi + (n+p)|D\psi|)((n-k)^+$$

$$+ (p-k)^+)\, dx = J_1 + J_2 + J_3,$$

(6.67)

in which ϕ stands for an L^2 function of the form

$$\phi = \rho_1(1+n+p) + g_0(|Dn| + |Dp|).$$

Set

$$A_k = \{x \in \Omega | n > k\},\, B_k = \{x \in \Omega | p > k\}$$

and let

$$\omega_k = |A_k| + |B_k|.$$

We estimate the three terms J_1, J_2, J_3 as follows:

$$J_1 \leq C\|f\|_{L^\ell}(\|p\|_{L^6} + \|n\|_{L^6})(\omega_k)^{(2\ell-3)/3\ell},$$
$$J_2 \leq C\|\phi\|_{L^2}(\|(n-k)^+\|_{L^6} + \|(p-k)^+\|_{L^6})(\omega_k)^{1/3},$$

and we can use the Poincaré inequality to assert that

$$J_2 \leq C\|\phi\|_{L^2}(\|D((n-k)^+)\|_{L^2} + \|D((p-k)^+)\|_{L^2})(\omega_k)^{1/3}.$$

Finally,

$$J_3 \leq C\|(n+p)^{1/2}D\psi\|_{L^2}(\|(n-k)^+\|_{L^6} + \|(p-k)^+\|_{L^6})(\|p\|_{L^6} + \|n\|_{L^6})(\omega_k)^{1/4}.$$

We next use the Poincaré inequality to give

$$\|(n-k)^+\|_{L^6} + \|(p-k)^+\|_{L^6} \leq c_0(\|D((n-k)^+)\|_{L^2} + \|D((p-k)^+)\|_{L^2}).$$

Including this on the right-hand side of J_2, J_3, and collecting the estimates of J_1, J_2, J_3 on the right-hand side of (6.67), we obtain

$$\int_\Omega (|D((n-k)^+)|^2 + |D((p-k)^+)|^2)\, dx \leq C(\omega_k)^{\min(\frac{1}{2}, \frac{2\ell-3}{3\ell})}.$$

Referring to formula (2.28), we see that the number p entering in this formula is such that

$$\frac{p-2}{2p} = \min\left(\frac{1}{4}, \frac{2\ell-3}{6\ell}\right);$$

hence $p = 4$ or $p = 6\ell/(\ell+3)$, and from the assumption on ℓ, this number is larger than $N = 2, 3$. Then the result follows from applying Lemma 2.9 (see end of Section 2.3).

- **Existence for g Bounded**

We assume here that

$$g \leq g_0 \text{ constant.}$$

Then under this additional assumption we prove the theorem.

1. Approximation

We prove the existence by an approximation procedure. Define

$$\phi_\delta(x) = \frac{x}{1 + \delta x}, \quad x \geq 0, \quad 0 < \delta \leq 1,$$

and set

$$r_\delta(s, t) = r(\phi_\delta(s), \phi_\delta(t)), \quad s, t \geq 0.$$

We prove that there exists a solution in $(H^1(\Omega))^3$ of

$$\begin{aligned}
-\mu_n \text{ div } (Dn - \phi_\delta(n^+)D\psi) &= r_\delta(n^+, p^+)(1 - \phi_\delta(n^+)\phi_\delta(p^+)) \\
&\quad + g(n, p, Dn, Dp, D\psi), \\
-\mu_p \text{ div } (Dp + \phi_\delta(p^+)D\psi) &= r_\delta(n^+, p^+)(1 - \phi_\delta(n^+)\phi_\delta(p^+)), \quad (6.68) \\
&\quad + g(n, p, Dn, Dp, D\psi), \\
-\Delta\psi &= \phi_\delta(p^+) - \phi_\delta(n^+) + f,
\end{aligned}$$

with the same boundary conditions as the original problem (6.52). To prove the existence of n, p, ψ (of course, depending on δ), we just notice that from (6.68) there exists an a priori estimate in $(H^1(\Omega))^3$ depending on δ, very easily obtained by writing the first two equations as

$$-\mu_n \Delta n = \text{ right-hand side,}$$

$$-\mu_p \Delta p = \text{ right-hand side,}$$

and checking that the right-hand sides have a norm in $(H^1_{\Gamma_D})'$ that is bounded by a fixed constant. This leads immediately to a fixed point argument in L^2, using the compact injection of H^1 into L^2 and the Leray–Schauder fixed point theorem. Hence the existence of a solution to (6.68) is obtained. By testing with n^-, p^-, we see immediately that they are 0. Hence we have proven the existence of a solution in $(H^1(\Omega))^3$, n, p, ψ, with $p, n \geq 0$ of

$$\begin{aligned}
-\mu_n \text{ div } (Dn - \phi_\delta(n)D\psi) &= r_\delta(n, p)(1 - \phi_\delta(n)\phi_\delta(p)) \\
&\quad + g(n, p, Dn, Dp, D\psi), \\
-\mu_p \text{ div } (Dp + \phi_\delta(p)D\psi) &= r_\delta(n, p)(1 - \phi_\delta(n)\phi_\delta(p)) \quad (6.69) \\
&\quad + g(n, p, Dn, Dp, D\psi), \\
-\Delta\psi &= \phi_\delta(p) - \phi_\delta(n) + f.
\end{aligned}$$

2. Estimate in H^1 independent of δ

We then obtain estimates that do not depend on δ. We test the first two equations in (6.69) with

$$\frac{1}{\mu_n}(\phi_\delta(n) - \phi_\delta(n_0)), \frac{1}{\mu_p}(\phi_\delta(p) - \phi_\delta(p_0))$$

and the third one with

$$\frac{1}{2}\left[(\phi_\delta(p))^2 - (\phi_\delta(n))^2 - (\phi_\delta(p_0))^2 + (\phi_\delta(n_0))^2\right].$$

Combining, we obtain

$$\int_\Omega (|Dn|^2 \phi'_\delta(n) + |Dp|^2 \phi'_\delta(p))\, dx$$

$$+ \int_\Omega r_\delta \phi_\delta(n) \phi_\delta(p) \left(\frac{1}{\mu_n}\phi_\delta(n) + \frac{1}{\mu_p}\phi_\delta(p)\right) dx$$

$$+ \int_\Omega D\psi.\frac{1}{2}D((\phi_\delta(p_0))^2 - (\phi_\delta(n_0))^2)\, dx \; + \int_\Omega (\phi_\delta(p) - \phi_\delta(n) + f)$$

$$+ \frac{1}{2}\left[(\phi_\delta(p))^2 - (\phi_\delta(n))^2 - (\phi_\delta(p_0))^2 + (\phi_\delta(n_0))^2\right] dx$$

$$= \int_\Omega [(Dn - \phi_\delta(n)D\psi)Dn_0\phi'_\delta(n_0) - (Dp - \phi_\delta(p)D\psi)Dp_0\phi'_\delta(p_0)]\, dx$$

$$+ \int_\Omega (g + r_\delta)\left[\frac{1}{\mu_n}(\phi_\delta(n) - \phi_\delta(n_0)) + \frac{1}{\mu_p}(\phi_\delta(p) - \phi_\delta(p_0))\right] dx$$

$$+ \int_\Omega r_\delta \phi_\delta(n)\phi_\delta(p)\left(\frac{1}{\mu_n}\phi_\delta(n_0) + \frac{1}{\mu_p}\phi_\delta(p_0)\right) dx.$$

$$(6.70)$$

Proceeding as for the a priori estimates, we can derive from (6.70) the inequality

$$\beta_0 \int_\Omega (\phi_\delta(n) + \phi_\delta(p))^3\, dx \le \frac{27}{4}\beta_0|\Omega|$$

$$+ \frac{1}{2}\int_\Omega \left[\frac{(\phi'_\delta(n_0))^2}{\phi'_\delta(n)}|Dn_0|^2 + \frac{(\phi'_\delta(p_0))^2}{\phi'_\delta(p)}|Dp_0|^2\right] dx$$

$$+ \int_\Omega D\psi.[(\phi_\delta(n_0) - \phi_\delta(n))\phi'_\delta(n_0)Dn_0 - (\phi_\delta(p_0) - \phi_\delta(p))\phi'_\delta(p_0)Dp_0]\, dx$$

$$+ \int_\Omega (\phi_\delta(n) + \phi_\delta(p))\left[\frac{1}{2}(\phi_\delta(n_0) + \phi_\delta(p_0))^2 + \frac{g_0}{\min(\mu_n, \mu_p)} + \beta_1\right] dx$$

$$+ \int_\Omega \frac{1}{2}f((\phi_\delta(p_0))^2 - (\phi_\delta(n_0))^2)\, dx + \int_\Omega \left[(\phi_\delta(n) + \phi_\delta(p))^2\left(\frac{1}{2}|f| + \beta_1\right)\right.$$

$$\left. + \beta_1 \phi_\delta(n)\phi_\delta(p)(\phi_\delta(n_0) + \phi_\delta(p_0))(1 + \phi_\delta(n_0) + \phi_\delta(p_0))\right] dx.$$

$$(6.71)$$

From the third equation in (6.69) we can write

$$\int_\Omega D\psi.D(\psi - \psi_0)\, dx = \int_\Omega (\phi_\delta(p) - \phi_\delta(n) + f)(\psi - \psi_0)\, dx,$$

and thus (as is easily checked)

$$\int_{\Omega} D\psi.[(\phi_{\delta}(n_0) - \phi_{\delta}(n))\phi'_{\delta}(n_0)Dn_0 - (\phi_{\delta}(p_0) - \phi_{\delta}(p))\phi'_{\delta}(p_0)Dp_0]\,dx$$

$$\leq \int_{\Omega} (\phi_{\delta}(n) + \phi_{\delta}(p)))|D\psi_0|(|Dn_0| + |Dp_0|)\,dx$$

$$+2\int_{\Omega} (\phi_{\delta}(n) + \phi_{\delta}(p))^2(c_0^2 + |Dn_0|^2 + |Dp_0|^2)\,dx$$

$$+\int_{\Omega} [(\phi_{\delta}(n_0) + \phi_{\delta}(p_0)))|D\psi_0|(|Dn_0| + |Dp_0|) + |D\psi_0|^2 + 2c_0^2|f|^2$$

$$+2(\phi_{\delta}(n_0) + \phi_{\delta}(p_0))^2(|Dn_0|^2 + |Dp_0|^2)]\,dx.$$

We use this last estimate in (6.71); from Young's inequality we can derive

$$\frac{\beta_0}{2} \int_{\Omega} (\phi_{\delta}(n) + \phi_{\delta}(p))^3\,dx \leq \delta^2 \int_{\Omega} (n^2|Dn_0|^2 + p^2|Dp_0|^2)\,dx + C,$$

where C does not depend on δ and depends on

$$g_0,\ \beta_1,\ \|f\|_{L^3},\ \|\psi_0\|,\ \|p_0\|_{W^{1,6}},\ \|n_0\|_{W^{1,6}}.$$

Using the Poincaré inequality, we can write

$$\int_{\Omega} (\phi_{\delta}(n) + \phi_{\delta}(p))^3\,dx \leq \gamma_0\delta^2 \int_{\Omega} (|D(n - n_0)|^2 + |D(p - p_0)|^2)\,dx + C, \quad (6.72)$$

where

$$\gamma_0 = \frac{4}{\beta_0}c_0(\|Dn_0\|_{L^4} + \|Dp_0\|_{L^4})^{1/2}.$$

Then define

$$\Phi_{\delta}(x) = \int_0^x \phi_{\delta}(y)\,dy.$$

Note that

$$(\phi_{\delta}(x) - \phi_{\delta}(x'))(\Phi_{\delta}(x) - \Phi_{\delta}(x')) \geq 0,$$

$$\Phi_{\delta}(x) \leq x\phi_{\delta}(x). \quad (6.73)$$

We then test the two first equations in (6.69) with $(n - n_0)/\mu_n, (p - p_0)/\mu_p$, and the third one with

$$\Phi_{\delta}(p) - \Phi_{\delta}(n) - \Phi_{\delta}(p_0) + \Phi_{\delta}(n_0).$$

Combining, we obtain

$$\int_{\Omega} (|D(n - n_0)|^2 + |D(p - p_0)|^2)\, dx + \int_{\Omega} (\Phi_\delta(p)$$
$$-\Phi_\delta(n) - \Phi_\delta(p_0) + \Phi_\delta(n_0))(\phi_\delta(p) - \phi_\delta(n) + f)\, dx$$
$$= \int_{\Omega} D\psi.[(\phi_\delta(p) - \phi_\delta(p_0))Dp_0 - (\phi_\delta(n) - \phi_\delta(n_0))Dn_0]\, dx$$
$$+ \int_{\Omega} (Dn_0.D(n - n_0) + Dp_0.D(p - p_0))\, dx$$
$$+ \int_{\Omega} (r_\delta(1 - \phi_\delta(p)\phi_\delta(n)) + g) \left(\frac{1}{\mu_n}(n - n_0) + \frac{1}{\mu_p}(p - p_0) \right) dx$$

$$(6.74)$$

and recall that

$$\|D(\psi - \psi_0)\|_{L^2} \le \|D\psi_0\|_{L^2} + c_0\|\phi_\delta(p) + \phi_\delta(n) + |f|\,\|_{L^2}. \qquad (6.75)$$

Making use of (6.73) and

$$\|n - n_0\|_{L^2} + \|p - p_0\|_{L^2} \le c_0(\|D(n - n_0)\|_{L^2} + \|D(p - p_0)\|_{L^2})$$

we can conveniently majorize (6.74) using (6.75), getting (after some easy steps)

$$\int_{\Omega} (|D(n - n_0)|^2 + |D(p - p_0)|^2)\, dx \le C \int_{\Omega} (\phi_\delta(n) + \phi_\delta(p))^3\, dx + C. \quad (6.76)$$

Then combine (6.76) with (6.72), taking $\delta \le \delta_0$ sufficiently small to obtain

$$\int_{\Omega} (|Dn_\delta|^2 + |Dp_\delta|^2 + |D\psi_\delta|^2)\, dx \le C,$$

where the constant does not depend on δ.

3. Estimate in L^∞ independent of δ

We know that n_δ, p_δ remain bounded in $L^6(\Omega)$. Take $k \ge k_0$, the constant defined in deriving the a priori estimate in L^∞ (see above). Then we test the two first equations (6.69) with $\frac{1}{\mu_n}(n-k)^+, \frac{1}{\mu_p}(p-k)^+$, and the third equation with $(\Phi_\delta(p) - \Phi_\delta(k))^+ - (\Phi_\delta(n) - \Phi_\delta(k))^+$. Combining we obtain

$$\int_{\Omega} (|D((n - k)^+)|^2 + |D((p - k)^+)|^2)\, dx$$
$$+ \int_{\Omega} (\phi_\delta(p) - \phi_\delta(n) + f)[(\Phi_\delta(p) - \Phi_\delta(k))^+ - (\Phi_\delta(n) - \Phi_\delta(k))^+]\, dx$$
$$= \int_{\Omega} (r_\delta(1 - \phi_\delta(p)\phi_\delta(n)) + g) \left(\frac{1}{\mu_n}(n - k)^+ + \frac{1}{\mu_p}(p - k)^+ \right) dx,$$

and we can then majorize as follows:

$$\int_{\Omega} (|D((n - k)^+)|^2 + |D((p - k)^+)|^2)\, dx$$
$$\le C \int_{\Omega} |f|(n + p)^2\, dx + C \int_{\Omega} (n + p + 1)[(n - k)^+ + (p - k)^+]\, dx.$$

Since n, p are bounded in L^6 and f is in L^ℓ, we deduce using the Poincaré inequality that

$$\int_\Omega (|D((n-k)^+)|^2 + |D((p-k)^+)|^2)\, dx \leq C\omega_k^{\min(\frac{1}{2}, \frac{2\ell-3}{3\ell})},$$

where ω_k has the same definition as in the a priori estimate, and C does not depend on δ. The result follows as in the a priori estimate.

4. Convergence as δ tends to 0
Pick a subsequence that is weakly convergent in $(H^1)^3$ and weakly $*$ convergent in $(L^\infty)^3$. Then it is standard to check first that ψ_δ converges strongly in H^1 and then that n_δ, p_δ converge strongly in H^1. Then we may as well assume that

$$n_\delta, Dn_\delta \to n, Dn \text{ a.e. in } \Omega$$

and similar considerations for p_δ, ψ_δ. Passing to the limit as δ tends to 0 in the variational formulation of (6.69) is fairly easy, and the result is obtained.

• **Existence in the General Case**
We replace g by

$$g_\epsilon = \frac{g}{1 + \epsilon g},$$

and we are back in the bounded g case. We deduce the existence of a solution for the problem (6.51), (6.52), with g replaced by g_ϵ. Since g_ϵ satisfies the same assumptions as g uniformly in ϵ, the a priori estimates are legitimate and yield immediately that the solution $n_\epsilon, p_\epsilon, \psi_\epsilon$ remains bounded in $H^1 \cap L^\infty$. Therefore, extracting a weakly convergent subsequence, we can see again that first ψ_ϵ then n_ϵ, p_ϵ converge strongly in H^1. So we may assume pointwise convergence a.e. of the functions as well as the gradients. Then

$$g_\epsilon(n_\epsilon, p_\epsilon, Dn_\epsilon, Dp_\epsilon, D\psi_\epsilon) \to g(n, p, Dn, Dp, D\psi) \text{ a.e. as } \epsilon \to 0.$$

From Vitali's theorem, the convergence also takes place in L^2. This permits us to pass to the limit in the variational formulation and to obtain the result.

6.4 Impact Ionization Without Recombination

6.4.1 Statement of the Problem

We shall consider here a case of pure impact ionization. As already pointed out, the Auger term allows one to prove a priori estimates in order to model avalanche effects. It is thus of interest to consider models without this effect. We present here such a case, although slightly academic in view of our assumptions on the mobilities μ_n, μ_p. So we write the conduction current densities equations as

$$- \text{div } J_n = g,$$
$$- \text{div } J_p = -g.$$

We recall the link with the quasi-Fermi potentials

$$J_n = -\mu_n \, nD\phi_n,$$
$$J_p = -\mu_p \, nD\phi_p,$$

and the dependence of n, p with respect to the potentials (including the electrostatic potential ψ)

$$n = F_n(\psi - \phi_n), \quad p = F_p(\phi_p - \psi).$$

Here it is very important to take different functions F_n, F_p, and more precisely to assume the particular structure

$$F_n(s) = e^{s\mu_n}, \quad F_p(s) = e^{s\mu_p}.$$

Then we have

$$J_n = Dn - n\mu_n D\psi, \quad J_p = -(Dp + p\mu_p D\psi).$$

Therefore, the equations become

$$-\Delta n + \mu_n \text{ div } (nD\psi) = g,$$
$$-\Delta p - \mu_p \text{ div } (pD\psi) = g, \tag{6.77}$$
$$-\Delta\psi = p - n + f,$$

$$n = n_0, \ p = p_0, \ \psi = \psi_0 \text{ on } \Gamma_D,$$
$$\frac{\partial n}{\partial \nu} = \frac{\partial p}{\partial \nu} = \frac{\partial \psi}{\partial \nu} = 0 \text{ on } \Gamma_N. \tag{6.78}$$

The special structure leading to the same coefficient in front of $\Delta n, \Delta p$ is crucial in order to take advantage of the fact that $p - n$ appears in the third equation.

We assume

$$f \in L^\ell(\Omega), \quad \ell > N = 2, 3,$$
$$n_0, p_0, \psi_0 \in H^{1,s}(\Omega), \tag{6.79}$$
$$s > 2, \text{ if } N = 2, \quad s = \frac{4N}{N+2} = \frac{12}{5} \text{ if } N = 3,$$
$$\text{ess sup}_{\Gamma_D} n_0, \text{ ess sup}_{\Gamma_D} p_0 < \infty, \tag{6.80}$$
$$n_0, p_0 \geq 0.$$

Then we state the following result.

Theorem 6.7. *We assume (6.54), (6.55), (6.60), (6.79), (6.80), and the smallness condition*

$$\|D\psi_0\|_{L^s} \leq \tau_0, \text{ sufficiently small.} \tag{6.81}$$

Then there exists a weak solution of (6.79), (6.80) in $H^1 \cap L^\infty(\Omega)$.

Remark 6.8. Theorem 6.7 is due to J. Frehse and J. Naumann [28].

6.4.2 Proof of Theorem 6.7

We shall prove only the a priori estimates. The proof itself can be carried over exactly as for Theorem 6.5.

• A Priori Estimates in H^1

1. Basic inequalities
From the first two equations in (6.77), we obtain

$$-\Delta(n-p) + \text{div}\,((n\mu_n + p\mu_p)D\psi) = 0,$$

which we test with $\psi - \psi_0$. At the same time, we test the third equation with $n - p - n_0 + p_0$. Combining results yields

$$
\begin{aligned}
\int_\Omega (n\mu_n + p\mu_p)|D\psi|^2\,dx &+ \int_\Omega (p-n)^2\,dx \\
&= \int_\Omega (n\mu_n + p\mu_p)D\psi.D\psi_0\,dx - \int_\Omega D\psi_0.D(n-p)\,dx \\
&\quad + \int_\Omega D\psi.D(n_0 - p_0)\,dx + \int_\Omega (p-n)(p_0 - n_0)\,dx \\
&\quad + \int_\Omega f(n-p-n_0+p_0)\,dx.
\end{aligned}
\tag{6.82}
$$

Next we test the first two equations in (6.77) respectively with $n - n_0, p - p_0$, and the third equation with $\frac{1}{2}(p^2 - n^2 - p_0^2 + n_0^2)$. Combining results yields

$$
\begin{aligned}
\frac{1}{\mu_n}\int_\Omega |Dn|^2\,dx &+ \frac{1}{\mu_p}\int_\Omega |Dp|^2\,dx + \frac{1}{2}\int_\Omega (p-n)(p^2 - n^2)\,dx \\
&= \frac{1}{2}\int_\Omega (p-n)(p_0^2 - n_0^2)\,dx + \int_\Omega D\psi.((p-p_0)Dp_0 - (n-n_0)Dn_0)\,dx \\
&\quad - \frac{1}{2}\int_\Omega f(p^2 - n^2 - p_0^2 + n_0^2)\,dx + \int_\Omega g\left(\frac{1}{\mu_n}(n-n_0) + \frac{1}{\mu_p}(p-p_0)\right)dx.
\end{aligned}
$$

Using the assumption (6.60) we majorize as follows:

$$
\begin{aligned}
\frac{1}{\max(\mu_n, \mu_p)}\int_\Omega (|Dn|^2 + |Dp|^2)\,dx &\le \int_\Omega |D\psi|.[(p_0 + p)|Dp_0| \\
&\quad + (n_0 + n)|Dn_0| + \frac{g_0}{\min(\mu_n, \mu_p)}(n+p)^2]\,dx \\
&\quad + \frac{g_0}{\min(\mu_n, \mu_p)}\int_\Omega (n+p)(|Dn| + |Dp|)\,dx \\
&\quad - \frac{1}{2}\int_\Omega f(p^2 - n^2 - p_0^2 + n_0^2)\,dx + \int_\Omega (p-n)(p_0^2 - n_0^2)\,dx.
\end{aligned}
\tag{6.83}
$$

We note that from the assumption (6.79), $p_0|Dp_0| + n_0|Dn_0| \in L^2$, $p_0, n_0 \in L^{12}$. Moreover,

$$\int_\Omega |D\psi| |Dp_0| p \, dx \le \frac{1}{2} \int_\Omega p|D\psi|^2 + \frac{1}{2} \int_\Omega p|Dp_0|^2$$

$$\le \frac{1}{2} \int_\Omega p|D\psi|^2 + \|p\|_{L^6} \|Dp_0\|_{L^s}^2,$$

and from the Poincaré inequality this last expression is majorized by

$$\frac{1}{2} \int_\Omega p|D\psi|^2 \, dx + c_0 \|Dp\|_{L^2} \|Dp_0\|_{L^s}^2 + (\|p_0\|_{L^6} + c_0\|Dp_0\|_{L^2})\|Dp_0\|_{L^s}^2.$$

From the third equation tested with $\psi - \psi_0$, we have

$$\|D\psi\|_{L^2} \le 2\|D\psi_0\|_{L^2} + c_0(\|f\|_{L^2} + \|p - n\|_{L^2}).$$

Finally,

$$\int_\Omega |f| |p^2 - n^2| \, dx \le \|f\|_{L^r} \|p - n\|_{L^2} \|p + n\|_{L^{2r/(r-2)}}$$

$$\le \|f\|_{L^r} \|p - n\|_{L^2} (\|p_0 + n_0\|_{L^{2r/(r-2)}} + c_0(\|Dp\|_{L^2} + \|Dn\|_{L^2})).$$

From now on we designate by C, constants depending on the data f, n_0, p_0, ψ_0 and their norms in the spaces of definition $L^r, W^{1,s}$ (at this stage only the H^1 norm of ψ_0 enters), and of course on the parameters. The smallness condition does not make an appearance yet. We deduce from (6.83) and the preceding estimates that

$$\|Dn\|_{L^2}^2 + \|Dp\|_{L^2}^2 \le C + C\|p - n\|_{L^2}^2 + C \int_\Omega (n + p)(|Dn| + |Dp|) \, dx$$

$$+ C \int_\Omega |D\psi|(n + p)^2 \, dx + C \int_\Omega |D\psi|^2(n + p) \, dx. \tag{6.84}$$

Now use

$$\int_\Omega |D\psi|(n + p)^2 \, dx \le \left(\int_\Omega |D\psi|^2(n + p) \, dx \right)^{1/2} \|n + p\|_{L^3}^{3/2}$$

$$\le \sqrt{2} \left(\int_\Omega |D\psi|^2(n + p) \, dx \right)^{1/2} \|n_0 + p_0\|_{L^3}^{3/2}$$

$$+ c_0\sqrt{2} \left(\int_\Omega |D\psi|^2(n + p) \, dx \right)^{1/2}$$

$$(\|D(n - n_0)\|_{L^2} + \|D(p - p_0)\|_{L^2})^{3/2}.$$

$$\int_\Omega (n + p)(|Dn| + |Dp|) \, dx$$

$$\le (\|Dn\|_{L^2} + \|Dp\|_{L^2})\|n + p\|_{L^3}^{3/4} \left(\int_\Omega (n + p) \, dx \right)$$

$$\leq (\|Dn\|_{L^2} + \|Dp\|_{L^2})\|n_0 + p_0\|_{L^3}^{3/4} \left(\int_\Omega (n+p)\, dx\right)$$

$$+ (\|Dn\|_{L^2} + \|Dp\|_{L^2})^{7/8} \left(\int_\Omega (n+p)\, dx\right).$$

Collecting results in (6.84) and making use of Young's inequality to absorb the term $\|Dn\|_{L^2} + \|Dp\|_{L^2}$ on the right-hand side into the left-hand side yields

$$\|Dn\|_{L^2}^2 + \|Dp\|_{L^2}^2 \leq C + C\|p - n\|_{L^2}^2 C \int_\Omega |D\psi|^2 (n+p)\, dx$$

$$+ C \left(\int_\Omega |D\psi|^2 (n+p)\, dx\right)^2 + C \left(\int_\Omega (n+p)\, dx\right)^2.$$
(6.85)

Now if we go back to (6.82), we check easily that

$$\int_\Omega |D\psi|^2 (n+p)\, dx + \|p - n\|_{L^2}^2 \leq C + \|D\psi_0\|_{L^2}(\|Dn\|_{L^2} + \|Dp\|_{L^2})$$

$$+ C \int_\Omega |D\psi_0|^2 (n+p)\, dx,$$

and thus also

$$\int_\Omega |D\psi|^2 (n+p)\, dx + \|p - n\|_{L^2}^2 \leq C + C\|D\psi_0\|_{L^3}^2 (\|Dn\|_{L^2} + \|Dp\|_{L^2}). \quad (6.86)$$

If we compare (6.85) and (6.86), we see that if there were no term $(\int_\Omega (n+p)\, dx)^2$ the result would have been a straightforward consequence of the smallness condition. So estimating $(\int_\Omega (n+p)\, dx)^2$ now becomes crucial.

2. Estimating the term $(\int_\Omega (n+p)\, dx)^2$
Note that $n+p$ satisfies the following conditions:

$$|-\Delta(n+p) + \text{div}((n\mu_n - p\mu_p)D\psi)| \leq g_0(|Dn| + |Dp| + (n+p)|D\psi|)\text{a.e.},$$

$$n+p = n_0 + p_0 \text{ on } \Gamma_D, \quad \frac{\partial(n+p)}{\partial\nu} = 0 \text{ on } \Gamma_N.$$

The purpose of the following trick is to be able to treat this inequality like an equation (this is very close to ideas of F. Tomi, [102]). We need some preliminaries.

Let $\sigma \in (L^\infty(\Omega))^N$ (we shall need $|\sigma(x)| \leq 2g_0$) and consider the linear variational problem

$$\int_\Omega D\Phi.Dz\, dx - \int_\Omega \Phi\sigma.Dz\, dx = \int_\Omega z\, dx,$$
$$\forall z \in H^1(\Omega), \quad z = 0 \text{ on } \Gamma_D, \qquad (6.87)$$
$$\Phi \in H^1(\Omega), \quad \Phi = 0 \text{ on } \Gamma_D.$$

Then this problem has a unique solution in $H^1 \cap L^\infty(\Omega)$ (this follows from Theorem 2.14). Moreover,

$$\Phi \geq 0.$$

This can be obtained by duality and using the maximum principle for the dual problem (adaptation of Theorem 2.16), or more directly by using the test function

$$z = \frac{\Phi^-}{1 + t\Phi^-}, \quad t > 0,$$

from which we obtain

$$\int_\Omega \frac{|D(\Phi^-)|^2}{(1 + \Phi^-)^2} \, dx < \frac{C}{t^2}.$$

Hence from the Poincaré inequality

$$\int_\Omega (\log(1 + t\Phi^-))^2 \, dx < C, \quad \forall t > 0.$$

This implies $\Phi^- = 0$.

We next test (6.87) with $z = n + p - (n_0 + p_0)$, and the first two equations (6.77) with Φ. Combining, we obtain

$$\int_\Omega (n + p - n_0 - p_0) \, dx + \int_\Omega \Phi\sigma.D(n + p - n_0 - p_0) \, dx$$
$$+ \int_\Omega D(n_0 + p_0).D\Phi \, dx + \int_\Omega D\psi.D\Phi(p\mu_p - n\mu_n) \, dx = 2 \int_\Omega g\Phi \, dx.$$
$$(6.88)$$

But

$$2\int_\Omega g\Phi dx \leq 2g_0 \int_\Omega \Phi(|D(n + p)| + |D(n - p)| + (n + p)|D\psi|) \, dx,$$

where we have used the fact that

$$|Dn| + |Dp| \leq |D(n + p)| + |D(n - p)|.$$

We now make precise our choice of σ. We pick

$$\sigma = \begin{cases} 2g_0 \dfrac{D(n + p)}{|D(n + p)|} & \text{if } D(n + p) \neq 0, \\ 0 & \text{if } D(n + p) = 0. \end{cases}$$

With this (ad hoc) choice we deduce from (6.88) that

$$\int_\Omega (n + p) \, dx \leq \int_\Omega (n_0 + p_0) \, dx + \int_\Omega \Phi\sigma.D(n_0 + p_0) \, dx$$
$$- \int_\Omega D(n_0 + p_0).D\Phi \, dx - \int_\Omega D\psi.D\Phi(p\mu_p - n\mu_n) \, dx$$
$$+ 2g_0 \int_\Omega \Phi(|D(n - p)| + (n + p)|D\psi|) \, dx,$$
$$(6.89)$$

and it follows that

$$\int_\Omega (n+p)\, dx \leq C + C\|D(n-p)\|_{L^1} + C \int_\Omega (n+p)|D\Psi|^2\, dx$$
$$+ C \left(\int_\Omega (n+p)|D\Phi|^2\, dx\right)^{1/2} \left(\int_\Omega (n+p)|D\psi|^2\, dx\right)^{1/2}.$$
$$(6.90)$$

Now if in (6.87) we take $z = (n+p)\Phi$, we get

$$\int_\Omega (n+p)|D\Phi|^2\, dx + \int_\Omega \Phi\, D\Phi . D(n+p)\, dx$$
$$- \int_\Omega \Phi\sigma.((n+p)D\Phi + \Phi D(n+p))\, dx = \int_\Omega (n+p)\Phi\, dx.$$

Hence

$$\int_\Omega (n+p)|D\Phi|^2\, dx \leq C\|D(n+p)\|_{L^2} + C \int_\Omega (p+n)\, dx,$$

which yields in (6.90)

$$\int_\Omega (n+p)\, dx \leq C + C\|D(n-p)\|_{L^1} + C \int_\Omega (n+p)|D\psi|^2\, dx$$
$$+ C\|D(n+p)\|_{L^2}^{1/2} \left(\int_\Omega (n+p)|D\psi|^2\, dx\right)^{1/2}.$$
$$(6.91)$$

There remains in (6.91) the problem of estimating $\|D(n-p)\|_{L^1}$. In fact, we shall use the obvious relation

$$\|D(n-p)\|_{L^1} \leq \int_{\Omega_0} |D(n-p)|\, dx + |\Omega - \Omega_0|^{1/2}\|D(n+p)\|_{L^2},$$

which is valid for any $\Omega_0 \subset \Omega$, the point being that majorizing just by $\|D(n+p)\|_{L^2}$ is not good enough, whereas putting a parameter in front that we can make as small as we wish will be sufficient. So it remains to estimate

$$\int_{\Omega_0} |D(n-p)|\, dx, \quad \overline{\Omega_0} \subset \Omega.$$

Recall that $n - p$ satisfies

$$-\Delta(n-p) + \text{div}\, ((n\mu_n + p\mu_p)D\Psi) = 0.$$

Pick a smooth cutoff function ζ such that

$$0 \leq \zeta \leq 1, \zeta = 1 \text{ on } \overline{\Omega_0}, \text{ with compact support in } \Omega.$$

The function $(n-p)\zeta$ satisfies

$$-\Delta((n-p)\zeta) = -\text{div}\,((n\mu_n + p\mu_p)\zeta\, D\Psi + 2(n-p)D\zeta)$$
$$+ D\zeta.D\psi(n\mu_n + p\mu_p) + (n-p)\Delta\zeta$$

with Dirichlet boundary conditions. Pick any $1 < \lambda \le \frac{12}{7}$, so that

$$\frac{\lambda}{2 - \lambda} \le \frac{2n}{n - 2}$$

is satisfied. Then the right-hand side of the preceding equation belongs to $W^{-1,\lambda}$. We rely on the fact that $-\Delta$ with Dirichlet boundary conditions is an isomorphism from $W^{1,\lambda}$ to $W^{-1,\lambda}$ to write

$$\le C(\|n - p\|_{L^\lambda} + \|(n + p)|D\psi|\,\|_{L^\lambda}),$$

and from the condition on λ and the Poincaré inequality we get,

$$\|D((n - p)\varsigma)\|_{L^\lambda} \le C\|D(n + p)\|_{L^2}^{1/2} \left(\int_\Omega (n + p)|D\Psi|^2\,dx \right)^{1/2}$$

$$+C \int_\Omega (n + p)|D\Psi|^2\,dx + C\|n - p\|_{L^2} + C.$$

Using this last estimate in (6.91) yields

$$\int_\Omega (n + p)\,dx \le C + C|\Omega - \Omega_0|^{1/2}\|D(n + p)\|_{L^2} + C \int_\Omega (n + p)|D\Psi|^2\,dx$$

$$+C\|n - p\|_{L^2} + C\|D(n + p)\|_{L^2}^{1/2} \left(\int_\Omega (n + p)|D\Psi|^2\,dx \right)^{1/2},$$

and thus turning back to (6.85) we obtain

$$\|Dn\|_{L^2}^2 + \|Dp\|_{L^2}^2 \le C + C|\Omega - \Omega_0|(\|Dn\|_{L^2}^2 + \|Dp\|_{L^2}^2)$$

$$+C \left(\int_\Omega |D\psi|^2(n + p)\,dx \right)^2$$

$$+C\|n - p\|_{L^2}^2 + C \left(\int_\Omega |D\psi|^2(n + p)\,dx \right)$$

$$\|D(n + p)\|_{L^2}.$$

Choosing Ω_0 sufficiently large and making use of (6.86) and the smallness condition we are done.

• A Priori Estimates in L^∞

This is now routine. We again consider

$$k_0 = \max(\text{ess sup}_{\Gamma_D} n_0, \text{ess sup}_{\Gamma_D} p_0)$$

and obtain for $k > k_0$,

$$\int_\Omega (|D((n - k)^+)|^2 + |D((p - k)^+)|^2)\,dx$$

$$+\frac{1}{2} \int_\Omega (p - n + f)[(p^2 - k^2)^+ - (n^2 - k^2)^+]\,dx$$

$$= \int_\Omega g \left(\frac{1}{\mu_n}(n - k)^+ + \frac{1}{\mu_p}(p - k)^+ \right)\,dx,$$

and we proceed as for the a priori estimates in Theorem 6.5.

The proof is complete.

7. Stationary Navier–Stokes Equations

7.1 Introduction

We study here the regularity of solutions of the stationary Navier–Stokes equations. For the usual Navier–Stokes equations for incompressible fluids on a bounded domain, where the time variable intervenes, the issue of regularity of weak solutions is an open problem in dimension $n = 3$ (see J. Leray [68] or O.A. Ladyzenskaya, V.A. Solonnikov, N.N. Ural'tseva [67] for the background). Only partial regularity is available; see L. Caffarelli, R. Kohn, L. Nirenberg [9].

However, in the stationary case the regularity is easily obtained by a bootstrap argument. This has motivated mathematicians to consider the regularity problem in higher dimensions. Although at first sight it might seem purely academic, this can provide new methods that might also be helpful for the nonstationary problem. In fact, experience (and considerations concerning scaling invariance) shows that the nonstationary case in 3 dimensions has a strong analogy with the stationary case in 5 dimensions. The time corresponds to two space variables.

The case $n = 4$ was solved by C. Gerhard [38] and M. Giaquinta, G. Modica [45]. For higher dimensions, H. Sohr [94] obtained regularity in the stationary case under the additional assumption that $u \in L^n$. A convenient reference is [36]. Partial regularity for $n = 5$, the analogue of the above result [9], was obtained by M. Struwe [98]. In a series of papers J. Frehse and M. Růžička [29], [30], [31], [32], [33] proved the existence of a regular (in the interior) solution to the stationary Navier–Stokes equation in 5 and 6 dimensions, and in 15 dimensions in the periodic case. It is likely that there is *no* restriction with respect to the dimension, at least in the periodic case. Furthermore, they obtained that *every* solution $u \in L^q, q \geq \max(4, \frac{n}{2})$ to the Dirichlet problem is regular (in the periodic case $q \geq \max(4, \frac{n}{4})$).

Based on the weighted estimates [29], M. Struwe [99] obtained independently the existence of a solution for the case $n = 5$, with basic domain R^n. It is interesting to note that methods based on hole-filling and Green function lead to regularity, in the spirit of this book, which explains why we include it. Concerning nonuniqueness see V.I. Youdovich [107].

We shall consider first the regularity issue for "maximum-like solutions" in any dimension, then prove the existence of "maximum-like solutions" with some restrictions on the dimension.

7.2 Regularity of "Maximum-Like Solutions"

7.2.1 Setting of the Problem

Let Ω be a smooth bounded domain of R^n. Also let

$$f \in (L^\infty(\Omega))^n. \tag{7.1}$$

We consider a vector function $u = (u^1, \ldots, u^n)$, the velocity of the flow, and a scalar function p, the pressure, such that

$$u \in (H_0^1(\Omega))^n, \ \operatorname{div} u = 0, \ \ p \in W^{1, \frac{n}{n-1}}(\Omega), \tag{7.2}$$

and

$$-\Delta u + u.Du + Dp = f, \tag{7.3}$$

where this last relation holds in a weak sense. The existence of a weak solution is more or less a routine argument, via Ritz–Galerkin approximations. Note that by Sobolev embedding theorems,

$$u \in (L^{2n/(n-2)}(\Omega))^n.$$

So

$$u.Du \in L^{n/(n-1)}(\Omega).$$

Then (7.3) means that for every $\phi \in H_0^1 \cap L^\infty(\Omega)$ we have

$$\int_\Omega D_j u^i D_j \phi \, dx + \int_\Omega u^j D_j u^i \phi \, dx + \int_\Omega D_i p \phi \, dx = \int_\Omega f^i \phi \, dx. \tag{7.4}$$

Note that from equation (7.3), $\Delta u \in (L^{n/(n-1)}(\Omega))^n$. The pressure equation is derived from (7.4) by replacing ϕ by $D_i \psi$ and summing up, where first $\psi \in C_0^\infty(\Omega)$ and then $\psi \in W_0^{1,n} \cap L^\infty(\Omega)$. We obtain

$$\int_\Omega D_j p D_j \psi \, dx = \int_\Omega D_i u^j D_j u^i \psi \, dx + \int_\Omega f^i D_i \psi \, dx. \tag{7.5}$$

Define

$$w = p + \frac{|u|^2}{2}.$$

This quantity is called "head pressure" and has some physical importance. In particular, the fact that we shall prove an upper bound for this quantity reflects, to a certain extent, Bernoulli's paradox, namely that if u becomes large, the pressure has to become small.

Note that $w \in L^{n/(n-2)}$. We define a "maximum-like solution" as a weak solution u such that

$$w^+ \in L^q_{\text{loc}}(\Omega), \quad q > \frac{n}{2}. \tag{7.6}$$

We begin by proving some regularity properties of "maximum-like solutions." If $q = \infty$, we call u a "maximum solution."

7.2.2 Some Regularity Properties of "Maximum-Like Solutions"

We are going to prove *weighted* estimates for "maximum-like" solutions.

Let Ω_0 and Ω_1 be two smooth subdomains of Ω such that

$$\overline{\Omega_0} \subset \Omega_1; \quad \overline{\Omega_1} \subset \Omega.$$

We consider a smooth cutoff function ζ such that

$$0 \le \zeta \le 1 \quad \zeta = 1 \text{ on } \Omega_1; \quad \zeta \text{ has compact support in } \Omega.$$

We now pick any point $x_0 \in \Omega_0$. Take $h > 0$ (which will tend to 0). We set

$$\Gamma = \Gamma_h^{x_0} = \begin{cases} \dfrac{1}{n-4} \dfrac{1}{(|x-x_0|^2 + h^2)^{(n-4)/2}}, & \text{if } n \ge 5, \\[2ex] \log \dfrac{(\text{diam } \Omega + h)}{(|x-x_0|^2 + h^2)^{1/2}}, & \text{if } n = 4. \end{cases}$$

Then we have

$$D_i \Gamma = -\frac{(x_i - x_{0,i})}{(|x-x_0|^2 + h^2)^{(n-2)/2}},$$

$$D_i D_j \Gamma = -\frac{\delta_{ij}}{(|x-x_0|^2 + h^2)^{(n-2)/2}} + \frac{(n-2)(x_i - x_{0,i})(x_j - x_{0,j})}{(|x-x_0|^2 + h^2)^{n/2}},$$

and finally

$$\Delta \Gamma = -\frac{2}{(|x-x_0|^2 + h^2)^{(n-2)/2}} - \frac{(n-2)h^2}{(|x-x_0|^2 + h^2)^{n/2}}.$$

We also notice that

$$|D_i \Gamma| \le C_n \frac{1}{|x-x_0|^{n-3}}, \tag{7.7}$$

$$|\Delta \Gamma| \le C_n \frac{1}{|x-x_0|^{n-2}}, \tag{7.8}$$

where the constants C_n do not depend on h or x_0.

Our first result is the following lemma.

Lemma 7.1. *We have the estimates*

$$\int_\Omega \frac{(u.(x-x_0))^2}{|x-x_0|^n}\zeta^2\,dx \le C_n(\Omega_0,\zeta),\quad \forall x_0\in\Omega_0, \tag{7.9}$$

$$\int_\Omega \frac{|w|}{|x-x_0|^{n-2}}\zeta^2\,dx \le C_n(\Omega_0,\zeta),\quad \forall x_0\in\Omega_0, \tag{7.10}$$

where the constant does not depend on x_0 but depends on the norm of ζw^+ in $L^q(\Omega)$.

Proof. We can test (7.3) with $-\zeta^2 D_i\Gamma$, and after summing up we obtain

$$\int_\Omega (p\Delta\Gamma + u^i u^j D_i D_j\Gamma)\zeta^2\,dx = -\int_\Omega \zeta^2 f^i D_i\Gamma\,dx$$

$$+ \int_\Omega D_j\zeta^2(\Delta u^j\Gamma - pD_j\Gamma - u^j u^i D_i\Gamma)\,dx. \tag{7.11}$$

Since $D_j\zeta^2$ vanishes on Ω_1, the second integral on the right-hand side is clearly bounded independently of h and x_0. The first integral on the right-hand side is also bounded, since f is bounded and thanks to the estimates (7.7), (7.8). To evaluate the left-hand side, we first write

$$p = 2w^+ - 2|w| - \frac{|u|^2}{2}$$

and notice that

$$u^i u^j D_i D_j\Gamma - \frac{|u|^2}{2}\Delta\Gamma = \frac{(n-2)[(u.(x-x_0))^2 + h^2|u|^2]}{(|x-x_0|^2 + h^2)^{n/2}}.$$

Therefore, we get from (7.11)

$$-\int_\Omega |w|\zeta^2\Delta\Gamma\,dx + C_n\int_\Omega \frac{(u.(x-x_0))^2}{(|x-x_0|^2 + h^2)^{n/2}}\,dx$$

$$\le -2\int_\Omega w^+\zeta^2\Delta\Gamma\,dx + C_n. \tag{7.12}$$

Thanks to the estimate (7.7) and the assumption (7.6), the right-hand side of (7.12) is bounded independently of h. Using

$$-\Delta\Gamma \ge \frac{2}{(|x-x_0|^2 + h^2)^{(n-2)/2}}$$

and applying Fatou's lemma we immediately obtain the result.   ◇

The next lemma deserves a proof only for $n \ge 5$.

Lemma 7.2. *We have the properties*

$$w \in L^2_{\text{loc}}(\Omega), \tag{7.13}$$

$$|u|^2, p \in L^s_{\text{loc}}(\Omega), \quad \forall s \in [1, 2). \tag{7.14}$$

In fact, for $n = 4$, from the information

$$p, |u|^2 \in L^{n/(n-2)}$$

we already know more than this.

Proof of Lemma 7.2. There are several steps:
1. *We demonstrate the property*

$$\int_\Omega \frac{|p| + |u|^2}{|x - x_0|^s} \zeta^2 \, dx \leq C_s(\Omega_0, \zeta), \quad \forall x_0 \in \Omega_0, \quad 2 < s < n - 2. \tag{7.15}$$

The proof is very similar to that of Lemma 7.1 (we just replace Γ by $(|x - x_0|^2 + h^2)^{-(s-2)/2}$). We use the fact that

$$(|x - x_0|^2 + h^2)^{-s/2} \leq C|x - x_0|^{-(n-2)},$$

which is true for $x \in \Omega$; and using Lemma 7.1 we assert that

$$\int_\Omega \frac{(u.(x - x_0))^2}{(|x - x_0|^2 + h^2)^{(s+2)/2}} \zeta^2 \, dx \leq C_s(\Omega_0, \zeta), \quad \forall x_0 \in \Omega_0,$$

and

$$\int_\Omega \frac{|w|}{(|x - x_0|^2 + h^2)^{s/2}} \zeta^2 \, dx \leq C_s(\Omega_0, \zeta), \quad \forall x_0 \in \Omega_0.$$

From the corresponding relation to (7.11), we deduce easily that

$$\left| \int_\Omega p\zeta^2 \left[\frac{n - 2 - s}{(|x - x_0|^2 + h^2)^{s/2}} + \frac{sh^2}{(|x - x_0|^2 + h^2)^{s/2}} \right] dx \right| \leq C_s(\Omega_0, \zeta).$$

Recalling that from the assumption on maximum solutions $p^+ \in L^q_{\text{loc}}$, we deduce easily the estimate (7.15).

2. *Proof of (7.14)*
Introduce a mollifier

$$(R_\epsilon z)(x) = \frac{1}{\epsilon^n} \int_{R^n} \beta\left(\frac{x - y}{\epsilon}\right) z(y) \, dy,$$

where

$$\beta(x) = \begin{cases} \exp \dfrac{1}{|x|^2 - 1} & \text{if } |x| < 1, \\ 0 & \text{if } |x| \geq 1. \end{cases}$$

We shall write z_ϵ instead of $R_\epsilon z$, to simplify the writing. Let $1 < r < 2$ and

$$\phi(x) = \phi_{h,\epsilon}(x) = \int_\Omega \frac{|p(y)|^{r-1} \operatorname{sgn} p_\epsilon(y)}{(|x-y|^2 + h^2)^{(n-2)/2}} \zeta^2(y)\, dy. \tag{7.16}$$

If we pick

$$n - 2 > s > n - \frac{2}{r-1},$$

then from Hölder's inequality we can majorize

$$|\phi(x)| \le C_s \left(\int_\Omega \frac{|p(y)|}{|x-y|^s} \zeta^2(y)\, dy \right)^{1/(r-1)},$$

and thus from (7.15),

$$|\phi(x)| \le C_s(\Omega_0, \zeta), \ \forall x \in \Omega_0, \tag{7.17}$$

and also since

$$|D\phi(x)| \le C \int_\Omega \frac{|p(y)|^{r-1}}{|x-y|^{n-1}}\, dy,$$

we have

$$\|D\phi\|_{L^1} \le C. \tag{7.18}$$

None of the constants depend on ϵ or h.

Let τ be a smooth cutoff function with compact support in Ω_0. We can take $\psi = (\tau^2 \phi)_\epsilon$ in the pressure equation (7.5). This is possible, at least for ϵ sufficiently small, since ψ will also have compact support in Ω. We deduce, permuting the mollifiers, that

$$-\int_\Omega p_\epsilon \tau^2 \Delta\phi(x)\, dx = -\int_\Omega (p_\epsilon \Delta(\tau^2) + 2Dp_\epsilon . D\tau^2)\phi\, dx$$

$$+ \int_\Omega (D_i u^j D_j u^i)_\epsilon \tau^2 \phi\, dx + \int_\Omega (f^i)_\epsilon D_i(\tau^2 \phi)\, dx.$$

From (7.17), (7.18) we clearly get

$$\left| -\int_\Omega p_\epsilon \tau^2 \Delta\phi(x)\, dx \right| \le C(\Omega_0),$$

where the constant does not depend on h or ϵ. Using

$$-\Delta\phi(x) = n(n-2)h^2 \int_\Omega \frac{|p(y)|^{r-1} \operatorname{sgn} p_\epsilon(y)}{(|x-y|^2 + h^2)^{(n+2)/2}} \zeta^2(y)\, dy$$

we obtain $|J_{h,\epsilon}| \le C$, where

$$J_{h,\epsilon} = n(n-2)h^2 \int_\Omega \int_\Omega p_\epsilon(x) \frac{|p(y)|^{r-1} \operatorname{sgn} p_\epsilon(y)}{(|x-y|^2 + h^2)^{(n+2)/2}} \zeta^2(y)\tau^2(x)\, dx\, dy.$$

Writing

$$J_{h,\epsilon} = J_{h,\epsilon}^1 + J_{h,\epsilon}^2,$$

where

$$J_{h,\epsilon}^1 = n(n-2)h^2 \int_\Omega \int_\Omega \frac{(p_\epsilon(x) - p_\epsilon(y))|p(y)|^{r-1} \operatorname{sgn} p_\epsilon(y)}{(|x-y|^2 + h^2)^{(n+2)/2}} \zeta^2(y)\tau^2(x) \, dx \, dy$$

and

$$J_{h,\epsilon}^2 = n(n-2)h^2 \int_\Omega \int_\Omega \frac{|p_\epsilon(y)||p(y)|^{r-1}}{(|x-y|^2 + h^2)^{(n+2)/2}} \zeta^2(y)\tau^2(x) \, dx \, dy,$$

it is easy to check that

$$J_{h,\epsilon}^1 \to 0, \text{ as } h \to 0,$$

$$J_{h,\epsilon}^2 \to n(n-2)\varpi_n \int_\Omega |p_\epsilon(x)||p(x)|^{r-1}\tau^2(x) \, dx \text{ as } h \to 0.$$

Therefore, we have proved that

$$\int_\Omega |p_\epsilon(x)||p(x)|^{r-1}\tau^2(x) \, dx \leq C.$$

Letting $\epsilon \to 0$, using Fatou's lemma we obtain

$$\int_\Omega |p(x)|^r \tau^2(x) \, dx \leq C.$$

Hence clearly

$$p \in L_{\text{loc}}^s(\Omega), \quad \forall s \in [1,2).$$

Noting that

$$|u|^2 \leq 2w^+ - 2p$$

we obtain immediately from the assumption on maximum solutions that

$$|u|^2 \in L_{\text{loc}}^s(\Omega), \quad \forall s \in [1,2). \tag{7.19}$$

3 Proof of (7.13):
In the definition of $\phi_{h,\epsilon}(x)$, (7.16), we replace $|p(y)|^{r-1}$, by $|w(y)|$. We can proceed with the same proof, using now (7.10) instead of (7.15). We then arrive at

$$|p||w| \in L_{\text{loc}}^1(\Omega). \tag{7.20}$$

Since we know

$$w^2 \leq wp + w^+ \frac{|u|^2}{2},$$

by referring to (7.19), (7.20), and the assumption on w^+, we obtain property (7.13). The proof of the lemma is complete. $\diamond$

Now introduce the function g, a solution of

$$-\Delta g = w \text{ in } \Omega,$$
$$g = 0 \text{ on } \partial\Omega. \tag{7.21}$$

This function will help to overcome the square growth of the vorticity term $u.Du$.

Clearly, g is defined uniquely as an element of $W^{2,\frac{n}{n-2}} \cap W_0^{1,\frac{n}{n-1}}(\Omega)$. Moreover, from (7.13),

$$g \in H^2_{\text{loc}}(\Omega).$$

We can say more.

Lemma 7.3. *We have the properties*

$$g \in L^\infty_{\text{loc}}(\Omega) \tag{7.22}$$

and

$$\int_\Omega \frac{|Dg|^2}{|x - x_0|^{n-2}} \zeta^2 \, dx \leq C_n(\Omega_0, \zeta), \quad \forall x_0 \in \Omega_0. \tag{7.23}$$

Proof. Let τ be a smooth cutoff function with compact support in Ω_0. Let us consider, for $x_0 \in \Omega_0$,

$$G = G_h^{x_0}(x) = \frac{1}{(|x - x_0|^2 + h^2)^{(n-2)/2}}$$

and test (7.21) with $G\tau^{2n}$. We get

$$-\int_\Omega g\Delta(G\tau^{2n}) \, dx = \int_\Omega wG\tau^{2n} \, dx,$$

and thanks to (7.10), the right-hand side is bounded, so we have

$$\left| \int_\Omega g\Delta(G\tau^{2n}) \, dx \right| \leq C(\Omega_0).$$

Next we see that

$$\left| \int_\Omega gDG.D\tau^{2n} \, dx \right| + \left| \int_\Omega gG\Delta\tau^{2n} \, dx \right|$$
$$\leq C \int_\Omega \frac{|g|\tau^{2n-1}}{|x - x_0|^{n-1}} \, dx + C \int_\Omega \frac{|g|\tau^{2n-2}}{|x - x_0|^{n-2}} \, dx.$$

Noting that from Sobolev embedding theorems $g \in L^{n/(n-4)}(\Omega)$, we obtain

$$h^2 \left| \int_\Omega \frac{g\tau^{2n}}{(|x - x_0|^2 + h^2)^{(n+2)/2}} \, dx \right|$$
$$\leq C + C(\|g\tau^{2n}\|_{L^\infty(\Omega_0)})^{(2n-1)/2n} + C(\|g\tau^{2n}\|_{L^\infty(\Omega_0)})^{(2n-2)/2n}.$$

But letting h tend to 0 and using (7.5) we deduce that a.e. x_0,

$$|g\tau^{2n}(x_0)| \le C + C(\|g\tau^{2n}\|_{L^\infty(\Omega_0)})^{(2n-1)/2n}.$$

Taking the ess sup on x_0, we deduce at once

$$\|g\tau^{2n}\|_{L^\infty(\Omega_0)} < \infty.$$

Thus, (7.22) has been proven.

We proceed to prove (7.23). We test this time with $gG\zeta^2$ and obtain

$$\int_\Omega |Dg|^2 G\zeta^2 \, dx - \int_\Omega \frac{g^2}{2}\zeta^2 \Delta G \, dx = \int_\Omega gwG\zeta^2 \, dx$$
$$+ \int_\Omega \frac{g^2}{2}(\Delta\zeta^2 G + 2DG.D\zeta^2) \, dx.$$

In the preceding equality, the second integral on the left-hand side is positive and can be eliminated. Thanks to (7.22) and (7.10), the first integral on the right-hand side is bounded. As far as the last integral is concerned, in fact the range of integration is $\Omega - \Omega_1$, and thus it is easily seen to be bounded. Hence

$$\int_\Omega |Dg|^2 G\zeta^2 \, dx \le C.$$

Letting h tend to 0, the result (7.23) is proven. $\Diamond$

7.2.3 The Navier–Stokes Inequality

By the Navier–Stokes inequality, we mean the relation

$$\int_\Omega D_j u^i D_j(u^i\gamma) \, dx \le \int_\Omega w\, u.D\gamma \, dx$$
$$+ \int_\Omega f.u\gamma \, dx, \; \forall\gamma \in C_0^\infty(\Omega), \quad \gamma \ge 0. \tag{7.24}$$

Note that all the terms in (7.24) make sense for "maximum-like solutions," as a consequence of the regularity properties proved in the preceding section. Unfortunately, it is not clear that the Navier–Stokes inequality is satisfied for the "maximum-like solutions." To obtain it formally (and actually, this even gives equality) we just take $\phi = u^i\gamma$ in (7.4) and sum up. But this is not, in general, legitimate (in fact, for $n = 4$, this does become legitimate, since $u^i\gamma$ belongs to $L^4(\Omega)$ and $u.Du + Dp$ to $L^{4/3}(\Omega)$). We shall see later on that for "maximum solutions," i.e., $q = \infty$ in (7.6), then the Navier–Stokes inequality is satisfied. Our objective here is to show that the Navier–Stokes inequality is sufficient to obtain full regularity of "maximum-like solutions."

We begin by giving some estimates in addition to those proven in the previous sections.

Lemma 7.4. *Consider a "maximum-like solution" satisfying the NS inequality. Then one has*

$$\int_\Omega \frac{|u|^2}{|x-x_0|^{n-2}}\zeta^2\,dx \le C_n(\Omega_0,\zeta),\quad \forall x_0\in\Omega_0,\tag{7.25}$$

$$\int_\Omega \frac{|Du|^2}{|x-x_0|^{n-4}}\zeta^2\,dx \le C_n(\Omega_0,\zeta),\quad \forall x_0\in\Omega_0,\quad n\ge 5,\tag{7.26}$$

$$\int_\Omega |Du|^2\log\frac{\operatorname{diam}\Omega}{|x-x_0|}\,\zeta^2\,dx \le C(\Omega_0,\zeta),\quad \forall x_0\in\Omega_0,\quad n=4.$$

Proof. We take $\gamma=\Gamma\zeta^2$ in the NS inequality. We obtain, by integration by parts,

$$\int_\Omega |Du|^2\Gamma\zeta^2\,dx - \int_\Omega \frac{|u|^2}{2}\Delta(\Gamma\zeta^2)\,dx \le \int_\Omega wu.D(\Gamma\zeta^2)\,dx + \int_\Omega f.u\Gamma\zeta^2\,dx.$$

We perform the calculations only for $n\ge 5$ (the case $n=4$ can be treated similarly). We replace w by $-\Delta g$ and again perform integration by parts. Recalling the formulas for the derivatives of Γ, we arrive at

$$\int_\Omega |Du|^2\Gamma\zeta^2\,dx + \int_\Omega |u|^2(n-4)\frac{\Gamma}{|x-x_0|^2+h^2}\zeta^2\,dx$$
$$\le \int_\Omega \zeta^2 D_j g D_j u^i D_i\Gamma\,dx + \int_\Omega \zeta^2 D_j g u^i D_i D_j\Gamma\,dx + \int_\Omega \zeta^2\Gamma f.u\,dx$$
$$+ \int_{\Omega-\Omega_1} \frac{|u|^2}{2}(2D\zeta^2.D\Gamma + \Gamma\Delta\zeta^2)\,dx + \int_{\Omega-\Omega_1}\Gamma D_j g D_j u^i D_i\zeta^2\,dx$$
$$+ \int_{\Omega-\Omega_1} D_j g u^i(D_i\zeta^2 D_j\Gamma + D_j\zeta^2 D_i\Gamma + D_i D_j\zeta^2\Gamma)\,dx.\tag{7.27}$$

All the integrals over $\Omega-\Omega_1$ are easily seen to be bounded (independently of h, x_0) by a constant $C(\Omega_0,\zeta)$. The three integrals involving ζ^2 on the right-hand side are treated by Young's inequality, absorbing terms in $|u|^2$ and $|Du|^2$ by the left-hand side. We thus obtain the following integrals:

$$\int_\Omega \zeta^2\Gamma|f|^2(|x-x_0|^2+h^2)\,dx,$$

$$\int_\Omega \zeta^2|Dg|^2\frac{|D\Gamma|^2}{\Gamma}\,dx,$$

$$\int_\Omega \zeta^2|Dg|^2\frac{|D^2\Gamma|^2}{\Gamma}(|x-x_0|^2+h^2)\,dx.$$

Using

$$|D\Gamma|^2 \le C\frac{\Gamma^2}{|x-x_0|^2+h^2},$$

$$|D^2\Gamma|^2 \le C\frac{\Gamma^2}{(|x-x_0|^2+h^2)^2},$$

and the property (7.23), we obtain that these three integrals are bounded by constants of the type $C(\Omega_0, \zeta)$. Therefore, the left-hand side of (7.27) is also bounded by such a constant. Letting h tend to 0 and using Fatou's lemma, the results (7.25), (7.26) are obtained.   ◇

7.2.4 Hole-Filling

Consider now, attached to any $x_0 \in \Omega_0$, the ball $B_R(x_0)$. Assume $R \leq R_0 = \frac{1}{2}$ dist $(\Omega_0, \partial\Omega_1)$, so $B_{2R}(x_0) \subset \Omega_1$. Let τ_R be the usual cutoff function, i.e., τ_R is smooth, equals 1 on B_R, has support in B_{2R}, and takes values in $[0, 1]$ such that

$$|D\tau_R| \leq \frac{C}{R}; \quad |D^2\tau_R| \leq \frac{C}{R^2}.$$

We shall prove some crucial estimates, which are to some extent local counterparts of (7.9), (7.10), (7.23), (7.25), (7.26).

Lemma 7.5. *We have the following inequalities:*

$$\int_{B_R} \frac{|w|}{|x - x_0|^{n-2}} \, dx + \int_{B_R} \frac{(u.(x - x_0))^2}{|x - x_0|^n} \, dx$$
$$\leq C \int_{B_{2R}-B_R} \frac{|w|}{|x - x_0|^{n-2}} \, dx + C \int_{B_{2R}-B_R} \frac{|u|^2}{|x - x_0|^{n-2}} \, dx + CR^\gamma, \tag{7.28}$$

$$\int_{B_R} \frac{|Dg|^2}{|x - x_0|^{n-2}} \, dx \leq C \int_{B_{2R}-B_R} \frac{|Dg|^2}{|x - x_0|^{n-2}} \, dx$$
$$+ C \int_{B_{2R}} \frac{|w|}{|x - x_0|^{n-2}} \, dx, \tag{7.29}$$

$$\int_{B_R} \frac{|Du|^2}{|x - x_0|^{n-4}} \, dx + \int_{B_R} \frac{|u|^2}{|x - x_0|^{n-2}} \, dx$$
$$\leq C \int_{B_{2R}-B_R} \frac{|u|^2}{|x - x_0|^{n-2}} \, dx + C \int_{B_{2R}} \frac{|Dg|^2}{|x - x_0|^{n-2}} \, dx + CR^6, \tag{7.30}$$

where $\gamma = 2 - \frac{n}{q}$, and the constants do not depend or R, x_0, but of course do depend on Ω_0, Ω_1. Note also that (7.30) is written for $n \geq 5$. For $n = 4$, one substitutes $\log \dfrac{\text{diam } \Omega}{|x - x_0|}$ for $\dfrac{1}{|x - x_0|^{n-4}}$.

Proof.
1. Proof of (7.28)

In the pressure equation (7.5) take $\psi = \Gamma\tau_R^2$. We obtain

$$-\int_\Omega p\Delta(\Gamma\tau_R^2) \, dx = \int_\Omega u^i u^j D_i D_j(\Gamma\tau_R^2) \, dx + \int_\Omega f.D(\Gamma\tau_R^2) \, dx.$$

Again replacing p by $2w^+ - |w| - |u|^2/2$ we get

$$\int_\Omega \tau_R^2 \left[-|w|\Delta\Gamma + \left(u^i u^j D_i D_j \Gamma - \frac{|u|^2}{2}\Delta\Gamma \right) \right] dx$$
$$= \int_\Omega \left(|w| + \frac{|u|^2}{2} \right) (2D\tau_R^2.D\Gamma + \Gamma\Delta\tau_R^2) \, dx \tag{7.31}$$
$$- \int_\Omega (2(u.D\tau_R^2)(u.D\Gamma) + \Gamma u^i u^j D_i D_j \tau_R^2) \, dx$$
$$+ 2\int_\Omega w^+ \Delta(\tau_R^2\Gamma) \, dx - \int_\Omega f.D(\tau_R^2\Gamma) \, dx.$$

The two first integrals on the right-hand side of (7.31) are in fact integrals on $B_{2R} - B_R$, since derivatives of τ_R are involved. Moreover, using the relations

$$|x - x_0| \geq R \geq \frac{|x - x_0|}{2},$$

which hold on $B_{2R} - B_R$, we majorize these integrals easily by

$$C \int_{B_{2R} - B_R} \frac{|w|}{|x - x_0|^{n-2}} \, dx + C \int_{B_{2R} - B_R} \frac{|u|^{n-2}}{|x - x_0|^{n-2}} \, dx.$$

Similarly, we check that

$$\left| \int_\Omega w^+ \Delta(\tau_R^2\Gamma) \, dx \right| \leq CR^\gamma, \quad \gamma = 2 - \frac{n}{q},$$
$$\left| \int_\Omega f.D(\tau_R^2\Gamma) \, dx \right| \leq CR^3.$$

Recalling then the expression for

$$u^i u^j D_i D_j \Gamma - \frac{|u|^2}{2}\Delta\Gamma,$$

letting h tend to 0, and referring to Fatou's lemma, we obtain (7.28).

2. Proof of (7.29)
We refer to the equation defining g, (7.21). Let g_R be the average of g on $B_{2R} - B_R$. We test (7.21) with $(g - g_R)G\tau_R^2$. We obtain (by neglecting a positive term arising from $-\Delta G$)

$$\int_\Omega |Dg|^2\tau_R^2 G \, dx \leq -2\int_\Omega (g - g_R)GDg.D\tau_R^2 \, dx$$
$$- \frac{1}{2}\int_\Omega (g - g_R)^2 G\Delta\tau_R^2 \, dx + \int_\Omega w(g - g_R)\tau_R^2 G \, dx.$$

Since $B_{2R} \subset \Omega_1$, we can use the fact that g is locally bounded to majorize the last integral by

$$C \int_{B_{2R}} \frac{|w|}{|x - x_0|^{n-2}} \, dx.$$

The two first integrals on the right-hand side hold on $B_{2R} - B_R$, thanks to the presence of derivatives of τ_R. This time we use

$$R^2 \leq |x - x_0|^2 + h^2,$$

which holds on $B_{2R} - B_R$, and write

$$\left| \int_{\Omega} (g - g_R)^2 G \Delta \tau_R^2 \, dx \right| \leq \frac{C'}{R^n} \int_{B_{2R} - B_R} (g - g_R)^2 \, dx$$

$$\leq \frac{C}{R^{n-2}} \int_{B_{2R} - B_R} |Dg|^2 \, dx$$

using the Poincaré inequality. The other term is treated similarly. Letting h tend to 0 and using Fatou's lemma, (7.29) is obtained.

3. Proof of (7.30)
This time we use the NS inequality, in which we take $\gamma = \Gamma \tau_R^2$. We then perform computations similar to those of Lemma 7.4 to yield

$$\int_{\Omega} |Du|^2 \Gamma \tau_R^2 \, dx + \int_{\Omega} |u|^2 (n-4) \frac{\Gamma}{|x - x_0|^2 + h^2} \tau_R^2 \, dx$$

$$\leq \int_{\Omega} \tau_R^2 D_j g D_j u^i D_i \Gamma \, dx + \int_{\Omega} \tau_R^2 D_j g u^i D_i D_j \Gamma \, dx + \int_{\Omega} \tau_R^2 \Gamma f.u \, dx$$

$$+ \int_{B_{2R} - B_R} \frac{|u|^2}{2} (2D\tau_R^2.D\Gamma + \Gamma \Delta \tau_R^2) \, dx + \int_{B_{2R} - B_R} \Gamma D_j g D_j u^i D_i \tau_R^2 \, dx$$

$$+ \int_{B_{2R} - B_R} D_j g u^i (D_i \tau_R^2 D_j \Gamma + D_j \tau_R^2 D_i \Gamma + D_i D_j \tau_R^2 \Gamma) \, dx.$$

The two integrals on the right-hand side involving $D_j g D_j u^i$ are treated by Young's inequality, absorbing the gradient into the left-hand side. This produces the following estimate on the right-hand side:

$$C \int_{B_{2R}} \frac{|Dg|^2}{|x - x_0|^{n-2}} \, dx.$$

The integral involving $|u|^2$ is estimated by

$$C \int_{B_{2R} - B_R} \frac{|u|^2}{|x - x_0|^{n-2}} \, dx.$$

The integral involving $D_j g u^i D_i D_j \tau_R^2 \Gamma$ is estimated by

$$C \int_{B_{2R} - B_R} \frac{|u|^2}{|x - x_0|^{n-2}} \, dx + C \int_{B_{2R} - B_R} \frac{|Dg|^2}{|x - x_0|^{n-2}} \, dx.$$

All the remaining integrals involving $D_j g u^i$ are treated by Young's inequality, absorbing the term in u into the left-hand side. This again produces

$$C \int_{B_{2R}} \frac{|Dg|^2}{|x - x_0|^{n-2}} \, dx.$$

The term involving $f.u$ is also treated by Young's inequality, absorbing u into the left-hand side. This produces a CR^6 term. Collecting results, we get (7.30). The proof is complete. ◇

We then state the key result, which will be a consequence of hole-filling.

Theorem 7.6. *Consider a "maximum-like solution" satisfying the NS inequality (see (7.6), (7.24)). Assume (7.1). Let Ω_0, Ω_1 be subdomains of Ω such that*

$$\overline{\Omega_0} \subset \Omega_1; \quad \overline{\Omega_1} \subset \Omega.$$

Then

$$\int_{B_R(x_0)} \frac{|Du|^2}{|x - x_0|^{n-4}} \, dx \le C(\Omega_0, \Omega_1) R^{2\delta}, \quad 0 < 2\delta < \gamma = 2 - \frac{n}{q}, \tag{7.32}$$

$$\forall x_0 \in \Omega_0, \quad R \le R_0 = \frac{1}{2} \text{dist} \, (\Omega_0, \partial\Omega_1).$$

For $n = 4$, (7.32) is valid, although we could, in fact, introduce the log, which would give a sharper estimate.

Proof. Let us set

$$z = \frac{|Du|^2}{|x - x_0|^{n-4}} + \frac{|u|^2}{|x - x_0|^{n-2}} + \frac{|Dg|^2}{|x - x_0|^{n-2}} + \frac{|w|}{|x - x_0|^{n-2}}.$$

Then combining the relations (7.28), (7.29), (7.30), we can write

$$\int_{B_R} z \, dx \le C \int_{B_{2R} - B_R} z \, dx + CR^\gamma, \quad \forall R \le R_0.$$

Moreover, we have

$$\int_{B_{R_0}} z \, dx \le C_0,$$

which is a consequence of (7.10), (7.23), (7.25), (7.26). But then the hole-filling technique applies (see Section 1.2.3). We know that

$$\delta < \delta_0 = \frac{1}{2} \min \left(\gamma, \log_2 \frac{1 + C}{C} \right).$$

The result follows. ◇

It is remarkable to use the above function z for the hole-filling procedure. It is customary to use the hole-filling only with terms of the type $\int |Du|^2 G dx$.

7.2.5 Full Regularity

We can now prove the following main result on regularity for "maximum-like solutions" of Navier–Stokes equations satisfying the NS inequality.

Theorem 7.7. *We make the same assumptions as in Theorem 7.6. Then the solution u, p satisfies*

$$u \in \left(W_{\mathrm{loc}}^{2,s}(\Omega)\right)^n, \quad p \in W_{\mathrm{loc}}^{1,s}(\Omega), \quad \forall 1 < s < \infty.$$

Proof. From (7.32) we deduce that

$$u \in W_{\mathrm{loc}}^{1,2,n-4+2\delta}(\Omega),$$

and thus from the embedding property of Morrey spaces we have

$$u \in L_{\mathrm{loc}}^{2\frac{2-\delta}{1-\delta},n-4+2\delta}(\Omega).$$

Note that the Morrey exponent $n - 4 + 2\delta$ is preserved in this embedding. For $n = 4$, this implies that $u \in L_{\mathrm{loc}}^{n+\epsilon}(\Omega)$, and a bootstrap argument can be performed to deduce full regularity. Assume then $n \geq 5$. Set

$$\ell_0 = 2, \quad s_0 = 2\frac{2-\delta}{1-\delta}, \quad r_0 = 2\frac{2-\delta}{3-2\delta}.$$

Then

$$u.Du \in L_{\mathrm{loc}}^{r_0,n-4+2\delta}(\Omega).$$

Therefore, from linear regularity theory,

$$u \in W_{\mathrm{loc}}^{2,r_0,n-4+2\delta}(\Omega), \quad p \in W_{\mathrm{loc}}^{1,r_0,n-4+2\delta}(\Omega),$$

and applying again the embedding property, we have

$$u \in W_{\mathrm{loc}}^{1,\ell_1,n-4+2\delta}(\Omega)$$

with

$$\ell_1 = \frac{(4-2\delta)r_0}{4-2\delta-r_0} = \frac{2-\delta}{1-\delta}.$$

So progress has been made, since $\ell_1 > \ell_0$. If $\ell_1 \geq 4 - 2\delta$, then u belongs to any L^p and we are done. If not, we proceed with the embedding procedure. Hence

$$u \in L_{\mathrm{loc}}^{s_1,n-4+2\delta}(\Omega)$$

with

$$s_1 = \frac{\ell_1(4-2\delta)}{4-2\delta-\ell_1}$$

and

$$u.Du \in L_{loc}^{r_1, n-4+2\delta}(\Omega)$$

with

$$r_1 = \frac{2s_1}{2 + s_1}.$$

We thus can define sequences ℓ_k, s_k, r_k, by the formulas

$$\ell_k = \frac{(4 - 2\delta)r_{k-1}}{4 - 2\delta - r_{k-1}},$$

$$s_k = \frac{\ell_k(4 - 2\delta)}{4 - 2\delta - \ell_k},$$

$$r_k = \frac{2s_k}{2 + s_k},$$

as long as $r_{k-1} < 4 - 2\delta$. In fact,

$$\frac{1}{\ell_k} = \frac{1}{\ell_{k-1}} - \frac{\delta}{4 - 2\delta},$$

and after a finite number of steps we obtain that u is any L^s, for any s. So we have full regularity.   ◇

7.3 Maximum Solutions and the NS Inequality

7.3.1 Notation and Setup

We consider here a weak solution u, p of (7.2) such that

$$u \in (H_0^1(\Omega))^n, \ \text{div } u = 0, \tag{7.33}$$

$$p \in W^{1, \frac{n}{n-1}}(\Omega), \tag{7.34}$$

$$-\Delta u + u.Du + Dp = f, \tag{7.35}$$

$$-\Delta p = Du \circ Du - \text{div } f, \tag{7.36}$$

where we have set

$$Du \circ Du = D_i u^j D_j u^i.$$

Consider

$$w = p + \frac{|u|^2}{2}.$$

We assume (maximum solution) that

$$w \leq c_0. \tag{7.37}$$

Then we have the following result.

Theorem 7.8. *A maximum solution of the Navier–Stokes equations satisfies*

$$\int_\Omega Dw.D\gamma\, dx + \int_\Omega |Du|^2\gamma\, dx$$

$$\leq \int_\Omega Du \circ Du\gamma\, dx + \int_\Omega wu.D\gamma\, dx + \int_\Omega f.u\gamma\, dx + \int_\Omega f.D\gamma\, dx \tag{7.38}$$

for any $\gamma \in C_0^\infty(\Omega), \gamma \geq 0$.

The Navier–Stokes inequality is an easy consequence of (7.38). Indeed, we just test the pressure equation with γ and subtract the result from (7.38).

7.3.2 Proof of Theorem 7.8

Formally, we want to test (7.35) with $u\gamma$ and (7.36) with γ. To make this formal calculation legitimate we shall use several approximations. We introduce first the mollified versions of the Navier–Stokes equations, namely,

$$-\Delta u_\epsilon + (u.Du)_\epsilon + Dp_\epsilon = f_\epsilon, \tag{7.39}$$

$$-\Delta p_\epsilon = (Du \circ Du)_\epsilon - \operatorname{div} f_\epsilon, \tag{7.40}$$

where ϵ denotes the parameter of the mollifier (see proof of Lemma 7.2). We let

$$w^\epsilon = p_\epsilon + \frac{1}{2}|u_\epsilon|^2.$$

Of course, w^ϵ is not the mollifier of w. However, it is easy to check that

$$w^\epsilon \leq c_0.$$

1. Preliminaries
Introduce the functions

$$g^\rho(r) = \frac{1}{1 + \rho r^2},$$

$$\ell^\delta(s) = \frac{1}{1 - \delta s}, \quad s \leq c_0, \quad \delta \leq \frac{1}{2c_0}.$$

Next define

$$G(x) = G^{\epsilon,\rho,\delta}(x) = g^\rho(|u_\epsilon(x)|)\ell^\delta(w^\epsilon(x)). \tag{7.41}$$

We can then multiply (7.39) by $Gu_\epsilon\gamma$ and (7.40) by $G\gamma$. Performing integrations by parts and adding up, we obtain

$$\int_\Omega Dw^\epsilon.D(G\gamma)\, dx + \int_\Omega |Du_\epsilon|^2 G\gamma\, dx$$

$$+ \int_\Omega (u.Du)_\epsilon u_\epsilon G\gamma\, dx + \int_\Omega u_\epsilon.Dp_\epsilon G\gamma\, dx \tag{7.42}$$

$$= \int_\Omega (Du \circ Du)_\epsilon G\gamma\, dx + \int_\Omega f_\epsilon.u_\epsilon G\gamma\, dx + \int_\Omega f_\epsilon.D(G\gamma)\, dx.$$

To deduce the basic inequality from (7.42), we treat the terms involving DG as follows:

$$\int_\Omega (Dw^\epsilon - f_\epsilon).DG\gamma \, dx \geq -\frac{2\rho^2}{\delta} \int_\Omega (g^\rho)^3 |u_\epsilon|^2 |Du_\epsilon|^2 \, dx$$

$$-\frac{\delta}{2} \int_\Omega G\ell^\delta |f_\epsilon|^2 \, dx + 2\rho \int_\Omega Gg^\rho u_\epsilon.Du_\epsilon.f_\epsilon \, dx.$$

Thus we state the basic inequality

$$\int_\Omega GDw^\epsilon.D\gamma \, dx + \int_\Omega |Du_\epsilon|^2 G\gamma \, dx$$

$$+ \int_\Omega (u.Du)_\epsilon u_\epsilon G\gamma \, dx + \int_\Omega u_\epsilon.Dp_\epsilon G\gamma \, dx$$

$$\leq \int_\Omega (Du \circ Du)_\epsilon G\gamma \, dx + \int_\Omega f_\epsilon.u_\epsilon G\gamma \, dx + \int_\Omega Gf_\epsilon.D\gamma \, dx \qquad (7.43)$$

$$+ \frac{2\rho^2}{\delta} \int_\Omega (g^\rho)^3 |u_\epsilon|^2 |Du_\epsilon|^2 \, dx + \frac{\delta}{2} \int_\Omega G\ell^\delta |f_\epsilon|^2 \, dx$$

$$- 2\rho \int_\Omega Gg^\rho u_\epsilon.Du_\epsilon.f_\epsilon \, dx.$$

2. Letting ϵ tend to 0

For fixed ρ, δ, we can let ϵ tend to 0 in (7.43) and obtain

$$\int_\Omega GDw.D\gamma \, dx + \int_\Omega |Du|^2 G\gamma \, dx + \int_\Omega u.DwG\gamma \, dx$$

$$\leq \int_\Omega Du \circ DuG\gamma \, dx + \int_\Omega f.uG\gamma \, dx + \int_\Omega Gf.D\gamma \, dx \qquad (7.44)$$

$$+ \frac{2\rho^2}{\delta} \int_\Omega (g^\rho)^3 |u|^2 |Du|^2 \, dx + \frac{\delta}{2} \int_\Omega G\ell^\delta |f|^2 \, dx$$

$$- 2\rho \int_\Omega Gg^\rho u.Du.f \, dx,$$

where it is clear that the arguments inside g^ρ, ℓ^δ, G are now u, w. Note that a term $\int_\Omega u.DwG\gamma \, dx$ has been obtained. This is the key term, where after integration by parts, we take advantage of the fact that div $u = 0$. We write it as

$$\int_\Omega u.DwG\gamma \, dx = \frac{1}{\delta} \int_\Omega \log(1 - \delta w) \, u.[D\gamma g^\rho - 2\rho u.Du(g^\rho)^2 \gamma] \, dx,$$

and thus (7.44) yields

$$\int_\Omega GDw.D\gamma \, dx + \int_\Omega |Du|^2 G\gamma \, dx + \frac{1}{\delta} \int_\Omega \log(1 - \delta w) \, u.D\gamma g^\rho \, dx$$

$$\leq \frac{2\rho}{\delta} \int_\Omega \log(1 - \delta w) \, (u.Du).u(g^\rho)^2 \gamma \, dx + \int_\Omega Du \circ DuG\gamma \, dx$$

$$+ \int_\Omega f.uG\gamma \, dx + \int_\Omega Gf.D\gamma \, dx + \frac{2\rho^2}{\delta} \int_\Omega (g^\rho)^3 |u|^2 |Du|^2 \, dx \qquad (7.45)$$

$$+ \frac{\delta}{2} \int_\Omega G\ell^\delta |f|^2 \, dx - 2\rho \int_\Omega Gg^\rho u.Du.f \, dx.$$

3. Letting ρ tend to 0

We let ρ tend to 0 in (7.45). By Lebesgue's theorem we obtain

$$\int_\Omega \ell^\delta Dw.D\gamma \, dx + \int_\Omega |Du|^2 \ell^\delta \gamma \, dx + \frac{1}{\delta} \int_\Omega \log(1 - \delta w) \, u.D\gamma \, dx$$
$$\leq \int_\Omega Du \circ Du \, \ell^\delta \gamma \, dx \qquad (7.46)$$
$$+ \int_\Omega f.u \, \ell^\delta \gamma \, dx + \int_\Omega \ell^\delta f.D\gamma \, dx + \frac{\delta}{2} \int_\Omega (\ell^\delta)^2 |f|^2 \, dx.$$

4. Letting δ tend to 0

We finally let δ tend to 0 in (7.46). The only term that warrants some care is

$$J^\delta = \frac{1}{\delta} \int_\Omega \log(1 - \delta w) \, u.D\gamma \, dx.$$

Note that

$$\frac{1}{\delta} \log(1 - \delta w) u.D\gamma \to -wu.D\gamma \quad \text{a.e. as } \delta \to 0.$$

Moreover, since

$$0 \leq \frac{1}{\delta} \log(1 - \delta w) \leq |w| \quad \text{for } w \leq 0$$

and

$$\left| \frac{1}{\delta} \log(1 - \delta w) \right| \leq 2c_0, \quad \forall 0 \leq \delta \leq \frac{1}{2c_0}, \quad 0 \leq w \leq c_0,$$

the integrand in J_δ is majorized by $C(|w| + 2c_0)|u|$, which is integrable. Hence from Lebesgue's dominated convergence theorem, we get

$$J^\delta \to - \int_\Omega wu.D\gamma \, dx.$$

From (7.46) we obtain (7.38). The proof is complete. $\Diamond$

7.4 Existence of a Regular Solution for $n \leq 5$

Up to here, we have proved that maximum-like solutions are regular in any dimension. The problem we consider now concerns finding maximum-like solutions.

We begin with some preliminaries concerning appropriate Green functions.

7.4.1 Green Function Associated with Incompressible Flows

Let $u \in (H_0^1(\Omega))^n$ such that div $u = 0$. Let $x_0 \in \Omega$, and let $\delta_h^{x_0}(x)$ be an approximation of the Dirac measure in x_0, namely

$$\delta_h^{x_0}(x) = \frac{1}{|B_h(x_0)|} \mathrm{I\!I}_{B_h(x_0)}(x), \quad h < \mathrm{dist}\,(x_0, \partial\Omega).$$

We consider $G_h^{x_0} = G$ to be defined by

$$-\Delta G_h - u.DG_h = \delta_h,$$
$$G_h = 0 \text{ on } \partial\Omega. \tag{7.47}$$

We shall prove only a priori estimates, some of which will depend on h. In fact, the complete proof requires us first to approximate u by a smooth divergence free flow, and proceed from there. We skip this step.

Theorem 7.9. *One has the estimates*

$$0 \le G_h \le C_h, \quad \|G_h\|_{H_0^1(\Omega)} \le C_h,$$

where the constant C_h depends only on h and not on u or x_0.

Proof. The second estimate is obtained just by testing with G_h. The positivity is obtained by testing with $(G_h)^-$. The L^∞ bound is obtained by the Moser iteration technique (see Section 2.3.1).   ◇

We next obtain estimates in higher Sobolev spaces, depending this time not only on h, but also on the norm of u in H_0^1.

Theorem 7.10. *One has the properties*

$$\|G_h\|_{H_{\text{loc}}^2}, \quad \|G_h\|_{W_{\text{loc}}^{1,4}} \le C_h, \tag{7.48}$$

where the local constant C_h depends on h and on the H_0^1 norm of u, but not on x_0.

Proof. Let ζ be a smooth cutoff function with compact support in Ω. We test with $-D_j(\zeta^2 D_j G)$ and obtain, after some integration by parts,

$$\int_\Omega \zeta^2 \|D^2 G\|^2 \, dx + \int_\Omega D_i D_j G D_i \zeta^2 D_j G \, dx + \int_\Omega u^i D_i \zeta^2 \frac{|DG|^2}{2} \, dx$$
$$= \int_\Omega \zeta^2 D_j u^i D_i G D_j G \, dx + \int_\Omega \delta_h D_j(\zeta^2 D_j G) \, dx. \tag{7.49}$$

Also,

$$\int_\Omega u^i D_i \zeta^2 \frac{|DG|^2}{2} \, dx$$
$$= -\frac{1}{2} \int_\Omega (D_i D_j \zeta^2 u^i D_j G + D_i \zeta^2 D_j u^i D_j G + D_i \zeta^2 u^i D_j D_j G) G \, dx,$$

which, thanks to the estimates proved in Theorem 7.9 together with (7.49), yields

$$\int_\Omega \zeta^2 \|D^2 G\|^2 \, dx \le \frac{C(h, \|u\|_{H_0^1}, \zeta)}{\delta} + \delta \int_\Omega \zeta^4 |DG|^4 \, dx, \tag{7.50}$$

where δ is arbitrary. Now using

$$\int_\Omega \zeta^4 |DG|^4 \, dx$$
$$= -2 \int_\Omega G[\zeta^4 \Delta G |DG|^2 + 2\zeta^4 D_i G D_j G D_i D_j G + |DG|^2 D_i G D_i \zeta^4] \, dx$$

we obtain

$$\int_\Omega \zeta^4 |DG|^4 \, dx = C(h) \int_\Omega \zeta^2 \|D^2 G\|^2 \, dx + C(h, \zeta).$$

This last estimate can be used in (7.50) with an appropriate δ to derive the result. $\diamond$

Finally, we prove L^q-estimates independent of h. They are basically the same as those of the Green function for the Laplace operator.

Theorem 7.11. *One has the estimates*

$$\|G_h\|_{L^p} \leq C_p, \quad \forall p < \frac{n}{n-2}, \tag{7.51}$$

$$\|DG_h\|_{L^q} \leq C_q, \quad \forall q < \frac{n}{n-1}, \tag{7.52}$$

where the constants are independent of h, u, x_0.

Proof. We test equation (7.47) with $G(1 + G^r)^{(-1)/r}$ and obtain

$$\int_\Omega \frac{|DG|^2}{(1 + G^r)^{(1+r)/r}} \, dx = \int_\Omega \delta_h \frac{G}{(1 + G^r)^{1/r}} \, dx \leq 1.$$

Next we have for any $\lambda > 0$,

$$\leq c \left(\int_\Omega |DG|^{n\lambda/(n+\lambda)} \, dx \right)^{(n+\lambda)/n\lambda},$$

and for $\lambda < \frac{2n}{n-2}$,

$$\left(\int_\Omega |G|^\lambda \, dx \right)^{1/\lambda}$$
$$\leq c \left(\int_\Omega \frac{|DG|^2}{(1 + G^r)^{(1+r)/r}} \, dx \right)^{1/2} \left(\int_\Omega (1 + G^r)^{\frac{n\lambda(r+1)}{r(2n+2\lambda-n\lambda)}} \, dx \right)^{(2n+2\lambda-n\lambda)/2n\lambda}$$

Noting that

$$\frac{2n + 2\lambda - n\lambda}{2n\lambda} < \frac{1}{\lambda}$$

we can guarantee that $\|G\|_{L^\lambda}$ is bounded, for

$$\lambda = \frac{n\lambda(r+1)}{2n + 2\lambda - n\lambda},$$

which implies $r < 1$ and

$$\lambda = \frac{n(1-r)}{n-2}.$$

This implies (7.51).

Next, taking $\mu < 2$, we can write

$$\left(\int_\Omega |DG|^\mu \, dx\right)^{1/\mu} \leq \left(\int_\Omega \frac{|DG|^2}{(1+G^r)^{(1+r)/r}} \, dx\right)^{1/2}$$
$$\left(\int_\Omega (1+G^r)^{\frac{\mu}{r(2-\mu)}} \, dx\right)^{(2-\mu)/2\mu},$$

and from (7.51) it is bounded if $\mu < n/(n-1)$. Hence (7.52) is obtained.

$\diamond$

We finally give a result that we can prove only when $n \leq 5$ (for $n = 6$, see [33] for a dimensional reduction argument). Basically, it says that away from the singularity x_0 the function G_h remains bounded independently of h.

Theorem 7.12. Let $R_0 < \frac{1}{2}\mathrm{diam}\,\Omega$ be given and $x_0, \xi_0 \in \Omega$, such that

$$|x_0 - \xi_0| > 2R_0.$$

Let $B_{R_0}(\xi_0)$ be the ball of center ξ_0 and radius R_0. Then one has the estimate

$$\|G_h\|_{L^\infty(B_{R_0}(\xi_0))} \leq C_{R_0}, \quad h < |x_0 - \xi_0| - 2R_0. \tag{7.53}$$

Proof.
1. *A general relation*
Consider a smooth function ζ with

$$0 \leq \zeta \leq 1, \quad \zeta\delta_h = 0. \tag{7.54}$$

Then we have, for $s \geq 1$,

$$\int_\Omega |D(\zeta G^{s/2})|^2 \, dx \leq 2 \int_\Omega G^s (\Delta\zeta^2 + |D\zeta|^2 - u.D\zeta^2) \, dx. \tag{7.55}$$

This relation is simply obtained by testing (7.47) with $G^{s-1}\zeta^2$, making use of the main assumption (7.54), and performing some easy computations. In fact, since $s - 1$ can be less than 1, and the gradient has to be calculated, one can begin with $(\epsilon + G)^{s-1}\zeta^2$, and then let ϵ tend to 0.

2. *Preparing for local Moser iteration*
Take $R \leq R_0/2$, and consider the balls $B_{R_0+R}(\xi_0)$ and $B_{R_0+2R}(\xi_0)$. Define a function ζ such that

$$\zeta = 1 \text{ on } B_{R_0+R}(\xi_0),$$
$$\zeta = 0 \text{ outside } B_{R_0+2R}(\xi_0),$$
$$0 \leq \zeta \leq 1; \; \zeta \text{ smooth},$$
$$|D\zeta| \leq \frac{C_{R_0}}{R}; \; |D^2\zeta| \leq \frac{C_{R_0}}{R^2}.$$

Pick $h < |x_0 - \xi_0| - 2R_0$. Then $B_h(x_0) \cap B_{R_0+2R}(\xi_0) = \emptyset$, and hence ζ satisfies (7.54).

We then apply (7.55), from which we deduce

$$\int_{\Omega} |D(\zeta G^{s/2})|^2 \, dx \leq C_{R_0} \frac{1}{R^2} \int_{B_{R_0+2R} \cap \Omega} G^s (1 + |u|) \, dx.$$

Since $|u| \in L^{2n/(n-2)}$, we deduce

$$\int_{\Omega} |D(\zeta G^{s/2})|^2 \, dx \leq C(R_0, \|u\|_{H_0^1}) \frac{1}{R^2} \left(\int_{B_{R_0+2R} \cap \Omega} G^{2ns/(n+2)} \, dx \right)^{(n+2)/2n}.$$

Using the Poincaré inequality for the left-hand side, we obtain

$$\left(\int_{B_{R_0+R} \cap \Omega} G^{ns/(n-2)} \, dx \right)^{(n-2)/n}$$

$$\leq C(R_0, \|u\|_{H_0^1}) \frac{1}{R^2} \left(\int_{B_{R_0+2R} \cap \Omega} G^{2ns/(n+2)} \, dx \right)^{(n+2)/2n}. \qquad (7.56)$$

The restriction on the dimension will now come from the condition

$$\frac{ns}{n-2} > \frac{2ns}{n+2},$$

which is required for applying Moser's iteration technique. Hence we need $n < 6$.

3. Applying Moser's iteration technique
We need a local variant of what has been presented in Section 2.3.1. First we deduce from (7.56) that

$$\|G\|_{L^{ns/(n-2)}(B_{R_0+R})} \leq C^{1/s} \left(\frac{1}{R} \right)^{2/s} \|G\|_{L^{2ns/(n+2)}(B_{R_0+2R})}. \qquad (7.57)$$

We now apply (7.57) with

$$R = R_{j+1} = \frac{R_0}{2^{j+1}}; \quad s = s_{j+1} = s_0 a^{j+1},$$

where

$$a = \frac{n+2}{2(n-2)} > 1.$$

Also, set

$$z_j = \|G\|_{L^{ns_j/(n-2)}(B_{R_0+R_j})}.$$

Then (7.57) reads

$$z_{j+1} \leq C^{1/(s_{j+1})} \left(\frac{1}{R_{j+1}} \right)^{2/(s_{j+1})} z_j.$$

Furthermore,

$$z_0 = \|G\|_{L^{ns_0/(n-2)}(B_{2R_0})},$$

and from (7.57), applied with $s = s_0$, $R = R_0$,

$$z_0 \leq \|G\|_{L^{2ns_0/(n+2)}(B_{3R_0})}.$$

But then we can choose $s_0 > 1$ such that

$$\frac{2ns_0}{n+2} < \frac{n}{n-2}.$$

Using then the result (7.51), we see that z_0 is finite. Since $a > 1$, the sequence z_j is bounded. Letting j tend to ∞, we obtain the result (7.53).   ◊

7.4.2 Approximation

Assume $n \leq 5$. We approximate the Navier–Stokes equations as follows:

$$u \in (H_0^1(\Omega))^n, \ \text{div} \ u = 0, \tag{7.58}$$

$$p \in W^{1, \frac{n}{(n-1)}}(\Omega), \tag{7.59}$$

$$-\Delta u + u.Du + \epsilon u|u|^2 + Dp = f. \tag{7.60}$$

The solution also satisfies estimates depending on ϵ:

$$\|u\|_{L^4(\Omega)} \leq \frac{K}{\epsilon^{1/4}}, \tag{7.61}$$

$$\epsilon \||u|u|^2\|_{L^{4/3}(\Omega)} \leq K\epsilon^{1/4}. \tag{7.62}$$

Moreover,

$$\|u\|_{H^{2,4/3}}, \|p\|_{W^{1,4/3}} \leq C(\epsilon). \tag{7.63}$$

We shall not detail the proof of results (7.58) to (7.63). They are obtained by applying the standard Galerkin method. The pressure equation is derived in a weak form as

$$-\Delta p = Du \circ Du + \text{div} \ (\epsilon u|u|^2 - f), \tag{7.64}$$

and we can use any function in $L^\infty \cap W_0^{1,4}(\Omega)$ as a test function.

7.4.3 Proof of Existence of a Maximum Solution for $n \leq 5$

We state our main result.

Theorem 7.13. *Assume Ω to be smooth and bounded, and $f \in (L^\infty(\Omega))^n$. If $n = 5$, there exists a weak solution of the stationary Navier–Stokes equations that is a maximum solution, hence satisfying (7.33), (7.34), (7.35), (7.36), (7.37). Consequently, this is also a fully regular solution, i.e.,*

$$u \in \left(W_{loc}^{2,s}(\Omega) \right)^n, \quad p \in W_{loc}^{1,s}(\Omega), \quad \forall 1 < s < \infty.$$

Proof. The existence of a weak solution is easily obtained from the approximation procedure u^ϵ, p^ϵ described above. The important point is to check that this approximation leads to a maximum solution. This is where the Green function intervenes. We of course consider the Green function $G = G_h$ associated with u^ϵ, but all the estimates obtained in Section 7.4.1 will be independent of ϵ, since we know that they depend at most on the H_0^1 norm of u. Moreover, thanks in particular to (7.48), we can take $uG\zeta^2$ as a test function in (7.60) and $G\zeta^2$ in the pressure equation (7.64), where ζ is a smooth function with compact support in Ω. Collecting results, we obtain

$$\int_\Omega Dw.D(G\zeta^2)\, dx + \int_\Omega |Du|^2 G\zeta^2\, dx + \int_\Omega u.DwG\zeta^2\, dx$$
$$+\epsilon \int_\Omega |u|^4 G\zeta^2\, dx + \epsilon \int_\Omega |u|^2 u.D(G\zeta^2)\, dx$$
$$= \int_\Omega Du \circ DuG\zeta^2\, dx + \int_\Omega f.uG\zeta^2\, dx + \int_\Omega f.D(G\zeta^2)\, dx.$$

At this stage we notice that

$$Du \circ Du - |Du|^2 \leq 0,$$

and hence from the positivity of the Green function we arrive at the inequality

$$\int_\Omega Dw.D(G\zeta^2)\, dx + \int_\Omega u.DwG\zeta^2\, dx$$
$$\leq -\epsilon \int_\Omega |u|^2 u.D(G\zeta^2)\, dx + \int_\Omega f.uG\zeta^2\, dx + \int_\Omega f.D(G\zeta^2)\, dx. \tag{7.65}$$

Since $w \in W^{1,\frac{4}{3}}$, we can derive from the equation of the Green function (7.47) that

$$\int_\Omega DG.D(w\zeta^2)\, dx + \int_\Omega Gu.D(w\zeta^2)\, dx = \int_\Omega \delta_h w\zeta^2\, dx. \tag{7.66}$$

Combining (7.65), (7.66) we obtain after rearrangements

$$\int_\Omega \delta_h w \zeta^2 \, dx \le -2 \int_\Omega GDw.D\zeta^2 \, dx - \int_\Omega wG\Delta(\zeta^2) \, dx$$
$$+ \int_\Omega wGu.D\zeta^2 \, dx - \epsilon \int_\Omega |u|^2 u.D(G\zeta^2) \, dx$$
$$+ \int_\Omega f.uG\zeta^2 \, dx + \int_\Omega f.D(G\zeta^2) \, dx,$$

and we pass to the limit for fixed h, as ϵ tends to 0. We obtain

$$\int_\Omega \delta_h w \zeta^2 \, dx \le -2 \int_\Omega GDw.D\zeta^2 \, dx - \int_\Omega wG\Delta(\zeta^2) \, dx$$

$$\tag{7.67}$$

$$+ \int_\Omega wGu.D\zeta^2 \, dx + \int_\Omega f.uG\zeta^2 \, dx + \int_\Omega f.D(G\zeta^2) \, dx,$$

where now u, w refers to the weak solution obtained in the limit. Note in particular that since $u \in L^{2n/(n-2)}$, $w \in L^{n/(n-2)}$, and G is bounded (for fixed h), the integral

$$\int_\Omega wGu.D\zeta^2 \, dx$$

is well-defined whenever $n \le 6$. Now we want to let h tend to 0. We cannot take arbitrary ζ. So we set

$$\Omega_0 = \{x| \text{ dist } (x, \partial\Omega) < \rho\},$$
$$\Omega_1 = \left\{x| \text{ dist } (x, \partial\Omega) < \frac{\rho}{2}\right\},$$

and take ζ to be a smooth cutoff function that takes the value 1 on Ω_1 and has compact support in Ω. We take $x_0 \in \Omega_0$. We can assert that on $\Omega - \Omega_1$, $G_h^{x_0}$ is bounded independently of h and $x_0 \in \Omega_0$. This follows from the fact that this set can be covered by a finite number of balls whose centers ξ_0 are such that

$$|x_0 - \xi_0| > \frac{\rho}{2},$$

and we can apply Theorem 7.12 to each of these balls. Then the right-hand side of (7.67) is bounded as h tends to 0. Note that for the integral

$$\int_\Omega f.uG\zeta^2 \, dx$$

the boundedness follows from (7.51), taking account of the fact that $n < 6$. Next, as h tends to 0,

$$\int_\Omega \delta_h w \zeta^2 \, dx \to w(x_0) \text{ a.e.,}$$

and thus we have obtained

$$w(x_0) \le C, \text{ a.e. } x_0 \in \Omega_0,$$

which proves that the weak solution is a maximum solution. The proof is complete.   ◊

7.5 Periodic Case: Existence of a Regular Solution for $n < 10$

7.5.1 Approximation

In fact, the latest result is $n = 15$.

We consider here $\Omega = (0, L)^n$ and denote the sides by

$$\Gamma_j = \partial\Omega \cap \{x_j = 0\}, \quad \Gamma_{j+n} = \partial\Omega \cap \{x_j = L\}.$$

We assume

$$f \in L^\infty(\Omega), \quad \int_\Omega f \, dx = 0. \tag{7.68}$$

The Navier–Stokes system reads

$$
\begin{aligned}
&u \in (H^1(\Omega))^n, \ p \in W^{1,\frac{n}{n-1}}(\Omega), \\
&u|_{\Gamma_j} = u|_{\Gamma_{j+n}}; \quad p|_{\Gamma_j} = p|_{\Gamma_{j+n}}, \\
&\frac{\partial u}{\partial x_k}\Big|_{\Gamma_j} = \frac{\partial u}{\partial x_k}\Big|_{\Gamma_{j+n}}, \ \forall j, k = 1, \dots, n, \\
&-\Delta u + u.Du + Dp = f, \\
&\text{div } u = 0, \\
&\int_\Omega u \, dx = 0.
\end{aligned}
\tag{7.69}
$$

We will construct a (weak) solution of (7.69) by the approximation procedure

$$
\begin{aligned}
&u \in (H^1(\Omega))^n \cap L^4(\Omega), \ p \in W^{1,\frac{4}{3}}(\Omega), \\
&u|_{\Gamma_j} = u|_{\Gamma_{j+n}}; \quad p|_{\Gamma_j} = p|_{\Gamma_{j+n}}, \\
&\frac{\partial u}{\partial x_k}\Big|_{\Gamma_j} = \frac{\partial u}{\partial x_k}\Big|_{\Gamma_{j+n}}, \ \forall j, k = 1, \dots, n, \\
&-\Delta u + u.Du + \epsilon|u|^2 u + Dp = f + \rho_\epsilon, \\
&\text{div } u = 0, \\
&\int_\Omega u \, dx = 0, \quad \rho_\epsilon = \frac{1}{|\Omega|}\int_\Omega \epsilon|u|^2 u \, dx.
\end{aligned}
\tag{7.70}
$$

The approximate solution is itself obtained by a standard Galerkin method, as in the Dirichlet case. By letting ϵ tend to 0 in the approximate solution, one obtains a weak solution of (7.69).

Note that we do not know whether the more natural approximation

$$-\Delta u + \epsilon\Delta^2 u + Dp = f$$

leads to a maximum solution.

7.5.2 A Specific Green Function

Comparing the present case with the Dirichlet case, the important point will be to show that the approximation procedure leads to a weak solution that is a "maximum-like solutions" satisfying the NS inequality (provided that $n < 10$). In the Dirichlet case (for $n = 5$) we showed that the approximation led to a maximum solution by using a specific Green function. In this case we use a specific Green function, of a different kind.

Let u, p be the solution of (7.69). Define

$$w = \frac{u^2}{2} + p,$$

and let $G = G^{\epsilon,\delta}$ be the solution of

$$
\begin{aligned}
&G \in H^2(\Omega) \cap W^{1,4}(\Omega) \cap L^\infty(\Omega), \\
&G|_{\Gamma_j} = G|_{\Gamma_{j+n}}; \quad \frac{\partial G}{\partial x_k}|_{\Gamma_j} = \frac{\partial G}{\partial x_k}|_{\Gamma_{j+n}}, \quad \forall j, k = 1, \ldots, n, \\
&-\Delta G - u.DG = \frac{w_+^s}{(1 + \delta w_+)^s} - \mu, \\
&\int_\Omega G \, dx = 0, \quad \mu = \frac{1}{|\Omega|} \int_\Omega \frac{w_+^s}{(1 + \delta w_+)^s} \, dx,
\end{aligned}
\tag{7.71}
$$

where $s \in R^+$ will be specified later, and $\delta > 0$ will tend to 0. The existence of G is obtained by first approximating u by a smooth function. We do not give the details of this step.

We begin with a lemma.

Lemma 7.14. *If $-1 < \lambda < 0$, then one has the estimate*

$$
\|G\|_{L^{\frac{n(\lambda+2)}{n-2}}} \leq c_0^2 \frac{(\lambda+2)^2}{2(\lambda+1)} \left\| \frac{w_+^s}{(1+\delta w_+)^s} \right\|_{\frac{n(\lambda+2)}{n+2(\lambda+1)}},
\tag{7.72}
$$

$$
\|DG\|_{L^1} \leq c_0^2 \frac{\lambda+2}{\lambda+1} |\Omega|^{\frac{n(\lambda+1)-\lambda}{n(\lambda+2)}} \left\| \frac{w_+^s}{(1+\delta w_+)^s} \right\|_{\frac{n(\lambda+2)}{n+2(\lambda+1)}},
\tag{7.73}
$$

where c_0 is the Poincaré constant.

Proof. We test (7.71) with $G|G|^\lambda$ and obtain

$$
(\lambda+1) \int_\Omega |DG|^2 |G|^\lambda \, dx = \int_\Omega \left(\frac{w_+^s}{(1+\delta w_+)^s} - \mu \right) G|G|^\lambda \, dx
$$

or

$$
\frac{4(\lambda+1)}{(\lambda+2)^2} \int_\Omega \left| D|G|^{\lambda+2/2} \right|^2 \, dx = \int_\Omega \left(\frac{w_+^s}{(1+\delta w_+)^s} - \mu \right) G|G|^\lambda \, dx.
$$

Using the Poincaré inequality yields

$$\frac{4(\lambda+1)}{(\lambda+2)^2}\left(\int_\Omega |G|^{\frac{n(\lambda+2)}{n-2}}\right)^{(n-2)/n} \leq c_0^2 \int_\Omega \frac{w_+^s}{(1+\delta w_+)^s}|G|^{\lambda+1}\,dx$$

$$+\frac{c_0^2}{|\Omega|}\int_\Omega \frac{w_+^s}{(1+\delta w_+)^s}\,dx \int_\Omega |G|^{\lambda+1}\,dx.$$

Using Hölder's inequality on the right-hand side, we deduce (7.72).
    Next we have

$$\int_\Omega |DG|\,dx \leq \left(\int_\Omega |DG|^2|G|^\lambda\,dx\right)^{1/2}\left(\int_\Omega |G|^{-\lambda}\,dx\right)^{1/2}.$$

Recall that from above,

$$\int_\Omega |DG|^2|G|^\lambda\,dx \leq \frac{2c_0^2}{\lambda+1}\left\||G|^{\lambda+1}\right\|_{\frac{n(\lambda+2)}{(\lambda+1)(n-2)}}\left\|\frac{w_+^s}{(1+\delta w_+)^s}\right\|_{\frac{n(\lambda+2)}{n+2(\lambda+1)}}.$$

This, combined with (7.72), implies (after easy calculations)

$$\int_\Omega |DG|^2|G|^\lambda\,dx \leq 2^{-\lambda}(c_0^2)^{\lambda+2}\frac{(\lambda+2)^{2(\lambda+1)}}{(\lambda+1)^{\lambda+2}}\left(\left\|\frac{w_+^s}{(1+\delta w_+)^s}\right\|_{\frac{n(\lambda+2)}{n+2(\lambda+1)}}\right)^{\lambda+2}.$$

Then

$$\int_\Omega |G|^{-\lambda}\,dx \leq \left(\int_\Omega |G|^{\frac{n(\lambda+2)}{n-2}}\,dx\right)^{\frac{-\lambda(n-2)}{n(\lambda+2)}}|\Omega|^{\frac{2n(\lambda+1)-2\lambda}{n(\lambda+2)}}.$$

Using (7.72) again to estimate the right-hand side and collecting results, we obtain (7.73).   ◇

We then prove the following result for the negative part of G.

Lemma 7.15.

$$\|G^-\|_{L^\infty} \leq C_p\|G^-\|_{L^p}, \quad \forall p > 1, \tag{7.74}$$

where C is independent of ϵ, δ.

Proof. We first notice that

$$\inf \frac{\int_\Omega |D((z^-)^q)|^2\,dx}{\int_\Omega ((z^-)^q)^2\,dx} > 0, \quad q \geq 1,$$

where the infimum is taken among all periodic z such that

$$\int_\Omega z\,dx = 0.$$

This can be checked by a standard argument, assuming the contrary and showing that it leads to a contradiction.

We now test (7.71) with $-(G^-)^\beta$, where $\beta > 1$. We obtain, neglecting a negative term on the right-hand side,

$$\beta \int_\Omega (G^-)^{\beta-1} |DG^-|^2 \, dx \leq |\mu| \int_\Omega (G^-)^\beta \, dx.$$

Hence

$$\int_\Omega |D((G^-)^{(\beta+1)/2})|^2 \, dx \leq |\mu|\beta \int_\Omega (G^-)^\beta \, dx.$$

Moreover, from what we said at the beginning, we can assert that

$$\int_\Omega |D((G^-)^{(\beta+1)/2})|^2 \, dx \geq c^2 \int_\Omega ((G^-)^{(\beta+1)/2})^2 \, dx.$$

From these two inequalities and the Sobolev embedding theorems, we deduce that

$$\left(\int_\Omega ((G^-)^{(\beta+1)/2})^{2n/(n-2)} \, dx \right)^{(n-2)/2n} \leq c_0 \left(|\mu|\beta \int_\Omega (G^-)^\beta \, dx \right)^{1/2}.$$

Let us set

$$H = \frac{G^-}{(1 + |\mu|)}.$$

Then also

$$\left(\int_\Omega H^{\frac{(\beta+1)n}{n-2}} \, dx \right)^{(n-2)/2n} \leq c_0 \beta^{1/2} \left(\int_\Omega H^\beta \, dx \right)^{1/2},$$

and we may assume that $c_0 > 1$.

We now are in place to apply Moser's iteration technique. Consider the sequence

$$\beta_{j+1} = (\beta_j + 1) \frac{n}{n-2}, \quad \beta_0 > 1, \quad \beta_0 \text{ given},$$

and set

$$z_j = \max(1, \|H\|_{\beta_j}).$$

Then from the above estimate, it follows that

$$z_{j+1} \leq (c_0^2 \beta_j)^{\frac{1}{\beta_{j+1}}} z_j,$$

and thus

$$z_j \leq C z_0 \quad \forall j.$$

Letting j tend to ∞ we get

$$\|H\|_\infty \leq C z_0,$$

and the result (7.74) follows at once.   ◇

7.5.3 Main Results

We begin by showing that the approximation (7.70) leads to a "maximum-like" solution; namely, we have the following result.

Lemma 7.16. *The weak solution u, p of (7.69) satisfies the estimate*

$$\|w_+\|_{L^q} \le c_q , \ \forall q < \infty, \ if \ n \le 6.$$
$$\|w_+\|_{L^{2n/(n-6)}} \le C, \ if \ n > 6. \tag{7.75}$$

Proof. Recall from (7.70) that

$$-\Delta u^\epsilon + u^\epsilon.Du^\epsilon + \epsilon|u^\epsilon|^2 u^\epsilon + Dp^\epsilon = f + \rho_\epsilon,$$

so that the pressure equation becomes

$$-\Delta p^\epsilon = Du^\epsilon \circ Du^\epsilon - \ \text{div} \ f + \ \text{div} \ (\epsilon|u^\epsilon|^2 u^\epsilon).$$

Testing the first relation with $u^\epsilon G^\epsilon$ (recall that $G^\epsilon = G^{\epsilon,\delta}$), and the pressure equation with G^ϵ, yields (after adding up; see also (7.4.3)),

$$\int_\Omega Dw^\epsilon.DG^\epsilon \, dx - \int_\Omega w^\epsilon u^\epsilon.DG^\epsilon \, dx + \int_\Omega \epsilon G^\epsilon |u^\epsilon|^4 \, dx$$
$$= \int_\Omega G^\epsilon (Du^\epsilon \circ Du^\epsilon - |Du^\epsilon|^2) \, dx + \int_\Omega f.DG^\epsilon \, dx + \int_\Omega u^\epsilon.fG^\epsilon \, dx$$
$$+ \int_\Omega u^\epsilon.\rho^\epsilon G^\epsilon \, dx - \int_\Omega \epsilon|u^\epsilon|^2 u^\epsilon.DG^\epsilon \, dx,$$

from which the following inequality follows:

$$\int_\Omega Dw^\epsilon.DG^\epsilon \, dx - \int_\Omega w^\epsilon u^\epsilon.DG^\epsilon \, dx \le \int_\Omega \epsilon G^\epsilon_- |u^\epsilon|^4 \, dx$$
$$- \int_\Omega G^\epsilon_- (Du^\epsilon \circ Du^\epsilon - |Du^\epsilon|^2) \, dx + \int_\Omega f.DG^\epsilon \, dx + \int_\Omega u^\epsilon.fG^\epsilon \, dx$$
$$+ \int_\Omega u^\epsilon.\rho^\epsilon G^\epsilon \, dx - \int_\Omega \epsilon|u^\epsilon|^2 u^\epsilon.DG^\epsilon \, dx.$$

Now using the equation of G^ϵ, (7.71), we get

$$\int_\Omega \frac{(w^\epsilon_+)^{s+1}}{(1 + \delta w^\epsilon_+)^s} \, dx \le \mu_{\epsilon,\delta} \int_\Omega w^\epsilon \, dx + \int_\Omega \epsilon G^\epsilon_- |u^\epsilon|^4 \, dx$$
$$- \int_\Omega G^\epsilon_- (Du^\epsilon \circ Du^\epsilon - |Du^\epsilon|^2) \, dx + \int_\Omega f.DG^\epsilon \, dx + \int_\Omega u^\epsilon.fG^\epsilon \, dx \tag{7.76}$$
$$+ \int_\Omega u^\epsilon.\rho^\epsilon G^\epsilon \, dx - \int_\Omega \epsilon|u^\epsilon|^2 u^\epsilon.DG^\epsilon \, dx.$$

We then use the estimates

$$\int_\Omega \epsilon |u^\epsilon|^4 \, dx \le C,$$

$$\int_\Omega |Du^\epsilon|^2 \, dx \le C,$$

$$\left(\int_\Omega \epsilon |u^\epsilon|^3 \, dx \right)^{4/3} \le C\epsilon^{1/4}, \tag{7.77}$$

$$\int_\Omega |w^\epsilon|^{n/(n-1)} \, dx \le C,$$

$$\int_\Omega |u^\epsilon|^{2n/(n-1)} \, dx \le C,$$

in (7.76) to deduce that

$$\int_\Omega \frac{(w_+^\epsilon)^{s+1}}{(1 + \delta w_+^\epsilon)^s} \, dx \le C \int_\Omega \frac{(w_+^\epsilon)^s}{(1 + \delta w_+^\epsilon)^s} \, dx + C \int_\Omega G_-^\epsilon \, dx$$

$$+ C \int_\Omega |DG^\epsilon| \, dx + C \left(\int_\Omega |G^\epsilon|^{2n/(n+2)} \, dx \right)^{(n+2)/2n} + \int_\Omega u^\epsilon . \rho^\epsilon G^\epsilon \, dx$$

$$- \int_\Omega \epsilon |u^\epsilon|^2 u^\epsilon . DG^\epsilon \, dx.$$

$$\tag{7.78}$$

Note first that

$$\int_\Omega \frac{(w_+^\epsilon)^s}{(1 + \delta w_+^\epsilon)^s} \, dx$$

$$\le C \left(\int_\Omega \frac{(w_+^\epsilon)^{s+1}}{(1 + \delta w_+^\epsilon)^s} \, dx \right)^{s/(s+1)} \left(\int_\Omega \frac{1}{(1 + \delta w_+^\epsilon)^s} \, dx \right)^{1/(s+1)}.$$

Therefore, the first term on the right-hand side of (7.78) can be absorbed by the left-hand side. Moreover, we use (7.72), (7.73), (7.74) to deduce from (7.78), provided that

$$\frac{(\lambda + 2)n}{n - 2} \ge \frac{2n}{n + 2}, \tag{7.79}$$

that

$$\int_\Omega \frac{(w_+^\epsilon)^{s+1}}{(1 + \delta w_+^\epsilon)^s} \, dx \le C_\lambda \left(\int_\Omega \left(\frac{(w_+^\epsilon)^s}{(1 + \delta w_+^\epsilon)^s} \right)^{\frac{n(\lambda+2)}{n+2(\lambda+1)}} \right)^{\frac{n+2(\lambda+1)}{n(\lambda+2)}} \, dx$$

$$+ \int_\Omega u^\epsilon . \rho^\epsilon G^\epsilon \, dx - \int_\Omega \epsilon |u^\epsilon|^2 u^\epsilon . DG^\epsilon \, dx. \tag{7.80}$$

Again using (7.77), and noting that for δ fixed, $G^{\epsilon,\delta}$ remains bounded in $W^{1,4}$, we can let ϵ tend to 0 in (7.80) to obtain

$$\int_\Omega \frac{(w_+)^{s+1}}{(1 + \delta w_+)^{s+1}} \, dx \le C_\lambda \left(\int_\Omega \left(\frac{(w_+)^s}{(1 + \delta w_+)^s} \right)^{\frac{n(\lambda+2)}{n+2(\lambda+1)}} \right)^{\frac{n+2(\lambda+1)}{n(\lambda+2)}} \, dx, \tag{7.81}$$

where w corresponds to the weak solution of (7.69). We then choose

$$s + 1 = \frac{n(\lambda + 2)}{n + 2(\lambda + 1)} s.$$

Hence

$$s = \frac{n + 2(\lambda + 1)}{(n - 2)(\lambda + 1)}.$$

After letting $\delta \to 0$ in (7.81) and using Fatou's lemma, this implies

$$\|w_+\|_{L^{\frac{n(\lambda + 2)}{(n-2)(\lambda + 1)}}} < \infty.$$

Since $\frac{n(\lambda + 2)}{(n-2)(\lambda + 1)}$ decreases as λ increases, we have to take λ as small as possible, and given the constraint (7.79) we must have

$$\lambda > -1, \quad \lambda \geq -\frac{8}{n + 2}.$$

Therefore, if $n \leq 6$, we may take $\lambda > -1$ arbitrary, and if $n > 6$, we have to take $\lambda = -8/(n + 2)$. Then (7.75) follows easily.   ◇

We then state the main result.

Theorem 7.17. *If (7.68) holds, and $n < 10$, then the weak solution of (7.69), obtained as limit of the approximation procedure (7.70), is regular in the sense that*

$$u \in (W^{2,p}(\Omega))^n, \quad p \in W^{1,p}(\Omega), \quad \forall 1 < p < \infty.$$

Proof. Since the solution is periodic, it is sufficient to prove local regularity in any bounded domain. From Theorem 7.7, we know, then, that it is enough to show that the weak solution u, p is a "maximum-like solutions" satisfying the NS inequality. From (7.6) and (7.75), we see that it is a "Maximum-Like Solutions" when $n \leq 6$, and for $n > 6$, we need

$$\frac{2n}{n - 6} > \frac{n}{2}.$$

But we have $n < 10$. So we know that (see Lemma 7.2)

$$w \in L^2(\Omega).$$

To show that the NS inequality (7.24) is satisfied, consider the approximate problem (7.70); it is easy to check that it satisfies the NS inequality, and that we can pass to the limit, which leads to the result. The proof is complete.

8. Strongly Coupled Elliptic Systems

8.1 Introduction

We shall consider systems of the type

$$-D_i a_i^\nu (x, Du) + f^\nu = 0, \quad \nu = 1, \ldots, N, \tag{8.1}$$

with

$$u = (u^1, \ldots, u^N).$$

For the study of interior regularity, we shall restrict ourselves for simplicity to the case of Dirichlet conditions

$$u = g, \text{ on } \partial\Omega, \tag{8.2}$$

where, to fix ideas, Ω is a smooth, bounded domain of R^n.

However, in practical applications, mixed boundary conditions are more realistic. We consider them at the end of this chapter.

Such systems are motivated in particular by the theory of nonlinear elasticity. Consider Lamé–Navier equations (see Remark 9.4)

$$-\mu \Delta u - (\lambda + \mu) D \text{ div } u + f = 0,$$

which describe the displacement $u(x)$ of the point x of an elastic body Ω under a volume force f. We shall consider a nonlinear version of this system, namely,

$$-D_i((\mu(Du)D_i u^\nu) + (\lambda(Du) + \mu(Du))D^\nu u^i) + f^\nu = 0, \tag{8.3}$$

which is an example of (8.1), with

$$a_i^\nu (x, Du) = \mu(Du)D_i u^\nu + (\lambda(Du) + \mu(Du))D^\nu u^i.$$

The model (8.3) corresponds to the physical situation of small displacements that are not infinitesimal. The case of large displacements leading to variational integrals with quasi-convex functionals are not considered here.

Existence and uniqueness of a solution of the system (8.1), (8.2) may be obtained easily, whenever we can use the theory of monotone operators. If we assume, for $s \geq 2$,

$a_i^\nu(x, p)$ is Caratheodory on $R^n \times R^{nN}$,
$|a_i^\nu(x, p)| \le \beta[K + |p|^{s-1}]$,
$(a_i^\nu(x, p) - a_i^\nu(x, \tilde{p})).(p_i^\nu - \tilde{p}_i^\nu) > 0, \quad \forall p \ne \tilde{p},$
$a_i^\nu(x, p).p_i^\nu \ge \alpha|p|^s,$

and

$$f \in (L^s(\Omega))^N, \quad g \text{ smooth},$$

then from the theory of monotone operators (see J.L. Lions [70]), we can assert that there exists one and only one solution of (8.1), (8.2) in $(W^{1,s}(\Omega))^N$.

8.2 H^2_{loc} and Meyers's Regularity Results

With natural regularity assumptions, we shall see that we can, in a straightforward way, get H^2_{loc} and Meyers's regularity for the gradient. We follow the approach of N.G. Meyers, E. Elcrat [77], taking comments of M. Giaquinta, G. Modica [43], [44] into account.

Let us assume that a_i^ν is C^1 and let us introduce the notation

$$a_{i,x_j}^\nu = \frac{\partial a_i^\nu}{\partial x_i},$$

$$a_{ij}^{\nu\,\mu} = \frac{\partial a_i^\nu}{\partial p_j^\mu},$$

where $i, j = 1, \ldots, n$, $\mu, \nu = 1, \ldots, N$.

We assume

$$\begin{aligned}
|a_i^\nu(x, p)| &\le b_i^\nu(x) + C|p|, \quad b_i^\nu(x) \in L^2(\Omega), \\
|a_{i,x_j}^\nu(x, p)| &\le b_{i,x_j}^\nu(x) + C|p|, \quad b_{i,x_j}^\nu(x) \in L^2(\Omega), \\
k_0|\xi|^2 &\le a_{ij}^{\nu\,\mu}(x, p)\xi_i^\nu \xi_j^\mu \le k_1|\xi|^2, \quad \forall \xi \in R^{nN}, \ k_0 > 0,
\end{aligned}$$ (8.4)

and also

$$f^\nu(x) \in L^2(\Omega).$$ (8.5)

We remark that by changing f^ν into $f^\nu - \sum_i a_{i,x_i}^\nu$, we may without loss of generality assume that

$$a_i^\nu(x, 0) = 0.$$ (8.6)

Our objective is to check the following classical theorem.

Theorem 8.1. *If we assume that (8.4) and (8.5) hold, then the solution of (8.1), (8.2) satisfies*

$$u \in (H^2)^N_{\mathrm{loc}}(\Omega).$$

Proof. We first introduce the subset Ω_ρ of Ω whose distance from the boundary of Ω is larger than ρ. Let ζ be a smooth cutoff function whose support is in Ω_ρ, is equal to 1 on $\Omega_{2\rho}$, and takes values in $[0, 1]$. We consider the finite difference operator

$$\Delta_z \phi(x) = \phi(x + z) - \phi(x).$$

Now choose $0 < h < \rho$, then the function

$$\Delta_{-he_k}(\zeta^2 \Delta_{he_k} u^\nu)$$

vanishes on the boundary, and is a possible test function for (8.1). Hence we have

$$\int_\Omega \Delta_{-he_k}(\zeta^2 \Delta_{he_k} u^\nu) D_i a_i^\nu(x, Du)\, dx = \int_\Omega \Delta_{-he_k}(\zeta^2 \Delta_{he_k} u^\nu) f^\nu\, dx.$$

We note that $D_i(\zeta^2 \Delta_{he_k} u^\nu)$ has support in Ω_ρ, and so we can perform integration by parts and interchange finite differences to obtain

$$-\int_\Omega D_i(\zeta^2 \Delta_{he_k} u^\nu)\, \Delta_{he_k} a_i^\nu(x, Du)\, dx = \int_\Omega \Delta_{-he_k}(\zeta^2 \Delta_{he_k} u^\nu) f^\nu\, dx. \quad (8.7)$$

From assumptions (8.4), we can write

$$\Delta_{he_k} a_i^\nu = h \int_0^1 a_{i,x_k}^\nu(x + \theta h e_k, Du(x + h e_k))\, d\theta$$

$$+ \int_0^1 a_{i\,j}^{\nu\,\mu}(x, Du + \theta \Delta_{he_k} Du) \Delta_{he_k} D_j u^\mu\, d\theta.$$

Set

$$m_{i\,j}^{\nu\,\mu} = \int_0^1 a_{i\,j}^{\nu\,\mu}(x, Du + \theta \Delta_{he_k} Du)\, d\theta.$$

Using this expression in (8.7), and making use of our assumptions, we have

$$k_0 \int_\Omega \zeta^2 |\Delta_{he_k} Du|^2\, dx \le Ch \int_\Omega \zeta^2 |\Delta_{he_k} Du|$$

$$\left(\int_0^1 \sum_{i,\nu} b_{i,x_k}^\nu(x + \theta h e_k)\, d\theta + |Du(x + h e_k)| \right) dx$$

$$+ C \int_\Omega \zeta |D\zeta| |\Delta_{he_k} u| \left(|\Delta_{he_k} Du| + h \right.$$

$$\left(\int_0^1 \sum_{i,\nu} b_{i,x_k}^\nu(x + \theta h e_k)\, d\theta \right.$$

$$\left. \left. + |Du(x + h e_k)| \right) \right) dx - \int_\Omega \Delta_{-he_k}(\zeta^2 \Delta_{he_k} u^\nu) f^\nu\, dx.$$

From assumptions (8.4), (8.5) we obtain

$$\frac{k_0}{2} \int_\Omega \zeta^2 |\Delta_{he_k} Du|^2 \, dx \leq Ch^2 + I,$$

where we have set

$$I = - \int_\Omega \Delta_{-he_k} (\zeta^2 \Delta_{he_k} u^\nu) f^\nu \, dx.$$

Now we may write

$$I = h \int_0^1 \left[\int_{\Omega - \theta h e_k} f^\nu (x + \theta h e_k) D_k (\zeta^2 \Delta_{he_k} u^\nu)(x) \, dx \right] d\theta,$$

and it follows that

$$|I| \leq C \frac{h^2}{\eta} + \eta \int_\Omega \zeta^2 |\Delta_{he_k} Du|^2 \, dx,$$

for any η. Picking η sufficiently small and collecting results yields

$$\int_\Omega \zeta^2 |\Delta_{he_k} Du|^2 \, dx \leq Ch^2.$$

Letting h tend to 0, we obtain the result.   ◇

For the Dirichlet problem the H^2 regularity holds up to the boundary when the domain is smooth:

Theorem 8.2. *If we assume (8.4), (8.5), with $g = 0$ and a smooth domain Ω, then the solution of (8.1), (8.2) satisfies*

$$u \in (H^2)^N(\Omega). \tag{8.8}$$

Proof. We only sketch the proof, which is a standard extension of that of Theorem 8.1. We use local charts, and after straightening the boundary, we reduce the problem to the following one. The domain is

$$\Omega = \tilde{\Omega} \cap \{x_n > 0\},$$
$$\tilde{\Omega} - \Omega = (\tilde{\Omega} \cap \{x_n = 0\}) \cup \{(x_t, -x_n) | x = (x_t, x_n) \in \Omega\},$$

in which $t = 1, \ldots, n - 1$ and where $\tilde{\Omega}$ is a bounded smooth domain of R^n. One introduces the sets $\tilde{\Omega}_\rho$, and ζ the smooth cutoff function whose support is in $\tilde{\Omega}_\rho$, which is equal to 1 on $\tilde{\Omega}_{2\rho}$, and $0 \leq \zeta \leq 1$. Of course,

$$\zeta = 1 \text{ on } \tilde{\Omega}_{2\rho} \cap \{x_n = 0\}.$$

For $k = 1, \ldots, n-1$, one can use the test function

$$\Delta_{-he_k}(\zeta^2 \Delta_{he_k} u^\nu)$$

and proceed exactly as in the proof of Theorem 8.1. In particular, it follows that

$$\int_{\{\tilde{\Omega}_{2\rho} \cap \{x_n > 0\}\}} |D_k Du|^2 \, dx \leq C_\rho. \tag{8.9}$$

To estimate $D_n D_n u$ one uses equations (8.1). Namely, one has

$$a_{n\,n}^{\nu\,\mu}(x, Du) D_n D_n u^\mu = f^\nu - a_{i\,,x_i}^\nu(x, Du)$$

$$- \sum_{i=1}^{n} \sum_{j=1}^{n-1} a_{i\,j}^{\nu\,\mu}(x, Du) D_i D_j u^\mu - \sum_{i=1}^{n-1} a_{i\,n}^{\nu\,\mu}(x, Du) D_i D_n u^\mu.$$

Since the matrix $a_{n\,n}^{\nu\,\mu}$ is uniformly coercive, it follows easily from the above relations that (8.9) also holds for $k = n$. Thus we have the result.   ◇

In the case of a scalar equation, the regularity $H^2(\Omega)$ is valid for *convex* bounded domains (see J. Kadlec [61]). The property has been extended to some systems (see J. Saranen [89]).

Unfortunately, in the nonlinear case $n \geq 3$, one cannot estimate the higher derivatives of the solution, although in view of the following theorem, one would expect that $D^2 u$ is still in some Nikolskii space, i.e.,

$$\sup_{h>0} |D^2 u(x + he) - D^2 u(x)|_{L^2} \leq K.$$

We indeed can complete Theorem 9.2 with Meyers's regularity result. We assume

$$b_{i,x_k}^\nu, f^\nu \in L^{2n/(n-2)}(\Omega). \tag{8.10}$$

Then we have the following result:

Theorem 8.3. *If we assume (8.4), (8.10), then the solution of (8.1), (8.2) satisfies*

$$u \in (W^{2,p_0})_{\text{loc}}^N(\Omega) \tag{8.11}$$

for some $2 < p_0 < \frac{2n}{n-2}$.

Proof. Take a ball $B_{2R}(x_0)$ whose closure is contained in Ω. Let τ_R be the usual cutoff function, whose support is in $B_{2R}(x_0)$, is equal to 1 in $B_R(x_0)$, and satisfies

$$0 \leq \tau_R \leq 1, \quad |D\tau_R| \leq \frac{C}{R}.$$

We denote by $c_{k,R}^\nu$ the average of $D_k u^\nu$ on B_{2R}. From Theorem 8.1 we can use

$$D_k(\tau_R^2(D_k u^\nu - c_{k,R}^\nu))$$

as a test function in (8.1). We get the relation

$$\int_\Omega D_k a_i^\nu(x, Du) \, D_i(\tau_R^2(D_k u^\nu - c_{k,R}^\nu)) \, dx = \int_\Omega f^\nu D_k(\tau_R^2(D_k u^\nu - c_{k,R}^\nu)) \, dx,$$

which, if we perform the differentiations, amounts to

$$\int_\Omega (a_{i,x_k}^\nu + a_{ij}^{\nu\mu} D_j D_k u^\mu)[\tau_R^2 D_i D_k u^\nu + 2\tau_R D_i \tau_R (D_k u^\nu - c_{k,R}^\nu)] \, dx$$
$$= \int_\Omega f^\nu D_k(\tau_R^2(D_k u^\nu - c_{k,R}^\nu)) \, dx.$$

From the assumptions, it follows that

$$k_0 \int_{B_R} |D_k Du|^2 \, dx \le C \int_{B_{2R}} \left(\sum_{i,\nu} (b_{i,x_k}^\nu)^2 + |Du|^2 + \sum_\nu (f^\nu)^2 \right) dx$$
$$+ \frac{C}{R^2} \int_{B_{2R}} |D_k u - c_{k,R}|^2 \, dx.$$

Using the Poincaré inequality, we get

$$k_0 \int_{B_R} |D_k Du|^2 \, dx \le C \int_{B_{2R}} \left(\sum_{i,\nu} (b_{i,x_k}^\nu)^2 + |Du|^2 + \sum_\nu (f^\nu)^2 \right) dx$$
$$+ \frac{C}{R^2} \left(\int_{B_{2R}} |D_k Du|^{2n/(n+2)} \, dx \right)^{(n+2)/n}.$$

We thus have obtained condition (1.88) applied to the function $(\sum_\nu (D_k u^\nu)^2)^{1/2}$. We can conveniently extend this function as an H^1 function, since we are looking for only a local result. Using Theorem 8.1 and Sobolev embedding, we have $Du \in L_{\text{loc}}^{2n/(n-2)}(\Omega)$. Then from assumption (8.10), the condition (1.89) is also satisfied. Thus applying Theorem 1.15, the result follows. ◇

If the dimension is 2, then the preceding Meyers's result yields $C^{1+\delta}$ local regularity. More precisely, we state the following theorem.

Theorem 8.4. *If we assume $n = 2$, (8.4), and*

$$b_{i,x_k}^\nu, f^\nu \in L_{\text{loc}}^\infty(\Omega), \tag{8.12}$$

then the solution of (8.1), (8.2) satisfies

$$u \in (C^{1+\delta})_{\text{loc}}^N(\Omega). \tag{8.13}$$

This is an obvious consequence of Theorem 8.3 and the Sobolev embedding theorem. We have

$$\delta = 1 - \frac{2}{p_0}.$$

$\diamond$

Having obtained $C^{1+\delta}$ regularity, one may apply the linear theory to derive higher regularity. Thus in the case of dimension 2, there is a theory providing full regularity if the data are smooth. For $n \geq 3$, there are counterexamples; see Section 8.7. However, one can emphasize a comment by J. Nečàs that for $n = 3, 4$ in the variational case the problem is still open.

8.3 Hölder Regularity

Under the assumptions (8.4) the solution of (8.1) is, in general, not Hölder. A. Koshelev [63] has given additional assumptions under which Hölder regularity can be proven. We shall present next the main ideas and results of his theory.

8.3.1 Preliminaries

We consider the system (8.1), (8.2), with

$$g = 0$$

to simplify. We make assumptions (8.4), (8.5) and the convenient change to have also (8.6). We define

$$L^\nu(u) = -D_i a_i^\nu(x, Du) + f^\nu.$$

We shall construct the solution of

$$L^\nu(u) = 0$$

by the following iteration:

$$-\Delta u_{k+1}^\nu = -\Delta u_k^\nu - \epsilon L^\nu(u_k),$$
$$u_{k+1}^\nu = 0, \text{ on } \partial\Omega, \tag{8.14}$$

for some convenient ϵ to be made precise later. We take for initial function

$$u_0 = 0.$$

The choice of ϵ will be made so that the iterative process above will converge, geometrically. We first notice that we can write

$$-\Delta(u_{k+1}^\nu - u_k^\nu) = -\Delta(u_k^\nu - u_{k-1}^\nu) + \epsilon D_i \left(a_{ij,k}^{\nu\,\mu} D_j(u_k^\mu - u_{k-1}^\mu) \right),$$
$$u_{k+1}^\nu - u_k^\nu = 0, \text{on } \partial\Omega, \tag{8.15}$$

where we have set

$$a_{ij,k}^{\nu\mu}(x) = \int_0^1 a_{ij}^{\nu\mu}(x, Du_{k-1} + \theta(Du_k - Du_{k-1}))d\theta.$$

We then write (8.15) under variational formulation as follows:

$$\int_\Omega D(u_{k+1}^\nu - u_k^\nu)Dv^\nu\, dx = \int_\Omega D(u_k^\nu - u_{k-1}^\nu)Dv^\nu\, dx$$
$$\tag{8.16}$$
$$-\epsilon \int_\Omega a_{ij,k}^{\nu\mu}(x)D_j(u_k^\mu - u_{k-1}^\mu)D_i v^\nu\, dx,$$

where the functions v^ν are in $H^1(\Omega)$. We use next a matrix form, introducing

$$U_k = D_i u_k^\nu, \quad V = D_i v^\nu,$$

which are vectors of R^{nN}. Similarly, we use the notation

$$A_k(x) = a_{ij,k}^{\nu\mu}(x),$$

which belongs to the family

$$A(x,p) = a_{ij}^{\nu\mu}(x,p).$$

With this notation, (8.16) becomes

$$\int_\Omega (U_{k+1} - U_k)V\, dx = \int_\Omega (I - \epsilon A_k)(U_k - U_{k-1})V\, dx. \tag{8.17}$$

Let us consider the matrix $A = A(x,p)$, and define its symmetric and antisymmetric parts

$$A^+ = \frac{A + A^*}{2}, \quad A^- = \frac{A - A^*}{2}.$$

Thanks to assumption (8.4) the eigenvalues $\lambda_i = \lambda_i(x,p)$ of A^+ satisfy

$$k_0 \le \lambda_i \le k_1.$$

So we can introduce the numbers

$$\lambda = \inf_{i,x,p} \lambda_i(x,p), \quad \Lambda = \sup_{i,x,p} \lambda_i(x,p).$$

Let us also consider the matrix

$$C = A^*A - (A^+)^2$$

and define

$$\sigma = \sup_{\xi,x,p} \frac{\xi^*C(x,p)\xi}{|\xi|^2}.$$

Since the elements of the matrix A are bounded, the value of σ is finite. Moreover, it is easy to check that

$$\sigma > -\Lambda^2.$$

We begin with a lemma.

Lemma 8.5. *We have the relation*

$$\sup_{x,p} \|I - \epsilon A(x,p)\|^2 \leq K_\epsilon^2 = \max\left\{(1-\epsilon\lambda)^2, (1-\epsilon\Lambda)^2\right\} + \epsilon^2\sigma.$$

Proof. It is easy to check that

$$\|I - A(x,p)\|^2 \leq \|I - \epsilon A^+(x,p)\|^2 + \epsilon^2\sigma$$
$$\leq \sup_{i,x,p}(1 - \epsilon\lambda_i(x,p))^2 + \epsilon^2\sigma,$$

and the result follows. ◇

It is important now to check that the minimum of K_ϵ^2 in ϵ is attained and is strictly less than 1. First we notice that

$$K_\epsilon^2 = \begin{cases} (1-\epsilon\lambda)^2 + \epsilon^2\sigma & \text{if } 0 \leq \epsilon \leq \dfrac{2}{\lambda+\Lambda} = \epsilon^*, \\ (1-\epsilon\Lambda)^2 + \epsilon^2\sigma & \text{if } \epsilon \geq \epsilon^*. \end{cases}$$

We then state the following result, whose proof is left as an exercise.

Proposition 8.6. *We have*

$$K^2 = \inf_\epsilon K_\epsilon^2 = \begin{cases} \dfrac{\sigma}{\sigma+\lambda^2} & \text{if } \sigma \geq \dfrac{\lambda(\Lambda-\lambda)}{2}, \\ \dfrac{(\Lambda-\lambda)^2 + 4\sigma}{(\Lambda+\lambda)^2} & \text{if } \sigma \leq \dfrac{\lambda(\Lambda-\lambda)}{2}. \end{cases} \tag{8.18}$$

Remark 8.7. When the matrix A is symmetric, $\sigma = 0$, and

$$K = \frac{\Lambda-\lambda}{\Lambda+\lambda}.$$

From now on, ϵ is chosen such that

$$K_\epsilon = K.$$

Turning now to equation (8.17), and chosing

$$V = U_{k+1} - U_k,$$

we deduce easily

$$\|U_{k+1} - U_k\|_{(L^2(\Omega))^{nN}} \leq K^k \|U_1\|_{(L^2(\Omega))^{nN}}. \tag{8.19}$$

Of course, we obtain that the sequence u_k satisfies

$$u_k \to u \text{ in } (H^1(\Omega))^N \text{ as } k \to \infty.$$

Remark 8.8. When additional assumptions are made (see below) the above iteration will have convergence properties in other spaces, in particular in a space implying the convergence in C^α.

8.3.2 Representation Using Spherical Functions

Spherical functions will represent an important tool in the following. Let us consider the unit sphere of R^n, and let

$$\theta = (\theta_1, \ldots, \theta_{n-1})$$

be the spherical system of coordinates of a point in the unit sphere. Let Δ' be the Laplace–Beltrami operator on the unit sphere S. The eigenvalues of this operator are the values $s(s + n - 1), s = 0, 1, \ldots$. The corresponding eigenfunctions denoted by $Y_{s,i}, i = 1, \ldots, k_s$ satisfy

$$\Delta' Y_{s,i} = s(s + n - 1) Y_{s,i}.$$

Note that

$$Y_{0,1} = Y_0 = \text{constant} = \frac{1}{\sqrt{|S|}}, \quad k_0 = 1.$$

So by the orthogonality condition,

$$\int_S Y_{s,i}(\theta) d\theta = 0, \ \forall s \geq 1.$$

Next, let us consider a ball $B_{\delta_0}(x_0)$. We can use the system of spherical coordinates (r, θ) to represent the points of this ball. Then we can expand functions defined on the ball $B_{\delta_0}(x_0)$ as a series

$$\phi(x) = \phi(r, \theta) = \sum_{s=0}^{+\infty} \sum_{i=1}^{k_s} \phi_{s,i}(r) Y_{s,i}(\theta).$$

The functions $\phi_{s,i}(r)$ are defined by the formulas

$$\phi_{s,i}(r) = \int_S \phi(r, \theta) Y_{s,i}(\theta) dS.$$

Since

$$\phi(0, \theta) = \phi(x_0),$$

we have from the above

$$\phi_{s,i}(0) = 0 \quad \forall s \geq 1.$$

For a function in $L^2(B_{\delta_0}(x_0))$, we have the identity

$$\int_{B_{\delta_0}(x_0)} |\phi(x)|^2 \, dx = \sum_{s,i} \int_0^{\delta_0} |\phi_{s,i}(r)|^2 r^{n-1} \, dr.$$

Next we state the following result:

Proposition 8.9. *If we assume $\phi \in H^1(B_{\delta_0}(x_0))$, then one has the identity*

$$\int_{B_{\delta_0}(x_0)} |D\phi(x)|^2 \, dx = \sum_{s,i} \int_0^{\delta_0} \left[|\phi_{s,i}(r)'|^2 + \frac{s(s+n-2)}{r^2} |\phi_{s,i}(r)|^2 \right] r^{n-1} \, dr,$$

(8.20)

where $\displaystyle\sum_{s,i} = \sum_{s=0}^{+\infty} \sum_{i=1}^{k_s}.$

Proof. We need to recall the expression of the Laplacian in spherical coordinates; namely,

$$\Delta = r^{1-n} \frac{\partial}{\partial r} \left(r^{n-1} \frac{\partial}{\partial r} \right) - \frac{1}{r^2} \Delta'.$$

We begin by proving that (8.20) holds when ϕ is smooth. Indeed, let $\xi(r)$ be any C^1 function that vanishes for $r = \delta_0$ and $r = 0$. We shall check that

$$\int_{B_{\delta_0}(x_0)} |D\phi(x)|^2 \xi^2(r) \, dx$$
$$= \sum_{s,i} \int_0^{\delta_0} \left[|\phi_{s,i}(r)'|^2 + \frac{s(s+n-2)}{r^2} |\phi_{s,i}(r)|^2 \right] r^{n-1} \xi^2(r) \, dr.$$

(8.21)

If this is true, (8.20) will follow easily by taking an increasing sequence of functions ξ converging a.e. to 1, using the monotone convergence property of integrals. Now, thanks to the smoothness of ϕ, the left-hand side of (8.21) becomes

$$= -\int_{B_{\delta_0}(x_0)} \phi(\Delta\phi\xi^2 + 2D\phi(x)D\xi\xi) \, dx$$
$$= -\int_{B_{\delta_0}(x_0)} \phi\left(\Delta\phi\xi^2 + 2\xi\xi' \frac{\partial\phi}{\partial r} \right) dx,$$

and using the representation formula we obtain

$$\int_{B_{\delta_0}(x_0)} |D\phi(x)|^2 \xi^2(r) \, dx = -\sum_{s,i} \int_0^{\delta_0} \phi_{s,i}(r^{n-1}\phi'_{s,i})' \xi^2 dr$$
$$+ \sum_{s,i} \int_0^{\delta_0} \frac{s(s+n-2)}{r^2} \phi_{s,i}^2 r^{n-1} \xi^2 dr$$
$$- \sum_{s,i} \int_0^{\delta_0} 2\xi\xi' \phi_{s,i}\phi'_{s,i} r^{n-1} dr.$$

Performing an integration by parts, one obtains (8.21). To prove the result for H^1, we proceed by approximation. Let us write, to simplify notation,

$$\|\phi\|^2 = \int_{B_{\delta_0}(x_0)} |D\phi(x)|^2 \, dx.$$

and

$$\|\phi\|^2 = \sum_{s,i} \int_0^{\delta_0} \left[|\phi_{s,i}(r)'|^2 + \frac{s(s+n-2)}{r^2} |\phi_{s,i}(r)|^2 \right] r^{n-1} \, dr.$$

These seminorms coincide for $\phi \in C^\infty(\overline{B_{\delta_0}(x_0)})$. Of course, the first is finite for any $\phi \in H^1(B_{\delta_0}(x_0))$. We first check that the second is also finite, and moreover,

$$\|\|\phi\|\| \le \|\phi\|. \tag{8.22}$$

We use the notation

$$\|\|\phi\|\|_S^2 = \sum_{s=0}^S \sum_1^{k_s} \int_0^{\delta_0} \left[|\phi_{s,i}(r)'|^2 + \frac{s(s+n-2)}{r^2} |\phi_{s,i}(r)|^2 \right] r^{n-1} \, dr.$$

Now let ϕ^k be a sequence of smooth functions converging to ϕ in H^1. Since it is a Cauchy sequence in H^1 and the functions are smooth, we assert easily that

$$\phi_{s,i}^k(r)' r^{(n-1)/2} \quad \text{and} \quad \phi_{s,i}^k(r) r^{(n-3)/2}$$

converge in $L^2(0, \delta_0)$. It easy to check that the limits are necessarily

$$\phi_{s,i}(r)' r^{(n-1)/2} \quad \text{and} \quad \phi_{s,i}(r) r^{(n-3)/2}.$$

Hence for any S,

$$\|\|\phi^k\|\|_S \to \|\|\phi\|\|_S.$$

On the other hand,

$$\|\|\phi^k\|\|_S \le \|\|\phi^k\|\| = \|\phi^k\|,$$

and letting $k \to \infty$ yields

$$\|\|\phi\|\|_S \le \|\phi\|$$

and letting S go to infinity, the result (8.22) follows. Let us define finally

$$\phi_S = \sum_{s=0}^S \sum_{i=1}^{k_s} \phi_{s,i}(r) Y_{s,i}(\theta).$$

Clearly,

$$\|\|\phi\|\|_S = \|\|\phi_S\|\|.$$

Considering the smooth approximation ϕ^k, we then can write

$$\|\|\phi^k\|\|_S = \|\|\phi_S^k\|\| = \|\phi_S^k\|,$$

and thus the sequence ϕ_S^k is also Cauchy in H^1, and thus converges to a limit, which is necessarily ϕ_S, since for the norm $\|\| \ \|\|$ it already converges to it. Therefore, necessarily

$$\|\|\phi_S\|\| = \|\phi_S\|.$$

If we now let S tend to infinity, we notice that ϕ_S remains in a bounded subset of H^1, and thus at least for a subsequence it converges weakly to a limit, which is necessarily ϕ, since it converges to it in L^2. From the lower semicontinuity of the norm it follows that

$$\|\phi\| \leq \liminf \|\!|\phi|\!\|_S \leq \|\!|\phi|\!\|.$$

Comparing to (8.22) the result follows.   ◇

In a similar way one can assert the formula

$$\int_{B_{\delta_0}(x_0)} |D\phi(x)|^2 r^\gamma \, dx$$

$$= \sum_{s,i} \int_0^{\delta_0} \left[|\phi_{s,i}(r)'|^2 + \frac{s(s+n-2)}{r^2} |\phi_{s,i}(r)|^2 \right] r^{\gamma+n-1} \, dr$$

(8.23)

for a real γ when the left-hand side is finite.

8.3.3 Statement of the Main Result

We shall make the following assumptions on the constant K defined in equation (8.18). We assume that there exists γ such that

$$-\frac{1}{2}\left(n - 2 + \sqrt{n^2 + 4n - 4}\right) \leq \gamma < -(n-2)$$

(8.24)

and

$$K_0 = K \frac{\sqrt{1 - \gamma \frac{n-2}{n-1}}}{1 - \frac{\gamma(\gamma+n-2)}{2(n-1)}} < 1.$$

(8.25)

We also assume

$$f^\nu(x) \in L^p(\Omega), \quad p \text{ sufficiently large.}$$

(8.26)

The main result is then stated as follows.

Theorem 8.10. *If we assume (8.4), (8.24), (8.25), and (8.26), then the solution of (8.1) (with $g = 0$) satisfies*

$$u \in ((C^\alpha_{\text{loc}}(\Omega))^N$$

with

$$\alpha = \frac{2 - n - \gamma}{2}.$$

Remark 8.11. With the choice of γ in the interval (8.24), one has

$$1 - \frac{\gamma(\gamma + n - 2)}{2(n - 1)} > 0.$$

Hence the definition of K_0 makes sense.

To prepare for the proof, we shall introduce some notation related to the use of Spherical functions. Let us define, for functions of one variable r, the expressions

$$I_s(\phi, \psi) = \int_0^{\delta_0} \left[\phi' \psi' + \frac{s(s + n - 2)}{r^2} \phi \psi \right] r^{n-1} \, dr$$

and

$$J_{\gamma,s}(\psi) = \int_0^{\delta_0} \left[\psi'^2 + \frac{s(s + n - 2)}{r^2} \psi^2 \right] r^{-\gamma + n - 1} \, dr.$$

We denote by $H^1_{\gamma,s}(0, \delta_0)$ the Hilbert space of functions ψ such that

$$J_{\gamma,s}(\psi) < \infty$$

with

$$\|\psi\|_{\gamma,s} = \sqrt{J_{\gamma,s}(\psi)}.$$

Note that $I_s(\phi, \psi)$ is a duality form between $H^1_{\gamma,s}(0, \delta_0)$ and $H^1_{-\gamma,s}(0, \delta_0)$. With this notation, considering an expression like (8.23), one has

$$\int_{B_{\delta_0}(x_0)} |D\phi(x)|^2 r^\gamma \, dx = \sum_{s,i} J_{-\gamma,s}(\phi_{s,i}).$$

Now pick, for $\gamma < 0$,

$$\phi \in H^1_{-\gamma,s}(0, \delta_0)$$

and let us introduce the cutoff function

$$\zeta(r) = \begin{cases} 1 & \text{for } 0 \le r \le \dfrac{\delta_0}{2}, \\ 0 \le \zeta \le 1, & \zeta \text{ smooth}, |\zeta'| \le \dfrac{C}{\delta_0}, \\ 0 & \text{for } \dfrac{3\delta_0}{4} \le r \le \delta_0. \end{cases} \qquad (8.27)$$

We shall derive useful estimates for the quantities

$$I_{\gamma,s}(\phi) = I_s(\phi, r^\gamma \phi \zeta),$$
$$K_{\gamma,s}(\phi) = J_{\gamma,s}(r^\gamma \phi \zeta).$$

We then state the following result.

Lemma 8.12. *If we assume*

$$\phi \in H^1_{-\gamma,s}(0, \delta_0), \quad \phi(0) = 0,$$

then for $s \geq 1$, one has the estimates

$$I_{\gamma,s}(\phi) \geq \left(1 - \frac{\gamma(\gamma+n-2)}{2(n-1)} - \eta\right) J_{-\gamma,s}(\phi) - C_\gamma \frac{\delta_0^\gamma}{\eta} J_{0,s}(\phi), \quad (8.28)$$

$$K_{\gamma,s}(\phi) \leq \left(1 - \frac{n-2}{n-1}\gamma + \eta\right) J_{-\gamma,s}(\phi) + C_\gamma \frac{\delta_0^\gamma}{\eta} J_{0,s}(\phi), \quad (8.29)$$

where $0 < \eta \leq 1$ can be arbitrarily small and C is independent of s, x_0, δ_0 and depends only on γ.

Proof. Performing a straightforward computation based on integration by parts, thanks to the fact that $\phi(0) = 0, \zeta(\delta_0) = 0$, one obtains the formula

$$I_{\gamma,s}(\phi) = \int_0^{\delta_0} \left(\phi'^2 + \left[s(s+n-2) - \frac{\gamma(\gamma+n-2)}{2}\right]\frac{\phi^2}{r^2}\right) r^{\gamma+n-1}\zeta\, dr$$
$$+ \int_0^{\delta_0} \phi\phi'\zeta' r^{\gamma+n-1}\, dr - \frac{\gamma}{2}\int_0^{\delta_0} \phi^2\zeta' r^{\gamma+n-2}\, dr.$$

The estimate (8.28) follows easily, making use of Hölder's inequality in the second integral and taking account of the fact that $\zeta = 1, \zeta' = 0$ for $0 < r < \delta_0/2$.

Similarly, one gets the formula

$$K_{\gamma,s}(\phi) = \int_0^{\delta_0} \left(\phi'^2 + \left[s(s+n-2) - \frac{\gamma(\gamma+n-2)}{2}\right]\frac{\phi^2}{r^2}\right) r^{\gamma+n-1}\zeta^2\, dr$$
$$+ \int_0^{\delta_0} [\phi\phi'(\zeta^2)' + \phi^2\zeta'^2] r^{\gamma+n-1}\, dr,$$

and the estimate (8.29) follows in the same way. ◇

We apply the preceding lemma, as follows. Pick $u \in (H^1(B_{\delta_0}(x_0)))^N$ (we consider vector functions for later application, but the treatment is identical to the case of scalar functions), and assume that

$$\int_{B_{\delta_0}(x_0)} |Du|^2 r^\gamma\, dx < \infty.$$

We consider the expansion of u in terms of Spherical functions; hence

$$u^\nu(x) = \sum_{s,i} u_{s,i}^\nu(r) Y_{s,i}(\theta),$$

and we have, see (8.23),

$$\int_{B_{\delta_0}(x_0)} |Du(x)|^2 r^\gamma\, dx$$
$$= \sum_{s,i} \int_0^{\delta_0} \left[|u_{s,i}(r)'|^2 + \frac{s(s+n-2)}{r^2}|u_{s,i}(r)|^2\right] r^{\gamma+n-1}\, dr.$$

We associate to u a function v, defined by its expansion

$$v^\nu(x) = \sum_{s,i} v^\nu_{s,i}(r)Y_{s,i}(\theta),$$

where

$$v^\nu_0(r) = v^\nu_{0,1}(r) = -\int_r^{\delta_0} (u^\nu_0)'(r)\rho^\gamma \, d\rho,$$

and

$$v^\nu_{s,i}(r) = r^\gamma v^\nu_{s,i}(r)\zeta(r).$$

As in Proposition 8.9, we can write

$$\int_{B_{\delta_0}(x_0)} Du.Dv \, dx = \sum_{s,i} I_s(u_{s,i}, v_{s,i})$$

and

$$\int_{B_{\delta_0}(x_0)} |Dv|^2 r^{-\gamma} \, dx = \sum_{s,i} J_{\gamma,s}(v_{s,i}),$$

and all these expressions are well-defined for u. Clearly,

$$I_0(u_0, v_0) = J_{-\gamma,0}(u_0)$$

and for $s \geq 1$, we have

$$I_s(u_{s,i}, v_{s,i}) = I_{\gamma,s}(u_{s,i}).$$

Therefore,

$$\int_{B_{\delta_0}(x_0)} Du.Dv \, dx = J_{-\gamma,0}(u_0) + \sum_{s\geq 1,i} I_{\gamma,s}(u_{s,i})$$

and similarly

$$\int_{B_{\delta_0}(x_0)} |Dv|^2 r^{-\gamma} \, dx = J_{-\gamma,0}(u_0) + \sum_{s\geq 1,i} K_{\gamma,s}(u_{s,i}).$$

We can then use the estimates (8.28), (8.29) (noticing that $u_{s,i}(0) = 0$ for $s \geq 1$) to write

$$\int_{B_{\delta_0}(x_0)} Du.Dv \, dx \geq \left(1 - \frac{\gamma(\gamma + n - 2)}{2(n-1)} - \eta\right) \sum_{s,i} J_{-\gamma,s}(u_{s,i})$$

$$-C_\gamma \frac{\delta_0^\gamma}{\eta} \sum_{s,i} J_{0,s}(u_{s,i})$$

and

$$\int_{B_{\delta_0}(x_0)} |Dv|^2 r^{-\gamma} \, dx \leq \left(1 - \frac{n-2}{n-1}\gamma + \eta\right) \sum_{s,i} J_{-\gamma,s}(u_{s,i}) + C_\gamma \frac{\delta_0^\gamma}{\eta} \sum_{s,i} J_{0,s}(u_{s,i}),$$

which also implies the following estimates:

$$\int_{B_{\delta_0}(x_0)} Du \cdot Dv \, dx$$
$$\geq \left(1 - \frac{\gamma(\gamma + n - 2)}{2(n-1)} - \eta\right) \int_{B_{\delta_0}(x_0)} |Du|^2 r^\gamma \, dx - C_\gamma \frac{\delta_0^\gamma}{\eta} \int_{B_{\delta_0}(x_0)} |Du|^2 \, dx,$$
$$(8.30)$$

$$\int_{B_{\delta_0}(x_0)} |Dv|^2 r^{-\gamma} \, dx$$
$$\leq \left(1 - \frac{n-2}{n-1}\gamma + \eta\right) \int_{B_{\delta_0}(x_0)} |Du|^2 r^\gamma \, dx + C_\gamma \frac{\delta_0^\gamma}{\eta} \int_{B_{\delta_0}(x_0)} |Du|^2 \, dx.$$
$$(8.31)$$

We are ready now to give the

Proof of Theorem 8.10.
Let us take any subdomain $\tilde{\Omega}$ such that

$$\overline{\tilde{\Omega}} \subset \Omega.$$

We fix δ_0 such that

$$\overline{B_{\delta_0}(\xi)} \subset \Omega, \quad \forall \xi \in \tilde{\Omega}.$$

We now pick any $x_0 \in \tilde{\Omega}$ and consider the sequence u_k introduced in (8.14) that converges in $H^1(\Omega)$ to the solution u of (8.1). Note that all u_k belong to $(W^{2,p}(\Omega))^N$ and thus, since p is large enough, we may assume

$$\int_{B_{\delta_0}(x_0)} |Du_k|^2 r^\gamma \, dx < \infty.$$

Recalling the variational formulation (8.16), and picking a test function whose support is in $B_{\delta_0}(x_0)$, we can write

$$\int_{B_{\delta_0}(x_0)} D(u_{k+1}^\nu - u_k^\nu) Dv^\nu \, dx = \int_{B_{\delta_0}(x_0)} D(u_k^\nu - u_{k-1}^\nu) Dv^\nu \, dx$$
$$-\epsilon \int_{B_{\delta_0}(x_0)} a_{ij,k}^{\nu\,\mu}(x) D_j(u_k^\mu - u_{k-1}^\mu) D_i v^\nu \, dx,$$
$$(8.32)$$

and using the constant K defined in (8.18) we obtain

$$\int_{B_{\delta_0}(x_0)} D(u_{k+1} - u_k) Dv \, dx$$
$$\leq K \left(\int_{B_{\delta_0}(x_0)} |D(u_k - u_{k-1})|^2 r^\gamma \, dx\right)^{1/2} \left(\int_{B_{\delta_0}(x_0)} |Dv|^2 r^{-\gamma} \, dx\right)^{1/2}.$$
$$(8.33)$$

We next consider estimates (8.30), (8.31) with

$$u = u_{k+1} - u_k.$$

The corresponding v vanishes at the boundary of $B_{\delta_0}(x_0)$, and thus we can use it in (8.32). Therefore, (8.33) as well as (8.30), (8.31) are valid. Hence we have

$$\left(1 - \frac{\gamma(\gamma + n - 2)}{2(n-1)} - \eta\right) \int_{B_{\delta_0}(x_0)} |D(u_{k+1} - u_k)|^2 r^\gamma \, dx$$

$$-C_\gamma \frac{\delta_0^\gamma}{\eta} \int_{B_{\delta_0}(x_0)} |D(u_{k+1} - u_k)|^2 \, dx$$

$$\leq K \left(\int_{B_{\delta_0}(x_0)} |D(u_k - u_{k-1})|^2 r^\gamma \, dx\right)^{1/2}$$

$$\left[\left(1 - \frac{n-2}{n-1}\gamma + \eta\right)^{1/2} \left(\int_{B_{\delta_0}(x_0)} |D(u_{k+1} - u_k)|^2 r^\gamma \, dx\right)^{1/2}\right.$$

$$\left. + C_\gamma \frac{\delta_0^{\gamma/2}}{\eta^{1/2}} \left(\int_{B_{\delta_0}(x_0)} |D(u_{k+1} - u_k)|^2 \, dx\right)^{1/2}\right].$$

Let us use the notation

$$\sigma_k^2 = \int_{B_{\delta_0}(x_0)} |D(u_k - u_{k-1})|^2 r^\gamma \, dx$$

and choose η sufficiently small so that, defining the constant

$$K_1 = K \frac{(1 - \frac{n-2}{n-2}\gamma + \eta)^{1/2}}{1 - \frac{\gamma(\gamma+n-1)}{2(n-1)} - \eta},$$

we still have $K_1 < 1$. Therefore, the previous estimate amounts to

$$\sigma_{k+1}^2 \leq K_1 \sigma_{k+1} \sigma_k + C_1 \sigma_k K^k + C_2 K^{2k}.$$

The constants depend on δ_0, γ. The terms K^k, K^{2k} are obtained as in (8.19). From the next lemma we can infer that

$$\sigma_k \leq C K_2^k$$

with $K_2 < 1$. Collecting results we have proven that the limit u satisfies

$$\int_{B_{\delta_0}(x_0)} |Du|^2 r^\gamma \, dx < C_{\delta_0, \gamma},$$

the constant being independent of x_0. Setting

$$\alpha = \frac{2 - n - \gamma}{2}$$

we deduce that the Morrey norm (see Chapter 1) satisfies

$$\||u|\|_{\alpha,\Omega}^2 = \sup_{\substack{x_0,R \\ B_R(x_0) \in \tilde{\Omega}}} \frac{1}{R^{n-2+2\alpha}} \int_{B_R(x_0)} |Du|^2 \, dx \leq C.$$

The proof is now complete.   ◇

It remains to prove the following lemma.

Lemma 8.13. *Let us consider a sequence of positive numbers σ_k such that*

$$\sigma_{k+1}^2 \leq K_1 \sigma_{k+1} \sigma_k + C_1 \sigma_k K^k + C_2 K^{2k},$$

where

$$0 \leq K_1 < 1, \quad 0 \leq K < 1.$$

Then there exists K_2, with $0 \leq K_2 < 1$, such that

$$\sigma_k \leq C_3 K_2^k.$$

Proof. We can write

$$\sigma_{k+1}^2 \leq \frac{K_1}{2}(\sigma_{k+1}^2 + \sigma_k^2) + \eta \sigma_k^2 + \frac{C}{\eta} K^{2k},$$

and setting

$$t = \frac{K_1 + 2\eta}{2 - K_1},$$

we can find η such that $t < 1$. Possibly majorizing K by a number q still less than 1, we can finally write

$$\sigma_{k+1} \leq \sigma_k t + C q^k$$

with

$$0 \leq t < q < 1.$$

Iterating we get

$$\sigma_{k+1} \leq \sigma_0 t^{k+1} + C q^k \sum_{j=0}^{k} \left(\frac{t}{q}\right)^{k-j},$$

and since $t < q$, the result follows immediately.   ◇

8.3.4 Additional Remarks

One can derive an immediate consequence of estimates (8.30), (8.31) that in fact has been implicitly used in the proof of Theorem 8.10. Let us consider the problem

$$\int_\Omega Du.Dv \, dx = \int_\Omega z.Dv \, dx, \quad \forall v \in H_0^1(\Omega), \quad u \in H_0^1(\Omega), \qquad (8.34)$$

and the subdomain $\tilde{\Omega}$ of points of Ω whose distance to the boundary of Ω is larger that δ_0. Then one has the following result.

Proposition 8.14. *For any point $x_0 \in \tilde{\Omega}$, one has the estimate*

$$\left(1 - \frac{\gamma(\gamma + n - 2)}{2(n-1)} - \eta\right) \int_{B_{\delta_0}(x_0)} |Du|^2 r^\gamma \, dx$$

$$\leq \left(\int_{B_{\delta_0}(x_0)} |z|^2 r^\gamma \, dx\right)^{1/2} \left[\left(1 - \frac{n-2}{n-1}\gamma + \eta\right)^{1/2} \left(\int_{B_{\delta_0}(x_0)} |Du|^2 r^\gamma \, dx\right)^{1/2}\right.$$

$$\left. + C \left(\int_{B_{\delta_0}(x_0)} |Du|^2 \, dx\right)^{1/2}\right] + C \int_{B_{\delta_0}(x_0)} |Du|^2 \, dx,$$

$$(8.35)$$

where the constant C depends on δ_0, γ, η, but not on x_0.

Proof. In fact, we use $(8.30), (8.31)$ and just notice that

$$\int_{B_{\delta_0}} Du.Dv \, dx = \int_{B_{\delta_0}} z.Dv \, dx,$$

since v vanishes at the boundary of B_{δ_0}. The rest of the proof is immediate, by Hölder's inequality and previous estimates. $\diamond$

One can easily derive the following form of (8.35) by modifying the constant η, which can be taken as small as needed:

$$\int_{B_{\delta_0}(x_0)} |Du|^2 r^\gamma \, dx$$

$$\leq \frac{1 - \frac{n-2}{n-1}\gamma + \eta}{\left(1 - \frac{\gamma(\gamma+n-2)}{2(n-1)}\right)^2} \int_{B_{\delta_0}(x_0)} |z|^2 r^\gamma \, dx + C \int_{B_{\delta_0}(x_0)} |Du|^2 \, dx,$$

$$(8.36)$$

where C depends on γ, δ_0, η, but not on x_0.

Let us now take

$$\gamma = 2 - n - \epsilon,$$

where ϵ is arbitrarily small. Replacing in (8.36) and using the fact that ϵ is small, we obtain the following form:

$$\int_{B_{\delta_0}(x_0)} |Du|^2 r^{2-n-\epsilon} \, dx \leq \left(1 + \frac{(n-2)^2}{n-1} + O(\epsilon)\right) \int_{B_{\delta_0}(x_0)} |z|^2 r^{2-n-\epsilon} \, dx$$

$$+ C(\epsilon) \int_{B_{\delta_0}(x_0)} |Du|^2 \, dx.$$

Koshelev has shown that the constant

$$1 + \frac{(n-2)^2}{n-1}$$

is sharp, by constructing the following example. Define

$$z_{\epsilon,i}(x) = |x|^{-1+\epsilon}\left(\delta_{in} + (1-n)\frac{x_i x_n}{|x|^2}\right).$$

Set

$$u_\epsilon(x) = ax_n|x|^{-1+\epsilon}.$$

For an appropriate choice of a, one has the relation

$$\Delta u_\epsilon = \operatorname{div} z_\epsilon.$$

For that purpose, one must take

$$a = a_\epsilon = \frac{n^2 - 3n + 3 + \epsilon(n-2)}{(1-\epsilon)(n-1+\epsilon)}.$$

Let B be the ball of center the origin and radius 1. Noting the formula

$$n\int_B x_n^2|x|^{-2-n+\epsilon}\,dx = \int_B |x|^{-n+\epsilon}\,dx$$

one easily computes

$$\int_B |z_\epsilon|^2|x|^{2-n-\epsilon}\,dx = \frac{|S|}{\epsilon}\frac{n^2 - 3n + 3}{n}$$

and

$$\int_B |Du_\epsilon|^2|x|^{2-n-\epsilon}\,dx = a_\epsilon^2\frac{n + \epsilon^2 - 1}{n}\frac{|S|}{\epsilon},$$

where $|S|$ represents the area of the unit sphere in R^n.

Therefore, one can write (as is easily seen)

$$\int_B |Du_\epsilon|^2|x|^{2-n-\epsilon}\,dx = \left(1 + \frac{(n-2)^2}{n-1} + O(\epsilon)\right)\int_B |z_\epsilon|^2|x|^{2-n-\epsilon}\,dx.$$

This proves the sharpness of the constant, since u_ϵ can always be considered as the restriction of a Dirichlet problem in a sufficiently large domain Ω.

8.3.5 Hölder's Continuity up to the Boundary

Our objective is to prove that in the case of a smooth boundary one can extend Hölder's continuity up to the boundary. Namely, we have the following theorem:

Theorem 8.15. *If we assume (8.4), (8.24), (8.25), (8.26), and that the domain is smooth, then the solution of (8.1), with $g = 0$, satisfies*

$$u \in \left(C^\alpha_{\mathrm{loc}}(\overline{\Omega})\right)^N \tag{8.37}$$

with

$$\alpha = \frac{2 - n - \gamma}{2}. \tag{8.38}$$

Before giving the proof, we need some preliminaries. First, using local charts, we can reduce the domain to

$$\Omega = \tilde{\Omega} \cap \{x_n > 0\},$$
$$\tilde{\Omega} - \Omega = (\tilde{\Omega} \cap \{x_n = 0\}) \cup \{(x_t, -x_n) | x = (x_t, x_n) \in \Omega\},$$

in which $t = 1, \ldots, n - 1$ and where $\tilde{\Omega}$ is a bounded smooth domain of R^n.

Next, let us consider the subdomain $\tilde{\Omega}_{\delta_0}$ of points of $\tilde{\Omega}$ that are at a distance from the boundary larger than δ_0. We shall consider balls $B_{\delta_0}(x_0)$ for $x_0 \in \tilde{\Omega}_{\delta_0}$ and define, if $x_{0,n} \geq 0$,

$$B_{\delta_0}(x_0)^+ = B_{\delta_0}(x_0) \cap \{x_n > 0\}.$$

Similarly, we define, if $x_{0,n} < 0$,

$$B_{\delta_0}(x_0)^- = B_{\delta_0}(x_0) \cap \{x_n < 0\}.$$

By definition, we have, if $x_{0,n} < 0$,

$$x \in B_{\delta_0}(x_0)^- \to x^* \in B_{\delta_0}(x_0^*)^+,$$

where x^* is the symmetric of x with respect to the hyperplan $\{x_n = 0\}$. Note that the indices $+, -$ make sense only when $x_{0,n} > 0$ or $x_{0,n} < 0$, respectively.

Moreover, it is easy to check that

$$B_{\delta_0}(x_0) \subset B_{\delta_0}(x_0)^+ \cup B_{\delta_0}(x_0^*)^-$$

if $x_{0\,n} > 0$, and a similar inclusion when $x_{0\,n} < 0$.

In fact, in view of the iteration we use (as is noted in Section 8.3.4), most of the proof relies on studying the auxiliary scalar problem

$$\int_\Omega Du.Dv\,dx = \int_\Omega z.Dv\,dx, \quad \forall v \in H_0^1(\Omega), \quad u \in H_0^1(\Omega) \qquad (8.39)$$

We first extend the variational problem (8.39) to $\tilde{\Omega}$, as follows. Define, for $x_n < 0$,

$$z^i(x_t, x_n) = -z^i(x_t, -x_n), \quad \forall i = 1, \ldots, n - 1,$$
$$z^n(x_t, x_n) = z^n(x_t, -x_n),$$

and extend u by antisymmetry:

$$u(x_t, x_n) = -u(x_t, -x_n).$$

Then it is easy to check that u is the solution of the variational problem

$$\int_{\tilde{\Omega}} Du.Dv\,dx = \int_{\tilde{\Omega}} z.Dv\,dx,$$
$$\forall v \in H_0^1(\tilde{\Omega}), \quad v = 0 \text{ on } \tilde{\Omega} \cap \{x_n = 0\}, \qquad (8.40)$$
$$u \in H_0^1(\tilde{\Omega}), \quad u = 0 \text{ on } \tilde{\Omega} \cap \{x_n = 0\}.$$

We then want to prove the following proposition, interesting in itself.

Proposition 8.16. *Pick* $x_0 \in \tilde{\Omega}_{\delta_0}, x_{0,n} \geq 0$,*then one has*

$$
\left(1 - \frac{\gamma(\gamma + n - 2)}{2(n-1)} - \eta\right) \int_{B_{\delta_0}(x_0)^+ \cup B_{\delta_0}(x_0^*)^-} |Du|^2 r^\gamma \, dx
$$

$$
\leq \left(\int_{B_{\delta_0}(x_0)^+ \cup B_{\delta_0}(x_0^*)^-} |z|^2 r^\gamma \, dx\right)^{1/2}
$$

$$
\left[\left(1 - \frac{n-2}{n-1}\gamma + \eta\right)^{1/2} \left(\int_{B_{\delta_0}(x_0)^+ \cup B_{\delta_0}(x_0^*)^-} |Du|^2 r^\gamma \, dx\right)^{1/2} \right. \tag{8.41}
$$

$$
\left. + C \left(\int_{B_{\delta_0}(x_0)^+} (|Du|^2 + |u|^2 + |z|^2) \, dx\right)^{1/2}\right]
$$

$$
+ C \int_{B_{\delta_0}(x_0)^+} (|Du|^2 + |u|^2 + |z|^2) \, dx.
$$

where the constants C *depend on* γ, δ_0, η, *but not on* x_0.
A similar relation holds for $x_0 \in \tilde{\Omega}_{\delta_0}, x_{0,n} < 0$.

Proof of Theorem 8.15.
We can consider a vector version of (8.39), (8.41), where u is a vector and z is a matrix. We then apply it to our iteration, with

$$
u = u^{k+1} - u^k,
$$
$$
z = (I - \epsilon A_k) D(u^k - u^{k-1}).
$$

With the choice of ϵ, by applying (8.41) we are back in the situation of Theorem 8.10. Hence we prove similarly for the limit u that

$$
\int_{B_{\delta_0}(x_0)^+ \cup B_{\delta_0}(x_0^*)^-} |Du|^2 r^\gamma \, dx < C_{\delta_0, \gamma}.
$$

Therefore, it follows also that

$$
\int_{B_{\delta_0}(x_0)} |Du|^2 r^\gamma \, dx < C_{\delta_0, \gamma},
$$

which now holds for any point in $\tilde{\Omega}$. Hence, we can conclude that u is Hölder on $\tilde{\Omega}$. This concludes the proof.   ◇

Proof of Proposition 8.16.
Because of the interior boundary condition, we cannot apply directly the methods of Section 8.3.3; in particular, the test function used there will not satisfy the boundary condition at $x_n = 0$.

1. Explicit formula

Pick $x_0 \in \bar{\Omega}_{\delta_0 \cap \{x_n \geq 0\}}$ and consider the ball $B_{\delta_0}(x_0)$. We shall give an explicit formula for $u(x)$ at points $x \in B_{\delta_0/2}(x_0)^+ \cup B_{\delta_0/2}(x_0^*)^-$. Recall the cutoff function $\zeta(r)$ defined by (8.27). We shall define

$$\zeta_{x_0}(x) = \zeta(|x - x_0|)\mathbb{1}_{\{x_n \geq 0\}} + \zeta(|x - x_0^*|)\mathbb{1}_{\{x_n < 0\}},$$

which is continuous and whose derivatives exist except for the singularity $x_n = 0$.

Let us consider the Green function[1]

$$G(x, y) = \frac{1}{(n - 2)|S|} \left(\frac{1}{|x - y|^{n-2}} - \frac{1}{|x^* - y|^{n-2}} \right).$$

Note that

$$G(x, y) = 0 \quad \text{if } y_n = 0 \text{ or if } x_n = 0.$$

The function $\zeta_{x_0}(y)G(x, y)$ (where x is a parameter) is a possible test function in (8.40). Note that $\zeta_{x_0}(y)$ is symmetric, and $G(x, y)$ is antisymmetric with respect to y.

Hence we obtain

$$\int_{B_{\delta_0}(x_0)^+} Du\, D(\zeta G)\, dy = \int_{B_{\delta_0}(x_0)^+} z\, D(\zeta G)\, dy$$

and also

$$\int_{B_{\delta_0}(x_0^*)^-} Du\, D(\zeta G)\, dy = \int_{B_{\delta_0}(x_0^*)^-} z\, D(\zeta G)\, dy.$$

We will consider just the positive part; the negative part is treated similarly. Since ζ vanishes at the boundary of B_{δ_0} and u vanishes on $\{y_n = 0\}$, we can perform another integration by parts on the left-hand side. Of course, we have to take care of the singularity. In fact, we have

$$\int_{B_{\delta_0}^+} u(y)\Delta_y G(x, y)\zeta(y)\, dy = u(x)$$

for $x \in B_{\delta_0/2}(x_0)^+$. Therefore, we obtain the formula

$$u(x) = -\int_{B_{\delta_0}^+} \zeta z\, D_y G\, dy - \int_{B_{\delta_0/2,\delta_0}^+} u\, G \Delta_y \zeta\, dy$$
$$-2 \int_{B_{\delta_0/2,\delta_0}^+} u\, D_y \zeta\, D_y G\, dy - \int_{B_{\delta_0/2,\delta_0}^+} z\, D_y \zeta G\, dy,$$

where

$$B_{\delta_0/2,\delta_0}^+ = B_{\delta_0}^+ - B_{\delta_0/2}^+,$$

---

[1] we assume, to fix ideas, that $n \geq 3$.

since $\zeta = 1$ on $B_{\delta_0/2}$. Making G explicit, we can write

$$u(x) = -\frac{1}{(n-2)|S|} \int_{B_{\delta_0}^+} \zeta z \left(D_y|x-y|^{2-n} - D_y|x^*-y|^{2-n}\right) dy$$
$$+ \int_{B_{\delta_0/2,\delta_0}^+} H(x,y)z(y)\,dy + \int_{B_{\delta_0/2,\delta_0}^+} K(x,y)u(y)\,dy, \tag{8.42}$$

where

$$H(x,y) = \frac{1}{(n-2)|S|}[|x^*-y|^{2-n} - |x-y|^{2-n}]D_y\zeta_{x_0},$$

$$K(x,y) = \frac{1}{(n-2)|S|}[|x^*-y|^{2-n} - |x-y|^{2-n}]\Delta_y\zeta_{x_0}$$
$$+ 2\frac{1}{(n-2)|S|}D_y(|x^*-y|^{2-n} - |x-y|^{2-n})D_y\zeta_{x_0}.$$

For the time being, we consider these functions for $y_n > 0$, and the derivatives are legitimate. Of course, they make sense when $y_n < 0$ as well, and they will be used later in that context.

Now let us restrict x to

$$x \in B_{q\delta_0/2}(x_0)^+, \quad q < 1.$$

We notice that

$$x \in B_{q\delta_0/2}(x_0)^+, \quad y \in B_{\delta_0/2,\delta_0}^+ \to |x-y|, \quad |x^*-y| > \frac{\delta_0}{2}(1-q).$$

Indeed, we have

$$|x-y| > |y-x_0| - |x_0-x| \geq \frac{\delta_0}{2} - q\frac{\delta_0}{2}.$$

Next, we use the relation

$$|x^*-y|^2 = |x-y|^2 + 4x_n y_n > |x-y|^2,$$

since $x_n > 0$, to establish the second inequality.

It follows that $H(x,y)$, $K(x,y)$ are smooth functions. So the important term is

$$R(x,y) = -\frac{1}{(n-2)|S|} \int_{B_{\delta_0}^+} \zeta z \left(D_y|x-y|^{2-n} - D_y|x^*-y|^{2-n}\right) dy,$$

which we write as follows:

$$R(x) = \frac{1}{(n-2)|S|}\left[\sum_{i=1}^{n}\frac{\partial}{\partial x_i}\int_{B_{\delta_0}^+} \zeta z^i(y)|x-y|^{2-n}\,dy\right.$$
$$\left.-\sum_{i=1}^{n-1}\frac{\partial}{\partial x_i}\int_{B_{\delta_0}^+} \zeta z^i(y)|x^*-y|^{2-n}\,dy + \frac{\partial}{\partial x_n}\int_{B_{\delta_0}^+} \zeta z^n(y)|x^*-y|^{2-n}\,dy\right].$$

In view of the choice of the extension of z, we check easily that

$$R(x) = \frac{1}{(n-2)|S|} \sum_{i=1}^{n} \frac{\partial}{\partial x_i} \left(\int_{B_{\delta_0}(x_0)^+ \cup B_{\delta_0}(x_0^*)^-} \zeta_{x_0} z^i(y) |x-y|^{2-n} \, dy \right).$$

The function R makes sense again for

$$x \in B_{\delta_0}(x_0)^+ \cup B_{\delta_0}(x_0^*)^-$$

and

$$R(x^*) = -R(x).$$

Similarly, we can write (as is easily seen), when $x \in B_{\delta_0/2}(x_0^*)^-$,

$$u(x) = R(x) + \int_{B_{\delta_0/2, \delta_0}^-} H(x,y) z(y) \, dy + \int_{B_{\delta_0/2, \delta_0}^-} K(x,y) u(y) \, dy, \quad (8.43)$$

and H, K are smooth whenever $x \in B_{\delta_0 q/2}(x_0^*)^-$.

The purpose of the next step is to treat such singular integrals.

2. Treatment of a singular integral

Let us consider two domains ω and $\underline{\omega}$ of R^n, and assume that

$$\overline{\omega} \subset B_\beta(x_0), \quad \overline{B_{\beta_1}(x_0)} \subset \underline{\omega}$$

with $\beta < \beta_1$. We assume that the vector function

$$\phi(x) = (\phi^1(x), \dots, \phi^n(x))$$

satisfies

$$\int_{\underline{\omega}} |\phi|^2 r^\gamma \, dx < \infty.$$

We consider next the singular integral

$$J(\phi) = \frac{1}{(n-2)|S|} D_i \int_{\underline{\omega}} \phi^i(y) |x-y|^{2-n} \, dy.$$

Then one has the estimate

$$\left(1 - \frac{\gamma(\gamma + n - 2)}{2(n-1)} - \eta\right) \int_{\omega} |DJ|^2 r^\gamma \, dx$$

$$\leq \left(\int_{\underline{\omega}} |\phi|^2 r^\gamma \, dx \right)^{1/2} \left[\left(1 - \frac{n-2}{n-1} \gamma + \eta \right)^{1/2} \left(\int_{B_\beta(x_0)} |DJ|^2 r^\gamma \, dx \right)^{1/2} \right.$$

$$\left. + C \left(\int_{\underline{\omega}} |\phi|^2 \, dx \right)^{1/2} \right] + C \int_{\underline{\omega}} |\phi|^2 \, dx, \quad (8.44)$$

where the constant depends on $\beta, \beta_1, \gamma, \eta$ but not on x_0. To prove (8.44), we introduce the variational problem

$$\int_{B_{\beta_1}(x_0)} D\psi.Dv \, dx = \int_{B_{\beta_1}(x_0)} \phi.Dv \, dx \qquad (8.45)$$

with

$$\psi = 0 \text{ on } \partial B_{\beta_1}. \qquad (8.46)$$

We can give the explicit solution of (8.45) and (8.46) by the formula

$$\psi(x) = \frac{1}{(n-2)|S|} \int_{B_{\beta_1}} \text{div } \phi \left[|x-y|^{2-n} - \left(\frac{\beta_1}{|x-x_0|} \right)^{n-2} |x'-y|^{2-n} \right] dy, \qquad (8.47)$$

where x' denotes the conjugate to the point x, given by the formula

$$x' = x_0 + \frac{\beta_1^2}{|x-x_0|^2}(x-x_0).$$

Note the formula

$$\frac{|x-x_0|}{\beta_1}|x'-y| = \frac{|y-x_0|}{\beta_1}|x-y'|,$$

and thus the quantity within brackets on the right-hand side of (8.47) vanishes when x or y lies on the boundary of $B_{\beta_1}(x_0)$. Note also that

$$x \in B_\beta \rightarrow |x'-x_0| \geq \frac{\beta_1^2}{\beta}.$$

So for $x \in B_\beta$ we can write, using integration by parts and exchange of x and y derivatives,

$$\psi(x) = J(\phi)(x) + \int_\omega \phi(y).\xi_{x_0}(x,y) \, dy,$$

where $\xi_{x_0}(x,y)$ is a smooth function, given by

$$\xi_{x_0}(x,y) = \frac{1}{(n-2)|S|} \mathbb{1}_{\{\omega - B_{\beta_1}(x_0)\}} \frac{\partial}{\partial y_i} |x-y|^{2-n}$$

$$+ \frac{1}{(n-2)|S|} \mathbb{1}_{\{B_{\beta_1}(x_0)\}}(y) \left(\frac{\beta_1}{|x-x_0|} \right)^{n-2} \frac{\partial}{\partial y_i} |x'-y|^{2-n}.$$

For ψ, on the other hand, we can use the estimates following from (8.30) and (8.31), to yield

$$\left(1 - \frac{\gamma(\gamma + n - 2)}{2(n-1)} - \eta\right) \int_{B_{\beta_1}(x_0)} |D\psi|^2 r^\gamma \, dx$$

$$\leq \left(\int_{B_{\beta_1}(x_0)} |\phi|^2 r^\gamma \, dx\right)^{1/2} \left[\left(1 - \frac{n-2}{n-1}\gamma + \eta\right)^{1/2}\right.$$

$$\left. \left(\int_{B_{\beta_1}(x_0)} |D\psi|^2 r^\gamma \, dx\right)^{1/2} + C \left(\int_{B_{\beta_1}(x_0)} |D\psi|^2 \, dx\right)^{1/2}\right]$$

$$+ C \int_{B_{\beta_1}(x_0)} |D\psi|^2 \, dx.$$

Using equation (8.45), we may also write

$$\left(1 - \frac{\gamma(\gamma + n - 2)}{2(n-1)} - \eta\right) \int_{B_\beta(x_0)} |D\psi|^2 r^\gamma \, dx$$

$$\leq \left(\int_{B_{\beta_1}(x_0)} |\phi|^2 r^\gamma \, dx\right)^{1/2} \left[\left(1 - \frac{n-2}{n-1}\gamma + \eta\right)^{1/2}\right. \tag{8.48}$$

$$\left. \left(\int_{B_\beta(x_0)} |D\psi|^2 r^\gamma \, dx\right)^{1/2} + C \left(\int_{B_{\beta_1}(x_0)} |\phi|^2 \, dx\right)^{1/2}\right]$$

$$+ C \int_{B_{\beta_1}(x_0)} |\phi|^2 \, dx.$$

Since clearly

$$\int_{B_\beta(x_0)} |DJ|^2 r^\gamma \, dx \leq (1 + \eta) \int_{B_\beta(x_0)} |D\psi|^2 r^\gamma \, dx + C \int_\omega |\phi|^2 \, dx,$$

we can deduce from (8.48) the estimate (8.44) by possibly modifying η.

3. Final estimate

We consider two cases.

Case 1: $|x_0 - x_0^*| \geq \delta_0$.

In that case, it is easy to check that

$$B_{\delta_0/2}(x_0) = B^+_{\delta_0/2}(x_0).$$

Recalling the definition of $R(x)$, we can write

$$R(x) = \frac{1}{(n-2)|S|} \sum_{i=1}^n \frac{\partial}{\partial x_i} \left(\int_{B_{\delta_0}(x_0)} \zeta(|x_0 - y|) \mathbb{1}_{\{y_n > 0\}} z^i(y) |x - y|^{2-n} \, dy\right)$$

$$+ R_1(x)$$

with

$$R_1(x) = \frac{1}{(n-2)|S|} \sum_{i=1}^{n} \frac{\partial}{\partial x_i} \left(\int_{B_{\delta_0}(x_0^*)} \zeta(|x_0^* - y|) \, z^i(y)|x - y|^{2-n} \, dy \right).$$

Now for

$$x \in B_{q\delta_0/2}(x_0)^+ = B_{q\delta_0/2}(x_0)$$

and $y \in B_{\delta_0}^-(x_0^*)$, we have $|x - y| > \frac{\delta_0}{2}(1 - q)$; hence

$$R_1(x) = \int_{B_{\delta_0}^-(x_0^*)} L(x, y) z(y) \, dy$$

with $L(x, y)$ smooth.

On the other hand,

$$R(x) = J(\zeta \mathrm{II}_{\{y_n > 0\}} z)(x) + R_1(x)$$

with

$$\underline{\omega} = B_{\delta_0}(x_0).$$

The solution $u(x)$ of (8.39) can be represented as

$$u(x) = \int_{B_{\delta_0/2,\delta_0}^+} H(x, y) z(y) \, dy + \int_{B_{\delta_0/2,\delta_0}^+} K(x, y) u(y) \, dy$$

$$+ \int_{B_{\delta_0}^-(x_0^*)} L(x, y) z(y) \, dy + + J(\zeta \mathrm{II}_{\{y_n > 0\}} z)(x).$$

We pick

$$\omega = B_{q\delta_0/2}(x_0)$$

and use (8.44), with

$$\beta = \frac{q\delta_0}{2}, \quad \beta_1 = \delta_0.$$

Collecting results, we can write

$$\left(1 - \frac{\gamma(\gamma + n - 2)}{2(n-1)} - \eta \right) \int_{B_{\delta_0/2q}(x_0)^+} |Du|^2 r^\gamma \, dx$$

$$\leq \left(\int_{B_{\delta_0}(x_0)^+} |z|^2 r^\gamma \, dx \right)^{1/2} \left[\left(1 - \frac{n-2}{n-1}\gamma + \eta \right)^{1/2} \right.$$

$$\left(\int_{B_{\delta_0/2q}(x_0)^+} |Du|^2 r^\gamma \, dx \right)^{1/2}$$

$$\left. + C \left(\int_{B_{\delta_0}(x_0)^+} (|Du|^2 + |u|^2 + |z|^2) \, dx \right)^{1/2} \right]$$

$$+ C \int_{B_{\delta_0}(x_0)^+} (|Du|^2 + |u|^2 + |z|^2) \, dx.$$

Since

$$\int_{B_{\delta_0/2q}(x_0^*)^-} |Du|^2 r^\gamma \, dx \le C \int_{B_{\delta_0/2q}(x_0^*)^-} |Du|^2 \, dx,$$

we deduce also that

$$\left(1 - \frac{\gamma(\gamma + n - 2)}{2(n-1)} - \eta\right) \int_{B_{\delta_0/2q}(x_0) + \cup B_{\delta_0/2q}(x_0^*)^-} |Du|^2 r^\gamma \, dx$$

$$\le \left(\int_{B_{\delta_0}(x_0) + \cup B_{\delta_0}(x_0^*)^-} |z|^2 r^\gamma \, dx\right)^{1/2} \times \left[\left(1 - \frac{n-2}{n-1}\gamma + \eta\right)^{1/2}\right.$$

$$\left(\int_{B_{\delta_0/2q}(x_0) + \cup B_{\delta_0/2q}(x_0^*)^-} |Du|^2 r^\gamma \, dx\right)^{1/2}$$

$$+ C \left(\int_{B_{\delta_0}(x_0)+} (|Du|^2 + |u|^2 + |z|^2)\, dx\right)^{1/2}\right] \tag{8.49}$$

$$+ C \int_{B_{\delta_0}(x_0)+} (|Du|^2 + |u|^2 + |z|^2)\, dx,$$

from which (8.41) follows.

Case 2: $|x_0 - x_0^*| < \delta_0$.

This is the hardest case, since in this case writing $R(x)$ as the sum of a singular integral and a regular one is no longer valid. We then define

$$\omega = B_{\delta_0 q/2}(x_0)^+ \cup B_{\delta_0 q/2}(x_0^*)^-$$

and

$$\underline{\omega} = B_{\delta_0(q+1)}(x_0)^+ \cup B_{\delta_0(q+1)}(x_0^*)^-.$$

We shall use the fact

$$\omega \subset B_{\delta_0(1+\frac{q}{2})}(x_0) \subset B_{\delta_0(1+q)}(x_0) \subset \underline{\omega}.$$

We apply (8.44) to that situation, noting that

$$\beta = \delta_0\left(1 + \frac{q}{2}\right), \quad \beta_1 = \delta_0(1+q).$$

We thus obtain

$$\left(1 - \frac{\gamma(\gamma + n - 2)}{2(n-1)} - \eta\right) \int_\omega |DJ|^2 r^\gamma \, dx$$

$$\le \left(\int_{\underline{\omega}} |\phi|^2 r^\gamma \, dx\right)^{1/2} \left[\left(1 - \frac{n-2}{n-1}\gamma + \eta\right)^{1/2}\right.$$

$$\left.\left(\int_\omega |DJ|^2 r^\gamma \, dx\right)^{1/2} + C \left(\int_{\underline{\omega}} |\phi|^2 \, dx\right)^{1/2}\right] + C \int_{\underline{\omega}} |\phi|^2 \, dx. \tag{8.50}$$

Now from representation formulas (8.42), (8.43), we can write

$$u(x) = J(\zeta_0 z)(x) + \text{ regular integrals}$$

for

$$x \in B_{\delta_0/2q}(x_0)^+ \cup B_{\delta_0/2q}(x_0^*)^-.$$

Using (8.50), we again obtain (8.49) and thus also (8.41).
  The proof is complete.   ◇

8.4 $C^{1+\alpha}$ Regularity

A. Koshelev [63] has also shown that by strengthening the condition (8.25) it is possible to obtain $C^{1+\alpha}$ regularity. We shall present this theory in this section.

8.4.1 Auxiliary Inequalities

We now state the estimates established by H.O. Cordes [12].

Lemma 8.17. *Let* $-n < \gamma < 2 - n$, *and*

$$\alpha = 1 - \frac{n+\gamma}{2}, \quad 0 < \alpha < 1.$$

We consider a ball $B_\delta(x_0)$, *and a function* u *such that*

$$\int_{B_\delta(x_0)} |D^2 u|^2 r^\gamma \, dx < \infty,$$

where $r = |x - x_0|$ *and*

$$\begin{aligned} u(x_0) &= Du(x_0) = 0, \\ u\big|_{\partial B_\delta} &= Du\big|_{\partial B_\delta} = 0. \end{aligned}$$

Then the following inequalities hold:

$$\int_{B_\delta(x_0)} |u|^2 r^{\gamma-4} \, dx \leq M_1^2(\alpha) \int_{B_\delta(x_0)} |\Delta u|^2 r^\gamma \, dx, \qquad (8.51)$$

$$\int_{B_\delta(x_0)} |Du|^2 r^{\gamma-4} \, dx \leq M_2^2(\alpha) \int_{B_\delta(x_0)} |\Delta u|^2 r^\gamma \, dx, \qquad (8.52)$$

$$\int_{B_\delta(x_0)} |D^2 u|^2 r^{\gamma-4} \, dx \leq M_3^2(\alpha) \int_{B_\delta(x_0)} |\Delta u|^2 r^\gamma \, dx, \qquad (8.53)$$

with the values

$$M_1(\alpha) = \begin{cases} \dfrac{1}{(n+1)\alpha} & \text{if } \alpha \le \dfrac{-n + \sqrt{(n+1)^2 + 1}}{2}, \\[3mm] \dfrac{1}{(n+1+\alpha)(1-\alpha)} & \text{if } \alpha \ge \dfrac{-n + \sqrt{(n+1)^2 + 1}}{2}, \end{cases} \quad (8.54)$$

$$(M_2(\alpha))^2 = M_1(\alpha)[1 + (n+2\alpha)(1+\alpha)M_1(\alpha)], \tag{8.55}$$

$$(M_3(\alpha))^2 = 1 + (n - 2 + 2\alpha)\frac{(1+\alpha)^2 + n[2 - (1-\alpha)^2]}{(1-\alpha)^2(n+1+\alpha)^2}. \tag{8.56}$$

Proof. We can extend u by 0 outside the ball $B_\delta(x_0)$. First by performing integration by parts, we establish the following:

$$\int_{B_\delta} |Du|^2 r^{\gamma-2}\, dx + \int_{B_\delta} u\, \Delta u\, r^{\gamma-2}\, dx = \frac{(\gamma - 2)(\gamma + n - 4)}{2}\int_{B_\delta} |u|^2 r^{\gamma-4}\, dx, \tag{8.57}$$

$$\int_{B_\delta} (\Delta u)^2 r^\gamma\, dx - \int_{B_\delta} |D^2 u|^2 r^\gamma\, dx$$
$$= \gamma(\gamma - 2)\int_{B_\delta} \left(\frac{\partial u}{\partial r}\right)^2 r^{\gamma-2}\, dx - \gamma(n + \gamma - 3)\int_{B_\delta} |Du|^2 r^{\gamma-2}\, dx. \tag{8.58}$$

Next, we take the expansion of u with respect to Spherical functions in B_δ,

$$u(x) = \sum_{si} u_{si}(r) Y_{si}(\theta),$$

and note that in view of the boundary conditions,

$$u_{si}(0) = u'_{si}(0) = u_{si}(\delta) = u'_{si}(\delta)$$

and the functions $u_{si}(r)$ vanish for $r \ge \delta$. We can then compute the quantity

$$D = \int_{B_\delta} (\Delta u)^2 r^\gamma\, dx$$
$$= \sum_{si} \int_0^\infty \Bigg(|u''_{si}|^2 + \frac{|u'_{si}|^2}{r^2}[(n-1)^2 - (n-1)(\gamma + n - 2)$$
$$+ 2s(s + n - 2)] + \frac{u_{si}^2}{r^4}s(s + n - 2)[s(s + n - 2)$$
$$- (\gamma - 2)(\gamma + n - 4)]\Bigg) r^{\gamma+n-1}\, dr. \tag{8.59}$$

We next make use of Hardy inequality (1.10) with $p = 2$ and with s respectively equal to $3 - n - \gamma$ and to $5 - n - \gamma$, to claim that

$$\int_0^\infty |u'_{si}|^2 r^{n+\gamma-3}\, dr \le \frac{4}{(2 - n - \gamma)^2}\int_0^\infty |u''_{si}|^2 r^{n+\gamma-1}\, dr,$$

$$\int_0^\infty |u_{si}|^2 r^{n+\gamma-5}\, dr \le \frac{4}{(4 - n - \gamma)^2}\int_0^\infty |u'_{si}|^2 r^{n+\gamma-3}\, dr.$$

Hence, estimating the right-hand side of (8.59) we obtain, after some easy calculations,

$$D \geq \sum_{si} \int_0^\infty |u_{si}|^2 \left(\left[\left(s + \frac{n-2}{2} \right)^2 - \left(\frac{\gamma-2}{2} \right)^2 \right]^2 \right.$$
$$\left. + s(s+n-2)(n+\gamma-4)(\gamma-2) \right) r^{n+\gamma-5} \, dr.$$

Thus clearly

$$D \geq \min_{s=0,1,\dots} \left[\left(s + \frac{n-2}{2} \right)^2 - \left(\frac{\gamma-2}{2} \right)^2 \right]^2 \int_{B_\delta} |u|^2 r^{\gamma-4} \, dx.$$

The minimum is attained for $s = 1$ or $s = 2$. Replacing γ in terms of α, we check easily that

$$\frac{1}{M_1(\alpha)} = \min(\alpha(n+\alpha), \, (1-\alpha)(n+1+\alpha)).$$

The value of $M_1(\alpha)$ (8.54) is easily deduced.

We next use (8.57) and Hölder's inequality to obtain

$$\int_{B_\delta} |Du|^2 r^{\gamma-2} \, dx \leq \left(\int_{B_\delta} |u|^2 r^{\gamma-4} \, dx \right)^{1/2} \left(\int_{B_\delta} |\Delta u|^2 r^\gamma \, dx \right)^{1/2}$$
$$+ (n+2\alpha)(\alpha+1) \int_{B_\delta} |u|^2 r^{\gamma-4} \, dx.$$

From (8.51) we deduce (8.52), with the value (8.55) of $M_2(\alpha)$.

We proceed by noticing the formula

$$\int_{B_\delta} |Du|^2 r^{\gamma-2} \, dx = \sum_{si} \int_0^\infty \left(|u_{si}'|^2 + \frac{s(s+n-2)}{r^2} \right) r^{n+\gamma-3} \, dr$$

and use it in (8.58) to obtain

$$\int_{B_\delta} |D^2 u|^2 r^\gamma \, dx - \int_{B_\delta} (\Delta u)^2 r^\gamma \, dx$$
$$= \gamma \sum_{si} \int_0^\infty \left((n-1)|u_{si}'|^2 + (n+\gamma-3)s(s+n-2) \frac{|u_{si}|^2}{r^2} \right) r^{n+\gamma-3} \, dr.$$

On the right-hand side, the term corresponding to $s = 0$ is obviously negative. The term corresponding to $s = 1$ is also negative. Indeed, it is equal to

$$\gamma \int_0^\infty \left((n-1)|u_{1\,i}'|^2 + (n+\gamma-3)(n-1) \frac{|u_{1\,i}|^2}{r^2} \right) r^{n+\gamma-3} \, dr,$$

and using Hardy's inequality, this term is smaller or equal to

$$\gamma(n-1)\left(1+\frac{2}{n+\gamma-4}\right)^2\int_0^\infty |u'_{1i}|^2 r^{n+\gamma-3}\,dr \le 0.$$

Therefore, using again Hardy's inequality,

$$\int_{B_\delta}|D^2u|^2 r^\gamma\,dx - \int_{B_\delta}(\Delta u)^2 r^\gamma\,dx$$

$$\le \gamma\sum_{s\ge 2i}\int_0^\infty\left[(n-1)\frac{(\gamma+n-4)^2}{4}+(\gamma+n-3)\right.$$

$$\left. s(s+n-2)\right]|u_{si}|^2 r^{n+\gamma-5}\,dr.$$

Multiplying and dividing each term by

$$\left[\left(s+\frac{n-2}{2}\right)^2-\left(\frac{\gamma-2}{2}\right)^2\right]^2,$$

we assert that

$$\int_{B_\delta}|D^2u|^2 r^\gamma\,dx - \int_{B_\delta}(\Delta u)^2 r^\gamma\,dx$$

$$\le \gamma(\gamma+n-3)\sup_{s\ge 2}\frac{s(s+n-2)+(n-1)\frac{(\gamma+n-4)^2}{4(\gamma+n-3)}}{\left[\left(s+\frac{n-2}{2}\right)^2-\left(\frac{\gamma-2}{2}\right)^2\right]^2}\int_{B_\delta}(\Delta u)^2 r^\gamma\,dx.$$

It remains to check that the sup is attained at $s=2$, to finally obtain the value of $(M_3(\alpha))^2$, given by (8.56). ◊

We want to state a similar result, without the condition that the function and its gradient vanish at x_0 and on the boundary of $B_\delta(x_0)$. First we give the following property.

Lemma 8.18. *For any ϵ, the following estimates hold:*

$$|u(x_0)|^2 \le 2\left(\fint_{B_\epsilon(x_0)}|u|^2\,dx + \frac{\epsilon^{n+2\alpha}}{\alpha^2}\fint_{B_\epsilon(x_0)}|Du|^2 r^\gamma\,dx\right), \qquad (8.60)$$

$$|Du(x_0)|^2 \le 2\left(\fint_{B_\epsilon(x_0)}|Du|^2\,dx + \frac{\epsilon^{n+2\alpha}}{\alpha^2}\fint_{B_\epsilon(x_0)}|D^2u|^2 r^\gamma\,dx\right). \qquad (8.61)$$

Proof. We just indicate the proof of (8.60). We write, with $x = (r, \theta)$,

$$|u(x_0)|^2 \le 2 \left(|u(x)|^2 + \left| \int_0^r \frac{\partial u}{\partial \rho}(\rho, \theta) \, d\rho \right|^2 \right).$$

We integrate over $B_\epsilon(x_0)$, and write

$$|B_\epsilon| |u(x_0)|^2 \le 2 \left(\int_{B_\epsilon} |u(x)|^2 \, dx \right.$$
$$\left. + \int_S \int_0^\epsilon r^{n+2\alpha} \left| \int_0^r \frac{\partial u}{\partial \rho}(\rho, \theta) \, d\rho \right|^2 r^{n+\gamma-3} \, dr \, dS \right).$$

Using Hardy's inequality and majorizing yields

$$|B_\epsilon| |u(x_0)|^2 \le 2 \left(\int_{B_\epsilon} |u(x)|^2 \, dx + \frac{\epsilon^{n+2\alpha}}{\alpha^2} \int_{B_\epsilon} |Du|^2 r^\gamma \, dx \right),$$

which is exactly the result. $\diamond$

Lemma 8.19. *Suppose u satifies*

$$\int_{B_\delta(x_0)} |D^2 u|^2 r^\gamma \, dx < \infty.$$

Then one may write the estimate

$$\int_{B_{\delta/2}(x_0)} |D^2 u|^2 r^\gamma \, dx \le (M_3^2 + \eta) \int_{B_{\delta/2}(x_0)} |\Delta u|^2 r^\gamma \, dx + C \|u\|_{H^2(B_\delta)}^2, \quad (8.62)$$

where the constant depends only on δ, γ, η.

Proof. Let us consider the cutoff function ζ, introduced in (8.27), and set

$$\tilde{u}(x) = u(x) - u(x_0) - Du(x_0).(x - x_0).$$

Then $\zeta \tilde{u}$ satisfies the conditions of applicability of Lemma 8.17, and hence

$$\int_{B_\delta(x_0)} |D^2(\zeta \tilde{u})|^2 r^{\gamma-4} \, dx \le M_3^2(\alpha) \int_{B_\delta(x_0)} |\Delta(\zeta \tilde{u})|^2 r^\gamma \, dx.$$

Performing the differentiations, using the fact that $D\zeta, D^2\zeta$ vanish on $B_{\delta/2(x_0)}$, and making use of the estimates (8.60), (8.61), one obtains (8.62). $\diamond$

Of course, (8.62) is just an example of a series of estimates that can be written as a consequence of Lemma 8.17. In particular, later on we shall use the estimate

$$\int_{B_{\delta/2(x_0)}} |Du - Du(x_0)|^2 r^{\gamma-2} \, dx \le (M_2^2 + \eta) \int_{B_{\delta/2(x_0)}} |\Delta u|^2 r^\gamma \, dx + C \|u\|_{H^2(B_\delta)}^2.$$
$$(8.63)$$

8.4.2 Main Result

We can now state the main $C^{1+\alpha}_{\text{loc}}$ regularity result. We shall need the additional restrictive condition

$$K'_0 = K\left((1 + |\gamma|)M_3^2(\alpha) + |\gamma|\|\gamma - 4|M_2^2(\alpha)\right) < 1. \tag{8.64}$$

Theorem 8.20. *If we assume (8.4), (8.24), (8.25), (8.26), (8.64), then the solution of (8.1), with $g = 0$, satisfies*

$$u \in \left(C^{1+\alpha}_{\text{loc}}(\Omega)\right)^N$$

with

$$\alpha = \frac{2 - n - \gamma}{2}.$$

The proof will use the iterative process (8.14), introduced in order to construct the limit u. We know that it converges stongly in $H^1(\Omega)$. We first prove the following result.

Proposition 8.21. *We have the estimate, recalling the definition of the subset $\tilde{\Omega}$ (see the Proof of Theorem 8.10),*

$$\int_{\tilde{\Omega}} |D^2 u_k|^2\, dx \leq C, \tag{8.65}$$

where the constant depends only on $\tilde{\Omega}$.

Proof. Let us consider a cutoff function ψ that is equal to 1 in $\tilde{\Omega}$, has support in Ω, and is $C^1(\overline{\Omega})$. We test equation (8.14) with

$$-D_i(\psi^2 D_i u_{k+1}^\nu),$$

and it follows after integration by parts, and exchange of derivatives that

$$\int_\Omega D_i D_j u_{k+1}^\nu D_i(\psi^2 D_j u_{k+1}^\nu)\, dx = \int_\Omega D_i D_j u_k^\nu D_i(\psi^2 D_j u_{k+1}^\nu)\, dx$$

$$-\epsilon \int_\Omega D_j a_i^\nu(x, Du_k) D_i(\psi^2 D_j u_{k+1}^\nu)\, dx + \epsilon \int_\Omega f^\nu D_j(\psi^2 D_j u_{k+1}^\nu)\, dx.$$

Performing differentiation we obtain

$$\int_\Omega \psi^2 |D^2 u_{k+1}^\nu|^2\, dx = \int_\Omega \psi^2 D_i D_j u_{k+1}^\nu(\delta_{il}\delta_{\mu\nu} - \epsilon a_{il}^{\nu\mu}(x, Du_k) D_j D_l u_k^\mu)\, dx$$

$$+ \int_\Omega D_i \psi^2 D_j u_{k+1}^\nu(D_i D_j u_k^\nu - D_i D_j u_{k+1}^\nu)\, dx$$

$$-\epsilon \int_\Omega a_{i\,x_j}^\nu(x, Du_k)\left(\psi^2 D_i D_j u_{k+1}^\nu + D_i \psi^2 D_j u_{k+1}^\nu\right)\, dx$$

$$-\epsilon \int_\Omega a_{il}^{\nu\mu}(x, Du_k) D_j D_l u_k^\mu D_i \psi^2 D_j u_{k+1}^\nu\, dx$$

$$+\epsilon \int_\Omega f^\nu(D_j \psi^2 D_j u_{k+1}^\nu + \psi^2 \Delta u_{k+1}^\nu)\, dx.$$

Majorizing, using assumptions (8.4), (8.26), the definition of K, and the fact that u_k is bounded in H^1, we easily deduce

$$\int_\Omega \psi^2 |D^2 u_{k+1}^\nu|^2 \, dx \le \frac{K+\eta}{2} \int_\Omega \psi^2 (|D^2 u_{k+1}^\nu|^2 \, dx + |D^2 u_k^\nu|^2) \, dx + C.$$

From this iteration, noting that

$$\frac{K+\eta}{2-K-\eta} < 1$$

for sufficiently small η, one easily derives

$$\int_\Omega \psi^2 |D^2 u_k^\nu|^2 \, dx \le C,$$

and thus (8.65) is proven.   ◇

Proof of Theorem 8.20.
We pick $x_0 \in \tilde{\Omega}$, and assume that δ_0 is such that the ball $B_{\delta_0}(x_0)$ is contained in $\tilde{\Omega}$. We can, of course, assume that δ_0 is fixed, by restricting x_0 to a subset of $\tilde{\Omega}$. We next consider a function $\phi(r)$ such that

$$\phi(r) = r^\gamma, \quad \text{in } B_{\delta_0/2}(x_0),$$

$$\phi(r) \text{ has support in } B_{\delta_0}(x_0),$$

$$\phi(r) \text{ is smooth in } B_{\delta_0}(x_0) - B_{\delta_0/2}(x_0).$$

We then test (8.14) with $\phi(r)\Delta u_{k+1}^\nu$. To simplify a little bit, we assume for the calculation that $a_i^\nu(x,p)$ does not depend on x. We first get

$$\int_{B_{\delta_0}} \phi(\Delta u_{k+1}^\nu)^2 \, dx = \int_{B_{\delta_0}} \phi \Delta u_{k+1}^\nu D_i(D_i u_k^\nu - \epsilon a_i^\nu(Du_k)) \, dx$$

$$+ \epsilon \int_{B_{\delta_0}} f^\nu \phi \Delta u_{k+1}^\nu \, dx.$$

Before performing further integration by parts, we can replace Du_k, Du_{k+1}, $a(Du_k)$ in this expression by

$$Du_k - Du_k(x_0), Du_{k+1} - Du_{k+1}(x_0), a(Du_k) - a(Du_k)(x_0).$$

We define

$$a_{il,k}^{\nu\,\mu}(x) = a_{il}^{\nu\,\mu}(Du_k(x)),$$

$$\tilde{a}_{il,k}^{\nu\,\mu}(x) = \int_0^1 a_{il}^{\nu\,\mu}(Du_k(x_0) + \theta(Du_k(x) - Du_k(x_0))) \, d\theta.$$

With this notation, we obtain

$$\int_{B_{\delta_0}} \phi(\Delta u_{k+1}^\nu)^2\, dx$$

$$= \int_{B_{\delta_0}} D_i D_j \phi(D_j u_{k+1}^\nu - D_j u_{k+1}^\nu(x_0))(\delta_{il}\delta_{\mu\nu} - \epsilon \tilde{a}_{il,k}^{\nu\,\mu})$$
$$(D_l u_k^\mu - D_l u_k^\mu(x_0))\, dx$$

$$+ \int_{B_{\delta_0}} D_i \phi(D_j u_{k+1}^\nu - D_j u_{k+1}^\nu(x_0))(\delta_{il}\delta_{\mu\nu} - \epsilon a_{il,k}^{\nu\,\mu}) D_j D_l u_k^\mu\, dx$$

$$+ \int_{B_{\delta_0}} D_j \phi D_i D_j u_{k+1}^\nu (\delta_{il}\delta_{\mu\nu} - \epsilon \tilde{a}_{il,k}^{\nu\,\mu})(D_l u_k^\mu - D_l u_k^\mu(x_0))\, dx$$

$$+ \int_{B_{\delta_0}} \phi D_i D_j u_{k+1}^\nu (\delta_{il}\delta_{\mu\nu} - \epsilon a_{il,k}^{\nu\,\mu}) D_j D_l u_k^\mu\, dx + \epsilon \int_{B_{\delta_0}} f^\nu \phi \Delta u_{k+1}^\nu\, dx.$$

Now on $B_{\delta_0/2}$ we have

$$\phi(r) = r^\gamma, \quad D\phi(r) = \gamma r^{\gamma-1}\frac{x - x_0}{r},$$
$$D_i D_j \phi(r) = \gamma r^{\gamma-2}\left(\delta_{ij} + (\gamma - 2)\frac{(x_i - x_{0,i})(x_j - x_{0,j})}{r^2}\right).$$

Remembering that γ is negative and using Proposition 8.21, we derive the following estimate:

$$\int_{B_{\delta_0/2}} |\Delta u_{k+1}|^2 r^\gamma\, dx$$

$$\le K\gamma(\gamma - 3) \int_{B_{\delta_0/2}} |Du_k - Du_k(x_0)||Du_{k+1} - Du_{k+1}(x_0)| r^{\gamma-2}\, dx$$

$$+ K|\gamma| \int_{B_{\delta_0/2}} |Du_k - Du_k(x_0)||D^2 u_{k+1}| r^{\gamma-1}\, dx$$

$$+ K|\gamma| \int_{B_{\delta_0/2}} |Du_{k+1} - Du_{k+1}(x_0)||D^2 u_k| r^{\gamma-1}\, dx$$

$$+ K|\gamma| \int_{B_{\delta_0/2}} |D^2 u_{k+1}||D^2 u_k| r^\gamma\, dx + \eta \int_{B_{\delta_0/2}} |\Delta u_{k+1}|^2 r^\gamma\, dx + C.$$

Let us set

$$\lambda_k = \int_{B_{\delta_0/2}} |D^2 u_k|^2 r^\gamma\, dx,$$

$$\mu_k = \int_{B_{\delta_0/2}} |Du_k - Du_k(x_0)|^2 r^{\gamma-2}\, dx.$$

Using Hölder's inequality, we can write

$$(1 - \eta) \int_{B_{\delta_0/2}} |\Delta u_{k+1}^\nu|^2 r^\gamma \, dx$$

$$\leq K[\gamma(\gamma - 3)\sqrt{\mu_k \mu_{k+1}} + |\gamma|(\sqrt{\mu_k \lambda_{k+1}} + \sqrt{\lambda_k \mu_{k+1}}) + \sqrt{\lambda_k \lambda_{k+1}}] + C.$$

We shall make use of the following algebraic inequality:

$$[\gamma(\gamma - 3)\sqrt{\mu_k \mu_{k+1}} + |\gamma|(\sqrt{\mu_k \lambda_{k+1}} + \sqrt{\lambda_k \mu_{k+1}}) + \sqrt{\lambda_k \lambda_{k+1}}]^2$$

$$\leq ((1 + |\gamma|)\lambda_k + |\gamma||\gamma - 4|\mu_k)((1 + |\gamma|)\lambda_{k+1} + |\gamma||\gamma - 4|\mu_{k+1}).$$

We also apply (8.63). Collecting results, we finally get

$$\int_{B_{\delta_0/2}} |\Delta u_{k+1}|^2 r^\gamma \, dx$$

$$\leq (K + \eta) \left((1 + |\gamma|)M_3^2(\alpha) + |\gamma||\gamma - 4|M_2^2(\alpha) \right)$$

$$\left(\int_{B_{\delta_0/2}} |\Delta u_{k+1}|^2 r^\gamma \, dx + C \right)^{1/2}$$

$$\times \left(\int_{B_{\delta_0/2}} |\Delta u_k|^2 r^\gamma \, dx + C \right)^{1/2}.$$

Now setting

$$\sigma_k = \int_{B_{\delta_0/2}} |\Delta u_k|^2 r^\gamma \, dx$$

and using the assumption (8.64), we can write

$$\sigma_{k+1}^2 \leq \theta(\sigma_k + C)(\sigma_{k+1} + C)$$

with $\theta < 1$. From this inequality, it easy to deduce that σ_k is bounded, hence also

$$\int_{B_{\delta_0/2}} |\Delta u_k|^2 r^\gamma \, dx \leq C,$$

where the constant depends only on δ_0, γ and not on x_0, provided that the ball $B_{\delta_0(x_0)}$ is contained in $\tilde{\Omega}$. From weak convergence properties this implies in the limit

$$\int_{B_{\delta_0/2}} |\Delta u|^2 r^\gamma \, dx \leq C,$$

which concludes the proof.   ◇

8.5 Almost Everywhere Regularity

Apart from the case $n = 2$, one does not have $C^{1+\alpha}$ regularity for the solution of elliptic systems in general, unless stringent conditions are assumed; see Theorem 8.20. Nevertheless, as we shall see, the lack of regularity affects only isolated points or "small" sets, provided that one knows a priori that the solution is bounded.

8.5.1 Regularity on Neighborhoods of Lebesgue Points

To be more specific about where the regularity holds, we can state the following theorem.

Theorem 8.22. *We assume (8.4) and*

$$b^\nu_{i,x_k} \in L^\infty_{\text{loc}}(\Omega), \ f^\nu \in W^{1,\infty}_{\text{loc}}(\Omega), \tag{8.66}$$

$a^{\nu\mu}_{ij}(x,p), a^\nu_{i,x_k}(x,p)$ *are uniformly continuous (with modulus of continuity bounded on bounded arguments) and concave.*

$$\tag{8.67}$$

We also assume that the solution of (8.1), (8.2) satisfies

$$D_i u^\nu \in L^\infty_{\text{loc}}(\Omega), \quad \forall i, \nu. \tag{8.68}$$

Let $x_0 \in \Omega$ be a Lebesgue point of Du such that

$$\lim_{R \to 0} \fint_{B_R} |Du - \overline{Du_R}|^2 \, dx = 0,$$

where $\overline{Du_R}$ is the average of Du on B_R.
Then one has
$$Du \in C^\alpha(\mathcal{U}(x_0)),$$

where $\mathcal{U}(x_0)$ is a neighborhood of x_0.

We shall only sketch the proof of this result, since the property will be revisited in the next chapter, in a dual form, which will also provide a proof. We refer to M. Giaquinta [40] for related results. By dual version we mean the following: We perform a change of unknown functions by solving the system

$$a^\nu_i(x, Du) = \sigma^\nu_i.$$

From the theory of monotone operators, we can define Du uniquely in terms of σ. Hence we may write the system

$$D_i u^\nu = g^\nu_i(x, \sigma),$$
$$\text{div } \sigma^\nu = f^\nu,$$

where
$$\sigma = (\sigma_i^\nu); \quad \text{div } \sigma^\nu = D_i \sigma_i^\nu.$$

We shall study this system in the next chapter, in the simpler case where the function g_i^ν does not depend explicitly on x. We shall state a result in Section 9.9 asserting that σ is C^δ in some neighborhood of any Lebesgue point. This is clearly equivalent to the statement of Theorem 8.22.

8.5.2 Proof of Theorem 8.22

1. Let x_0 be a point of Ω and τ_R the usual cutoff function, which takes values in $[0, 1]$, is equal to 1 on $B_R(x_0)$, vanishes outside $B_{2R}(x_0)$, and satisfies

$$|D\tau_R| \le \frac{C}{R}.$$

We write the trivial relation

$$\int_\Omega D_i(D_k u^\nu - c_k^\nu) \tau_R^2 D_k a_i^\nu(x, Du)\, dx = \int_\Omega \tau_R^2 D_k a_i^\nu(x, Du) D_k D_i u^\nu\, dx,$$

where c_k^ν is an arbitrary constant.

Assuming that $B_{2R}(x_0) \subset \Omega$, which will be achieved for R sufficiently small, we may perform an integration by parts, and using equation (8.1), we obtain

$$-\int_\Omega (D_k u^\nu - c_k^\nu) D_i \tau_R^2 (a_{i,x_k}^\nu + a_{ij}^{\nu\mu} D_k D_j u^\mu)\, dx - \int_\Omega (D_k u^\nu - c_k^\nu) \tau_R^2 D_k f^\nu\, dx$$

$$= \int_\Omega \tau_R^2 (a_{i,x_k}^\nu + a_{ij}^{\nu\mu} D_k D_j u^\mu) D_k D_i u^\nu\, dx.$$

Defining

$$\Phi(x_0, R) = \frac{\int_{B_R} |D^2 u|^2\, dx}{R^{n-2}}, \tag{8.69}$$

it follows from the preceding calculation and assumptions (8.66), (8.68) that

$$\Phi(x_0, R) \le CR + K \fint_{B_{2R}} |Du - \overline{Du_{2R}}|^2\, dx,$$

and thus, when x_0 is a Lebesgue point of Du, we obtain

$$\Phi(x_0, R) \to 0 \text{ as } R \to 0.$$

2. Take Ω' open with
$$\overline{\Omega'} \subset \Omega.$$

Let ϵ be given and x_0 be a Lebesgue point of Du in Ω'. We first consider R_0 such that

$$B_{3R_0}(x_0) \subset \Omega'.$$

We shall consider $R < R_0$ fixed such that

$$\Phi(x_0, 2R) + R^2 < \frac{\epsilon}{2}.$$

We then consider a neighborhood $\mathcal{U}(x_0)$ of x_0 contained in Ω', possibly depending on R, such that

$$\Phi(\xi, 2R) + R^2 < \epsilon, \quad \forall \xi \in \mathcal{U}(x_0).$$

Finally, we consider $B_{R_0} \cap \mathcal{U}(x_0)$: for ξ in this neighbourhood we have, collecting results,

$$\Phi(\xi, 2R) + R^2 < \epsilon, \quad B_{2R}(\xi) \subset \Omega', \tag{8.70}$$

and thus also, from the local Meyers property (see Theorem 8.3),

$$\left(\fint_{B_R(\xi)} |D^2 u|^p \, dx \right)^{1/p} \leq K_p \left[\left(\fint_{B_{2R}(\xi)} |D^2 u|^2 \, dx \right)^{1/2} + C \right]. \tag{8.71}$$

3. We differentiate (8.1) and write

$$-D_i a_{i,x_k}^\nu(x, Du) - D_i(a_{ij}^{\nu\,\mu}(x, Du)D_j D_k u^\mu) + D_k f^\nu = 0. \tag{8.72}$$

We next introduce the auxiliary problem, with solution denoted by τ_k^ν,

$$\begin{aligned}
-D_i(a_{ij}^{\nu\,\mu}(\xi, \overline{Du_R})D_j \tau_k^\mu) &= 0, \\
D_k u^\nu - \tau_k^\nu|_{\partial B_R(\xi)} &= 0,
\end{aligned} \tag{8.73}$$

where $\overline{Du_R}$ denotes the average of Du on $B_R(\xi)$.

We test both (8.72) and (8.73), with $D_k u^\nu - \tau_k^\nu$, over $B_R(\xi)$. After easy manipulations, we obtain the following relation:

$$\int_{B_R(\xi)} a_{ij}^{\nu\,\mu}(\xi, \overline{Du_R})(D_j \tau_k^\mu - D_j D_k u^\mu)(D_i D_k u^\nu - D_i \tau_k^\nu) \, dx$$

$$= \int_{B_R(\xi)} (a_{ij}^{\nu\,\mu}(x, Du) - a_{ij}^{\nu\,\mu}(\xi, \overline{Du_R}))D_j D_k u^\mu(D_i D_k u^\nu - D_i \tau_k^\nu) \, dx$$

$$- \int_{B_R(\xi)} (f^\nu(x) - f^\nu(\xi))(\Delta u^\nu - D_k \tau_k^\nu) \, dx$$

$$+ \int_{B_R(\xi)} (a_{i,x_k}^\nu(x, Du) - a_{i,x_k}^\nu(\xi, \overline{Du_R}))(D_i D_k u^\nu - D_i \tau_k^\nu) \, dx.$$

Making use of the assumptions (8.66), (8.67), (8.68), we deduce easily the estimate

$$\int_{B_R(\xi)} |D(\tau - Du)|^2 \, dx \leq C \int_{B_R(\xi)} \omega_0(R^2 + |Du - \overline{Du_R}|^2)|D^2u|^2 \, dx$$

$$+ \int_{B_R(\xi)} \omega_1(R^2 + |Du - \overline{Du_R}|^2) \, dx,$$

where $\omega_0(t), \omega_1(t)$ are increasing concave functions, bounded on bounded sets, that are dependent on the modulus of continuity of the data.

Using Hölder's inequality, and the boundedness and concavity of ω_0, ω_1, we derive

$$\int_{B_R(\xi)} |D(\tau - Du)|^2 \, dx$$

$$\leq C|B_R(\xi)| \left(\omega_0 \left(R^2 + \fint_{B_R(\xi)} |Du - \overline{Du_R}|^2 \, dx \right) \right)^{(p-2)/p}$$

$$\left(\fint_{B_R(\xi)} |D^2u|^p \, dx \right)^{2/p} + C|B_R(\xi)|\omega_1 \left(R^2 + \fint_{B_R(\xi)} |Du - \overline{Du_R}|^2 \, dx \right).$$

Using the Poincaré inequality, and the local Meyers's property (8.71), we obtain finally

$$\int_{B_R(\xi)} |D(\tau - Du)|^2 \, dx \leq C \left(\omega_0(R^2 + \Phi(\xi, R)) \right)^{(p-2)/p}$$

$$\left(\int_{B_{2R}(\xi)} |D^2u|^2 \, dx + R^n \right) + CR^n \omega_1(R^2 + \Phi(\xi, R)) \qquad (8.74)$$

4. Returning to the auxiliary problem (8.73), we prove the following property of τ, by successive differentiation (the details of which are left to the reader). We refer also to Theorem 9.10:

$$\|D\tau\|^2_{L^\infty(B_{\frac{R}{2^{\lceil \frac{n}{2} \rceil +1}}}(\xi))} \leq \frac{K}{R^n} \int_{B_R(\xi)} |D\tau|^2 \, dx.$$

Next, if we consider

$$\rho \leq \frac{R}{2^{\lceil \frac{n}{2} \rceil +1}},$$

we deduce that

$$\frac{\int_{B_\rho(\xi)} |D\tau|^2 \, dx}{\rho^{n-2}} \leq K \frac{\rho^2}{R^2} \frac{\int_{B_R(\xi)} |D\tau|^2 \, dx}{R^{n-2}}. \qquad (8.75)$$

Noting that

$$\int_{B_{2\rho}(\xi)} |D^2u|^2 \, dx \leq 2 \int_{B_{2\rho}(\xi)} |D^2u - D\tau|^2 \, dx + 2 \int_{B_{2\rho}(\xi)} |D\tau|^2 \, dx,$$

picking

$$\rho \leq \frac{R}{2^{[\frac{n}{2}]+2}},$$

and using (8.75) and (8.74), we obtain, after some easy manipulations,

$$\Phi(\xi, 2\rho) \leq C \left(\frac{R}{\rho}\right)^{n-2} \left[(\omega_0(R^2 + \Phi(\xi, R)))^{(p-2)/p} (R^2 + \Phi(\xi, 2R)) \right.$$
$$\left. + R^2 \omega_1(R^2 + \Phi(\xi, R))\right] + C \frac{\rho^2}{R^2} \Phi(\xi, R).$$

Let us set

$$\psi(\xi, R) = \phi(\xi, 2R) + R^2.$$

Then the preceding estimate yields

$$\psi(\xi, \rho) \leq C \frac{\rho^2}{R^2} \psi(\xi, R) \left[1 + \left(\frac{R}{\rho}\right)^n ((\omega_0(\psi(\xi, R)))^{(p-2)/p} + \omega_1(\psi(\xi, R)))\right].$$
$$(8.76)$$

Write

$$\beta = \frac{\rho}{R} \leq \frac{1}{2^{[\frac{n}{2}]+2}}.$$

Then (8.76) reads

$$\psi(\xi, \beta R) \leq C\beta^2 \psi(\xi, R) \left[1 + \beta^{-n} ((\omega_0(\psi(\xi, R)))^{(p-2)/p} + \omega_1(\psi(\xi, R)))\right].$$

Note that thanks to (8.70), we have

$$\psi(\xi, R) \leq \epsilon.$$

Fix $0 < \alpha < 1$, and choose β such that

$$2C\beta^{2-2\alpha} \leq 1$$

and ϵ such that

$$\beta^{-n} ((\omega_0(\epsilon))^{(p-2)/p} + \omega_1(\epsilon)) \leq 1.$$

With these choices, we deduce that

$$\psi(\xi, \beta R) \leq \beta^{2\alpha} \psi(\xi, R). \qquad (8.77)$$

Note that ξ may not be a Lebesgue point. So we do not have a priori information on the value of $\psi(\xi, \beta R)$. But from (8.77) one proves, among other things, that

$$\psi(\xi, \beta R) \leq \epsilon,$$

and therefore it is legitimate to replace R by βR in the previous calculations. We can then iterate and obtain

$$\psi(\xi, \beta^k R) \le \beta^{2k\alpha} \psi(\xi, R).$$

Thus in fact,

$$\psi(\xi, \rho) \le C\rho^{2\alpha},$$

which, by Morrey's result, proves that

$$Du \in C^\alpha \text{ in } \mathcal{U}(x_0).$$

This completes the proof of Theorem 8.22. ◇

Remark 8.23. In contrast to the case of the scalar equation, we cannot state a $W^{2,p}_{\text{loc}}$ regularity result for any p larger than Meyers's exponent. On the other hand, under additional smoothness assumptions on the data, we can go a step further and obtain a $C^{2+\alpha}_{\text{loc}}$ regularity property. This will be done explicitly for the dual version in Section 9.8, and thus will not be done here.

8.6 Regularity in the Uhlenbeck Case

8.6.1 Setting of the Problem

It clearly follows from the previous section that if we find a situation where we can prove

$$Du \in \left(L^\infty_{\text{loc}}(\Omega)\right)^{n^2} \tag{8.78}$$

and

$$\Phi(x_0, R) \to 0 \text{ as } R \to 0 \tag{8.79}$$

for any $x_0 \in \Omega$, where $\Phi(x_0, R)$ has been defined in (8.69), then local $C^{1+\alpha}$ regularity will follow. One example of when this situation arises is known as the Uhlenbeck case, which we describe in the following.

We assume

$$a^\nu_i(x, p) = \gamma(|p|)p^\nu_i, \tag{8.80}$$

where the function γ satisfies (we consider only positive arguments)

$$\gamma_0 = \gamma(0) \le \gamma(s) \le \bar{\gamma}, \quad \gamma' > 0, \quad \gamma'' < 0. \tag{8.81}$$

Note that as a consequence of (8.81), we also have

$$s\gamma'(s) \le \bar{\gamma} - \gamma_0.$$

For the function (8.80) we have

$$a^\nu_{i,x_k}(x, p) = 0,$$
$$a^{\nu\mu}_{ij}(x, p) = \gamma(|p|)\delta_{ij}\delta_{\mu\nu} + \frac{\gamma'(|p|)}{|p|}p^\nu_i p^\mu_j.$$

It is then easy to check that assumptions (8.4) and (8.6) are satisfied. Therefore, if we consider the problem

$$-D_i(\gamma(|Du|)D_i u^\nu) = f^\nu,$$
$$u = (H_0^1(\Omega))^N, \tag{8.82}$$

with the further assumption

$$f^\nu \in L^{2n/(n-2)}(\Omega), \tag{8.83}$$

then we can assert that there exists a unique solution of (8.82) that satisfies

$$u^\nu \in W_{loc}^{2,p_0}(\Omega) \text{ for some } p_0, \quad 2 < p_0 < \frac{2n}{n-2}.$$

The fact that there exists a unique solution follows from the remark that u is the unique minimum of the functional

$$J(u) = \int_\Omega F(|Du|^2)\,dx - 2\int_\Omega f.u\,dx,$$

where

$$F(s) = \int_0^s \gamma(s)\,ds.$$

To state the regularity result, we need to strengthen assumption (8.83) into

$$f^\nu \in W_{loc}^{1,n}(\Omega). \tag{8.84}$$

We can then assert the following.

Theorem 8.24. *If we assume (8.81), (8.83), (8.84), then the solution of (8.82) satisfies*

$$u^\nu \in C_{loc}^{1+\delta}(\Omega).$$

8.6.2 Proof of Theorem 8.24

As discussed earlier it is sufficient to prove (8.78), (8.79).

Proof of (8.78)
Let θ be a smooth function, $0 \le \theta \le 1$, with compact support in Ω. We shall use Moser's technique.
   We test (8.82) with

$$-D_k(\theta^2|Du|^s D_k u^\nu).$$

After integration by parts, we arrive at

$$\int_\Omega D_k(\gamma(|Du|)D_i u^\nu)D_i \theta^2 |Du|^s D_k u^\nu\,dx$$
$$+ \int_\Omega \theta^2 D_k(\gamma(|Du|)D_i u^\nu)D_i(|Du|^s D_k u^\nu)\,dx = \int_\Omega \theta^2 D_k f^\nu |Du|^s D_k u^\nu\,dx.$$

Differentiating, we get

$$\int_\Omega \theta^2 |D^2 u|^2 |Du|^s \gamma(|Du|)\, dx$$

$$+ \int_\Omega \theta^2 \gamma'(|Du|)|Du|^{s-1} \left| D\, \frac{|Du|^2}{2} \right|^2 dx$$

$$+ s \int_\Omega \theta^2 \gamma(|Du|)|Du|^{s-2} \left| D\, \frac{|Du|^2}{2} \right|^2 dx$$

$$+ s \int_\Omega \theta^2 \gamma'(|Du|)|Du|^{s-3} Du^\nu . D\, \frac{|Du|^2}{2} Du^\nu . D\, \frac{|Du|^2}{2}\, dx$$

$$= \int_\Omega \theta^2 D_k f^\nu |Du|^s D_k u^\nu\, dx$$

$$- \int_\Omega D_i \theta^2 |Du|^s \left[\gamma(|Du|) D_i\, \frac{|Du|^2}{2} + \frac{\gamma'(|Du|)}{|Du|} D_i u^\nu D_k u^\nu D_k\, \frac{|Du|^2}{2} \right] dx.$$

Given the assumptions on γ, we deduce that

$$\frac{s}{2}\gamma_0 \int_\Omega \theta^2 |Du|^{s-2} \left| D\, \frac{|Du|^2}{2} \right|^2 dx$$

$$\leq \int_\Omega \theta^2 |Df||Du|^{s+1}\, dx + \frac{4\bar\gamma}{s} \int_\Omega |D\theta|^2 |Du|^{s+2}\, dx.$$

This estimate can also been written in the form

$$\frac{s\gamma_0}{(s+2)^2} \int_\Omega \left| D\, \theta |Du|^{(s+2)/2} \right|^2 dx$$

$$\leq \int_\Omega \theta^2 |Df||Du|^{s+1}\, dx + \left(\frac{4\bar\gamma}{s} + \frac{2s\gamma_0}{(s+2)^2} \right) \int_\Omega |D\theta|^2 |Du|^{s+2}\, dx.$$

We now choose a ball

$$B_{2R_0}(\xi) \subset \Omega$$

and assume $R \leq R_0/2$. We consider a smooth function $\theta = \theta_R$ that takes values in $[0, 1]$, equals 1 on $B_{R_0+R}(\xi)$, and vanishes outside $B_{R_0+2R}(\xi)$. Applying the above estimate with this function and using the Poincaré inequality yields

$$\frac{1}{s+2} \left(\int_{B_{R_0+R}} |Du|^{\frac{(s+2)n}{n-2}}\, dx \right)^{(n-2)/n}$$

$$\leq C \int_{B_{R_0+2R}} |Df||Du|^{s+1}\, dx + \frac{C}{R^2} \int_{B_{R_0+2R}} |Du|^{s+2}\, dx$$

$$\leq C_{R_0} \left(\int_{B_{R_0+2R}} |Df|^n \right)^{1/n} \left(\int_{B_{R_0+2R}} |Du|^{\frac{(s+2)n}{n-1}} \, dx \right)^{\frac{(s+1)(n-1)}{(s+2)n}}$$

$$+ \frac{C_{R_0}}{R^2} \left(\int_{B_{R_0+2R}} |Du|^{\frac{(s+2)n}{n-1}} \, dx \right)^{(n-1)/n}.$$

Setting

$$\zeta_{\lambda,R} = \max(1, \| \, |Du| \, \|_{L^\lambda(B_{R_0+R})}),$$

we have proven the following estimate:

$$\zeta_{\frac{(s+2)n}{n-2},R} \leq \left(C_{R_0} \frac{s+2}{R^2} \right)^{1/(s+2)} \zeta_{\frac{(s+2)n}{n-1},2R}.$$

We can then proceed as in Moser's technique. Define the sequences s_j and R_j by

$$s_j = 2 \left(\left(\frac{n-1}{n-2} \right)^j - 1 \right), \quad R_j = \frac{R_0}{2^j},$$

and set

$$z_j = \zeta_{\frac{(s_j+2)n}{n-2},R_j}.$$

We deduce as usual

$$z_j \leq \Pi_{k=0}^j \left(C_{R_0} (4a)^{j+1} \right)^{a^{\frac{-(j+1)}{2}}} z_0$$

with

$$a = \frac{n-1}{n-2},$$

and z_0 is bounded. So z_j is bounded for any j. Letting j go to ∞, we obtain

$$\| |Du| \|_{L^\infty(B_{R_0})} \leq C_{R_0}.$$

So (8.78) has been proven.

Proof of (8.79)
We use the same function θ as above and we now test (8.82) with

$$-D_k(\theta^2 G^{x_0} D_k u^\nu),$$

where G^{x_0} is some Green function attached to a fixed arbitrary point x_0 (which we shall specify later on). We interchange integration by parts and obtain

$$\int_\Omega D_k(\gamma(|Du|)D_i u^\nu)D_i \theta^2 G^{x_0} D_k u^\nu \, dx$$

$$+ \int_\Omega \theta^2 D_k(\gamma(|Du|)D_i u^\nu)D_i(G^{x_0} D_k u^\nu) \, dx = \int_\Omega \theta^2 D_k f^\nu G^{x_0} D_k u^\nu \, dx.$$

We then differentiate the above expressions; first we introduce the symmetric matrix

$$a_{ki} = \gamma(|Du|)\delta_{ki} + \frac{\gamma'(|Du|)}{|Du|}D_i u^\nu D_k u^\nu.$$

Thanks to the properties of the function γ, we clearly have

$$\gamma_0|\xi|^2 \leq a_{k\,i}\xi_i\xi_k \leq (2\bar{\gamma} - \gamma_0)|\xi|^2.$$

The Green function is then defined relative to this matrix, the point x_0, and a domain Q larger than Ω, with Dirichlet conditions on the boundary. It will enjoy the standard properties of Green functions, namely

$$G^{x_0} \in L^{\frac{n}{n-2}-\delta}(Q),$$

$$\frac{c_0}{|x - x_0|^{n-2}} \leq G^{x_0} \leq \frac{c_1}{|x - x_0|^{n-2}},$$

where the constants c_0, c_1 depend only on γ_0, γ_1, and the estimate holds in a neighborhood of x_0 whose closure is in Q. In particular, it holds on $\bar{\Omega}$. After some easy calculations we arrive at the equation

$$\int_\Omega \theta^2 a_{k\,i} D_i G^{x_0} D_k \frac{|Du|^2}{2}\,dx$$

$$+ \int_\Omega \theta^2 G^{x_0}\left[|D^2u|^2\gamma(|Du|) + \frac{\gamma'(|Du|)}{|Du|}\left|D\frac{|Du|^2}{2}\right|^2\right]dx$$

$$= 2\int_\Omega \theta D_i\theta G^{x_0}\left[\gamma(|Du|)D_i\frac{|Du|^2}{2} + \frac{\gamma'(|Du|)}{|Du|}D_iu^\nu D_k u^\nu D_k\frac{|Du|^2}{2}\,dx\right]dx$$

$$+ \int_\Omega \theta^2 G^{x_0} D_k u^\nu D_k f^\nu\,dx.$$

Therefore, we also have

$$\int_\Omega \theta^2 a_{k\,i} D_i G^{x_0} D_k \frac{|Du|^2}{2}\,dx$$

$$+ \frac{1}{2}\int_\Omega \theta^2 G^{x_0}\left[|D^2u|^2\gamma(|Du|) + \frac{\gamma'(|Du|)}{|Du|}\left|D\frac{|Du|^2}{2}\right|^2\right]dx$$

$$\leq 2\int_\Omega |D\theta|^2 G^{x_0}|Du|^2(\gamma(|Du|) + \gamma'(|Du|)|Du|)\,dx$$

$$+ \int_\Omega \theta^2 G^{x_0}|Du||Df|\,dx.$$

In particular, it follows that

$$\int_\Omega \theta^2 G^{x_0}|D^2u|^2\,dx \leq C_\theta.$$

According to the above estimates on the Green function, it follows that

$$\int_\Omega \theta^2 \frac{|D^2 u|^2}{|x - x_0|^{n-2}} \, dx \le C_\theta.$$

Therefore,

$$\int_{B_R(x_0)} \theta^2 \frac{|D^2 u|^2}{|x - x_0|^{n-2}} \, dx \to 0 \quad \text{as } R \to 0.$$

Hence also

$$\frac{\int_{B_R} |D^2 u|^2 \, dx}{R^{n-2}} \to 0 \quad \text{as } R \to 0,$$

and the proof of (8.79) is complete. $\diamond$

8.7 Counterexamples

In Chapter 5 we alluded to solutions of elliptic systems that were in $H^1 \cap L^\infty$ and not continuous. Famous counterexamples to the regularity of solutions of elliptic systems go back to E. De Giorgi [14] and E. Giusti, M. Miranda [47]. The De Giorgi counterexample is the following: For $n \ge 3$, consider the functional

$$J(u) = \int_\Omega \left[|Du|^2 + \left((n-2)\mathrm{div}u + n \frac{x_i x_\nu}{|x|^2} D_i u^\nu \right)^2 \right] dx$$

on $H^1(\Omega; R^n)$, where Ω is the unit ball.
    Then the vector function

$$u(x) = \frac{x}{|x|^b}$$

with

$$b = \frac{n}{2} \{ 1 - [(2n-2)^2 + 1]^{-1/2} \}$$

is the unique minimum of $J(u)$ among functions that are equal to x on the boundary. Note that $u(x)$ is not bounded. The Giusti–Miranda counterexample is the following: Consider the functional

$$J(u) = \int_\Omega \left[|Du|^2 + \left(\mathrm{div}u + \frac{4}{n-2} \frac{u^i u^\nu}{1 + |u|^2} D_i u^\nu \right)^2 \right] dx,$$

where again Ω is the unit ball. Then for n sufficiently large, $x/|x|$ is the unique minimum of $J(u)$ among functions in $H^1 \cap L^\infty(\Omega)$ that are equal to x on the boundary.
    Following A. Koshelev [63], one can also consider the functional (which extends that of De Giorgi) as follows:

$$J(u) = \int_{\Omega} \left[|Du|^2 + \left(c\,\mathrm{div}u + d\frac{x_i x_\nu}{|x|^2} D_i u^\nu \right)^2 \right] dx,$$

still for $u \in H^1(\Omega; R^n)$, where Ω is the unit ball and $u = x$ on the boundary. We shall find the values of c, d in order that the minimum of $J(u)$ be the function

$$\frac{x}{|x|}.$$

Writing the Euler equations yields

$$D_i a_i^\nu(x, Du) = 0$$

with

$$a_i^\nu(x, Du) = a_{ij}^{\nu\mu}(x) D_j u^\mu,$$

where

$$a_{ij}^{\nu\mu}(x) = \delta_{\nu\mu}\delta_{ij} + \left(c\delta_{j\mu} + d\frac{x_j x_\mu}{|x|^2} \right) \left(c\delta_{i\nu} + d\frac{x_i x_\nu}{|x|^2} \right).$$

A direct calculation shows that $x|x|^{-b}$ solves the Euler system with

$$b = \frac{n}{2} - \sqrt{\frac{n^2}{4} - \frac{d(n-1)(nc+d)}{1+(c+d)^2}}$$

and

$$b = 1 \text{ if } d = \frac{c^2+1}{(n-2)c}.$$

The solution is bounded, but not continuous at 0. We shall check that for an appropriate choice of c, the condition (8.25) is not satisfied for any $\gamma < 2-n$. This will prove that this condition is sharp for systems.

First, recalling the notions introduced in Section 8.3.1, we have

$$A(x, p) = a_{ij}^{\nu\mu}(x),$$

which is symmetric. Hence the constant K is given by

$$K = \frac{\Lambda - \lambda}{\Lambda + \lambda}$$

(see Remark 8.7). We next calculate easily

$$\lambda = 1, \quad \Lambda = 1 + c^2(n-1) + (c+d)^2.$$

Hence

$$K = \frac{c^2(n-1) + (c+d)^2}{2 + c^2(n-1) + (c+d)^2}.$$

We choose

$$c = (n-1)^{-1/2} \left[1 + \frac{(n-2)^2}{n-1} \right]^{-1/4}.$$

From this choice of c, and the corresponding value of d, one computes the value

$$K = \left[1 + \frac{(n-2)^2}{n-1} \right]^{-1/2}.$$

Writing

$$\gamma = 2 - n - \epsilon, \quad \epsilon > 0,$$

we obtain

$$K_0 = \frac{\sqrt{1 + \epsilon \frac{n-2}{n^2-3n+3}}}{1 - \frac{(n-2+\epsilon)\epsilon}{2(n-1)}},$$

which is strictly larger than 1 for any positive ϵ.

In the case of diagonal systems with quadratic growth Hamiltonians, we have already referred to the general lack of continuity properties of the solutions. Let us mention as a complement the following example due to J. Frehse [26]. Let $n = N = 2$, and

$$u^1(x) = \sin \log \log \frac{1}{|x|}, \quad u^2(x) = \cos \log \log \frac{1}{|x|}.$$

The pair u^1, u^2 belongs to $H^1 \cap L^\infty$ and is a solution of

$$-\Delta u^1 = 2 \frac{u^1 + u^2}{1 + |u|^2} |Du|^2,$$

$$-\Delta u^2 = 2 \frac{u^2 - u^1}{1 + |u|^2} |Du|^2.$$

We also mention a more recent example of a Lipschitz vector function that is the unique minimum of a smooth convex functional and that is not C^1. In this case, the dimension is $n \geq 5$ (the cases $n = 3, n = 4$ are open). This example is due to W. Hao, S. Leonardi, J. Nečas [54].

Consider

$$\Omega = \{x \in R^n \mid |x| < 1\}, \quad n \geq 5.$$

We work with functions

$$u : \Omega \to R^{n^2},$$

for which we need some notation. We write

$$u = (u_{ij}(x)), \quad i, j = 1, \dots, n,$$

and its gradient is denoted by Du. We introduce a combination of derivatives as follows:

$$\tau_i u = D_i u_{jj},$$

$$\nabla_i u = \frac{1}{2}(D_j(u_{ij} + u_{ji}) + \tau_i u),$$

$$\nabla_{ijk} u = \frac{1}{6}(D_k(u_{ij} + u_{ji}) + D_i(u_{jk} + u_{kj}) + D_j(u_{ki} + u_{ik}))$$

$$- \frac{2n-1}{3(n^2-1)}(\delta_{ij} \nabla_k u + \delta_{jk} \nabla_i u + \delta_{ki} \nabla_j u).$$

We define the function

$$\hat{u} = \frac{x_i x_j}{|x|} - \frac{1}{n}\delta_{ij}|x|.$$

This matrix-valued function will realize the unique global minimum of the functional

$$\Phi(u) = \frac{\alpha}{4}\int_\Omega |Du|^4 \, dx + \frac{\alpha_0}{2}\int_\Omega |Du|^2 \, dx$$

$$+ \frac{\alpha_1}{2}\int_\Omega \nabla_m u \nabla_m u \nabla_{ijk} u \nabla_{ijk} u \, dx$$

$$+ \frac{\alpha_2}{4}\int_\Omega (\nabla_{ijk} u \nabla_{ijk} u) \, dx + \frac{\alpha_3}{4}\int_\Omega (\nabla_i u \nabla_i u)^2 \, dx$$

$$+ \alpha_4 \int_\Omega \nabla_i u \nabla_j u \nabla_k u \nabla_{ijk} u) dx + \alpha_5 \int_\Omega \nabla_i \tau_i \, dx + \frac{\alpha_6}{2}\int_\Omega \tau_i \tau_i \, dx.$$

This functional is minimized on the set

$$\{u \in W^{1,4}(\Omega, R^{n^2}) : u = \hat{u} \text{ on } \partial\Omega\}.$$

The constants are adjusted in order that the gradient of Φ vanish for $u = \hat{u}$, and to obtain the strict convexity of the functional. This leads to extremely tedious calculations, and the condition $n \geq 5$ plays a fundamental role in achieving the convexity. We just give the values of the constants:

$$\alpha = \frac{n}{(2n^2 - n - 1)^3},$$

$$\alpha_0 = \left(\frac{1}{n^2-1}\right)^2,$$

$$\alpha_1 = \begin{cases} \dfrac{n^4}{(n^2+2)(n^2-1)} & \text{if } n \leq 50, \\ 0 & \text{if } n > 50, \end{cases}$$

$$\alpha_2 = n^2 - 1 - \left(\frac{n^2-1}{n}\right)^2 \alpha_1,$$

$$\alpha_3 = \left(\frac{n}{n^2-1}\right)^3 \left[-\alpha_1 \frac{n^2-1}{n} + 3n \right.$$

$$\left. + n(n^2 - n + 4)(\beta_1 + \beta_2) - \beta_3(n+1)\right],$$

$$\alpha_4 = \left(\frac{n}{n^2-1}\right)^2 (1+\beta_1+\beta_2)n,$$

$$\alpha_5 = \begin{cases} \dfrac{n-4+\beta\beta_3(n+2)}{2\beta(n-1)} & \text{if n} \le 50, \\[3mm] \dfrac{\beta\beta_3(n+2)}{2\beta(n-1)} & \text{if n} > 50, \end{cases}$$

$$\alpha_6 = \frac{3\alpha_5^2}{\alpha_0}n^3,$$

$$\beta = \frac{1-n}{n(n^3-n^2-n+1)},$$

$$\beta_1 = -\beta(n+2)\beta_3,$$

$$\beta_2 = 2\beta(n-1)\alpha_5,$$

$$\beta_3 = \alpha_0 + \left(\frac{1}{2n^2-n-1}\right)^2.$$

8.8 Regularity for Mixed Boundary Value Systems

8.8.1 Stating the Problem

We present here a problem studied by C. Ebmeyer and J. Frehse [20], where mixed boundary values are considered. To simplify the presentation and to concentrate on the main ideas, we shall restrict ourselves to the Zaremba problem on a cube. Thus we take

$$\Omega = \{|x_j| < R_0, \ j = 1,\ldots,n-1, 0 < x_n < 2R_0\},$$

and if $\Gamma = \partial\Omega$, we take

$$\Gamma_D = \{x|x_n = 0, 0 \le x_1 \le R_0\}$$

and

$$\Gamma_N = \Gamma - \Gamma_D.$$

We consider a function $a(x,p)$ where $x \in R^n, p \in R^{nN}$ that is C^2, and we set

$$a_{x_k} = \frac{\partial a}{\partial x_k}, \quad a_k^\nu = \frac{\partial a}{\partial p_k^\nu},$$

and as in Section 8.2,

$$a_{k,x_j}^\nu = \frac{\partial a_k^\nu}{\partial x_j},$$

$$a_{kj}^{\nu\mu} = \frac{\partial a_k^\nu}{\partial p_j^\mu},$$

where $j, k = 1,\ldots,n, \ \mu,\nu = 1,\ldots,N$. We assume

$$k_0|\xi|^2 \leq a_{kj}^{\nu\mu}(x,p)\xi_k^\nu\xi_j^\mu \leq k_1|\xi|^2, \forall \xi \in R^{nN}, \ k_0 > 0, \ a \geq 0,$$
$$|a_{x_k}(x,p)| \leq g_{x_k}(x) + C|p|^2, g_{x_k} \in L^1(\Omega),$$
$$|a_k^\nu(x,p)| \leq g_k^\nu(x) + C|p|, g_k^\nu(x) \in L^2(\Omega), \quad (8.85)$$
$$|a_{k,x_j}^\nu(x,p)| \leq g_{k,x_j}^\nu(x) + C|p|, g_{k,x_j}^\nu(x) \in L^2(\Omega),$$

and let also

$$f^\nu(x) \in L^2(\Omega). \quad (8.86)$$

We consider the system of nonlinear equations

$$-D_i a_i^\nu(x, Du) + f^\nu = 0, \quad \nu = 1, \dots, N, \quad (8.87)$$

and the mixed boundary values

$$u^\nu = 0 \text{ on } \Gamma_D, \quad a_i^\nu(x, Du)n_i(x) = 0, \quad x \in \Gamma_N, \quad (8.88)$$

where n represents the outward normal.

We recall the observation that by changing the definition of f^ν we may, without loss of generality, assume that

$$a_i^\nu(x, 0) = 0.$$

Thus all assumptions quoted in Section 8.1 for the existence and uniqueness of a solution of (8.87), (8.88) in $(H^1(\Omega))^N$ are satisfied. We have also, from Theorem 8.1,

$$u \in \left(H^2_{\text{loc}}(\Omega)\right)^N.$$

Our objective is to investigate with the use of fractional derivatives the possible regularity near the boundary $x_n = 0$. We prove the following result.

Theorem 8.25. *Given the assumptions (8.85), (8.86), the solution of (8.87), (8.88) satisfies*

$$u \in \left(W^{1+\frac{1-\delta}{2},2}(\Omega)\right)^N, \quad 0 < \delta < 1.$$

We refer to Section 1.1.1 for the definition of the Besov–Nikol'skĭi spaces.

In the linear theory, in particular in the linear case, much stronger results are available. Indeed, u can be represented as an asymptotic expansion, say

$$u = \tilde{u} + s_1 + \cdots + s_N,$$

where $\tilde{u}$ is smooth and the s_i carry singularities, which behave like $|x|^{\alpha_i}$. In the nonlinear and in the strongly coupled cases, one cannot expect such an expansion. The result of Theorem 8.25 in the above generality is optimal. See also [49], [13], [62].

8.8.2 Proof of Theorem 8.25

We shall use the finite difference operators

$$\Delta_z \phi(x) = \phi(x+z) - \phi(x).$$

Introduce the following subsets of Ω:

$$\Omega_\rho = \{|x_j| < R_0 - \rho, \, j = 1, \ldots, n-1, 0 < x_n < 2R_0 - \rho\}$$

with $0 < \rho \le R_0/3$.

Lemma 8.26. *Let ζ be a cutoff function such that*

$$0 \le \zeta \le 1, \zeta = 1 \text{ on } \Omega_{2\rho}, \quad \text{supp } \zeta \subset \Omega_\rho, \zeta \text{ smooth,}$$

and let $0 < h < \rho$. Then one has the estimate

$$\int_\Omega \frac{\zeta}{h} |\Delta_{he_k} Du|^2 \, dx \le C, \quad 1 \le k \le n-1. \tag{8.89}$$

Proof. The key point is that when $k \le n-1$,

$$\zeta \Delta_{he_k} u^\nu$$

is a possible test function for the ν equation. Indeed, it vanishes on the Dirichlet boundary, since from the form of the domain, $x + he_k$ remains in Γ_D when $x \in \Omega_\rho, x_n = 0$, and $x_1 \ge 0$. Therefore, we obtain

$$\int_\Omega D_i(\zeta \Delta_{he_k} u^\nu) a_i^\nu \, dx + \int_\Omega f^\nu \zeta \Delta_{he_k} u^\nu \, dx = 0.$$

Hence

$$\int_\Omega \zeta D_i \Delta_{he_k} u^\nu \, a_i^\nu \, dx + \int_\Omega D_i \zeta \, \Delta_{he_k} u^\nu \, a_i^\nu \, dx + \int_\Omega f^\nu \zeta \Delta_{he_k} u^\nu \, dx = 0. \tag{8.90}$$

When it simplifies notation, we shall write

$$\bar{u} = u(x + he_k).$$

From the Taylor expansion of the function a, we also have

$$a(x, D\bar{u}) = a(x, Du) + a_i^\nu(x, Du) \, \Delta_{he_k} D_i u^\nu \tag{8.91}$$
$$+ m_{ij}^{\nu\mu} \Delta_{he_k} D_i u^\nu \, \Delta_{he_k} D_j u^\mu$$

with

$$m_{ij}^{\nu\mu} = \int_0^1 (1-t) a_{ij}^{\nu\mu}(x, Du + t\Delta_{he_k} Du) \, dt.$$

Therefore, let

$$I = \int_{\Omega} \frac{\zeta}{h} m_{ij}^{\nu\mu} \Delta_{he_k} D_i u^{\nu} \Delta_{he_k} D_j u^{\mu} \, dx.$$

From assumptions (8.85) we have

$$I \geq k_0 \int_{\Omega} \frac{\zeta}{h} |\Delta_{he_k} Du|^2 \, dx.$$

On the other hand, from (8.91) we have

$$I = \int_{\Omega} \frac{\zeta}{h} (a(x, D\bar{u}) - a(x, Du)) dx - \int_{\Omega} \frac{\zeta}{h} a_i^{\nu}(x, Du) \, \Delta_{he_k} D_i u^{\nu} \, dx,$$

so using (8.90),

$$I = \int_{\Omega} \frac{\zeta}{h} (a(x, D\bar{u}) - a(x, Du)) dx + \int_{\Omega} \frac{D_i \zeta}{h} \Delta_{he_k} u^{\nu} a_i^{\nu} \, dx$$

$$+ \int_{\Omega} f^{\nu} \frac{\zeta}{h} \Delta_{he_k} u^{\nu} \, dx = 0.$$

We denote the various terms by

$$I = II + III + IV.$$

Since the functions u^{ν} are in $H^1(\Omega)$, the first easy observation, thanks to assumption (8.85), is that

$$|III| + |IV| \leq C.$$

Next write II as

$$II = \int_{\Omega} \frac{\zeta}{h} \Delta_{he_k} a(x, Du) \, dx - \int_{\Omega} \frac{\zeta}{h} (a(\bar{x}, D\bar{u}) - a(x, D\bar{u})) dx$$

$$= \int_{\Omega} \Delta_{he_k} \left(\frac{\zeta}{h} a(x, Du) \right) dx - \int_{\Omega} \frac{1}{h} \Delta_{he_k} \zeta \, a(\bar{x}, D\bar{u}) \, dx$$

$$- \int_{\Omega} \frac{\zeta}{h} (a(\bar{x}, D\bar{u}) - a(x, D\bar{u})) dx$$

and write

$$II = - \int_{\Omega} \frac{1}{h} \Delta_{he_k} \zeta \, a(\bar{x}, D\bar{u}) \, dx$$

$$- \int_{\Omega} \zeta \int_0^1 a_{x_k}(x + the_k, D\bar{u}) dt \, dx,$$

where we have made use of the fact that ζ vanishes on the sets $\Omega + he_k - \Omega \cap (\Omega + he_k)$, $\Omega - \Omega \cap (\Omega + he_k)$, since $h < \rho$. Using assumption (8.85) and the fact that the u^{ν} are in $H^1(\Omega)$, we have

$$|II| \leq C.$$

The proof is complete.

We deduce immediately the following corollary.

Corollary 8.27.

$$\int_\Omega \frac{\zeta}{h}|\Delta_{he_k}\, a_j^\nu(x, Du)|^2\, dx \le C, \quad 1 \le k \le n-1,$$

$$j = 1, \ldots, n, \quad \nu = 1, \ldots, N. \tag{8.92}$$

Proof. We use Taylor's formula

$$\Delta_{he_k}\, a_j^\nu(x, Du) = h\int_0^1 a_{j,x_k}^\nu(x + the_k, Du(x + he_k))dt$$

$$+ \int_0^1 a_{ji}^{\nu\mu}(x, Du + t\Delta_{he_k} Du)\Delta_{he_k} D_i u^\mu\, dt$$

and assumption (8.85). The result is then an immediate consequence of (8.89).
◇

To estimate the normal derivative, we use a result whose proof, relying on Fourier analysis, is postponed to the next section. To state this result we need some notation. Let us set

$$R = R_0 - 2\rho$$

and consider the cube

$$Q_\sigma = \{|x_i| < R, i = 1, \ldots, n-1, \sigma R < x_n < (\sigma + 2)R\},$$

where

$$0 < \sigma < \frac{2\rho}{R_0 - 2\rho}$$

and in fact, σ will tend to 0. At any rate,

$$Q_\sigma \subset \Omega_{2\rho}.$$

We also introduce the cube

$$Q_\sigma^* = \{|x_i| < R, i = 1, \ldots, n-1, \sigma R < x_n < (\sigma + 1)R\},$$

and we notice that since $\rho < R_0/3$,

$$Q_\sigma^* + he_n \subset Q_\sigma \quad \text{for } h < \rho.$$

The key result is stated now as follows.

Lemma 8.28. *Consider functions H, $G_k, k = 1, \ldots, n-1$, that are locally H^1 in Ω and satisfy*

$$D_n H = \sum_{k=1}^{n-1} D_k G_k + g \tag{8.93}$$

with $g \in L^2(\Omega)$. Assume that for $h < \rho$,

$$\frac{1}{h} \int_{Q_\sigma} |\Delta_{he_k} G_k|^2 \, dx \le C, \quad 1 \le k \le n-1, \tag{8.94}$$

and

$$\frac{1}{h} \int_{Q_\sigma} |\Delta_{he_j} H|^2 \, dx \le C, \quad 1 \le j \le n-1, \tag{8.95}$$

where the constants do not depend on σ. Then one has

$$\frac{1}{h^{1-\delta}} \int_{Q_\sigma^*} |\Delta_{he_n} H|^2 \, dx \le C_\delta \tag{8.96}$$

for any $0 < \delta < 1$, where C_δ does not depend on σ.

The proof is postponed to the next section. Instead, we show how to make use of the preceding results to complete the proof of Theorem 8.25.

Proof of Theorem 8.25.
We first apply Lemma 8.28, with

$$H = D_j u^\nu, \quad 1 \le j \le n-1,$$
$$G_j = D_n u^\nu, \quad G_k = 0 \quad \text{if } k \ne j, \quad g = 0.$$

Clearly, (8.93) holds. We then notice that $Q_\sigma \subset \Omega_{2\rho}$ and apply Lemma 8.26, using the fact that $\zeta = 1$ on Q_σ. Hence (8.94), (8.95) are also satisfied. Therefore, applying (8.96) we assert that

$$\frac{1}{h^{1-\delta}} \int_{Q_\sigma^*} |\Delta_{he_n} D_j u^\nu|^2 \, dx \le C_\delta, \quad \forall 1 \le j \le n-1. \tag{8.97}$$

We then apply Lemma 8.28 differently, with

$$H = a_n^\nu(x, Du),$$
$$G_k = -a_k^\nu(x, Du), g = f^\nu.$$

From the ν equation, (8.93) clearly follows. Using Corollary 8.27 and arguing as above, we see that again (8.94), (8.95) hold. Therefore, applying (8.96), we now obtain

$$\frac{1}{h^{1-\delta}} \int_{Q_\sigma^*} |\Delta_{he_n} a_n^\nu(x, Du)|^2 \, dx \le C_\delta. \tag{8.98}$$

Then, using the same Taylor expansion once again, we can write (as in the proof of the above corollary)

$$\int_0^1 a_{n\,n}^{\nu\,\mu}(x, Du + t\Delta_{he_n} Du)\Delta_{he_n} D_n u^\mu \, dt = \Delta_{he_n} a_n^\nu(x, Du)$$

$$-h \int_0^1 a_{n,x_n}^\nu(x + the_n, Du(x + he_n)) dt$$

$$-\sum_{i=1}^{n-1} \int_0^1 a_{n\,i}^{\nu\,\mu}(x, Du + t\Delta_{he_n} Du)\Delta_{he_n} D_i u^\mu \, dt.$$

Noting that the matrix

$$\int_0^1 a_{n\,n}^{\nu\,\mu}(x, Du + t\Delta_{he_n} Du)dt$$

is uniformly positive definite, and making use of the assumption (8.85), we deduce the inequality

$$\sum_\mu |\Delta_{he_n} D_n u^\mu|^2 \le C \left(\sum_\nu |\Delta_{he_n} a_n^\nu|^2 + h^2 \int_0^1 (g_{n,x_n}^\nu (x + the_n))^2 dt \right.$$

$$\left. \times\ |Du(x + he_n)|^2 + \sum_{i=1}^{n-1} \sum_\mu |\Delta_{he_n} D_i u^\mu|^2 \right).$$

Then, thanks to estimates (8.97) and (8.98), we obtain

$$\frac{1}{h^{1-\delta}} \int_{Q_\sigma^*} |\Delta_{he_n} D_n u^\mu|^2 \, dx \le C_\delta.$$

Therefore, collecting results, we have

$$\frac{1}{h^{1-\delta}} \int_{Q_\sigma^*} |\Delta_{he_j} Du|^2 \, dx \le C_\delta, \quad \forall j = 1, \ldots, n.$$

We can then let σ tend to 0 and obtain

$$\frac{1}{h^{1-\delta}} \int_{Q_0^*} |\Delta_{he_j} Du|^2 \, dx \le C_\delta, \quad \forall j = 1, \ldots, n.$$

Since Q_0^* differs from $\Omega_{2\rho}$ only on the part $R_0 - 2\rho \le x_n \le 2R_0 - 2\rho$, on which the solution is H^2 anyway, we can state

$$\frac{1}{h^{1-\delta}} \int_{\Omega_{2\rho}} |\Delta_{he_j} Du|^2 \, dx \le C_\delta, \quad \forall j = 1, \ldots, n, h < \rho.$$

Now if we consider any translation h, we can assert that

$$\frac{1}{|h|^{1-\delta}} \int_{\{x \in \Omega_{3\rho} | x+h \in \Omega\}} |\Delta_h Du|^2 \, dx \le C_\delta, \quad |h| < \rho.$$

Note that translations $\pm h\,e_j, j = 1, \ldots, n-1$, with $0 < h < \rho$ are permitted, whereas with general positive h, only $h\,e_n$ is permitted.

To treat the remainder $\Omega - \Omega_{3\rho}$, we proceed with a finite covering of Ω with cubes, and by extending the definition of the equations by reflection on the Neumann part of the boundary. The boundary $x_1 = R_0$ will produce a situation with mixed Dirichlet and Neumann conditions, which is treated as above. All together, we finally have

$$\frac{1}{|h|^{1-\delta}} \int_{\{x \in \Omega | x + h \in \Omega\}} |\Delta_h Du|^2 dx \leq C_\delta,$$

and from the definition of the Besov–Nikol'skiĭ spaces the conclusion is obtained.

Remark 8.29. Instead of taking f^ν as data, we could take the function

$$f^\nu + \sum_{i=1}^{n} D_i f_i^\nu$$

with

$$f_i^\nu \in H^1(\Omega),$$

and the Neumann boundary condition then becomes

$$(-a_i^\nu(x, Du) + f_i^\nu(x))n_i(x) = 0, \quad x \in \Gamma_N.$$

8.8.3 Proof of Lemma 8.28

It remains to give the proof of Lemma 8.28. The method will consist of extending the definition of H, G_{k}, g by reflection to construct periodic functions, and using specific results available for periodic functions, obtained by Fourier representation. For that purpose we shall need the cube

$$\hat{Q}_\sigma = \{x | -R < x_1 < 3R, \ldots, -R < x_{n-1} < 3R, (\sigma - 2)R < x_n < (\sigma + 2)R\},$$

which extends Q_σ and therefore is not included in $\Omega_{2\rho}$. We consider functions that coincide with H, G_{k}, g on Q_σ, and complete their definition on $\hat{Q}_\sigma$ by reflection, so they are periodic in all directions, with period $4R$. We call those extensions

$$\hat{H}, \hat{G}_k, \hat{g}.$$

They satisfy, by construction,

$$\hat{H}(x \pm 4Re_k) = \hat{H}(x), \forall k = 1, \ldots, n,$$
$$\hat{H}(x) = \hat{H}(x + 2(R - x_j)e_j) \text{ if } R < x_j < 3R,$$
$$\hat{H}(x) = \hat{H}(x + 2(\sigma R - x_n)e_n) \text{ if } (\sigma - 2)R < x_n < \sigma R,$$
$$\hat{H}(x) = H(x) \text{ on } Q_\sigma,$$

and similar formulas for $\hat{G}_k, \hat{g}$.

Define also

$$\chi(s) = \begin{cases} 1 & \text{if } -R < s < R, \\ -1 & \text{if } R < s < 3R, \end{cases}$$

and $\hat{\chi}$ its periodic extension. The functions $\hat{H}, \hat{G}_k$ are $H^1(R^n)$, and we have, obviously,

$$D_n \hat{H} = \hat{\chi}(x_n) \widehat{D_n H}.$$

Therefore, the relation (8.93) implies

$$D_n \hat{H} = \hat{\chi}(x_n) \left(\sum_{k=1}^{n-1} \widehat{D_k G_k} + \hat{g} \right). \tag{8.99}$$

It is now crucial to investigate what properties (8.94), (8.95) imply about the corresponding extensions $\hat{H}$, $\hat{G}_k$. We claim that for $h < R$,

$$\frac{1}{h} \int_{\hat{Q}_\sigma} |\Delta_{he_j} \hat{H}|^2 \, dx \leq C \frac{1}{h} \int_{Q_\sigma} |\Delta_{he_j} H|^2 \, dx$$

$$+ C \frac{1}{h} \int_0^1 \int_{Q_\sigma} |\Delta_{\theta h e_j} H|^2 \, dx \, d\theta. \tag{8.100}$$

It is easy to convince oneself that (8.100) amounts to the following one-dimensional property:

$$\int_{-R}^{3R} |\hat{H}(x+h) - \hat{H}(x)|^2 \, dx \leq C \int_{-R}^{R} |H(x+h) - H(x)|^2 \, dx$$

$$+ C \int_0^1 \int_{-R}^{R} |H(x+h\theta) - H(x)|^2 \, dx \, d\theta \tag{8.101}$$

for $h < R$. One may wonder why the values of H beyond R, which intervene in the right-hand side of (8.101) through $x + h$, play a role, since from the construction of $\hat{H}$ only the values of H between $-R$ and R are used in the left-hand side. As we shall see, more accurate estimates could not be written as conveniently.

Decomposing the integral of the left-hand side of (8.100), we check easily that from the definition of $\hat{H}$ we have

$$\int_{-R}^{3R} |\hat{H}(x+h) - \hat{H}(x)|^2 \, dx = 2 \int_{-R}^{R-h} |H(x+h) - H(x)|^2 \, dx$$

$$+ \int_{R-h}^{R} |H(2R - x - h) - H(x)|^2 \, dx$$

$$+ \int_{R-h}^{R} |H(x+h-2R) - H(-x)|^2 \, dx.$$

The first integral is controlled, clearly, by the first integral of the right-hand side of (8.101). The other two are treated in a similar way, which we explain only for the last one. We proceed by averaging the arguments. First we notice that

$$\int_{R-h}^{R} |H(x+h-2R) - \int_0^1 H(x+h-2R+h\theta) d\theta|^2 \, dx$$

$$\leq \int_0^1 \int_{-R}^{-R+h} |H(x+h\theta) - H(x)|^2 \, dx,$$

which is controlled by the second integral of the right-hand side of (8.101). The same occurs with the quantity

$$\int_{R-h}^{R} |H(-x) - \int_0^1 H(-x + h\theta)d\theta|^2 \, dx.$$

We then consider the quantity

$$\int_{R-h}^{R} \left| \int_0^1 (H(x + h - 2R + h\theta) - H(-x + h\theta))d\theta \right|^2 dx$$

$$\leq \int_0^1 \int_{R-h}^{R-h/2} |H(x + h - 2R + h\theta) - H(-x + h\theta)|^2 \, dxd\theta$$

$$+ \int_0^1 \int_{R-h/2}^{R} |H(x + h - 2R + h\theta) - H(-x + h\theta)|^2 \, dxd\theta$$

$$= \frac{1}{h} \int_{R-h}^{R-h/2} \left(\int_{x+h-2R}^{x+2h-2R} |H(\eta + 2R - h - 2x) - H(\eta)|^2 \, d\eta \right) dx$$

$$+ \frac{1}{h} \int_{R-h/2}^{R} \left(\int_{-x}^{-x+h} |H(\eta + 2x + h - 2R) - H(\eta)|^2 \, d\eta \right) dx$$

$$\leq \frac{1}{h} \int_{R-h}^{R-h/2} \int_{-R}^{R} |H(\eta + 2R - h - 2x) - H(\eta)|^2 \, d\eta \, dx$$

$$+ \frac{1}{h} \int_{R-h/2}^{R} \int_{-R}^{R} |H(\eta + 2x + h - 2R) - H(\eta)|^2 \, d\eta \, dx.$$

By making changes of variables $2R - h - 2x = h\theta$ in the first and $2x + h - 2R = h\theta$ in the second integrals, then changing η into x, we control these terms by the second integral of the right-hand side of (8.101). Collecting results, the estimate (8.101) follows.

For G_k a related, but not equivalent, property will be used. Namely,

$$h \int_{\hat{Q}_\sigma} \left| \int_0^1 \widehat{D_k G_k}(x + h\theta e_k) \, d\theta \right|^2 dx \leq C \frac{1}{h} \int_{Q_\sigma} |\Delta_{he_k} G_k|^2 \, dx \tag{8.102}$$
$$+ C \frac{1}{h} \int_0^1 \int_{Q_\sigma} |\Delta_{\theta he_k} G_k|^2 \, dx \, d\theta.$$

This also relies on a one-dimensional property, which we write as follows:

$$h^2 \int_{-R}^{3R} \left| \int_0^1 \widehat{G'}(x + h\theta)d\theta \right|^2 dx \leq C \int_{-R}^{R} |G(x + h) - G(x)|^2 \, dx \tag{8.103}$$
$$+ C \int_0^1 \int_{-R}^{R} |G(x + h\theta) - G(x)|^2 \, dx \, d\theta.$$

Checking (8.103) is a little bit more involved than (8.101), but the ideas are identical, so we omit the proof.

We make now use of the assumptions (8.94), (8.95) and of the properties (8.100), (8.102) to assert that for $h < \rho$, one has

$$h \int_{\hat{Q}_\sigma} \left| \int_0^1 \widehat{D_k G_k}(x + h\theta e_k)\, d\theta \right|^2 dx \le C \tag{8.104}$$

and

$$\frac{1}{h} \int_{\hat{Q}_\sigma} |\Delta_{he_j} \hat{H}|^2\, dx \le C, \quad \forall j = 1, \ldots, n - 1. \tag{8.105}$$

Thanks to the relation (8.99), we make use of (8.104) for each k to yield the assertion

$$h \int_{\hat{Q}_\sigma} \left| \int_0^1 \cdots \int_0^1 D_n \hat{H}(x + h(\theta_1 e_1 + \cdots + \theta_{n-1} e_{n-1})) d\theta_1 \cdots d\theta_{n-1} \right|^2 dx \le C. \tag{8.106}$$

We now are equipped to make use of the Fourier decomposition. Indeed, we have the Fourier property

$$\hat{H}(x) = \sum_{m = (m_1, \ldots, m_n)} \beta_m \exp \frac{\pi i}{2R}(m_1(x_1 - R) + \cdots + m_{n-1}(x_{n-1} - R)$$
$$+ m_n(x_n - \sigma R)).$$

We have, of course,

$$\sum_m |\beta_m|^2 = \left(\frac{1}{4R} \right)^n \int_{\hat{Q}_\sigma} |\hat{H}|^2 \le C.$$

But properties (8.105), (8.106) yield additional information. From (8.105), we easily deduce

$$\sum_m |\beta_m|^2 \frac{\sin^2(m_j h)}{h} \le C, \quad \forall h < \bar{\rho} = \frac{\pi \rho}{4R}, \tag{8.107}$$

and from (8.106) we get

$$h^{3-2n} \sum_m |\beta_m|^2 |m_n|^2 \prod_{k=1}^{n-1} \frac{\sin^2(m_k h)}{|m_k|^2} \le C, \quad \forall h < \bar{\rho} = \frac{\pi \rho}{4R}. \tag{8.108}$$

To proceed, we reason almost identically for both (8.107) and (8.108). The trickier one is (8.108), and we concentrate on it.

We can assert that (8.108) holds for $h = 2^{-j}$, for $j \ge j_0$; hence

$$2^{j(2n-3)} \sum_m |\beta_m|^2 |m_n|^2 \prod_{k=1}^{n-1} \frac{\sin^2\left(\frac{m_k}{2^j}\right)}{|m_k|^2} \le C.$$

Among other things, it follows that

$$I = \sum_{j=j_0}^{+\infty} \frac{2^{j(2n-3)}}{j^2} \sum_{\{m|2^{j-2}\pi < \max_{k=1}^{n-1} |m_k| \le 2^{j-1}\pi\}} |\beta_m|^2 |m_n|^2 \prod_{k=1}^{n-1} \frac{\sin^2\left(\frac{m_k}{2^j}\right)}{|m_k|^2} \le C.$$

Using the fact that $\sin x \ge \frac{x}{2}$, for $0 < x < \frac{\pi}{2}$, we also have

$$\sum_{j=j_0}^{+\infty} \frac{1}{j^2 2^j} \sum_{\{m|2^{j-2}\pi < \max_{k=1}^{n-1} |m_k| \le 2^{j-1}\pi\}} |\beta_m|^2 |m_n|^2 \le C.$$

Now

$$2^{j-2}\pi < \max_{k-1}^{n-1} |m_k| \Rightarrow \frac{2^{j(1+\delta)}}{\sum_{k=1}^{n-1} |m_k|^{1+\delta}} < 2^{1+\delta}.$$

Therefore, we can state that

$$\sum_{j=j_0}^{+\infty} \frac{2^{j\delta}}{j^2} \sum_{\{m|2^{j-2}\pi < \max_{k=1}^{n-1} |m_k| \le 2^{j-1}\pi\}} \frac{|\beta_m|^2 |m_n|^2}{\sum_{k=1}^{n-1} |m_k|^{1+\delta}} \le C.$$

Since for $\delta > 0$, $2^{j\delta}/j^2$ is bounded below, we have obtained

$$\sum_m \frac{|\beta_m|^2 |m_n|^2}{\sum_{k=1}^{n-1} |m_k|^{1+\delta}} \le C. \tag{8.109}$$

We claim that a similar argument, applied to (8.107), and used for any j, yields

$$\sum_m |\beta_m|^2 \sum_{k=1}^{n-1} |m_k|^{1-\delta} \le C. \tag{8.110}$$

From Young's inequality, we assert that

$$|m_n|^{1-\delta} \le \frac{1-\delta}{2} \frac{|m_n|^2}{\sum_{k=1}^{n-1} |m_k|^{1+\delta}} + \frac{1+\delta}{2} \sum_{k=1}^{n-1} |m_k|^{1-\delta},$$

and thus from (8.109) and (8.110) we obtain

$$\sum_m |\beta_m|^2 |m_n|^{1-\delta} \le C. \tag{8.111}$$

Then recalling the relation

$$4 \sum_m |\beta_m|^2 \sin^2\left(\frac{\pi}{4R} m_n h\right) = \left(\frac{1}{4R}\right)^n \int_{\hat{Q}_\sigma} |\Delta_{he_n} \hat{H}|^2 \, dx$$

and using $|\sin x| \le |x|$, it follows from (8.111) that

$$\frac{1}{h^{1-\delta}} \int_{\hat{Q}_\sigma} |\Delta_{he_n} \hat{H}|^2 \, dx \leq C.$$

Now if we take $x \in Q_\sigma^*$ and $h < \rho$, we have

$$\Delta_{he_n} \hat{H} = \Delta_{he_n} H,$$

and thus the result (8.96) is obtained, which concludes the proof of the lemma.

8.8.4 Further Regularity

We shall assume here that

$$u^\nu \in C^\alpha(\overline{\Omega}). \tag{8.112}$$

In the scalar case, it is easy to give sufficient conditions for which (8.112) holds. Namely, if the a_i^ν reduce to one a_i and if we assume

$$f(x), a_{i,x_i}(x, 0) \in L^p(\Omega), \quad p > \frac{n}{2},$$

then (8.112) holds. It is enough to notice that the problem reduces to a linear one, writing

$$a_{i\,k}(x) = \int_0^1 a_{i\,k}(x, \theta \, Du) \, d\theta$$

and

$$Au = -D_i(a_{i\,k}D_ku),$$

so one has

$$Au = -D_i a_i(x, Du) + a_{i,x_i}(x, 0) = -f + a_{i,x_i}(x, 0).$$

The linear theory (see Sections 2.3.1 and 2.3.2) yields the result (8.112).

Now from the Miranda–Nirenberg interpolation result (1.20) we can assert that

$$u^\nu \in W^{1,q}(\Omega), \quad \forall q < \frac{n(3 - \delta)}{n - \alpha(1 - \delta)}.$$

In particular, picking

$$\delta < \frac{3\alpha}{n + 3\alpha},$$

we obtain

$$u^\nu \in L^{3+\epsilon_0}(\Omega), \quad \epsilon_0 < \frac{3\alpha}{n - \alpha}. \tag{8.113}$$

We then derive a regularity result for the second derivative. More precisely, we have the following theorem.

Theorem 8.30. *We make the assumptions (8.85), (8.86), and assume that the solution of (8.87), (8.88) satisfies*

$$u \in (C^\alpha(\Omega))^N.$$

Then we also have

$$u \in (W^{2,p}(\Omega))^N$$

with

$$p < \frac{6 + 2\epsilon_0}{5 + \epsilon_0}, \quad \epsilon_0 < \frac{3\alpha}{n - \alpha}.$$

Consider the distance function defined by

$$d_\epsilon(x) = \|x_1\| + x_n - \epsilon\|$$

and notice that for any $\beta > 0$, one has

$$\int_\Omega d_\epsilon^{-2+\beta}(x) \mathbb{1}_{\{|x_1|>\epsilon\}} \, dx \le C. \tag{8.114}$$

We then introduce

$$w_\epsilon(x) = d_\epsilon^\gamma(x) \mathbb{1}_{\{|x_1|>\epsilon\}}$$

with

$$2 > \gamma > 1.$$

Then by this choice one has

$$w_\epsilon \in W^{1,\infty}(\Omega).$$

Moreover, we note also that

$$\frac{|Dw_\epsilon|}{w_\epsilon^{1/2}} \mathbb{1}_{\{|x_1|>\epsilon\}} = \gamma\sqrt{2} d_\epsilon(x)^{\gamma/(2-1)} \mathbb{1}_{\{|x_1|>\epsilon\}},$$

and thus from (8.114),

$$\int_\Omega \left(\frac{|Dw_\epsilon|}{w_\epsilon^{1/2}}\right)^\lambda \mathbb{1}_{\{|x_1|>\epsilon\}} \le C \quad \text{if } \lambda < \frac{4}{2 - \gamma}. \tag{8.115}$$

Similarly, we can assert that

$$\int_\Omega w_\epsilon^{-\mu} \mathbb{1}_{\{|x_1|>\epsilon\}} \le C \quad \text{if } \mu < \frac{2}{\gamma}. \tag{8.116}$$

We will make the choices of the various constants precise later on.

We now prove the following lemma.

Lemma 8.31.

$$\int_{\Omega_{2\rho}} w_0 |D^2 u|^2 \, dx \le C_\rho, \tag{8.117}$$

where Ω_ρ has been defined in Section 8.8.2.

Proof. Consider the function

$$-\Delta_{-he_k}(\zeta^2 w_\epsilon \Delta_{he_k} u^\nu), \quad 1 \le k \le n-1,$$

where ζ is the cutoff function introduced in the proof of Theorem 8.25, whose support is in Ω_ρ and equals 1 on $\Omega_{2\rho}$, and $h < \rho$. Then it is an admissible test function for equation (8.87), provided that $h < \epsilon$. One has to check that it vanishes for $x_0 = 0, x_1 \ge 0$. This is in question only when $0 < x_1 < h$, for which w_ϵ vanishes. We consider the integrals

$$I = \int_\Omega a_i^\nu D_i(\Delta_{-he_k}(\zeta^2 w_\epsilon \Delta_{he_k} u^\nu)) \, dx$$

and

$$II = -\int_\Omega f^\nu \Delta_{-he_k}(\zeta^2 w_\epsilon \Delta_{he_k} u^\nu) \, dx.$$

We have by (8.87)

$$I = II.$$

Observing that the support of ζ is contained in $\Omega - he_k$, we may write

$$I = \int_\Omega \Delta_{he_k} a_i^\nu D_i(\zeta^2 w_\epsilon \Delta_{he_k} u^\nu) \, dx = I_1 + I_2$$

with

$$I_1 = \int_\Omega \Delta_{he_k} a_i^\nu D_i(\Delta_{he_k} u^\nu) \zeta^2 w_\epsilon \, dx,$$

$$I_2 = \int_\Omega \Delta_{he_k} a_i^\nu \Delta_{he_k} u^\nu D_i(\zeta^2 w_\epsilon) \, dx.$$

On the other hand, we recall that, from assumption (8.85),

$$\Delta_{he_k} a_i^\nu = h \int_0^1 a_{i,x_k}^\nu(x + \theta he_k, Du(x + he_k)) \, d\theta$$

$$+ \int_0^1 a_{ij}^{\nu\mu}(x, Du + \theta \Delta_{he_k} Du) \Delta_{he_k} D_j u^\mu \, d\theta.$$

Using this expression and assumption (8.85) again, we obtain the estimates

$$I_1 \ge k_0 \int_\Omega |\Delta_{he_k} Du|^2 \zeta^2 w_\epsilon \, dx$$

$$-h \int_\Omega \left(\int_0^1 g_{i,x_k}^\nu(x + \theta he_k) d\theta + C|Du(x + he_k)| \right) |\Delta_{he_k} D_i u^\nu| \zeta^2 w_\epsilon \, dx$$

and thus

$$I_1 \geq \frac{k_0}{2} \int_\Omega |\Delta_{he_k} Du|^2 \zeta^2 w_\epsilon \, dx - Ch^2.$$

We proceed with

$$|I_2| \leq \eta \int_\Omega |\Delta_{he_k} Du|^2 \zeta^2 w_\epsilon \, dx$$

$$+ C\frac{h^2}{\eta} \int_{\Omega_\rho} \left|\frac{\Delta_{he_k} u}{h}\right|^2 \frac{|D(\zeta^2 w_\epsilon)|^2}{\zeta^2 w_\epsilon} \mathrm{II}_{\{|x_1|\epsilon\}} \, dx$$

$$+ h^2 \int_\Omega \left(\int_0^1 \sum_{i,\nu} g^\nu_{i,x_k}(x + \theta h e_k) \, d\theta + C|Du(x + he_k)|\right)$$

$$|D(\zeta^2 w_\epsilon)| \left|\frac{\Delta_{he_k} u}{h}\right| dx,$$

where η is arbitrary.

In estimating the $|I_2|$ terms, we notice first that

$$\frac{|D(\zeta^2 w_\epsilon)|^2}{\zeta^2 w_\epsilon} \leq C + \zeta\frac{|Dw_\epsilon|^2}{w_\epsilon},$$

and then, using the estimate (8.113) and Hölder's inequality, we get

$$\int_{\Omega_\rho} \left|\frac{\Delta_{he_k} u}{h}\right|^2 \frac{|D(\zeta^2 w_\epsilon)|^2}{\zeta^2 w_\epsilon} \mathrm{II}_{\{|x_1|\epsilon\}} \, dx \leq C\|Du\|^2_{L^{3+\epsilon_0}(\Omega)} + C$$

$$+ \int_\Omega \left(\frac{|Dw_\epsilon|}{w_\epsilon^{1/2}}\right)^{\frac{2(3+\epsilon_0)}{1+\epsilon_0}} \mathrm{II}_{\{|x_1|>\epsilon\}} \, dx.$$

Now using the statement (8.115), in which we choose

$$\lambda = \frac{2(3 + \epsilon_0)}{1 + \epsilon_0},$$

and letting

$$\gamma > \frac{4}{3 + \epsilon_0},$$

we find that the second integral in the estimate of $|I_2|$ is bounded. Collecting results, we claim that

$$|I_2| \leq \eta \int_\Omega |\Delta_{he_k} Du|^2 \zeta^2 w_\epsilon \, dx + C\frac{h^2}{\eta}.$$

This permits us to estimate I; specifically,

$$I \geq \frac{k_0}{4} \int_\Omega |\Delta_{he_k} Du|^2 \zeta^2 w_\epsilon \, dx - Ch^2. \tag{8.118}$$

We proceed to estimate II. We write

$$II = h \int_\Omega \int_0^1 f^\nu D_k \left(\zeta^2 w_\epsilon \Delta_{he_k} u^\nu \right) (x - \theta h e_k) \, dx$$

$$= \int_0^1 \left[\int_{\Omega - he_k} f^\nu (x + \theta h e_k) D_k \left(\zeta^2 w_\epsilon \Delta_{he_k} u^\nu \right) \, dx \right] d\theta.$$

Using the same techniques as before, we assert that

$$|II| \leq C \frac{h^2}{\eta} + \eta \int_\Omega |\Delta_{he_k} Du|^2 \zeta^2 w_\epsilon \, dx. \tag{8.119}$$

Then, combining (8.118) and (8.119) with $I = II$, and picking η sufficiently small, we get

$$\int_\Omega |\Delta_{he_k} Du|^2 \zeta^2 w_\epsilon \, dx \leq C h^2.$$

Dividing by h^2, letting h tend to 0, and then ϵ tend to 0, we have proven

$$\int_\Omega \zeta^2 w_0 |D_k Du|^2 \, dx, \quad \forall k = 1, \ldots, n - 1,$$

and in particular,

$$\int_{\Omega_{2\rho}} w_0 |D_k Du|^2 \, dx, \quad \forall k = 1, \ldots, n - 1. \tag{8.120}$$

To complete the proof of the lemma, it remains to estimate $D_n D_n u$. For that we use equation (8.87). We have

$$D_n a_n^\nu = f^\nu - \sum_{i=1}^{n-1} D_i a_i^\nu.$$

Making use of assumption (8.85) and differentiating, we can write

$$a_{n\,n}^{\nu\,\mu}(x, Du) D_n D_n u^\mu = f^\nu - a_{n,x_n}^\nu(x, Du)$$

$$- \sum_{j=1}^{n-1} a_{n\,j}^{\nu\,\mu}(x, Du) D_n D_j u^\mu$$

$$- \sum_{i=1}^{n-1} a_{i\,k}^{\nu\,\mu}(x, Du) D_i D_k u^\mu - \sum_{i=1}^{n-1} a_{i,x_i}^\nu(x, Du).$$

Using the fact that the matrix $a_{n\,n}^{\nu\,\mu}(x, Du)$ is uniformly coercive, we get

$$|D_n Du| \leq C \left(|f| + \sum_{i,\nu} g_{i,x_i}^\nu + |Du| + \sum_i |D_i Du| \right),$$

and using (8.120) and previous estimates, we obtain (8.117). Thus the proof of the lemma is complete.   ◇

Proof of Theorem 8.30.
We can now proceed with the proof of Theorem 8.30. We simply write

$$\int_{\Omega_{2\rho}} |D^2 u|^p \, dx \leq C \int_{\Omega_{2\rho}} w_0 |D^2 u|^2 \, dx + C \int_{\Omega_{2\rho}} w_0^{-p/(2-p)} \, dx,$$

and thus, making use of (8.116) and the preceding lemma, the right-hand side will be bounded if

$$\frac{p}{2-p} < \frac{2}{\gamma}$$

and we have

$$p < \frac{6 + 2\epsilon_0}{5 + \epsilon_0}.$$

Therefore, we have obtained

$$\int_{\Omega_{2\rho}} |D^2 u|^p \, dx \leq C.$$

We use a covering argument to get the result for Ω. The proof is complete.   ◇

8.8.5 Domain with a Corner. Mixed Boundary Conditions

We give some brief indications for the case of a domain with a corner. We suppose, to fix ideas, that

$$\Omega = \left\{ -\frac{x_n}{2} < x_1 < R_0, |x_2|, \ldots, |x_{n-1}| < R_0, 0 < x_n < 2R_0 \right\}$$

and suppose that we still have

$$\Gamma_D = \{x | x_n = 0, 0 \leq x_1 \leq R_0\}$$

and

$$\Gamma_N = \Gamma - \Gamma_D.$$

The definitions of $\Omega_\rho, Q_\sigma, Q_\sigma^*$ and of the cutoff function ζ are unchanged (see Section 8.8.2). We then claim that Theorem 8.25 remains valid. Indeed, consider Lemma 8.26. The same proof can be carried over. Questions arise only in treating the integral II. More precisely, we concentrate on the term

$$\int_\Omega \Delta_{he_k} \left(\frac{\zeta}{h} a(x, Du) \right) dx,$$

which was equal to 0 previously, but which is not in this case. In fact, this term equals

$$\int_{\Omega+he_k-\Omega\cap(\Omega+he_k)} \frac{\zeta}{h} a(x, Du)\, dx - \int_{\Omega-\Omega\cap(\Omega+he_k)} \frac{\zeta}{h} a(x, Du)\, dx.$$

The first integral is still 0, since ζ vanishes on the domain of integration. The second will not vanish when $k = 1$, but thanks to the fact that $a > 0$, it is negative. Hence, we may assert only

$$II \leq C.$$

But this is sufficient to carry over the proof. So we obtain (8.89) and (8.92). We then have to modify Lemma 8.28 slightly. The assumptions (8.94) and (8.95) become, from the definition of Ω,

$$\frac{1}{h} \int_{Q_\sigma \cap \Omega} |\Delta_{he_k} G_k|^2\, dx \leq C, \quad 1 \leq k \leq n-1,$$

$$\frac{1}{h} \int_{Q_\sigma \cap \Omega} |\Delta_{he_j} H|^2\, dx \leq C, \quad 1 \leq j \leq n-1,$$

where the constants do not depend on σ. Then we assert that

$$\frac{1}{h^{1-\delta}} \int_{Q_\sigma^* \cap \Omega} |\Delta_{he_n} H|^2\, dx \leq C_\delta \tag{8.121}$$

for any $0 < \delta < 1$, where C_δ does not depend on σ. The rest of the proof is unchanged. To check (8.121), we consider an extension of H outside Ω (i.e., when $x_1 + x_n/2 < 0$). We define

$$\tilde{H}(x) = \begin{cases} -5H(x - (2x_1 + x_n)e_1) + 10H\left(x - \left(\frac{3}{2}x_1 + \frac{3}{4}x_n\right)e_1\right) \\ \quad -4H\left(x - \left(\frac{5}{4}x_1 + \frac{5}{8}x_n\right)e_1\right) & \text{if } x_1 + \frac{x_n}{2} < 0, \\ = H(x), & \text{if } x_1 + \frac{x_n}{2} \geq 0. \end{cases}$$

The merit of this extension is first that it is C^1, and second that it allows us to assert the following relation:

$$D_n \tilde{H}(x) = \sum_{k=1}^{n-1} D_k \tilde{G}_k(x) + \tilde{g}(x),$$

where

$$\tilde{G}_1(x) = \begin{cases} -5H(x - (2x_1 + x_n)e_1) + 15H\left(x - \left(\frac{3}{2}x_1 + \frac{3}{4}x_n\right)e_1\right) \\ \quad -10H\left(x - \left(\frac{5}{4}x_1 + \frac{5}{8}x_n\right)e_1\right) \\ \quad +5G_1(x - (2x_1 + x_n)e_1) - 20G_1\left(x - \left(\frac{3}{2}x_1 + \frac{3}{4}x_n\right)e_1\right) \\ \quad +16G_1\left(x - \left(\frac{5}{4}x_1 + \frac{5}{8}x_n\right)e_1\right) & \text{if } x_1 + \frac{x_n}{2} < 0, \\ G_1(x) & \text{if } x_1 + \frac{x_n}{2} \geq 0 \end{cases}$$

$$\tilde{G}_k(x) = \begin{cases} -5G_k(x - (2x_1 + x_n)e_1) + 10G_k\left(x - \left(\frac{3}{2}x_1 + \frac{3}{4}x_n\right)e_1\right) \\ -4G_1\left(x - \left(\frac{5}{4}x_1 + \frac{5}{8}x_n\right)e_1\right) & \text{if } x_1 + \frac{x_n}{2} < 0, \\ \tilde{G}_k(x) = G_k(x) & \text{if } x_1 + \frac{x_n}{2} \geq 0, \quad 2 \leq k \leq n-1, \end{cases}$$

$$\tilde{g}(x) = \begin{cases} -5g(x - (2x_1 + x_n)e_1) + 10g\left(x - \left(\frac{3}{2}x_1 + \frac{3}{4}x_n\right)e_1\right) \\ -4g\left(x - \left(\frac{5}{4}x_1 + \frac{5}{8}x_n\right)e_1\right) & \text{if } x_1 + \frac{x_n}{2} < 0, \\ \tilde{g}(x) = g(x) & \text{if } x_1 + \frac{x_n}{2} \geq 0, \end{cases}$$

We may then apply Lemma 8.28, provided that we check the properties

$$\frac{1}{h}\int_{Q_\sigma} |\Delta_{he_k}\tilde{G}_k|^2\, dx \leq C, \quad 1 \leq k \leq n-1,$$

$$\frac{1}{h}\int_{Q_\sigma} |\Delta_{he_j}\tilde{H}|^2\, dx \leq C, \quad 1 \leq j \leq n-1,$$

where the constants do not depend on σ. We do not detail this. We just mention that the only problem arises on the strip $-h < x_1 + x_n/2 < 0$, where $\tilde{G}_k(x), \tilde{H}(x)$ and $\tilde{G}_k(x + he_k), \tilde{H}(x + he_j)$ are defined by different formulas. This is similar to the situation encountered in the proof of Lemma 8.28. The same method, in particular the use of the averaging procedure, is used. Therefore, we assert that

$$\frac{1}{h^{1-\delta}}\int_{Q_\sigma^*} |\Delta_{he_n}\tilde{H}|^2\, dx \leq C_\delta,$$

from which (8.121) follows immediately.   ◇

8.8.6 Domain with a Corner. Dirichlet Boundary Conditions

The results of this section are due to C. Ebmeyer [16], where a more general setup is presented. We consider the same domain as above,

$$\Omega = \left\{-\frac{x_n}{2} < x_1 < R_0, |x_2|, \ldots, |x_{n-1}| < R_0, 0 < x_n < 2R_0\right\},$$

but we suppose now that

$$\Gamma_D = \left\{x|x_1 + \frac{x_n}{2} = 0, 0 < x_n < 2R_0\right\} \cup \{x|x_n = 0, 0 \leq x_1 \leq R_0\}$$

and

$$\Gamma_N = \Gamma - \Gamma_D.$$

The definitions of $\Omega_\rho, Q_\sigma, Q_\sigma^*$ and of the cutoff function ζ are again unchanged (see Section 8.8.2). So the Dirichlet condition holds on the corner part of the boundary. We claim again that Theorem 8.25 remains valid. However, some substantial changes have to be made to obtain the result of Lemma 8.26. We shall need an extension of the function u and of $a(x, p)$ for $x_1 + x_n/2 < 0$. We define

$$u(x_1, \ldots, x_n) = 0$$
$$a(x_1, \ldots, x_n, p_1, \ldots, p_n)$$
$$= a(-x_1 - x_n, x_2, \ldots, x_n, p_1, -p_2, \ldots, -p_n - 1, p_1 - p_n)$$
$$\text{for } -R_0 < x_1 < -x_n/2, \quad |x_2|, \ldots, |x_{n-1}| < R_0, \quad 0 < x_n < 2R_0.$$

The introduction of these extensions is motivated by the fact that we shall work not only with $\zeta\Delta_{he_k}u^\nu$, but also with $\zeta\Delta_{-he_k}u^\nu$, which requires u to be defined when $x_1 + x_n/2 < 0$. Consider the function

$$\zeta\Delta_{-he_k}u^\nu.$$

We claim that it is an acceptable test function for the νth equation of (8.87). Indeed, we notice that

$$\zeta\Delta_{-he_k}u^\nu = \zeta(x)(u^\nu(x - he_k) - u^\nu(x))$$

vanishes on the Dirichlet part of the boundary. Using this test function yields

$$\int_\Omega a_i^\nu(x, Du)D_i(\zeta\Delta_{-he_k}u^\nu)\,dx + \int_\Omega f^\nu\zeta\Delta_{-he_k}u^\nu\,dx = 0.$$

Hence

$$\int_\Omega \zeta D_i\Delta_{-he_k}u^\nu\,a_i^\nu\,dx + \int_\Omega D_i\zeta\,\Delta_{-he_k}u^\nu\,a_i^\nu\,dx + \int_\Omega f^\nu\zeta\Delta_{-he_k}u^\nu\,dx = 0.$$
$$(8.122)$$

When it simplifies notation, we write

$$\underline{u}(x) = u(x - he_k).$$

We use the Taylor expansion of the function a to obtain

$$a(x, D\underline{u}) = a(x, Du) + a_i^\nu(x, Du)\,\Delta_{-he_k}D_iu^\nu$$
$$+ \tilde{m}_{ij}^{\nu\mu}\Delta_{-he_k}D_iu^\nu\,\Delta_{-he_k}D_ju^\mu$$
$$(8.123)$$

with

$$\tilde{m}_{ij}^{\nu\mu} = \int_0^1 (1-t)a_{ij}^{\nu\mu}(x, Du + t\Delta_{-he_k}Du)\,dt.$$

Then let

$$I = \int_\Omega \frac{\zeta}{h}\tilde{m}_{ij}^{\nu\mu}\Delta_{-he_k}D_iu^\nu\,\Delta_{-he_k}D_ju^\mu\,dx.$$

From the assumptions (8.85) we have

$$I \geq \frac{k_0}{2} \int_\Omega \frac{\zeta}{h} |\Delta_{-he_k} Du|^2 \, dx.$$

On the other hand, from (8.122) and (8.123) we also have

$$I = \int_\Omega \frac{\zeta}{h} \left(a(x, D\underline{u}) - a(x, Du) \right) dx$$

$$+ \int_\Omega \frac{D_i \zeta}{h} \Delta_{-he_k} u^\nu a_i^\nu \, dx + \int_\Omega f^\nu \frac{\zeta}{h} \Delta_{-he_k} u^\nu \, dx,$$

and we denote the various terms by

$$I = II + III + IV.$$

Since the functions u^ν are in $H^1(\Omega)$, the first easy observation, thanks to assumptions (8.85), (8.86), is that

$$|III| + |IV| \leq C.$$

Next write II as

$$II = \int_\Omega \frac{\zeta}{h} \Delta_{-he_k} a(x, Du) \, dx - \int_\Omega \frac{\zeta}{h} (a(\underline{x}, D\underline{u}) - a(x, D\underline{u})) dx$$

$$= \int_\Omega \Delta_{-he_k} \left(\frac{\zeta}{h} a(x, Du) \right) dx$$

$$- \int_\Omega \frac{1}{h} \Delta_{-he_k} \zeta \, a(\underline{x}, D\underline{u}) \, dx$$

$$- \int_\Omega \zeta \int_0^1 a_{x_k}(x - the_k, D\underline{u}) dt \, dx.$$

The integrals on the second and third lines are bounded, so we may write

$$II \leq C + \int_{\Omega - he_k - \Omega \cap (\Omega - he_k)} \frac{\zeta}{h} a(x, Du) \, dx$$

$$- \int_{\Omega - \Omega \cap (\Omega - he_k)} \frac{\zeta}{h} a(x, Du) \, dx.$$

We first check that ζ vanishes on the domain of integration of all of the above integrals when $k = 2, \ldots, n - 1$. The case $k = 1$ remains. The last integral is negative anyway, and we ignore it. The annoying term is the first one, which is positive and unbounded with respect to h. We check that thanks to the extension we have chosen, Du vanishes on the sets $\Omega - he_1 - \Omega \cap (\Omega - he_1)$. Thus due to the assumptions (8.85) we have

$$\int_{\Omega - he_1 - \Omega \cap (\Omega - he_1)} \frac{\zeta}{h} |a(x, Du)| \, dx \leq C \int_{\Omega - he_1 - \Omega \cap (\Omega - he_1)} \frac{\zeta}{h} \leq C.$$

This completes the proof of Lemma 8.26, adapted to the present situation. Then we can proceed as in Section 8.8.5, for mixed boundary conditions, with no major changes. The corresponding result to Theorem 8.25 can also be derived.

9. Dual Approach to Nonlinear Elliptic Systems

9.1 Introduction

One can get a quick idea of the dual approach by simply considering the Dirichlet problem

$$-\Delta u = f, \quad u|_\Gamma = 0$$

and setting

$$\sigma = -Du.$$

Then we get the system

$$\operatorname{div} \sigma = f, \quad \sigma = -Du, \quad u|_\Gamma = 0,$$

and σ appears as the solution of the variational problem

$$\min_{\operatorname{div} \sigma = f} \frac{1}{2} \int_\Omega |\sigma|^2 \, dx.$$

The issue is now to derive regularity results on σ instead of u. Of course, in the above example this idea does not offer much insight, but in many problems arising from elasticity and plasticity theory it turns out to be quite fruitful. In the theory of elastoplasticity one is interested in two physical quantities, the displacement vector $u(x)$, and the stress tensor $\sigma(x)$, where the variable x is in Ω, and the domain Ω is the region occupied by the elastic-plastic body. The vector u belongs to R^n, and σ is a symmetric matrix. The strain is a tensor $\epsilon(u)$ related to the vector of displacements by the formula

$$\epsilon_{i,j} = \frac{1}{2}(D_i u_j + D_j u_i).$$

In the so-called primal problem the unknown is the displacement u_i, whereas in the dual problem the unknown is the stress σ. Since we are mostly concerned with the theory of regularity, we shall not deal with the general theory of dual variational problems, for which we refer to R. Temam [101], G. Duvaut, J.L. Lions [15], I. Ekeland, R. Temam [23], J. Nečas, I. Hlaváček ([86]).

The dual approach is important, for instance, for finite element approximation to elliptic problems. Using a finite element approach analogously to the continuous dual approach, one stresses the fact that it is the unknown stress σ that has to be approximated accurately, and where error estimates are desired.

The dual formulation arose first in the theory of linear elasticity. M.E. Gurtin [53] remarks that the basic ideas go back to Cotterill (1865) and Donati (1890). The first proved theorems seem to have been given by Colonetti (1912) and Prange (1916). See [53] for more details on historical facts. A more modern, more general formulation of the dual problem in connection with PDEs is due to K. Friedrichs [34]. Friedrichs' work contains related, but more specialized, work of E. Trefftz [103]. See also E. Zeidler [108] for additional historical remarks. In many problems arising from elasticity and plasticity theory, this approach, conveniently used, turns out to be quite fruitful.

Partial regularity results for nonlinear elliptic systems were first obtained in 1968 in the papers of C. Morrey [79] and E. Giusti, M. Miranda [48]. Our point is that the dual formulation is also useful for that goal. Indeed, we were able to show in a number of cases that getting a regularity result on the dual problem can be easier than on the primal problem. The basic regularity issue concerns getting H_{loc}^1 for σ. In the Hencky model, one introduces a constraint on the size of the deviator of the stress, defined by

$$\sigma_D = \sigma - \frac{1}{n} I \operatorname{tr} \sigma,$$

and the constraint is

$$|\sigma_D| \leq \mu.$$

It took a long time until H_{loc}^1 was obtained by G.A. Seregin [93]. Seregin approximated the problem by a variational problem with the displacements as unknowns, and the stress, expressed by the displacements, is estimated uniformly, although the method leads to poor estimates. We proved the result directly on the dual problem in a previous work [6], through a penalized problem, the Norton–Hoff problem, obtaining at the same time a uniform (with respect to the penalty parameter) regularity result. Our approach led also in a straightforward way to a similar regularity result for the evolution version of the Hencky problem, which is the Prandtl–Reuss problem, whereas it is still open how to obtain it through a primal formulation. There are other situations where our approach brings clear advantages. Indeed, suppose that we add the constraint that the matrix σ is symmetric. Then the primal formulation introduces the strain, which is derived from the displacement through the formula

$$\epsilon(u) = \frac{1}{2}(Du + Du^T).$$

Usually the strain is just a measure, from which one gets only

$$u \in L^{n/(n-1)},$$

whereas the stress can still be regular.

9.2 Preliminaries

9.2.1 Notation

Let Ω be a bounded Lipschitz domain of R^n, whose boundary is denoted by Γ. The boundary will be divided in two parts, Γ_0 and Γ_1. We shall denote by $W^{1,p}_{\Gamma_0}(\Omega)$ and $H^1_{\Gamma_0}(\Omega)$ the closed subspaces of functions in $W^{1,p}(\Omega)$ and $H^1(\Omega)$, respectively, that vanish on Γ_0. We shall also use the spaces of vector-valued functions $\left(W^{1,p}_{\Gamma_0}(\Omega)\right)^n$, $\left(H^1_{\Gamma_0}(\Omega)\right)^n$. When $\Gamma_0 = \Gamma$, one writes $W^{1,p}_0(\Omega)$ and $H^1_0(\Omega)$, following the usual notation. When $\Gamma_0 \subset \Gamma$, we assume that the measure of Γ_0 is positive.

We next consider the space of $n \times n$ symmetric (or alternatively non symmetric) matrices whose elements are in L^p, denoted by $\mathcal{L}^p_{\text{sym}}$ (or $\mathcal{L}^p$) with the norm

$$\|\sigma\|_{\mathcal{L}^p} = |\,|\sigma|\,|_{L^p},$$

where on the right-hand side $|\sigma|$ designates the norm of the matrix σ,

$$|\sigma| = \left(\sum_{i,j=1}^{n} \sigma_{ij}^2\right)^{1/2}.$$

In the case $p = 2$, it corresponds to the the the Hilbert norm

$$\|\sigma\|_{\mathcal{L}^2} = \left(\sum_{i,j=1}^{n} \int_{\Omega} \sigma_{ij}^2 \, dx\right)^{1/2}.$$

It will be convenient to use the notation

$$\sigma.\tau = \sum_{i,j=1}^{n} \sigma_{ij}\tau_{ij}$$

to represent the scalar product of two matrices σ and τ, similar to the scalar product of vectors in R^n. We shall also use the notation

$$\text{div } \sigma = \sum_{i=1}^{n} D_i \sigma_{ij},$$

which is a vector, and

$$\nu.\sigma = \sum_{i=1}^{n} \nu_i \sigma_{ij},$$

where ν is the outward unit normal on the boundary Γ.

9.2.2 Properties of the Operators $\epsilon(u)$ and Du

In this subsection we summarize important properties of the operators

$$\epsilon(u) = \frac{1}{2}(Du + Du^*) \text{ and } Du.$$

Since the properties of $\epsilon(u)$ are harder to prove, we state only these; all of them are valid for Du also, and even some more that we shall not need.

The main property is Korn's inequality. Namely, for $1 < p < \infty$,

$$c_0\|u\|_{W_{\Gamma_0}^{1,p}} \leq \|\epsilon(u)\|_{\mathcal{L}^p} \leq c_1\|u\|_{W_{\Gamma_0}^{1,p}}, \tag{9.1}$$

where c_0, c_1 are positive constants depending only on Ω, Γ_0, p. The right-hand inequality is obvious; the left-hand one is Korn's inequality. It is important to stress that the cases $p = 1, p = \infty$ are excluded. When $p = 1$, one has nevertheless the estimate

$$\|u\|_{(L^{n/(n-1)})^n} \leq c_2\|\epsilon(u)\|_{\mathcal{L}^1},$$

where c_2 is a constant depending only on Ω, Γ_0. Note also that for distributions u such that $\epsilon(u) \in \mathcal{L}^1$, the trace of u on the boundary Γ is defined as an element of $(L^1(\Gamma))^n$, and thus the functions that vanish on the part Γ_0 are well-defined.

From (9.1) it follows that ϵ, considered as a map from $\left(W_{\Gamma_0}^{1,p}(\Omega)\right)^n$ into $\mathcal{L}_{\text{sym}}^p$, has a closed range. Thus, by the closed range theorem, we can assert that

$$\text{range } \epsilon = (\ker \epsilon^*)^\perp.$$

We make explicit the space $\ker \epsilon^*$. We have

$$\epsilon^* : \mathcal{L}_{\text{sym}}^q \to \left((W_{\Gamma_0}^{1,p})^n\right)^*.$$

It is easy to check that (in the distributional sense)

$$\tau \in \ker \epsilon^* \Rightarrow \text{div } \tau = 0.$$

Now, if

$$\tau \in \mathcal{L}_{\text{sym}}^q \text{ and } \text{div } \tau \in (L^q)^n,$$

it is possible to define the trace

$$\nu.\tau \in (W^{-1/2,q}(\Gamma))^n$$

using the notation defined in Section 9.2.1. One has Green's formula

$$\int_\Omega \epsilon(u).\tau \, dx + \int_\Omega u.\text{div } \tau \, dx = \int_\Gamma u.\nu.\tau \, d\Gamma$$

for any $u \in (W^{1,p})^n$. Collecting results, we can assert the following result.

Proposition 9.1. *If $\sigma \in \mathcal{L}_{\text{sym}}^p$ satisfies*

$$\int_\Omega \sigma.\tau \, dx = 0, \forall \tau \in \mathcal{L}_{\text{sym}}^q \text{ such that } \text{div } \tau = 0, \nu.\tau = 0 \text{ on } \Gamma_1,$$

then there exists one and only one $u \in (W_{\Gamma_0}^{1,p})^n$ such that

$$\sigma = \epsilon(u).$$

Remark 9.2. If in the statement of Proposition 9.1 we discard the symmetry condition, then we obtain

$$\sigma = Du.$$

We refer to the book of R. Temam [101] for details and proofs of the assertions of this section.

We shall also need the following result, valid for $1 < p < \infty$: If a distribution ϕ has all first derivatives $D_i\phi$ in $W^{-1,p}(\Omega)$, then it belongs to $L^p(\Omega)$.

9.3 Elasticity Models

9.3.1 Primal and Dual Problems

We consider a tensor function

$$A_{ij,hk} \in L^\infty \text{ such that } A_{ij,hk} = A_{ji,hk} = A_{ij,kh} = A_{hk,ij},$$

$$\sum_{i,j;h,k=1}^n A_{ij,hk}\tau_{ij}\tau_{hk} \geq \alpha|\tau|^2, \quad \forall \tau, \alpha > 0, \tag{9.2}$$

and functions

$$f_j \in L^2(\Omega), \quad j = 1, \dots, n.$$

The primal problem is the following. Define the functional

$$L(u) = \frac{1}{2}\int_\Omega (A^{-1})_{ij,hk}D_iu_jD_hu_k \, dx + \int_\Omega f_iu_i \, dx,$$

and we minimize $L(u)$ over $(H_{\Gamma_0}^1(\Omega))^n$. We leave the reader to check that the minimum u satisfies the relations

$$\epsilon(u) = A\sigma, \, u \in (H_{\Gamma_0}^1(\Omega))^n,$$
$$\sigma \in \mathcal{L}_{\text{sym}}^2, \quad \text{div } \sigma = f, \quad \nu.\sigma|_{\Gamma_1} = 0. \tag{9.3}$$

Remark 9.3. If we replace (9.2) by

$$A_{ij,hk} \in L^\infty \text{ such that } A_{ij,hk} = A_{hk,ij},$$

$$\sum_{i,j;h,k=1}^{n} A_{ij,hk} \tau_{ij} \tau_{hk} \ge \alpha |\tau|^2, \quad \forall \tau, \quad \alpha > 0,$$

so that some of the symmetry assumptions are relaxed, then (9.3) has to be replaced by

$$Du = A\sigma, \ u \in (H^1_{\Gamma_0}(\Omega))^n,$$
$$\sigma \in \mathcal{L}^2, \ \text{div } \sigma = f, \ \nu.\sigma|_{\Gamma_1} = 0. \tag{9.4}$$

Remark 9.4. The Lamé equations correspond to the tensor

$$(A^{-1})_{ij,hk} = \mu \delta_{ih} \delta_{jk} + (\lambda + \mu) \delta_{ij} \delta_{hk}.$$

Now we can view σ in (9.3), (9.4) as the solution of the following minimization problem, the dual problem:

We consider the set

$$\mathcal{K} = \{\sigma \in \mathcal{L}^2_{\text{sym}} | \text{ div } \sigma = f, \nu.\sigma = 0 \text{ on } \Gamma_1\}.$$

Define on $\mathcal{L}^2_{\text{sym}}$ the functional

$$J(\sigma) = \frac{1}{2} \int_\Omega A_{ij,hk} \sigma_{ij} \sigma_{hk} \, dx.$$

Then the dual problem corresponds to the variational problem

$$\text{minimize } J(\sigma) \text{ over } \mathcal{K}.$$

It is clear that the functional $J(\sigma)$ has a unique minimum on the nonempty closed convex subset of $\mathcal{L}^2_{\text{sym}}$, $\mathcal{M} \cap \mathcal{K}$. We shall call this minimum σ also. It is also easy to check that there exists u such that (9.3) is satisfied.

9.3.2 A Hybrid Model

There is a third variational problem that can also be introduced. It is a mix of the primal and dual problems and incorporates both u and σ. It is known as the Hellinger–Reissner principle (see [86]). Namely, one considers the functional

$$\mathcal{J}(\sigma, u) = \frac{1}{2} \int_\Omega A_{ij,hk} \sigma_{ij} \sigma_{hk} \, dx - \int_\Omega \sigma_{ij} D_i u_j \, dx - \int_\Omega f_i u_i \, dx$$

over

$$u \in (H^1_{\Gamma_0}(\Omega))^n, \quad \sigma \in \mathcal{L}^2_{\text{sym}},$$

and we look for

$$\max_u \min_\sigma \mathcal{J}(\sigma, u).$$

In fact, if we minimize in σ, for fixed u, we recover exactly the primal problem in u.

9.4 H^1_{loc} Theory for the Nonsymmetric Case

9.4.1 Presentation of the Problem

Consider a scalar function $G(\sigma)$ on $n \times n$ matrices such that

$$G(0) = 0, \ G \text{ is } C^1, g = G',$$
$$|g(\sigma)| \leq K|\sigma|(1 + |\sigma|^{p-2}),$$
$$(g(\sigma_1) - g(\sigma_2)).(\sigma_1 - \sigma_2) \geq \alpha|\sigma_1 - \sigma_2|^2 + (\beta(|\sigma_1|)\sigma_1 - \beta(|\sigma_2|)\sigma_2).(\sigma_1 - \sigma_2),$$
$$\beta : R^+ \to R^+, \quad \beta(0) = 0, \quad \beta'(x) \geq k(p-2)x^{p-3}, \quad p \geq 2.$$

$$(9.5)$$

Note that it immediately follows from (9.5) that

$$\frac{\alpha}{2}|\sigma|^2 + \frac{k}{p}|\sigma|^p \leq G(\sigma) \leq K \left(\frac{1}{2}|\sigma|^2 + \frac{1}{p}|\sigma|^p \right)$$

and that G is strictly convex. Define

$$\mathcal{K}^p = \{\sigma \in \mathcal{L}^p \mid \text{div } \sigma = f, \nu.\sigma = 0 \text{ on } \Gamma_1\}, \tag{9.6}$$

where

$$f \in (W^{1,p}(\Omega))^n \tag{9.7}$$

and the functional J is given by

$$J(\sigma) = \int_\Omega G(\sigma) \, dx.$$

We define the following minimization problem:

$$\text{Minimize } J(\sigma) \text{ over } \mathcal{K}^p.$$

It is clear that there exists a unique minimum (denoted also by σ), and from Proposition 9.1 and Remark 9.2 a unique u such that

$$Du = g(\sigma), \ u \in \left(W^{1,\frac{p}{p-1}}_{\Gamma_0}(\Omega) \right)^n,$$
$$\sigma \in \mathcal{L}^p, \ \text{div } \sigma = f, \ \nu.\sigma|_{\Gamma_1} = 0. \tag{9.8}$$

Remark 9.5. The primal problem can be written as follows

$$\text{div } g^{-1}(Du) = f, \quad u \in \left(W^{1,\frac{p}{p-1}}_{\Gamma_0}(\Omega) \right)^n,$$

where g^{-1} is the inverse, or as

$$(G''^{-1})_{ij,hk}(g^{-1}(Du))D_i D_h u_k = f_j.$$

9.4.2 H^1_{loc} Regularity

We state now the main result of this section. Since we state a local result, we use a smooth function θ with compact support in Ω, satisfying $0 \leq \theta \leq 1$.

Theorem 9.6. *If we assume (9.5) and (9.7), then the minimum σ satisfies*

$$\int_\Omega \theta^2 (1 + |\sigma|^{p-2}) |D\sigma|^2 \, dx \leq C_p. \tag{9.9}$$

We begin with a formal proof.

• Formal Proof

We test the first relation (9.8) with $-D_k(\theta^2 D_k \sigma)$ and perform two integrations by parts. Since we are proceeding formally, we may as well assume that G is C^2, in which case we may assert from (9.5) that

$$G''(\sigma) \geq (\alpha + \beta(|\sigma|))I.$$

Making use of the second relation in (9.8), we obtain

$$\int \theta^2 G''(\sigma) D_k \sigma . D_k \sigma \, dx = -\int g_{kj}(\sigma)(D_i \theta^2 D_k \sigma_{ij} + \theta^2 D_k f_j) \, dx. \tag{9.10}$$

Then from assumptions (9.5) and (9.7) we obtain, using Hölder's inequality,

$$\int \theta^2 (\alpha + k|\sigma|^{p-2}) |D\sigma|^2 \, dx \leq 4K^2 \int |D\theta|^2 \left(\frac{|\sigma|^2}{\alpha} + \frac{|\sigma|^p}{k} \right) dx$$
$$+ 2K \int \theta^2 (|\sigma| + |\sigma|^{p-1}) |Df| \, dx,$$

and the result follows.   ◇

To prove rigorously the estimate (9.9), one method is to proceed with finite differences instead of derivatives. To that end we introduce the operators

$$D^h_{k,+} \phi(x) = \frac{\phi(x + he_k) - \phi(x)}{h},$$

$$D^h_{k,-} \phi(x) = \frac{\phi(x) - \phi(x - he_k)}{h},$$

where e_k represents the kth unit vector. We note the following discrete integration by parts formula in R^n:

$$\int D^h_{k,-} \phi(x) \, \psi \, dx = -\int D^h_{k,+} \psi(x) \, \phi \, dx.$$

Also, we have the finite difference formula for composition of functions

$$D^h_{k,+}\psi(\sigma(x)) = \int_0^1 [D_\sigma\psi(\sigma(x) + \lambda h D^h_{k,+}\sigma(x))D^h_{k,+}\sigma(x)]\,d\lambda.$$

We want to obtain the discrete analogue of (9.10). We test the first relation (9.8) with $-D^h_{k,-}(\theta^2 D^h_{k,+}\sigma)$ and do a discrete integration by parts. We obtain

$$\int_\Omega \theta^2 \frac{g(\sigma + h D^h_{k,+}\sigma) - g(\sigma)}{h} D^h_{k,+}\sigma\,dx \tag{9.11}$$
$$= -\int_\Omega \int_0^1 g_{kj}(\sigma(x + \lambda h e_k))(D_i\theta^2 D^h_{k,+}\sigma_{ij} + \theta^2 D^h_{k,+}f_j)dx\,d\lambda.$$

Of course, (9.11) is now perfectly rigorous. If we use the assumptions (9.5) on g, we can assert that

$$\int_\Omega \theta^2(\alpha + k\min(|\sigma(x + he_k)|^{p-2}, |\sigma(x)|^{p-2}))|D^h_{k,+}\sigma|^2\,dx$$
$$\leq \int_\Omega \int_0^1 K(1 + |\sigma(x + \lambda h e_k)|^{p-2})|\sigma(x + \lambda h e_k)|(2\theta|D\theta|\|D^h_{k,+}\sigma|$$
$$+\theta^2|D^h_{k,+}f|)dx\,d\lambda.$$

But now we cannot use Hölder's inequality, since the functions having exponent $p - 2$ do not coincide. The only case in which we can immediately conclude the result is where $p = 2$. Indeed, we can then apply Hölder's inequality and obtain

$$\frac{\alpha + k}{2}\int_\Omega \theta^2|D^h_{k,+}\sigma|^2\,dx$$
$$\leq \int_\Omega \int_0^1 2K|\sigma(x + \lambda h e_k)|\left(K\frac{|D\theta|^2}{\alpha + k}|\sigma(x + \lambda h e_k)| + \theta^2|D^h_{k,+}f|\right)dx\,d\lambda.$$

The right-hand side is bounded in h, so (9.9) follows.

Proof of Theorem 9.6.
We know that the result is true whenever $p = 2$. The idea is then to approximate G by a functional that has at most quadratic growth. We begin by producing a suitable approximation.

• **Approximation**
We first define
$$\hat{G}(\sigma) = G(\sigma) - \frac{\alpha}{4}|\sigma|^2 - \frac{k}{2p}|\sigma|^p.$$

Clearly, $\hat{G}(\sigma)$ has the same properties as G; i.e., (9.5) holds with α and k replaced by $\alpha/2$ and $k/2$ on the left-hand side.

We then define

$$\hat{G}_N(\sigma)$$

$$= \begin{cases} \hat{G}(\sigma) & \text{if } |\sigma| \leq N, \\ \hat{G}\left(\dfrac{N\sigma}{|\sigma|}\right) + (\hat{G})'\left(\dfrac{N\sigma}{|\sigma|}\right)\left(\sigma - \dfrac{N\sigma}{|\sigma|}\right) + \gamma_N \left|\sigma - \dfrac{N\sigma}{|\sigma|}\right|^2 & \text{if } |\sigma| \geq N, \end{cases}$$

where the constant γ_N is to be chosen such that

$$\frac{\alpha + kN^{p-2}}{4} \geq \gamma_N \geq \frac{\alpha}{4}.$$

Consider the conjugate function

$$(\hat{G}_N)^*(\tau) = \sup {}_\sigma(\sigma.\tau - \hat{G}_N(\sigma)),$$

which is convex. To render it strictly convex we just add

$$j_N(\tau) = \epsilon_N |\tau|^{p/(p-1)},$$

where

$$\epsilon_N \to 0 \text{ as } N \to \infty.$$

Then consider the conjugate

$$((\hat{G}_N)^* + j_N)^*(\sigma) = \sup_\tau (\sigma.\tau - (\hat{G}_N)^*(\tau) - j_N(\tau)).$$

Finally, we take

$$G_N(\sigma) = ((\hat{G}_N)^* + j_N)^*(\sigma) + \frac{\alpha}{4}|\sigma|^2$$
$$+ \frac{k}{2p}|\sigma|^p \mathbb{1}_{\{|\sigma| \leq N\}} + \left(\frac{k}{4}N^{p-2}|\sigma|^2 + k\left(\frac{1}{2p} - \frac{1}{4}\right)N^p\right) \mathbb{1}_{\{|\sigma| > N\}}.$$

First we check several important properties of G_N.

Note that for $|\sigma| \geq 2N$,

$$\hat{G}_N(\sigma) \geq \frac{\gamma_N}{4}|\sigma|^2 \geq \frac{\alpha}{16}|\sigma|^2,$$

whereas for $N \leq |\sigma| \leq 2N$,

$$\hat{G}_N(\sigma) \geq \hat{G}\left(\frac{\sigma N}{|\sigma|}\right) = \frac{N}{|\sigma|}\int_0^1 \hat{G}'\left(\lambda\frac{N\sigma}{|\sigma|}\right)\sigma\, d\lambda.$$

Since (see (9.5))

$$\left(\frac{\alpha}{2} + \frac{k}{2}|\sigma|^{p-2}\right) \leq \hat{G}'(\sigma).\sigma,$$

we see in particular that

$$\hat{G}_N(\sigma) \geq \frac{\alpha}{4} N^2 \geq \frac{\alpha}{16} |\sigma|^2.$$

But for $|\sigma| \leq N$ we have

$$\hat{G}_N(\sigma) = \hat{G}(\sigma) \geq \frac{\alpha}{4} |\sigma|^2,$$

so it follows that for all values of the argument

$$\hat{G}_N(\sigma) \geq \frac{\alpha}{16} |\sigma|^2. \tag{9.12}$$

Then we write, for $|\sigma| \geq N$,

$$\hat{G}_N(\sigma) = \hat{G}(\sigma) - \left(1 - \frac{N}{|\sigma|}\right) \int_0^1 \left(\hat{G}'\left(\frac{N\sigma}{|\sigma|} + \lambda \left(\sigma - \frac{N\sigma}{|\sigma|} \right) \right) \right.$$
$$\left. - \hat{G}'\left(\frac{N\sigma}{|\sigma|} \right) \right) . \sigma\, d\lambda + \gamma_N \left| \sigma - \frac{N\sigma}{|\sigma|} \right|^2 .$$

From the estimates on $\hat{G}'$ it follows that

$$\hat{G}_N(\sigma) \leq \hat{G}(\sigma) - \left(\frac{\alpha}{4} + \frac{k}{4} N^{p-2} \right) \left| \sigma - \frac{N\sigma}{|\sigma|} \right|^2 + \gamma_N \left| \sigma - \frac{N\sigma}{|\sigma|} \right|^2 ,$$

and thus from the choice of γ_N we have

$$\hat{G}_N(\sigma) \leq \hat{G}(\sigma).$$

The function $\hat{G}_N(\sigma)$ is continuous. Therefore, considering the conjugate, which is convex (and hence locally Lipschitz), it immediately follows that

$$\hat{G}^*(\tau) \leq (\hat{G}_N)^*(\tau) \leq \frac{4|\tau|^2}{\alpha}.$$

The function $(\hat{G}_N)^* + j_N$ is strictly convex and coercive. Therefore, $\sigma.\tau - (\hat{G}_N)^*(\tau) - j_N(\tau)$ has a unique maximum in τ, for any σ, which we denote by $\tau_N(\sigma)$. It is a continuous function. From convex analysis (see [23]) it follows that

$$((\hat{G}_N)^* + j_N)^*(\sigma) \text{ is convex and } C^1.$$

We note that

$$(\hat{G}_N)^*(\tau) + j_N(\tau) \leq \left(\frac{4}{\alpha} + \epsilon_N \right) |\tau|^2 + \epsilon_N,$$

and so

$$\frac{\alpha}{16 + 4\alpha\epsilon_N} |\sigma|^2 - \epsilon_N \leq ((\hat{G}_N)^* + j_N)^*(\sigma) \leq (\hat{G})^{**}(\sigma) = \hat{G}(\sigma).$$

We then check the important property that $(\hat{G}_N)^*(\tau)$ is an extension of $(\hat{G})^*(\tau)$ in the following sense:

$$(\hat{G}_N)^*(\tau) = (\hat{G})^*(\tau) \text{ if } |\tau| \le \frac{N\alpha}{16}.$$

This follows from the fact that from the estimate (9.12),

$$(\hat{G}_N)^*(\tau) = \sup_{\{|\sigma| \le \frac{16}{\alpha}|\tau|\}} (\sigma.\tau - \hat{G}_N(\sigma)).$$

To derive a similar estimate for the second conjugate, consider $\tau(\sigma)$ which achieves the supremum in τ of the function

$$\sigma.\tau - (\hat{G})^*(\tau).$$

By convex analysis we have

$$\hat{G}'(\sigma) = \tau(\sigma).$$

Hence from the definition of $\hat{G}$,

$$|\tau(\sigma)| \le \left(K + \frac{\alpha}{2}\right)|\sigma| + \left(K + \frac{k}{2}\right)|\sigma|^{p-1}.$$

By definition

$$((\hat{G}_N)^* + j_N)^*(\sigma) \ge \sigma.\tau(\sigma) - (\hat{G}_N)^*(\tau(\sigma)) - j_N(\tau(\sigma)),$$

and thus if

$$\left(K + \frac{\alpha}{2}\right)|\sigma| + \left(K + \frac{k}{2}\right)|\sigma|^{p-1} \le \frac{N\alpha}{16},$$

we have

$$\begin{aligned} ((\hat{G}_N)^* + j_N)^*(\sigma) &\ge \sigma.\tau(\sigma) - (\hat{G})^*(\tau(\sigma)) - j_N(\tau(\sigma)) \\ &\ge \hat{G}(\sigma) - j_N(\tau(\sigma)) \\ &\ge \hat{G}(\sigma) - \epsilon_N K_p(|\sigma| + |\sigma|^{p-1})^{p/(p-1)}. \end{aligned}$$

Summarizing, we have

$$((\hat{G}_N)^* + j_N)^*(\sigma) \ge \hat{G}(\sigma) - \epsilon_N K_p(|\sigma| + |\sigma|^{p-1})^{p/(p-1)},$$
$$\forall \sigma, \text{ with } |\sigma| \le \rho_N,$$

where ρ_N is the solution of the equation

$$\left(K + \frac{\alpha}{2}\right)\rho + \left(K + \frac{k}{2}\right)\rho^{p-1} = \frac{N\alpha}{16}.$$

Among other things we have proved that

$$((\hat{G}_N)^* + j_N)^*(\sigma) \to \hat{G}(\sigma), \quad N \to \infty.$$

Of course, we can now state similar results for our final approximation $G_N(\sigma)$, namely (noting that $\rho_N < N$)

$G_N(\sigma)$ is strictly convex and C^1,

$$G_N(\sigma) \geq \frac{\alpha}{4} \frac{5 + \alpha \epsilon_N}{4 + \alpha \epsilon_N} |\sigma|^2$$

$$+ \frac{k}{2p} |\sigma|^p \mathbb{1}_{\{|\sigma| \leq N\}}$$

$$+ \left(\frac{k}{4} N^{p-2} |\sigma|^2 + k \left(\frac{1}{2p} - \frac{1}{4} \right) N^p \right) \mathbb{1}_{\{|\sigma| > N\}} - \epsilon_N,$$

$$G_N(\sigma) \leq G(\sigma),$$

$$G_N(\sigma) \geq G(\sigma) - \epsilon_N K_p(|\sigma| + |\sigma|^{p-1})^{p/(p-1)}, \quad \forall \sigma, |\sigma| \leq \rho_N. \quad (9.13)$$

Let us set

$$g_N(\sigma) = G'_N(\sigma).$$

Now

$$g_N(\sigma) = \frac{\alpha}{2} \sigma + \frac{k}{2} (\min(|\sigma|, N))^{p-2} \sigma + \tau_N(\sigma),$$

so using the fact that $\tau_N(\sigma)$ is monotone, we have

$$(g_N(\sigma_1) - g_N(\sigma_2)) \cdot (\sigma_1 - \sigma_2) \geq \frac{\alpha}{2} |\sigma_1 - \sigma_2|^2 + \frac{k}{2} (\min(|\sigma_1|, |\sigma_2|, N))^{p-2} |\sigma_1 - \sigma_2|^2. \quad (9.14)$$

We can derive an estimate from above on the norm of $g_N(\sigma)$, by first observing that

$$0 \leq \sigma \cdot \tau_N(\sigma) - (\hat{G}_N)^*(\tau_N(\sigma)).$$

After some easy (but tedious) calculations, relying on the fact that

$$\hat{G}_N(\sigma) \leq |\sigma|^2 (K_1 + K_2(\min(|\sigma|, N))^{p-2}),$$

we can show that the following estimates hold:

$$(\hat{G}_N)^*(\tau) \geq \frac{|\tau|^2}{4(K_1 + K_2 N^{p-2})} \quad \text{if } |\tau| > (2K_1 + pK_2) N^{p-1}$$

and

$$(\hat{G}_N)^*(\tau) \geq \frac{p-1}{p} \frac{|\tau|^{p/(p-1)}}{(2K_1 + pK_2)^{1/(p-1)}} - K_1 \frac{p-2}{p} \quad \text{if } |\tau| \leq (2K_1 + pK_2) N^{p-1}.$$

Collecting results, one can derive the following estimate:

$$|\tau_N(\sigma)| \leq K_p |\sigma| (1 + (\min(|\sigma|, N))^{p-2}) + K_p.$$

Given our choice of $g_N(\sigma)$ above, the same estimate holds:

$$|g_N(\sigma)| \le K_p|\sigma|(1 + (\min(|\sigma|, N))^{p-2}) + K_p. \tag{9.15}$$

Moreover, from the definition of g_N and the monotonicity of τ_N we obtain also

$$g_N(\sigma).\sigma \ge \frac{\alpha}{2}|\sigma|^2 + \frac{k}{2}(\min(|\sigma|, N))^{p-2}|\sigma|^2. \tag{9.16}$$

We need a final estimate, which will be obtained by noting that

$$0 \le \sigma.\tau_N(\sigma) - (\hat{G})^*(\tau_N(\sigma)),$$

since

$$(\hat{G}_N)^*(\tau) \ge (\hat{G})^*.$$

Therefore, using the previous estimate on $(\hat{G})^*$ we get

$$|\tau_N(\sigma)|^{p/(p-1)} \le K_p\tau_N(\sigma).\sigma + K_p.$$

The same estimate holds for g_N:

$$|g_N(\sigma)|^{p/(p-1)} \le K_p g_N(\sigma).\sigma + K_p. \tag{9.17}$$

It will also be useful to make the following observation, which is easily checked:

$$G(\sigma \mathbb{1}_{|\sigma| \le \rho_N}) \le G(\sigma)\mathbb{1}_{|\sigma| \le \rho_N}. \tag{9.18}$$

• **Approximate Optimization Problem**

We consider now the functional

$$J_N(\sigma) = \int_\Omega G_N(\sigma)\, dx$$

over $\mathcal{K}^2$. There exists a unique minimum σ^N of $J_N(\sigma)$. We study the limit of σ^N as N goes to infinity. At one point, σ denoted the minimum of the functional $J(\sigma)$, to save notation in a context where this was not too confusing. To avoid confusion now, we rename this minimum $\hat{\sigma}$. Then let

$$\tilde{\sigma}^N = \sigma^N \mathbb{1}_{|\sigma^N| \le \rho_N}.$$

We make use of properties (9.13) to assert that picking any $\sigma \in \mathcal{K}^p$,

$$J_N(\sigma^N) \le J_N(\sigma) \le J(\sigma)$$

and also

$$\int_\Omega |\sigma^N|^2\, dx \le C,$$

$$\int_\Omega |\sigma^N|^p \mathbb{1}_{|\sigma^N| \le N}\, dx \le C,$$

$$N^p \text{ meas } \{|\sigma^N| \ge N\} \le C.$$

In particular, σ^N remains in a bounded subset of $\mathcal{L}^2$, and $\tilde{\sigma}^N$ remains in a bounded subset of $\mathcal{L}^p$. Moreover,

$$\sigma^N - \tilde{\sigma}^N \to 0 \text{ in } \mathcal{L}^1.$$

On the other hand, using the last property (9.13), we can write

$$J_N(\sigma^N) \geq \int_\Omega G_N(\sigma^N)\, \mathbb{1}_{|\sigma^N| \leq \rho_N}\, dx + \frac{\alpha}{4} \int_\Omega |\sigma^N|^2 \mathbb{1}_{|\sigma^N| > \rho_N}\, dx$$

$$\geq \int_\Omega G(\sigma^N)\, \mathbb{1}_{|\sigma^N| \leq \rho_N}\, dx - C\,\epsilon^N,$$

and from (9.18),

$$J_N(\sigma^N) \geq J(\tilde{\sigma}^N) - C\,\epsilon^N.$$

If we now extract subsequences, still denoted σ^N, $\tilde{\sigma}^N$, that converge weakly in $\mathcal{L}^2$, $\mathcal{L}^p$, respectively, to the same limit σ^*, we deduce from the lower semicontinuity of J that

$$J(\sigma^*) \leq J(\sigma) \quad \forall \sigma,$$

and since $\sigma^* \in \mathcal{K}^p$, necessarily

$$\sigma^* = \hat{\sigma}.$$

Moreover,

$$J(\tilde{\sigma}^N) \to J(\hat{\sigma}),$$

$$\int_\Omega |\sigma^N|^2 \mathbb{1}_{|\sigma^N| > \rho_N}\, dx \to 0,$$

as $N \to \infty$.

The strong convergence of $\sigma^N \to \hat{\sigma}$ in $\mathcal{L}^2$ follows from monotonicity arguments, which we sketch briefly.

Set

$$\mathcal{A}(\sigma)(x) = \int_0^1 g(\lambda \sigma(x))\, d\lambda$$

defined on the Banach space $\mathcal{L}^p$. It maps $\mathcal{L}^p$ continuously into its dual $\mathcal{L}^{p/(p-1)}$, and is clearly a monotone operator. By definition,

$$J(\sigma) = \langle \mathcal{A}(\sigma), \sigma \rangle.$$

Therefore, the sequence $\tilde{\sigma}^N$ converges weakly in $\mathcal{L}^p$ to $\hat{\sigma}$ and

$$\langle \mathcal{A}(\tilde{\sigma}^N), \tilde{\sigma}^N \rangle \to \langle \mathcal{A}(\hat{\sigma}), \hat{\sigma} \rangle.$$

This implies (see [70]) that

$$\mathcal{A}(\tilde{\sigma}^N) \to \mathcal{A}(\hat{\sigma})$$

weakly, and therefore

$$\langle \mathcal{A}(\tilde{\sigma}^N) - \mathcal{A}(\hat{\sigma}), \tilde{\sigma}^N - \hat{\sigma} \rangle \to 0,$$

which implies the strong convergence.

• **End of Proof**

The approximation σ^N belongs to $\left(H^1_{\text{loc}}(\Omega)\right)^{n \times n}$. Going back to (9.11), we can write

$$\int_\Omega \theta^2 \frac{g_N(\sigma^N + hD^h_{k,+}\sigma^N) - g_N(\sigma^N)}{h} D^h_{k,+}\sigma \, dx$$

$$= -\int_\Omega \int_0^1 g_{N,kj}(\sigma^N(x + \lambda h e_k))(D_i\theta^2 D^h_{k,+}\sigma^N_{ij} + \theta^2 D^h_{k,+}f_j)dx\, d\lambda.$$

Using (9.14) we derive the inequality

$$\int_\Omega \theta^2 \left[\frac{\alpha}{2} + \frac{k}{2}(\min(|\sigma^N(x + he_k)|, |\sigma^N(x)|, N))^{p-2}\right] |D^h_{k,+}\sigma^N|^2 \, dx$$

$$\tag{9.19}$$

$$\leq -\int_\Omega \int_0^1 g_{N,kj}(\sigma^N(x + \lambda h e_k))(D_i\theta^2 D^h_{k,+}\sigma^N_{ij} + \theta^2 D^h_{k,+}f_j)dx\, d\lambda.$$

We then let h tend to 0 in (9.19). Using the fact that $\sigma^N \in H^1_{\text{loc}}(\Omega)$ and the estimate (9.15) we arrive at

$$\int_\Omega \theta^2 \left[\frac{\alpha}{2} + \frac{k}{2}(\min(|\sigma^N(x)|, N))^{p-2}\right] |D\sigma^N|^2 \, dx$$

$$\leq -\int_\Omega g_{N,kj}(\sigma^N)(D_i\theta^2 D_k\sigma^N_{ij} + \theta^2 D_k f_j)dx.$$

Using the estimate (9.15) for g_N and using Hölder's inequality to absorb the gradient into the left-hand side, we arrive at

$$\int_\Omega \theta^2[1 + (\min(|\sigma^N(x)|, N))^{p-2}]|D\sigma^N|^2 \, dx$$

$$\leq C \int_\Omega |\sigma^N|^2 (\min(|\sigma^N(x)|, N))^{p-2} \, dx + C + \int_\Omega |g_N(\sigma^N)|^{p/(p-1)} \, dx + C,$$

and thus from both estimates (9.16) and (9.17) it follows that

$$\int_\Omega \theta^2[1 + (\min(|\sigma^N(x)|, N))^{p-2}]|D\sigma^N|^2 \, dx \leq C + \int_\Omega g_N(\sigma^N).\sigma^N \, dx + C.$$

$$\tag{9.20}$$

To estimate the right-hand side of (9.20) we use the optimality condition

$$\int_\Omega g_N(\sigma^N).\sigma^N \, dx = \int_\Omega g_N(\sigma^N).\tau \, dx$$

for any $\tau \in \mathcal{K}^p$. Using (9.17) again and Hölder's inequality we find that the right-hand side of (9.20) is bounded independently of N. In particular, we have shown that

$$\int_\Omega \theta^2 [1 + |\tilde{\sigma}^N|^{p-2}]|D\tilde{\sigma}^N|^2 \, dx \leq C.$$

From Fatou's lemma, letting $N \to \infty$, and using the fact that $\tilde{\sigma}^N$ remains bounded in $\mathcal{L}^p$ and that we can assume that it converges pointwise to $\hat{\sigma}$, we obtain the same estimate with $\tilde{\sigma}^N$ replaced by $\hat{\sigma}$.

The proof is complete.   ◇

9.5 H^1_{loc} Theory for the Symmetric Case

9.5.1 Presentation of the Problem

We consider the same scalar function $G(\sigma)$ as before, but the argument σ is now a symmetric $n \times n$ matrix. We still assume (9.5), (9.7). Note that $g(\sigma)$ is a symmetric matrix. This time let

$$\mathcal{K}^p_{\text{sym}} = \{\sigma \in \mathcal{L}^p_{\text{sym}} \mid \text{div } \sigma = f, \nu.\sigma = 0 \text{ on } \Gamma_1\}$$

and the functional J be defined by

$$J(\sigma) = \int_\Omega G(\sigma) \, dx.$$

We consider the following minimization problem:

$$\text{Minimize } J(\sigma) \text{ over } \mathcal{K}^p_{\text{sym}}. \tag{9.21}$$

It is clear that there exists a unique minimum (denoted also by σ) and from Proposition 9.1 a unique u such that

$$\epsilon(u) = g(\sigma), \quad u \in (W^{1,\frac{p}{p-1}}_{\Gamma_0}(\Omega))^n,$$
$$\sigma \in \mathcal{L}^p_{\text{sym}}, \quad \text{div } \sigma = f, \quad \nu.\sigma|_{\Gamma_1} = 0.z$$

9.5.2 H^1_{loc} Regularity

We state the result corresponding to Theorem 9.6. We need also

$$\Delta f_j \in L^{np/(n+p)}(\Omega). \tag{9.22}$$

Theorem 9.7. *If we assume (9.5), (9.7), and (9.22), then the minimum* σ
of problem (9.21) satisfies

$$\int_{\Omega} \theta^2 (1 + |\sigma|^{p-2}) |D\sigma|^2 \, dx \leq C_p. \tag{9.23}$$

As we shall see, the proof is more involved than that of Theorem 9.6, although of course many steps are identical. We shall point out the main differences, and develop the part that requires specific treatment. In fact, in the formal proof one can understand the source of the additional difficulty.

● **Formal Proof**
Proceeding as in the nonsymmetric case we first obtain

$$\int \theta^2 G''(\sigma) D_k \sigma . D_k \sigma = - \int D_k u_j (D_i \theta^2 D_k \sigma_{ij} + \theta^2 D_k f_j) \, dx.$$

We want $\epsilon(u)$ to appear on the right-hand side, and so we write

$$- \int D_k u_j D_i \theta^2 D_k \sigma_{ij} \, dx = -2 \int \epsilon_{jk} D_i \theta^2 D_k \sigma_{ij} \, dx + \int D_j u_k D_i \theta^2 D_k \sigma_{ij} \, dx$$

$$= -2 \int g_{jk} D_i \theta^2 D_k \sigma_{ij} \, dx + \int D_j u_k D_i \theta^2 D_k \sigma_{ij} \, dx.$$

So, comparing with the nonsymmetric case, there is an additional term involving the vector u remaining on the right-hand side. We treat it as follows:

$$\int D_j u_k D_i \theta^2 D_k \sigma_{ij} \, dx = - \int_{\Omega} u_k [D_i \theta^2 D_k f_i + D_k \sigma_{ij} D_i D_j \theta^2] \, dx$$

$$= \int_{\Omega} [\text{div } u \, f_i D_i \theta^2 + u_k f_i D_i D_k \theta^2] \, dx$$

$$+ \int_{\Omega} [\text{div } u \sigma_{ij} D_i D_j \theta^2 + u_k \sigma_{ij} D_i D_j D_k \theta^2] \, dx.$$

We note that

$$\text{div } u = \text{tr } g.$$

Moreover, we write

$$- \int D_k u_j \, \theta^2 D_k f_j \, dx = \int_{\Omega} u_j (D_k \theta^2 D_k f_j + \theta^2 \triangle f_j) \, dx,$$

and collecting results, we arrive at the equality

$$\int \theta^2 G''(\sigma) D_k \sigma . D_k \sigma \, dx$$

$$= -2 \int g_{jk} D_i \theta^2 D_k \sigma_{ij} \, dx + \int \text{tr } g \, (f_i D_i \theta^2 + \sigma_{ij} D_i D_j \theta^2) \, dx \tag{9.24}$$

$$+ \int u_k [f_i D_i D_k \theta^2 + \sigma_{ij} D_i D_j D_k \theta^2 + D_j \theta^2 D_j f_k + \theta^2 \triangle f_k] \, dx$$

Although u remains, the formal proof can proceed, since u enters only in a term where there is no gradient of σ. This, however, will generate additional technical difficulties in the justification.

We proceed by writing the finite differences analogue of (9.24). After some steps, we obtain

$$
\int_\Omega \theta^2 \frac{g(\sigma + hD^h_{k,+}\sigma) - g(\sigma)}{h} D^h_{k,+}\sigma \, dx
$$

$$
= -2 \int_\Omega \int_0^1 g_{kj}(\sigma(x + \lambda he_k)) D_i\theta^2 D^h_{k,+}\sigma_{ij} \, dx
$$

$$
+ \int_\Omega \int_0^1 \int_0^1 g_{k\,k}(x + (\lambda - \mu)he_k) D_i D_j \theta^2(x - \mu he_k)\sigma_{ij} \, dx \, d\lambda \, d\mu
$$

$$
- \int_\Omega \int_0^1 u_k(x + \lambda he_k) D_j \theta^2 D^h_{k,+} f_j \, dx \, d\lambda + \int_\Omega u_k D^h_{j,-}(\theta^2 D^h_{j,+} f_k) \, dx
$$

$$
+ \int_\Omega \int_0^1 \int_0^1 u_k(x + (\lambda - \mu)he_k) D_i D_j D_k \theta^2(x - \mu he_k)\sigma_{ij} \, dx \, d\lambda \, d\mu.
$$

$$(9.25)$$

We can conclude as in the case $p = 2$.

Proof of Theorem 9.7 for $2 \le p \le p_0 = \frac{2n}{n-4}$.
Naturally, if $n \le 4$, this will imply that the result is completely proven, since in that case there is no restriction on p.

The idea is to proceed, as in the nonsymmetric case, by using our approximation G^N. So we write (9.25) for the corresponding pair σ^N, u^N. We obtain

$$
\int_\Omega \theta^2 \frac{g_N(\sigma^N + hD^h_{k,+}\sigma^N) - g_N(\sigma^N)}{h} D^h_{k,+}\sigma^N \, dx
$$

$$
= -2 \int_\Omega \int_0^1 g_{N,kj}(\sigma^N(x + \lambda he_k)) D_i\theta^2 D^h_{k,+}\sigma^N_{ij} \, dx
$$

$$
+ \int_\Omega \int_0^1 \int_0^1 g_{N,k\,k}(x + (\lambda - \mu)he_k) D_i D_j \theta^2(x - \mu he_k)\sigma^N_{ij} \, dx \, d\lambda \, d\mu
$$

$$
- \int_\Omega \int_0^1 u^N_k(x + \lambda he_k) D_j \theta^2 D^h_{k,+} f_j \, dx \, d\lambda + \int_\Omega u^N_k D^h_{j,-}(\theta^2 D^h_{j,+} f_k) \, dx
$$

$$
+ \int_\Omega \int_0^1 \int_0^1 u^N_k(x + (\lambda - \mu)he_k) D_i D_j D_k \theta^2(x - \mu he_k)\sigma_{ij} \, dx \, d\lambda \, d\mu.
$$

$$(9.26)$$

Of course, the convergence of the sequence σ^N to σ is the same as in the nonsymmetric case, since it relies only on the convergence of the functional to be minimized. We need now to obtain results on the convergence of u^N. But we can see, as in the nonsymmetric case, that $g^N(\sigma^N)$ remains in a bounded set of $\mathcal{L}^{p/(p-1)}_{\text{sym}}$. Since it is equal to $\epsilon(u^N)$, we can refer to Korn's inequality to claim that

$$\|u^N\|_{\left(W_{r_0}^{1,\frac{p}{p-1}}(\Omega)\right)^n} \le C,$$

and by Sobolev inequalities

$$\|u^N\|_{\left(L^{\frac{pn}{np-n-p}}(\Omega)\right)^n} \le C. \tag{9.27}$$

Using (9.14) as in the nonsymmetric case and then letting h go to 0 (N fixed) in (9.26) yields

$$
\int_\Omega \theta^2 \left[\frac{\alpha}{2} + \frac{k}{2}(\min(|\sigma^N(x)|, N))^{p-2}\right] |D\sigma^N|^2 \, dx
$$
$$
\le -2 \int_\Omega g_{N,kj}(\sigma^N) D_i \theta^2 D_k \sigma_{ij}^N \, dx + \int_\Omega \text{tr } g_N(\sigma^N) D_i D_j \theta^2 \sigma_{ij}^N \, dx
$$
$$
- \int_\Omega u_k^N D_j \theta^2 D_k f_j \, dx + \int_\Omega u_k^N D_j(\theta^2 D_j f_k) \, dx \tag{9.28}
$$
$$
+ \int_\Omega u_k^N D_i D_j D_k \theta^2 \, \sigma_{ij}^N \, dx.
$$

The difference between our situation and the nonsymmetric case lies in the last integral of the right-hand side, because the estimates on u^N and σ^N do not suffice. This is where the restriction on p will help. We first replace θ by θ^3 (which is, of course, possible) and note the crucial estimate

$$
\left|\int_\Omega u_k^N \sigma_{ij}^N D_i D_j D_k \theta^6 \, dx\right| \le C \int_\Omega |u^N| |\sigma^N| \theta^3 \, dx \le C \| \sigma^N \theta^3 \|_{\mathcal{L}_{\text{sym}}^{pn/(p+n)}}
$$
$$
\le C \| \sigma^N \theta^3 \|_{\mathcal{L}_{\text{sym}}^{2n/(n-2)}} \le C \| |D(\sigma_l \theta^3)| \|_{L^2},
$$

where we have used the restriction on p.

We use this estimate in (9.28). All other terms are easily estimated as in the nonsymmetric case; we need to use the assumptions (we especially need (9.22)). Finally, we arrive at an inequality of the form

$$
\int_\Omega \theta^6 (\min(|\sigma^N|, N))^{p-2} |D\sigma^N|^2 \, dx + \| |D(\sigma^N \theta^3)| \|_{L^2}^2 \le C \| |D(\sigma^N \theta^3)| \|_{L^2} + C,
$$

and we can conclude as in the nonsymmetric case.

Proof of Theorem 9.7 by a bootstrap argument.
We shall see that the result is also true with a weaker restriction on p, namely

$$
p_0 < p \le \frac{np_0}{n - 2 - p_0} = p_1.
$$

The argument can be reproduced, and in this way, by a bootstrap argument, the restriction on p can be waived.

We consider the function

$$G_{N,p_0}(\sigma) = ((\hat{G}_N)^* + j_N)^*(\sigma) + \frac{\alpha}{4}|\sigma|^2$$
$$+ \frac{k}{2p}|\sigma|^p \mathrm{1\!I}_{\{|\sigma|\leq N\}}$$
$$+ \left(\frac{k}{2p_0} N^{p-p_0} |\sigma|^{p_0} + k \left(\frac{1}{2p} - \frac{1}{2p_0} \right) N^p \right) \mathrm{1\!I}_{\{|\sigma|>N\}}$$

and the approximate optimization problem

$$J_{N,p_0}(\sigma) = \int_\Omega G_{N,p_0}(\sigma)\, dx$$

over $\mathcal{K}^{p_0}$.

Considering the gradient g_{N,p_0} of G_{N,p_0} we check the properties

$$(g_{N,p_0}(\sigma_1) - g_N(\sigma_2)).(\sigma_1 - \sigma_2) \geq \frac{\alpha}{2}|\sigma_1 - \sigma_2|^2$$
$$+ \frac{k}{2}\min(|\sigma_1|^{p_0-2}(|\sigma_1| \wedge N)^{p-p_0},\ |\sigma_2|^{p_0-2}(|\sigma_2| \wedge N)^{p-p_0})|\sigma_1 - \sigma_2|^2,$$

$$|g_{N,p_0}(\sigma)| \leq K_p|\sigma|(1 + |\sigma|^{p_0-2}\min(|\sigma|,N)^{p-p_0}) + K_p,$$

$$g_{N,p_0}(\sigma).\sigma \geq \frac{\alpha}{2}|\sigma|^2 + \frac{k}{2}|\sigma|^{p_0}(\min(|\sigma|,N)^{p-p_0},$$

$$|g_{N,p_0}(\sigma)|^{p/(p-1)} \leq K_p g_N(\sigma).\sigma + K_p.$$

If we again call σ^N the minimum of $J_{N,p_0}(\sigma)$, we again obtain convergence to σ (also earlier called $\hat{\sigma}$) in the space $\mathcal{L}^2_{\text{sym}}$ as N tends to infinity. We have

$$\frac{k}{2}\int_\Omega |\sigma^N|^{p_0}\min(|\sigma^N|,N)^{p-p_0}\, dx \leq \int_\Omega g_{N,p_0}(\sigma^N).\sigma^N\, dx \leq C,$$

from which it follows in particular that $|\sigma^N|$ remains bounded in $L^{p_0}(\Omega)$.

Moreover, we know from the value of p_0 and the first part of the proof that

$$\int_\Omega \theta^2 |\sigma^N|^{p_0-2}|D\sigma^N|^2\, dx < \infty$$

(we do not claim at this stage that it is uniform in N). We can then write the equivalent of (9.28):

$$\int_\Omega \theta^2 \left[\frac{\alpha}{2} + \frac{k}{2}|\sigma^N|^{p_0-2}(\min(|\sigma^N|,N))^{p-p_0} \right] |D\sigma^N|^2\, dx$$
$$\leq -2\int_\Omega g_{N,kj}(\sigma^N)D_i\theta^2 D_k\sigma^N_{ij}\, dx + \int_\Omega \mathrm{tr}\ g_N(\sigma^N)D_iD_j\theta^2\sigma^N_{ij}\, dx$$
$$- \int_\Omega u^N_k D_j\theta^2 D_k f_j\, dx + \int_\Omega u^N_k D_j(\theta^2 D_j f_k)\, dx$$
$$+ \int_\Omega u^N_k D_iD_jD_k\theta^2\ \sigma^N_{ij}\, dx.$$

$$(9.29)$$

We replace θ by $\theta^{\frac{3p_0}{2(p_0-1)}}$. To simplify the notation we write $s = \frac{3p_0}{2(p_0-1)}$. Consider again the crucial term

$$\left| \int_\Omega u_k^N \sigma_{ij}^N D_i D_j D_k \theta^{2s} \, dx \right| \le C \int_\Omega |u^N| |\sigma^N| \theta^{2s-3} \, dx$$

$$\le C \| |\sigma^N| \theta^{2s/p_0} \|_{L^{pn/(p+n)}(\Omega)}.$$

To get an estimate on $\| |\sigma^N| \theta^{2s/p_0} \|_{L^{pn/(p+n)}(\Omega)}$ that is uniform in N we first notice that the restriction on p implies

$$\frac{pn}{p+n} \le \frac{p_0 n}{n-2}.$$

Therefore, it is sufficient to estimate $\| |\sigma^N| \theta^{2s/p_0} \|_{L^{p_0 n/(n-2)}(\Omega)}$. Now we use the Sobolev–Poincaré inequality to write

$$\| |\sigma^N| \theta^{2s/p_0} \|_{L^{p_0 n/(n-2)}(\Omega)} \le C \left(\int_\Omega |D(|\sigma^N|^{p_0/2})|^2 \theta^{2s} \, dx \right)^{1/p_0} + C$$

$$= C \left(\int_\Omega |\sigma^N|^{p_0-2} |D\sigma^N|^2 \theta^{2s} \, dx \right)^{1/p_0} + C$$

$$\le C \left(\int_\Omega |D\sigma^N|^2 \theta^{2s} \, dx + \int_\Omega |\sigma^N|^{p_0-2} |D\sigma^N|^2 \right.$$

$$\left. (\min(|\sigma^N|, N))^{p-p_0} \theta^{2s} \, dx \right)^{1/p_0} + C,$$

which can be estimated by the left-hand side of (9.29). The other terms on the right-hand side are estimated as usual. Therefore, the result has been proven for $p \le p_1$.

Clearly, we can now proceed with a bootstrap argument, which permits us to waive the restrictions on p. Hence, the estimate (9.23) has been proven, and the proof of Theorem 9.7 is complete.   ◇

9.5.3 Reducing the Symmetric Case to the Nonsymmetric Case

We shall show here a useful technique, which in particular permits us to reduce the study of regularity in the symmetric case to that in the nonsymmetric case. We use the notation of the previous section, so g is symmetric and the gradient of G. We define, for a nonsymmetric matrix σ,

$$G^\delta(\sigma) = G(\sigma) + \frac{1}{2\delta} \sum_{ij} (\sigma_{ij} - \sigma_{ji})^2,$$

and the penalty δ will tend to 0. The gradient is given by

$$g^\delta(\sigma) = g(\sigma) + \frac{1}{\delta}(\sigma - \sigma^*).$$

We check immediately that

$$(g^\delta(\sigma^1) - g^\delta(\sigma^2)).(\sigma^1 - \sigma^2) \geq (y(\sigma^1) - y(\sigma^2)).(\sigma^1 - \sigma^2).$$

Let us consider the problem of minimizing the functional

$$J^\delta(\sigma) = \int_\Omega G^\delta(\sigma)\,dx$$

over $\mathcal{K}^p$ defined in (9.6). It is fairly easy to check that there exists a unique minimum σ^ν and that

$$\sigma^\nu \to \sigma \text{ in } \mathcal{L}^2 \text{ strongly and in } \mathcal{L}^p \text{ weakly,}$$

where σ is the solution of the problem (9.21). Now we can write down the optimality conditions

$$Du^\delta = g(\sigma^\delta) + \frac{1}{\delta}(\sigma^\delta - (\sigma^\delta)^*), \quad u^\delta \in \left(W^{1,\frac{p}{p-1}}_{\Gamma_0}(\Omega)\right)^n,$$
$$\sigma^\delta \in \mathcal{L}^p, \quad \text{div } \sigma^\delta = f, \quad \nu.\sigma^\delta|_{\Gamma_1} = 0.$$

Since

$$\epsilon(u^\delta) = g(\sigma^\delta),$$

we obtain that it remains in a bounded subset of $\mathcal{L}^{p/(p-1)}_{\text{sym}}$, and thus from Korn's inequality we derive

$$u^\delta \to u \text{ in } \left(W^{1,\frac{p}{p-1}}_{\Gamma_0}(\Omega)\right)^n \text{ weakly.}$$

Since we assume here that Theorem 9.6 is available for u^δ, we can write the inequality (see (9.10))

$$\int \theta^2(\alpha + k|\sigma^\delta|^{p-2})|D\sigma^\delta|^2\,dx \leq - \int g^\delta_{kj}(\sigma^\delta)(D_i\theta^2 D_k\sigma^\delta_{ij} + \theta^2 D_k f_j)\,dx.$$
$$\tag{9.30}$$

We have to estimate the right-hand side:

$$\text{RHS} = -2\int g_{kj}(\sigma^\delta)(D_i\theta^2 D_k\sigma^\delta_{ij} + \theta^2 D_k f_j)\,dx$$
$$+ \int_\Omega D_j u^\delta_k(D_i\theta^2 D_k\sigma^\delta_{ij} + \theta^2 D_k f_j)\,dx.$$

The only delicate term is

$$\int_\Omega D_j u_k^\delta D_i \theta^2 D_k \sigma_{ij}^\delta \, dx$$

$$= -\int_\Omega D_j g_{k\,k}(\sigma^\delta) D_i \theta^2 \sigma_{ij}^\delta \, dx - \int_\Omega D_j u_k^\delta D_i D_k \theta^2 \sigma_{ij}^\delta \, dx.$$

The latter two integrals are bounded by a constant, thanks to the available estimates. Therefore, the right-hand side of (9.30) is bounded by a constant as δ tends to 0. By Fatou's lemma the regularity of the limit σ follows.

9.6 L_{loc}^∞ Theory for the Nonsymmetric Uhlenbeck Case

9.6.1 Setting of the Problem and Statement of Results

We consider the problem (9.8). The Uhlenbeck case refers to

$$g(\sigma) = \beta(|\sigma|)\sigma, \tag{9.31}$$

where

$$\beta(0) = \alpha > 0, \quad \beta'(x) \geq k(p-2)x^{p-3}, \tag{9.32}$$

and

$$\beta(x) \leq K(1 + x^{p-2}). \tag{9.33}$$

So we can write the necessary conditions

$$Du = g(\sigma), \quad u \in \left(W_{\Gamma_0}^{1,\frac{p}{p-1}}(\Omega)\right)^n,$$
$$\sigma \in \mathcal{L}^p, \quad \text{div } \sigma = f, \quad \nu.\sigma|_{\Gamma_1} = 0. \tag{9.34}$$

We shall assume

$$f \in (W^{1,\max(p,n)}(\Omega))^n. \tag{9.35}$$

Note that thanks to Theorem 9.6 we have the local property

$$\int_\Omega \theta^2 (1 + |\sigma|^{p-2})|D\sigma|^2 \, dx \leq C_p \tag{9.36}$$

for any smooth function θ with compact support in Ω. Our objective is to prove the following result.

Theorem 9.8. *If we assume (9.31), (9.32), (9.33), and (9.35), then σ is locally bounded in Ω.*

9.6.2 Proof of Theorem 9.8

We shall adapt Moser's iteration technique to take account of the localization, as in Section 7.4. We prove that for any ball $B_{R_0}(\xi)$ such that

$$B_{2R_0}(\xi) \subset \Omega$$

we have

$$|\sigma(x)| \leq C_{R_0}, \quad x \in B_{R_0}(\xi). \tag{9.37}$$

Let $R \leq R_0/2$. We consider a smooth function $\theta = \theta_R$ that is equal to 1 on B_{R_0+R}, vanishes outside B_{R_0+2R}, and takes values in $[0,1]$ such that $|D\theta| \leq C/R$.

We test (9.34) with $-D_k(\theta^2 D_k(|\sigma|^s \sigma))$. After some calculations and double integrations by parts, we arrive at the following relation:

$$\int_\Omega \theta^2 \beta(|\sigma|)|\sigma|^s |D\sigma|^2 \, dx + \int_\Omega \theta^2 [(s+1)\beta'(|\sigma|)|\sigma|^{s-1}$$

$$+ s\beta(|\sigma|)|\sigma|^{s-2}] \left| D\frac{|\sigma|^2}{2} \right|^2 dx$$

$$= -\int_\Omega \theta^2 \beta(|\sigma|)|\sigma|^s \sigma_{kj} D_k f_j \, dx - \int_\Omega \beta(|\sigma|)|\sigma|^s \sigma_{kj} D_k \sigma_{ij} D_i \theta^2 \, dx$$

$$+ \int_\Omega s\theta^2 \beta'(|\sigma|)|\sigma|^{s-3} \sigma_{kj} \sigma_{ij} D_k \left(\frac{|\sigma|^2}{2} \right) D_i \left(\frac{|\sigma|^2}{2} \right) dx.$$

We can absorb the last integral on the right-hand side into the left-hand side. It follows that

$$\int_\Omega \theta^2 \beta(|\sigma|)|\sigma|^s |D\sigma|^2 \, dx + \int_\Omega \theta^2 |\sigma|^{s-2} [\beta'(|\sigma|)|\sigma| + s\beta(|\sigma|)] \left| D\frac{|\sigma|^2}{2} \right|^2 dx$$

$$\leq -\int_\Omega \theta^2 \beta(|\sigma|)|\sigma|^s \sigma_{kj} D_k f_j \, dx - \int_\Omega \beta(|\sigma|)|\sigma|^s \sigma_{kj} D_k \sigma_{ij} D_i \theta^2 \, dx.$$

Using Hölder's inequality to absorb the term involving $D_k \sigma_{ij}$ on the right-hand side into the left-hand side, we obtain

$$\int_\Omega s\theta^2 |\sigma|^{s-2} \beta(|\sigma|)(\sigma.D\sigma)^2 \, dx$$

$$\leq \int_\Omega \theta^2 \beta(|\sigma|)|\sigma|^{s+1} |Df| \, dx + \int_\Omega \beta(|\sigma|)|\sigma|^{s+2} |D\theta|^2 \, dx.$$

Using the assumptions on β we derive

$$ks \int_\Omega \theta^2 |\sigma|^{s+p-4} (\sigma.D\sigma)^2 \, dx$$

$$\leq K \int_\Omega (1 + |\sigma|^{p-2})|\sigma|^{s+1} (\theta^2 |Df| + |\sigma||D\theta|^2) \, dx,$$

which reads

$$\frac{2sk}{(s+p)^2} \int_\Omega |D\left(\theta|\sigma|^{(s+p)/2}\right)|^2 \, dx$$

$$\leq K \int_\Omega \theta^2 |Df|(|\sigma|^{s+1} + |\sigma|^{p+s-1}) \, dx$$

$$+ K \int_\Omega |D\theta|^2 |\sigma|^{s+2} \, dx + \left(K + \frac{4sk}{(s+p)^2}\right) \int_\Omega |D\theta|^2 |\sigma|^{s+p} \, dx.$$

From now on C designates not an explicit, but a generic, constant, possibly depending on R_0. Taking account of the estimate on $D\theta$ and applying the Poincaré inequality on the left-hand side (this is a standard step in Moser's technique) we obtain

$$\frac{1}{s+p}\left(\int_{B_{R_0+R}} |\sigma|^{\frac{(s+p)n}{n-2}} \, dx\right)^{(n-2)/n} \leq C \int_{B_{R_0+2R}} |Df|(|\sigma|^{s+1} + |\sigma|^{p+s-1}) \, dx$$

$$+ \frac{C}{R^2} \int_{B_{R_0+2R}} |\sigma|^{s+2} \, dx + \frac{C}{R^2} \int_{B_{R_0+2R}} |\sigma|^{s+p} \, dx.$$

$$(9.38)$$

We use the notation

$$\zeta_{\lambda,R} = \max\left(1, \||\sigma|\|_{L^\lambda(B_{R_0+R})}\right).$$

Then we can reinterpret the inequality (9.38) as follows:

$$\zeta_{\frac{(s+p)n}{n-2},R} \leq \left(\frac{C(s+p)}{R^2}\right)^{1/(s+p)} \zeta_{\frac{(s+p)n}{n-1},2R}.$$

Everything is then routine. Define the sequence

$$s_{j+1} + p = (s_j + p)\frac{n-1}{n-2}, \quad s_0 = 0,$$

$$R_{j+1} = \frac{R_0}{2^{j+1}},$$

and set

$$z_j = \zeta_{\frac{(s_j+p)n}{n-2},R_j}.$$

Thanks to the result (9.36) and using the Poincaré inequality, we see that z_0 is finite. The iteration technique applies, and we note that

$$s_j + p = pa^j, \quad a = \frac{n-1}{n-2}.$$

Therefore,

$$z_j \leq \prod_{k=0}^{j} (C\,(4a)^{j+1})^{\frac{a^{-(j+1)}}{p}},$$

and since $a > 1$, we get

$$z_j \leq C, \quad \forall j.$$

Letting j tend to ∞ we have obtained (9.37) and completed the proof.

9.7 $W_{\text{loc}}^{1,p}$ Theory for the Nonsymmetric Case

9.7.1 Assumptions and Results

We consider the situation of Section 9.4, with $p = 2$. In other words, we assume

$$G(0) = 0, \quad G \text{ is } C^2, \quad g = G', \quad g(0) = 0,$$
$$\alpha I \leq g'(\sigma) \leq M I. \tag{9.39}$$

Furthermore, we assume that

$$f \in \left(W^{1,\frac{2n}{n-2}}(\Omega) \right)^n. \tag{9.40}$$

We can assert that there exists a unique pair σ, u such that

$$Du = g(\sigma), \quad u \in (H^1_{\Gamma_0}(\Omega))^n,$$
$$\sigma \in \mathcal{L}^2, \quad \text{div}\,\sigma = f, \quad \nu.\sigma|_{\Gamma_1} = 0. \tag{9.41}$$

Furthermore, the regularity result of Theorem 9.6 applies to yield that

$$\sigma \in H^1_{\text{loc}}.$$

Our objective is to show a Meyers' type result:

Theorem 9.9. *If we assume (9.39) and (9.40), then the solution σ of (9.41) satisfies*

$$\sigma \in W_{\text{loc}}^{1,p}(\Omega)$$

with

$$2 \leq p \leq p_0, \quad 2 < p_0 < \frac{2n}{n-2}.$$

9.7.2 Proof of Theorem 9.9

We first notice that by the Sobolev embedding we can assert that

$$|\sigma| \in L_{\text{loc}}^{2n/(n-2)}(\Omega). \tag{9.42}$$

In order to make apparent a scaling related to the local character of the estimates, we make the following assumption about the form of the domain:

$$\overline{B_{R_0}(\xi)} \subset \Omega \subset B_{2R_0}(\xi)$$

for a convenient ξ and R_0. We shall write

$$\Omega' = B_{R_0}(\xi)$$

and consider a function ζ such that

$$\text{supp } \zeta = \Omega', \quad 0 \le \zeta \le 1, \quad \zeta \text{ smooth,}$$
$$|D\zeta| \le \frac{C}{R_0}.$$

We next consider balls $B_R = B_R(x_0)$, with $x_0 \in \Omega$. Note that from these choices we have

$$\Omega' \subset B_{3R_0}(x_0). \tag{9.43}$$

Let τ_R be a smooth function such that

$$0 \le \tau_R \le 1, \quad \tau_R = 1 \text{ on } B_R \text{ with support in } B_{2R} \text{ and } |D\tau_R| \le \frac{C}{R}.$$

We test the first equation (9.41) with

$$-D_k(\zeta^2 \tau_R^2 D_k \sigma),$$

and after some, by now standard, steps involving integration by parts we get

$$\int_\Omega \zeta^2 \tau_R^2 D_i g_{kj}(\sigma) \, D_k \sigma_{ij} \, dx = \int_\Omega \zeta^2 \tau_R^2 D_k g_{ij}(\sigma) D_k \sigma_{ij} \, dx.$$

We introduce a constant c_{kj} (which will be specified later), do an additional integration by parts, and make use of the second relation (9.41) to obtain

$$-\int_\Omega \zeta^2 \tau_R^2 (g_{kj}(\sigma) - c_{kj}) D_k f_j \, dx - \int_\Omega D_i(\zeta^2 \tau_R^2)(g_{kj}(\sigma) - c_{kj}) D_k \sigma_{ij} \, dx$$
$$= \int_\Omega \zeta^2 \tau_R^2 g'_{ij,hl} D_k \sigma_{hl} D_k \sigma_{ij} \, dx.$$

From this inequality we derive the estimate (denoting by K a generic constant)

$$
\int_\Omega \zeta^2 \tau_R^2 |D\sigma|^2 \, dx \le \frac{K}{R^2} \int_{\Omega \cap B_{2R}} \zeta^2 |g(\sigma) - c|^2 \, dx
$$

$$
+ K \int_\Omega \tau_R^2 \left[|D\zeta|^2 |g(\sigma) - c|^2 + \zeta^2 |Df||g(\sigma) - c| \right] \, dx,
$$

(9.44)

where c stands for the matrix c_{kj}. We now choose

$$
c = \frac{\int_{\Omega' \cap B_{2R}} g(\sigma) \, dx}{|\Omega' \cap B_{2R}|},
$$

so we may write (9.44) as follows:

$$
\int_\Omega \zeta^2 \tau_R^2 |D\sigma|^2 \, dx \le \frac{K}{R^2} \int_{\Omega \cap B_{2R}} \zeta^2 |g(\sigma) - c|^2 \, dx
$$

$$
+ \frac{K}{R_0^2} \int_{\Omega' \cap B_{2R}} (|\sigma|^2 + |Df|^2) \, dx.
$$

Let x^* be the minimum of ζ on $\Omega' \cap B_{2R}$. A first observation is to check the following inequality:

$$
\frac{1}{2R^2} \int_{\Omega' \cap B_{2R}} \zeta^2 |g(\sigma) - c|^2 \, dx \le \frac{\zeta^2(x^*)}{R^2} \int_{\Omega' \cap B_{2R}} |g(\sigma) - c|^2 \, dx
$$

$$
+ 48 \|D\zeta\|_{\infty, \Omega' \cap B_{2R}}^2,
$$

where ∞ refers to the L^∞ norm over $\Omega' \cap B_{2R}$. This is easily obtained by using Taylor's formula and Hölder's inequality. Using the Poincaré inequality we have

$$
\int_{\Omega' \cap B_{2R}} |g(\sigma) - c|^2 \, dx \le c_0 \left(\int_{\Omega' \cap B_{2R}} |Dg(\sigma)|^{2n/(n+2)} \, dx \right)^{(n+2)/n},
$$

and from the assumption (9.39) we get

$$
\int_{\Omega' \cap B_{2R}} |g(\sigma) - c|^2 \, dx \le K \left(\int_{\Omega' \cap B_{2R}} |D\sigma|^{2n/(n+2)} \, dx \right)^{(n+2)/n}.
$$

Collecting results and taking account of the fact that x^* is the minimum of ζ on $\Omega' \cap B_{2R}$, we also deduce that

$$
\int_{\Omega' \cap B_R} \zeta^2 |D\sigma|^2 \, dx \le \frac{K}{R^2} \left(\int_{\Omega' \cap B_{2R}} \zeta^{2n/(n+2)} |D\sigma|^{2n/(n+2)} \, dx \right)^{(n+2)/n}
$$

$$
+ \frac{K}{R_0^2} \int_{\Omega' \cap B_{2R}} (|\sigma|^2 + |Df|^2) \, dx.
$$

(9.45)

Setting

$$\phi = \left(\frac{|\sigma|^2 + |Df|^2}{R_0^2} \right)^{n/(n+2)} \mathrm{II}_{\Omega'}$$

and

$$z = \zeta^{2n/(n+2)} |D\sigma|^{2n/(n+2)} \mathrm{II}_{\Omega'},$$

then (9.45) becomes

$$\fint_{B_R} z^{(n+2)/n} \, dx \le K \left(\fint_{B_R} z \, dx \right)^{(n+2)/n} + K \fint_{B_R} \phi^{(n+2)/n} \, dx.$$

From (9.40) and (9.42) we see that

$$\phi \in L^{(n+2)/(n-2)}(R^n).$$

We can then use the "reverse Hölder's inequality" (see Section 1.2.1) to assert that there exists p with

$$2 \le p \le p_0, \quad 2 < p_0 < \frac{2n}{n-2},$$

not depending on R_0, such that for $R \le 3R_0/2$ (any finite upper bound is fine) we have

$$\left(\fint_{B_R} z^{\frac{n+2}{2n} p} \, dx \right)^{2n/(n+2)p}$$

$$\le K_p \left[\left(\fint_{B_{2R}} z^{(n+2)/n} \, dx \right)^{n/(n+2)} + \left(\fint_{B_{2R}} \phi^{\frac{n+2}{2n} p} \, dx \right)^{2n/(n+2)p} \right].$$

Thus from the definitions of z and ϕ as well as (9.43) we deduce the estimate

$$\fint_{\Omega' \cap B_R} \zeta^p |D\sigma|^p \, dx \le K_p \left[\left(\fint_{\Omega' \cap B_{2R}} \zeta^2 |D\sigma|^2 \, dx \right)^{p/2} \right.$$

$$\left. + \frac{1}{R_0^p} \fint_{\Omega' \cap B_{2R}} (|\sigma|^p + |Df|^p) \, dx \right], \tag{9.46}$$

and of course the proof of the theorem is complete. ◇

9.8 $C_{\text{loc}}^{1+\delta}$ Regularity for the Nonsymmetric Case

9.8.1 Setting of the Problem and Statement of Results

We consider the situation of Section 9.7.1, with additional conditions. More precisely, we assume

$$
\begin{aligned}
&g(\sigma) : R^{n \times n} \to R^{n \times n}, \\
&\alpha I \le g'(\sigma) \le MI, \\
&g' \text{ Lipschitz.}
\end{aligned}
\tag{9.47}
$$

We consider a pair σ, u such that

$$
\begin{aligned}
&Du = g(\sigma), \ u \in (H^1(\Omega))^n, \\
&\sigma \in \left(H_{\text{loc}}^1(\Omega)\right)^{n \times n}, \ \text{div } \sigma = 0.
\end{aligned}
\tag{9.48}
$$

In fact, if we compare to (9.41), we have taken $f = 0$. This assumption is made to simplify technicalities and can be removed. Since we are interested in a local regularity result, we do not specify boundary conditions, and thus we do not address the existence issue. We shall assume that

$$
\sigma \in C_{\text{loc}}^\delta(\Omega).
\tag{9.49}
$$

Our objective is to prove the following result

Theorem 9.10. *We assume (9.47) and consider a pair σ, u satisfying (9.48) and (9.49). Then*

$$
D\sigma \in C_{\text{loc}}^{\delta'}(\Omega), \quad \forall \delta' < \delta.
$$

Take an open set Ω' such that

$$
\overline{\Omega'} \subset \Omega.
$$

We pick R_0 such that

$$
B_{R_0}(x_0) \subset \Omega, \quad \forall x_0 \in \Omega'.
$$

Additional restrictions on the size of R_0 will be made later.

The result will be obtained using the Campanato spaces (see Section 1.1.1). In order to prove that

$$
D\sigma \in C^{\delta'}(\overline{\Omega'}), \quad \delta' < \delta,
\tag{9.50}
$$

we need to prove that the Campanato norm

$$
\sup_{x_0, R} \frac{\int_{B_R(x_0) \cap \Omega'} |D\sigma - D\sigma_{x_0, R}|^2 \, dx}{R^{n + 2\delta'}}
$$

is finite, where

$$D\sigma_{x_0,R} = \frac{\int_{B_R(x_0) \cap \Omega'} D\sigma \, dx}{|B_R(x_0) \cap \Omega'|}.$$

It is easy to check that we can take $x_0 \in \Omega'$ without loss of generality. Moreover, thanks to the second property (9.48), we can also assume witout loss of generality that $R \le R_0$. Define, then, for $x_0 \in \Omega'$ and $R \le R_0$,

$$\Phi(R) = \frac{\int_{B_R(x_0)} |D\sigma - \overline{D\sigma_R}|^2 \, dx}{R^{n+2\delta'}},$$

where

$$\overline{D\sigma_R} = \frac{\int_{B_R(x_0)} D\sigma \, dx}{|B_R(x_0)|}.$$

Since

$$\int_{B_R(x_0) \cap \Omega'} |D\sigma - D\sigma_{x_0,R}|^2 \, dx \le 4 \int_{B_R(x_0)} |D\sigma - \overline{D\sigma_R}|^2 \, dx,$$

it is then enough to prove that

$$\Phi(R) \le C \text{ for } R \le R_0, \quad x_0 \in \Omega', \tag{9.51}$$

where R_0 will be conveniently selected, and for any $\delta' < \delta$.

9.8.2 Preliminary Results

We consider an auxiliary problem as follows:

$$\begin{aligned} D_i \omega_j^k &= g'_{ij,hl}(\sigma(x_0)) \tau_{hl}^k, \quad \omega^k \in (H^1(B_R(x_0)))^n, \\ \tau^k &\in \mathcal{L}^2, \quad \text{div } \tau^k = 0, \\ \nu.(D_k\sigma - \tau^k)|_{\partial B_R(x_0)} &= 0. \end{aligned} \tag{9.52}$$

We shall write τ for the family of functions τ^k.

We next differentiate (9.48) with respect to k and obtain

$$D_i D_k u_j = g'_{ij,hl}(\sigma) D_k \sigma_{hl},$$

which we test by $(D_k\sigma - \tau^k)_{ij}$ on $B_R(x_0)$.

Using the summation convention to facilitate calculation, it is easy to deduce that

$$\int_{B_R(x_0)} g'(\sigma) D\sigma.(D\sigma - \tau) \, dx = 0.$$

Similarly, testing (9.52) with the same quantity yields

$$\int_{B_R(x_0)} g'(\sigma(x_0)) \tau.(D\sigma - \tau) \, dx = 0.$$

Combining the two relations we get

$$\int_{B_R(x_0)} g'(\sigma(x_0))(D\sigma - \tau).(D\sigma - \tau)\, dx$$
$$= \int_{B_R(x_0)} (g'(\sigma(x_0)) - g'(\sigma))D\sigma.(D\sigma - \tau)\, dx.$$

Using assumptions (9.47) and (9.49) we obtain the following estimate:

$$\int_{B_R(x_0)} |D\sigma - \tau|^2\, dx \leq CR^{2\delta} \int_{B_R(x_0)} |D\sigma|^2\, dx. \tag{9.53}$$

Take $\rho \leq R$ and let c^k_{ij} be an arbitrary set of constants. We can write

$$\int_{B_\rho(x_0)} |D\sigma - c|^2\, dx \leq 2\int_{B_\rho(x_0)} |D\sigma - \tau|^2\, dx + 2\int_{B_\rho(x_0)} |\tau - c|^2\, dx$$
$$\leq 2\int_{B_\rho(x_0)} |\tau - c|^2\, dx + 2CR^{2\delta}\int_{B_R(x_0)} |D\sigma|^2\, dx,$$

and since

$$\int_{B_\rho(x_0)} |D\sigma - \overline{D\sigma_\rho}|^2\, dx \leq \int_{B_\rho(x_0)} |D\sigma - c|^2\, dx,$$

we obtain the estimate

$$\int_{B_\rho(x_0)} |D\sigma - \overline{D\sigma_\rho}|^2\, dx \leq 2\int_{B_\rho(x_0)} |\tau - c|^2\, dx + 2CR^{2\delta}\int_{B_R(x_0)} |D\sigma|^2\, dx. \tag{9.54}$$

The next useful estimate concerns τ, itself a solution of (9.52). We shall first prove the following estimate:

$$\|D\tau\|^2_{L^\infty\left(B_{\frac{R}{2^{[\frac{n}{2}]+2}}}(x_0)\right)} \leq \frac{K}{R^n}\int_{B_{R/2}(x_0)} |D\tau|^2\, dx, \tag{9.55}$$

where the constant K is a generic constant depending only on n. The notation $[\frac{n}{2}]$ denotes the integer part of $\frac{n}{2}$. Set

$$A = g'(\sigma(x_0)), \quad \Pi = \nu.D_k\sigma$$

to simplify the notation, and write (9.52) suppressing the index k:

$$\begin{aligned} D_i\omega_j &= (A\tau)_{ij}, \\ \operatorname{div}\tau &= 0, \\ \nu.\tau|_{\partial B_R(x_0)} &= \Pi. \end{aligned} \tag{9.56}$$

Note that on $\partial B_R(x_0)$,

$$\nu = \frac{x - x_0}{R}, \quad |x - x_0| = R.$$

Introduce new functions

$$\tilde{\tau}(\xi) = \tau \left(x_0 + \frac{\xi R}{2^{[\frac{n}{2}]+2}} \right),$$

$$\tilde{\omega}(\xi) = \frac{2^{[\frac{n}{2}]+2}}{R} \omega \left(x_0 + \frac{\xi R}{2^{[\frac{n}{2}]+2}} \right),$$

with

$$|\xi| \le 2^{[\frac{n}{2}]+1}.$$

Then the pair $\tilde{\tau}, \tilde{\omega}$ is a solution of

$$\begin{aligned} D_{\xi_i}\tilde{\omega}_j(\xi) &= (A\tilde{\tau})_{ij}(\xi), \\ \operatorname{div} \tilde{\tau} &= 0, \\ \tilde{\nu}.\tilde{\tau}|_{\partial B_{2^{[\frac{n}{2}]+1}}(0)}(\xi) &= \nu.\tau|_{\partial B_{R/2}(x_0)} \left(x_0 + \frac{\xi R}{2^{[\frac{n}{2}]+2}} \right), \end{aligned} \quad (9.57)$$

where

$$\tilde{\nu} = \frac{\xi}{2^{[\frac{n}{2}]+1}} = \frac{2(x - x_0)}{R}, \quad \text{with } |x - x_0| = \frac{R}{2}.$$

Let us consider a smooth function such that

$$\theta = \begin{vmatrix} 1 & \text{on } B_1(0), \\ 0 & \text{outside } B_2(0). \end{vmatrix}$$

We differentiate (9.57) twice to obtain, inside the ball $B_{2^{[\frac{n}{2}]+1}}(0)$,

$$\begin{aligned} D_{\xi_i}D_{\xi_k}D_{\xi_l}\tilde{\omega}_j(\xi) &= (AD_{\xi_k}D_{\xi_l}\tilde{\tau})_{ij}(\xi), \\ \operatorname{div} D_{\xi_k}D_{\xi_l}\tilde{\tau} &= 0. \end{aligned}$$

We test the first relation with

$$\theta^2 \left(\frac{\xi}{2^{[\frac{n}{2}]}} \right) D_{\xi_k}D_{\xi_l}\tilde{\tau},$$

which vanishes on the boundary of $B_{2^{[\frac{n}{2}]+1}}(0)$. We integrate by parts and make use of the divergence condition and

$$D_{\xi_k}D_{\xi_l}\tilde{\omega}_j = (AD_{\xi_l}\tilde{\tau})_{kj}$$

to obtain (after easy calculations)

$$\int_{B_{2^{[\frac{n}{2}]}}(0)} |D^2\tilde{\tau}|^2 \, d\xi \le \frac{c}{2^{2[\frac{n}{2}]}} \int_{B_{2^{[\frac{n}{2}]+1}}(0)} |D\tilde{\tau}|^2 \, d\xi.$$

Differentiating again, and proceeding in the same way, we prove

$$\int_{B_{2^{[\frac{n}{2}]-1}}(0)} |D^3 \tilde{\tau}|^2 \, d\xi \leq \frac{c}{2^{2([\frac{n}{2}]-1)}} \int_{B_{2^{[\frac{n}{2}]}}(0)} |D^2 \tilde{\tau}|^2 \, d\xi.$$

We can continue as long as we like. In particular, we get

$$\int_{B_1(0)} |D^{[\frac{n}{2}]+2} \tilde{\tau}|^2 \, d\xi \leq \frac{c^{[\frac{n}{2}]+1}}{2^{[\frac{n}{2}]([\frac{n}{2}]+1)}} \int_{B_{2^{[\frac{n}{2}]+1}}(0)} |D\tilde{\tau}|^2 \, d\xi.$$

Noting that

$$\left[\frac{n}{2}\right] + 1 > \frac{n}{2},$$

it follows from Sobolev embedding theorems that

$$\|D\tilde{\tau}\|_{L^\infty(B_1(0))}^2 \leq K \int_{B_{2^{[\frac{n}{2}]+1}}(0)} |D\tilde{\tau}|^2 \, d\xi,$$

where K is a constant depending only on n. Returning to the function τ, we immediately get the estimate (9.55). It remains to notice the estimate

$$\int_{B_{R/2}(x_0)} |D\tau|^2 \, dx \leq \frac{K}{R^2} \int_{B_R(x_0)} |\tau - c|^2 \, dx, \tag{9.58}$$

where c is arbitrary. This can be obtained by operating directly on (9.56). Differentiating once yields

$$D_i D_k \omega_j = (A D_k \tau)_{ij},$$
$$\text{div } D_k \tau = 0,$$

and we test with $\theta^2 \left(\frac{2(x-x_0)}{R}\right) D_k \tau$. We obtain (using the divergence condition)

$$\int_{B_R(x_0)} A D_k \tau \, D_k \tau \theta_{R/2}^2 \, dx = -\int_{B_R(x_0)} (A(\tau - c))_{kj} D_i \theta_{R/2}^2 D_k \tau_{ij} \, dx,$$

where

$$\theta_{R/2}(x) = \theta \left(\frac{2(x-x_0)}{R}\right)$$

and where an arbitrary constant c has been introduced. The property (9.58) follows easily. Next, combining (9.55) and (9.58) we obtain

$$\|D\tau\|_{L^\infty \left(B_{\frac{R}{2^{[\frac{n}{2}]+2}}}(x_0)\right)}^2 \leq \frac{K}{R^{n+2}} \int_{B_R(x_0)} |\tau - c|^2 \, dx. \tag{9.59}$$

Pick any ρ. From the Poincaré inequality, we have

$$\frac{1}{\rho^2}\int_{B_\rho(x_0)}|\tau-\overline{\tau_\rho}|^2\,dx \le C\int_{B_\rho(x_0)}|D\tau|^2\,dx \le C\rho^n\|D\tau\|^2_{L^\infty(B_\rho(x_0))}.$$

Therefore, if

$$\rho \le \frac{R}{2^{[\frac{n}{2}]+2}}, \tag{9.60}$$

we have

$$\frac{1}{\rho^2}\int_{B_\rho(x_0)}|\tau-\overline{\tau_\rho}|^2\,dx \le C\rho^n\|D\tau\|^2_{L^\infty\left(B_{\frac{R}{2^{[\frac{n}{2}]+2}}}(x_0)\right)}.$$

Using (9.59) we get

$$\int_{B_\rho(x_0)}\frac{|\tau-\overline{\tau_\rho}|^2}{\rho^n}\,dx \le C\frac{\rho^2}{R^2}\int_{B_R(x_0)}\frac{|\tau-c|^2}{R^n}\,dx \tag{9.61}$$

for any ρ satisfying (9.60).

9.8.3 Proof of Theorem 9.10

Let $\delta' < \delta$. We first deduce from (9.61) that

$$\int_{B_\rho(x_0)}\frac{|\tau-\overline{\tau_\rho}|^2}{\rho^{n+2\delta'}}\,dx \le C\left(\frac{\rho}{R}\right)^{2-2\delta'}\int_{B_R(x_0)}\frac{|\tau-c|^2}{R^{n+2\delta'}}\,dx. \tag{9.62}$$

We go back to (9.54), and recalling (9.62), we assert that

$$\int_{B_\rho(x_0)}\frac{|D\sigma-\overline{D\sigma_\rho}|^2}{\rho^{n+2\delta'}}\,dx \le K\left(\frac{\rho}{R}\right)^{2-2\delta'}\int_{B_R(x_0)}\frac{|\tau-c|^2}{R^{n+2\delta'}}\,dx$$
$$+K\frac{R^{2\delta}}{\rho^{n+2\delta'}}\int_{B_R(x_0)}|D\sigma|^2\,dx$$

for ρ satisfying (9.60).

We take $c = \overline{D\sigma_R}$ and use (9.53) to derive

$$\int_{B_\rho(x_0)} \frac{|D\sigma - \overline{D\sigma_\rho}|^2}{\rho^{n+2\delta'}}\, dx$$

$$\leq K \left(\frac{\rho}{R}\right)^{2-2\delta'} \frac{\int_{B_R(x_0)}[R^{2\delta}|D\sigma|^2 + |D\sigma - \overline{D\sigma_R}|^2]dx}{R^{n+2\delta'}}\, dx \quad (9.63)$$

$$+K\frac{R^{2\delta}}{\rho^{n+2\delta'}}\int_{B_R(x_0)}|D\sigma|^2\, dx.$$

Set

$$\beta = \frac{\rho}{R} \leq \frac{1}{2^{[\frac{n}{2}]+2}}.$$

Recalling the notation $\Phi(R)$, it follows easily from (9.63) that

$$\Phi(\rho) \leq K\left(\beta^{2-2\delta'} + \frac{R^{2\delta}}{\beta^{n+2\delta'}}\right)\Phi(R) + K\frac{R^{2\delta-2\delta'}}{\beta^{n+2\delta'}}|\overline{D\sigma_R}|^2. \quad (9.64)$$

We now pick β such that

$$K\beta^{2-2\delta'} \leq \frac{1}{4}$$

and, of course, also satisfying

$$\beta \leq \frac{1}{2^{[\frac{n}{2}]+2}}.$$

We now choose R_0 as a function of β and impose the first (of two) restriction

$$\frac{K(R_0(\beta))^{2\delta}}{\beta^{n+2\delta'}} \leq \frac{1}{4}. \quad (9.65)$$

With these restrictions, for $R \leq R_0(\beta)$, we deduce from (9.64) that

$$\Phi(\beta R) \leq \frac{1}{2}\Phi(R) + K\left(\frac{R_0(\beta)^{2\delta-2\delta'}}{\beta^{n+2\delta'}}\right)|\overline{D\sigma_R}|^2. \quad (9.66)$$

We now define the sequence

$$R_i = R_0(\beta)\,\beta^i.$$

Consider $\overline{D\sigma_{R_i}}$; recalling its definition, we can write

$$|\overline{D\sigma_{R_i}}| \leq |\overline{D\sigma_{R_{i-1}}}| + \frac{1}{|B_{R_i}|^{1/2}}\left(\int_{B_{R_i}}|D\sigma - \overline{D\sigma_{R_{i-1}}}|^2\, dx\right)^{1/2},$$

and thus, as is easily checked,

$$|\overline{D\sigma_{R_i}}| \leq |\overline{D\sigma_{R_{i-1}}}| + KR_0(\beta)^{\delta'}\beta^{(i-1)\delta'-\frac{n}{2}}(\Phi(R_{i-1}))^{1/2}.$$

Adding up we get

$$|\overline{D\sigma_{R_i}}| \leq |\overline{D\sigma_{R_0(\beta)}}| + KR_0(\beta)^{\delta'}\sum_{j=0}^{i-1}\beta^{j\delta'-\frac{n}{2}}(\Phi(R_j))^{1/2}.$$

We go back to (9.66), which we apply with $R = R_i$. We obtain the inequality

$$\Phi(R_{i+1}) \leq \frac{1}{2}\Phi(R_i) + K_1\frac{R_0(\beta)^{2\delta}}{\beta^{2n+2\delta'}}\left(\sum_{j=0}^{i-1}\beta^{j\delta'}(\Phi(R_j))^{1/2}\right)^2 \tag{9.67}$$
$$+K|\overline{D\sigma_{R_0(\beta)}}|^2\frac{R_0(\beta)^{2\delta-2\delta'}}{\beta^{n+2\delta'}}.$$

If we set

$$S_i = \max_{j=0,\ldots,i-1}\Phi(R_j),$$

then we deduce from (9.67) that

$$\Phi(R_{i+1}) \leq S_{i+1}\left(\frac{1}{2} + K_1\frac{R_0(\beta)^{2\delta}}{\beta^{2n+2\delta'}}\left(\sum_{j=0}^{\infty}\beta^{j\delta'}\right)^2\right) +K|\overline{D\sigma_{R_0(\beta)}}|^2\frac{R_0(\beta)^{2\delta-2\delta'}}{\beta^{n+2\delta'}}.$$

From this estimate, it is also easy to deduce

$$S_{i+1} \leq S_{i+1}\left(\frac{1}{2} + K_1\frac{R_0(\beta)^{2\delta}}{\beta^{2n+2\delta'}}\left(\sum_{j=0}^{\infty}\beta^{j\delta'}\right)^2\right) + K|\overline{D\sigma_{R_0(\beta)}}|^2\frac{R_0(\beta)^{2\delta-2\delta'}}{\beta^{n+2\delta'}}.$$

We impose then a final restriction on $R_0(\beta)$, namely

$$K_1\frac{R_0(\beta)^{2\delta}}{\beta^{2n+2\delta'}}\left(\sum_{j=0}^{\infty}\beta^{j\delta'}\right)^2 < \frac{1}{4},$$

which in fact implies (9.65) for a convenient value of K_1. With this choice, we obtain easily from the above estimate on S_{i+1} that

$$S_{i+1} \leq K\frac{|\overline{D\sigma_{R_0(\beta)}}|^2}{R_0(\beta)^{2\delta'}}.$$

Therefore, the sequence $\Phi(R_i)$ is bounded in i.

Now if $R_{i+1} < R < R_i$, we first assert that

$$\Phi(R) \leq \frac{\int_{B_{R_i}(x_0)} |D\sigma - \overline{D\sigma_R}|^2 \, dx}{R_{i+1}^{n+2\delta'}}.$$

Then

$$\Phi(R) \leq \frac{2}{\beta^{n+2\delta'}} \left(\Phi(R_i) + \frac{|\overline{D\sigma_R} - \overline{D\sigma_{R_i}}|^2}{R_i^{2\delta'}} \right).$$

Moreover, as is easily seen,

$$\frac{|\overline{D\sigma_R} - \overline{D\sigma_{R_i}}|^2}{R_i^{2\delta'}} \leq \frac{1}{\beta^n} \Phi(R_i).$$

Collecting results we can assert that

$$\Phi(R) \leq \frac{2}{\beta^{n+2\delta'}} \left(1 + \frac{1}{\beta^n} \right) \Phi(R_i),$$

and thus the result (9.51) has been proven, and the proof of Theorem 9.10 is complete. $\diamond$

9.9 C^δ Regularity on Neighborhoods of Lebesgue Points for the Nonsymmetric Case

9.9.1 Setting of the Problem and Statement of Results

Our objective here is to prove a result that is similar to Theorem 9.10, weakening assumption (9.49) into

$$\sigma \in L^\infty_{\mathrm{loc}}(\Omega). \tag{9.68}$$

We shall also weaken the assumption (9.47) into

$g(\sigma) : R^{n \times n} \to R^{n \times n}$,
$\alpha I \leq g'(\sigma) \leq MI$,
g' uniformly continuous
(with modulus of continuity bounded on bounded arguments) and concave.
$$\tag{9.69}$$
We shall see that a result analogous to Theorem 9.10 holds, although only on neighborhoods of Lebesgue points (partial regularity).

Theorem 9.11. *We assume (9.69), and we consider a pair σ, u satisfying (9.48) and (9.68). Then for each Lebesgue point x_0 of σ, we have $\sigma \in C^\delta(\mathcal{U}(x_0))$, where $\mathcal{U}(x_0)$ is some neighborhood of x_0.*

We proceed as in Theorem 9.10, introducing an open set Ω' with

$$\overline{\Omega'} \subset \Omega.$$

We again pick R_0 such that

$$B_{R_0}(x_0) \subset \Omega, \quad \forall x_0 \in \Omega',$$

with additional restrictions on the size of R_0 to be given later.
We can assert, thanks to assumption (9.68), that

$$\sigma \text{ is bounded on } \Omega'.$$

We may also assert that the local Gehring property holds, see (9.46):

$$\left(\fint_{\Omega' \cap B_R} |D\sigma|^p \, dx \right)^{1/p} \leq K_p \left[\left(\fint_{\Omega' \cap B_{2R}} |D\sigma|^2 \, dx \right)^{1/2} + C \right] \tag{9.70}$$

with

$$2 \leq p \leq p_0, \quad 2 < p_0 < \frac{2n}{n-2}.$$

In the proof we use the notation

$$\Phi(\xi, R) = \frac{\int_{B_R(\xi)} |D\sigma|^2 \, dx}{R^{n-2}}.$$

We make use of the following properties of the function $\Phi(\xi, R)$: if ξ is a Lebesgue point of σ, then $\Phi(\xi, R)$ tends to 0 as $R \to 0$, and $\Phi(\xi, R)$ is continuous in ξ, for fixed R.
     To check the first property, we consider (9.44) and we choose $c = g(\sigma(x_0))$, $\zeta = 1$ on Ω'. Recalling that $f = 0$, we get

$$\Phi(x_0, R) \leq \frac{K}{R^n} \int_{\Omega \cap B_{2R}} |g(\sigma) - g(\sigma(x_0))|^2 \, dx,$$

and the result follows from the definition of Lebesgue points. The second property is easy to check.

9.9.2 Proof of Theorem 9.11

Let ϵ be given. Let x_0 be a Lebesgue point of σ in Ω'. We first consider R_0 such that

$$B_{3R_0}(x_0) \subset \Omega'.$$

We shall consider $R < R_0$ fixed such that

$$\Phi(x_0, 2R) + R^2 < \frac{\epsilon}{2}.$$

We then consider a neighborhood $\mathcal{U}(x_0)$ of x_0 contained in Ω', possibly depending on R, such that

$$\Phi(\xi, 2R) + R^2 < \epsilon, \quad \forall \xi \in \mathcal{U}(x_0).$$

Finally, we consider $B_{R_0} \cap \mathcal{U}(x_0)$.

For ξ in this neighborhood we have, collecting results,

$$\Phi(\xi, 2R) + R^2 < \epsilon, \quad B_{2R}(\xi) \subset \Omega', \tag{9.71}$$

and thus also, see (9.70),

$$\left(\fint_{B_R(\xi)} |D\sigma|^p \, dx \right)^{1/p} \leq K_p \left[\left(\fint_{B_{2R}(\xi)} |D\sigma|^2 \, dx \right)^{1/2} + C \right].$$

Denoting by $\bar{\sigma}_R$ the average of σ on $B_R(\xi)$, we use the following property:

$$|g'(\sigma(x)) - g'(\bar{\sigma}_R)|^2 \leq \omega(|\sigma(x) - \bar{\sigma}_R|^2), \quad \forall x \in B_R(\xi), \tag{9.72}$$

and

$$\omega \text{ is bounded and concave}, \quad \omega(t) \to 0 \text{ as } t \to 0,$$

which is an easy consequence of the last part of (9.69).

We next introduce an auxiliary problem as in the proof of Theorem 9.10:

$$\begin{aligned} D_i w_j^k &= g'_{ij,hl}(\bar{\sigma}_R)\tau_{hl}^k, \quad w^k \in (H^1(B_R(\xi)))^n, \\ \tau^k &\in \mathcal{L}^2, \quad \operatorname{div} \tau^k = 0, \\ \nu.(D_k \sigma &- \tau^k)|_{\partial B_R(\xi)} = 0. \end{aligned}$$

We prove as in Theorem 9.10 the following property of τ:

$$\|\tau\|^2_{L^\infty(B_{\frac{R}{2^{[\frac{n}{2}]+1}}}(\xi))} \leq \frac{K}{R^n} \int_{B_R(\xi)} |\tau|^2 \, dx.$$

Note that we need an estimate on τ and not on $D\tau$ here, but the method to get it is the same. Next, if we consider

$$\rho \leq \frac{R}{2^{[\frac{n}{2}]+1}},$$

we deduce easily

$$\frac{\int_{B_\rho(\xi)} |\tau|^2 \, dx}{\rho^{n-2}} \leq K \frac{\rho^2}{R^2} \frac{\int_{B_R(\xi)} |\tau|^2 \, dx}{R^{n-2}}. \tag{9.73}$$

Proceeding as in Section 9.8.2, we check

$$\int_{B_R(\xi)} |D\sigma - \tau|^2 \, dx \le C \int_{B_R(\xi)} |g'(\bar{\sigma}_R) - g'(\sigma)|^2 |D\sigma|^2 \, dx,$$

and using (9.72) we get

$$\int_{B_R(\xi)} |D\sigma - \tau|^2 \, dx \le C \int_{B_R(\xi)} (\omega(|\bar{\sigma}_R - \sigma|^2)) |D\sigma|^2 \, dx.$$

From Hölder's inequality, we obtain

$$\int_{B_R(\xi)} |D\sigma - \tau|^2 \, dx$$

$$\le C \left(\int_{B_R(\xi)} (\omega(|\bar{\sigma}_R - \sigma|^2))^{p/(p-2)} \, dx \right)^{(p-2)/p} \left(\int_{B_R(\xi)} |D\sigma|^p \, dx \right)^{2/p}.$$

Hence we also have

$$\int_{B_R(\xi)} |D\sigma - \tau|^2 \, dx$$

$$\le C|B_R(\xi)| \left(\fint_{B_R(\xi)} (\omega(|\bar{\sigma}_R - \sigma|^2))^{p/(p-2)} \, dx \right)^{(p-2)/p} \left(\fint_{B_R(\xi)} |D\sigma|^p \, dx \right)^{2/p}.$$

Using the fact that ω is bounded yields

$$\int_{B_R(\xi)} |D\sigma - \tau|^2 \, dx$$

$$\le C|B_R(\xi)| \left(\fint_{B_R(\xi)} \omega(|\bar{\sigma}_R - \sigma|^2) \, dx \right)^{(p-2)/p} \left(\fint_{B_R(\xi)} |D\sigma|^p \, dx \right)^{2/p},$$

and from the concavity of ω,

$$\int_{B_R(\xi)} |D\sigma - \tau|^2 \, dx$$

$$\le C|B_R(\xi)| \left(\omega \left(\fint_{B_R(\xi)} |\bar{\sigma}_R - \sigma|^2 \, dx \right) \right)^{(p-2)/p} \left(\fint_{B_R(\xi)} |D\sigma|^p \, dx \right)^{2/p}.$$

Using the local Gehring property (9.70) and the Poincaré inequality, we obtain (generically using $\omega(x)$ for $\omega(Cx)$, where C is a positive constant)

$$\int_{B_R(\xi)} |D\sigma - \tau|^2 \, dx \tag{9.74}$$

$$\le C \left(\omega \left(\frac{\int_{B_R(\xi)} |D\sigma|^2 \, dx}{R^{n-2}} \right) \right)^{(p-2)/p} \left(\int_{B_{2R}(\xi)} |D\sigma|^2 \, dx + R^n \right).$$

Proceeding as in the proof of Theorem 9.10, we write

$$\int_{B_{2\rho}(\xi)} |D\sigma|^2 \, dx \le 2 \int_{B_{2\rho}(\xi)} |D\sigma - \tau|^2 \, dx + 2 \int_{B_{2\rho}(\xi)} |\tau|^2 \, dx.$$

Using (9.74) and (9.73) and collecting results, we get

$$
\frac{\int_{B_{2\rho}(\xi)} |D\sigma|^2 \, dx}{\rho^{n-2}}
$$

$$
\le C \left(\omega \left(\frac{\int_{B_R(\xi)} |D\sigma|^2 \, dx}{R^{n-2}} \right) \right)^{(p-2)/p} \left(\frac{\int_{B_{2R}(\xi)} |D\sigma|^2 \, dx}{R^{n-2}} + R^2 \right) \left(\frac{R}{\rho} \right)^{n-2}
$$

$$
+ C \frac{\rho^2}{R^2} \frac{\int_{B_R(\xi)} |D\sigma|^2 \, dx}{R^{n-2}}.
$$

(9.75)

Let

$$\psi(\xi, R) = \phi(\xi, 2R) + R^2.$$

We derive easily from (9.75) the following relation:

$$\psi(\xi, \rho) \le C \frac{\rho^2}{R^2} \psi(\xi, R) \left(1 + \left(\frac{R}{\rho} \right)^n (\omega(\psi(\xi, R)))^{(p-2)/p} \right). \qquad (9.76)$$

Let

$$\beta = \frac{\rho}{R} \le \frac{1}{2^{[\frac{n}{2}]+1}}.$$

Then (9.76) reads

$$\psi(\xi, \beta R) \le C \beta^2 \psi(\xi, R) \left(1 + \beta^{-n} (\omega(\psi(\xi, R)))^{(p-2)/p} \right).$$

Note that thanks to (9.71), we have

$$\psi(\xi, R) \le \epsilon.$$

Fix $0 < \delta < 1$. Choose β such that

$$2C\beta^{2-2\delta} \le 1$$

and ϵ such that

$$\beta^{-n} (\omega(\epsilon))^{(p-2)/p} \le 1.$$

With these choices, we deduce

$$\psi(\xi, \beta R) \le \beta^{2\delta} \psi(\xi, R). \qquad (9.77)$$

Note that ξ may not be a Lebesgue point, so we do not have a priori information on the value of $\psi(\xi, \beta R)$. But from (9.77), one proves among other things that

$$\psi(\xi, \beta R) \le \epsilon,$$

and therefore it is legitimate to replace R by βR in the previous calculations. We can then iterate and obtain

$$\psi(\xi, \beta^k R) \le \beta^{2k\delta} \psi(\xi, R).$$

Thus in fact,

$$\psi(\xi, \rho) \le C\rho^{2\delta},$$

which proves, using Morrey's result, that

$$\sigma \in C^\delta \text{ in } \mathcal{U}(x_0).$$

This completes the proof of Theorem 9.11.   ◇

Remark 9.12. We refer to M. Giaquinta [40] for a similar argument developed in the primal formulation of the nonlinear system. See also Section 8.5.

Remark 9.13. From the C^δ regularity in $\mathcal{U}(x_0)$ we can show that

$$D\sigma \in C^{\delta'} \text{ in } \mathcal{U}'(x_0)$$

in a smaller neighborhood, provided that we assume (9.47) and we apply Theorem 9.10.

9.9.3 Additional Results in the Uhlenbeck Case

In the Uhlenbeck case one can prove that every point is a Lebesgue point, which then improves what has been obtained in the previous section.

We recall that the Uhlenbeck case corresponds to (see Section 9.6.1)

$$g(\sigma) = \beta(|\sigma|)\sigma,$$

where

$$\beta(0) = \alpha > 0 \beta'(x) \ge k(p-2)x^{p-3}$$

and

$$\beta(x) \le K(1 + x^{p-2}).$$

We state the following result.

Theorem 9.14. *If we make the same assumptions as in Theorem 9.8, then every point in Ω is a Lebesgue point of σ.*

Proof. We prove that if θ is a smooth function such that $0 \le \theta \le 1$ and has compact support in Ω, then

$$\int \theta^2 \frac{|D\sigma|^2}{|x - x_0|^{n-2}} \, dx \le C. \tag{9.78}$$

This implies the result, since for any fixed $x_0 \in \Omega$ and $B_R(x_0) \subset \Omega$,

$$\int_{B_R(x_0)} \frac{|D\sigma|^2}{R^{n-2}}\,dx \leq \int_{B_R(x_0)} \theta^2 \frac{|D\sigma|^2}{|x-x_0|^{n-2}}\,dx$$

for a convenient function θ. Thus the left-hand side tends to 0 as $R \to 0$. To prove (9.78) we introduce a specific Green function. We define the matrix

$$a_{ij}(x) = (\beta'(|\sigma|)|\sigma| + \beta(|\sigma|))\delta_{ij} - \beta'(|\sigma|)\frac{\sigma_{ik}\sigma_{jk}}{|\sigma|},$$

where, of course, σ is evaluated at point x. This matrix is symmetric and coercive, since

$$a_{ij}(x)\xi_i\xi_j \geq \beta(|\sigma|)|\xi|^2 \geq \alpha|\xi|^2.$$

We consider the Green function G^{x_0} related to this matrix at any point x_0. Then G has the standard properties. Hence to prove (9.78) it is sufficient to prove

$$\int \theta^2 |D\sigma|^2 G^{x_0}\,dx \leq C. \tag{9.79}$$

Starting with

$$D_j u_k = \beta(|\sigma|)\sigma_{jk}$$

and testing with

$$-D_i(\theta^2 D_j(\sigma_{ik}G))$$

we get

$$\int \theta^2 \beta'(|\sigma|)\frac{(\sigma.D_j\sigma)^2}{|\sigma|}G\,dx + \int \theta^2 \beta'(|\sigma|)\sigma.D_j\sigma\, D_jG\,|\sigma|\,dx$$
$$+ \int \theta^2 \beta(|\sigma|)|D\sigma|^2 G\,dx + \int \theta^2 \beta(|\sigma|)\sigma.D_j\sigma\, D_jG\,dx$$
$$= -\int \beta(|\sigma|)\sigma_{jk}D_i\theta^2 D_j(\sigma_{ik}G)\,dx - \int \theta^2\beta(|\sigma|)\sigma_{jk}D_j(\sigma_{ik}D_iG)\,dx$$
$$- \int \theta^2\beta(|\sigma|)\sigma_{jk}D_j(f_kG)\,dx.$$

After easy calculations, and making use of the definition of a_{ij}, we obtain

$$\int \theta^2 \beta'(|\sigma|)\frac{(\sigma.D_j\sigma)^2}{|\sigma|}G\,dx + \int \theta^2\beta(|\sigma|)|D\sigma|^2 G\,dx$$
$$+ \int \theta^2 a_{ij}D_jG\, D_i\left(\frac{1}{2}|\sigma|^2\right)\,dx$$
$$= -\int \beta(|\sigma|)\sigma_{jk}D_i\theta^2 D_j\sigma_{ik}\,G\,dx - \int \theta^2\beta(|\sigma|)\sigma_{jk}D_j f_k G\,dx.$$

Thus clearly,

$$\frac{1}{2}\int \theta^2\beta(|\sigma|)|D\sigma|^2 G\,dx \leq 2\int \beta(|\sigma|)|D\theta|^2|\sigma|^2 G\,dx + \int \theta^2\beta(|\sigma|)|\sigma|\,|Df|G\,dx.$$

Recalling that σ is bounded, and making use of the properties of Green functions, the estimate (9.79) follows. $\Diamond$

Theorem 9.15. *If we make the same assumptions as in Theorem 9.8, then*

$$\sigma \in C^{1+\delta}_{\text{loc}}(\Omega).$$

Proof. First, we know from Theorem 9.8 that σ is locally bounded. The result is then a consequence of Theorems 9.14 and 9.11. Indeed, to each point of Ω we can associate a neighborhood in which the property is true. Taking a compact subset of Ω, we find a finite covering, and thus the result follows.

$\Diamond$

10. Nonlinear Elliptic Systems Arising from plasticity Theory

10.1 Introduction

In this chapter (which is a continuation of Chapter 9 and will rely heavily on it) we are interested in specific models of plasticity. We shall need to attach to tensor σ its deviator

$$\sigma_D = \sigma - 1/nI \operatorname{tr} \sigma,$$

which has trace 0. We will consider two models of plasticity, the Hencky model, which is a model of perfect plasticity where,

$$|\sigma_D| \leq \mu$$

(μ is a given constant), and the Norton–Hoff model, which is an approximation to the Hencky model, where the constraint of perfect plasticity is relaxed with a penalty term. In fact, the Norton–Hoff model will be a particular case of the models considered in Chapter 9, but we shall consider a sequence of these models. Again these models are formulated as variational problems in which the unknown is the stress tensor and the displacement is recovered indirectly. The convergence of the approximation is very natural in the context of variational problems and follows from general penalty methods (see R. Temam [101], G. Duvaut, J.L. Lions [15], J.L. Lions [70], P. Le Tallec [69]).

We develop here the regularity properties of the solution of both the approximation and the limit problems. In fact, we show regularity properties of the solution of the approximation problem (the Norton–Hoff model) with estimates that are uniform with respect to the parameter of approximation. Therefore, it gives in the limit the regularity of the solution of the Hencky model. We refer also to the work of G.A. Seregin [93], where he proceeds differently, starting with the variational problem of the displacement, obtaining the stress by duality and proving estimates on the stress from estimates on the displacement. Our method has the advantage of being more direct and applicable to the Prandtl–Reuss model with simple modification, and also gives useful properties of the Norton–Hoff approximation.

10.2 Description of Models

10.2.1 Spaces $U(\Omega)$, $\Sigma(\Omega)$

Let Ω be a bounded Lipschitz domain of R^n, whose boundary is denoted by Γ. We refer to the notation of Section 9.2. We shall need to work with specific spaces related to the operator ϵ when not much regularity is available, so we review some important spaces in that context, referring again to R. Temam [101] for details and proofs. We consider the space of Radon measures on Ω. It is the subspace of distributions μ such that

$$\sup_{\{\phi \in C_0^\infty(\Omega)| \ |\phi(x)| \le 1\}} \langle \mu, \phi \rangle < \infty$$

endowed with the above norm. It is a Banach space, denoted by $M^1(\Omega)$. If μ belongs to $M^1(\Omega)$, then $\langle \mu, \phi \rangle$ has a meaning for $\phi \in L^\infty(\Omega)$, and one also uses the notation

$$\langle \mu, \phi \rangle = \int_\Omega \phi \, d\mu.$$

We shall consider the space of symmetric matrices whose elements belong to $M^1(\Omega)$, denoted by $\mathcal{M}_{sym}^1$.

Definition 10.1. Let

$$U(\Omega) = \{u \in (L^1(\Omega))^n | \epsilon(u) \in \mathcal{M}_{sym}^1, \ \mathrm{div}\,(u) \in L^2(\Omega)\}$$

equipped with the natural norm

$$\|u\|_{U(\Omega)} = \|u\|_{(L^1(\Omega))^n} + \sum_{ij=1}^n \|\epsilon_{ij}(u)\|_{M^1(\Omega)} + \|\mathrm{div}\,u\|_{L^2(\Omega)}.$$

In fact,

$$u \in (\mathcal{D}'(\Omega))^n \text{ and } \epsilon(u) \in \mathcal{M}_{sym}^1 \Rightarrow u \in (L^{n/(n-1)}(\Omega))^n,$$

and the injection of $U(\Omega)$ in $(L^p(\Omega))^n$ is compact for $1 \le p < n/(n-1)$. The weak-star topology on $U(\Omega)$ is defined as follows:

$$u_m \to u \text{ in } (L^1(\Omega))^n,$$
$$\mathrm{div}\,u_m \to \mathrm{div}\,u \text{ in } L^2(\Omega) \text{ weakly},$$
$$\epsilon(u_m) \to \epsilon(u) \text{ in } \mathcal{M}_{sym}^1 \text{ weakly},$$

and from bounded sequences in $U(\Omega)$, one can extract weakly convergent subsequences.

The trace of an element of $U(\Omega)$ can be defined as an element of $(L^1(\Gamma))^n$. We turn now to the space $\Sigma(\Omega)$.

Definition 10.2. Let

$$\Sigma(\Omega) = \{\sigma \in \mathcal{L}^2_{\text{sym}} | \text{ div } \sigma \in (L^{n+1}(\Omega))^n, \sigma_D \in \mathcal{L}^\infty_{\text{sym}}\}$$

equipped with the natural norm

$$\|\sigma\|_{\Sigma(\Omega)} = \|\sigma\|_{\mathcal{L}^2_{\text{sym}}} + \|\text{div } \sigma\|_{(L^{n+1}(\Omega))^n} + \|\sigma_D\|_{\mathcal{L}^\infty_{\text{sym}}}.$$

We have

$$\Sigma(\Omega) \subset \mathcal{L}^p_{\text{sym}}, \ \forall 1 \le p < \infty, \text{ with continuous injection}$$

The trace $\nu.\sigma$ of an element of $\Sigma(\Omega)$ can be defined as an element of $(H^{-1/2}(\Gamma))^n$. Consider now $\sigma \in \Sigma(\Omega), u \in U(\Omega)$. Then $\sigma_D.\epsilon_D(u)$ belongs to $M^1(\Omega)$, with the estimate

$$\|\sigma_D.\epsilon_D(u)\|_{M^1(\Omega)} \le \|\sigma_D\|_{\mathcal{L}^\infty_{\text{sym}}} \|\epsilon_D(u)\|_{M^1(\Omega)}.$$

It follows that $\sigma.\epsilon(u)$ also belongs to $M^1(\Omega)$, with the formula

$$\sigma.\epsilon(u) = \sigma_D.\epsilon_D(u) + \frac{1}{n} \text{div } u.\text{tr } \sigma.$$

Now for $\phi \in \mathcal{C}_0^\infty(\Omega)$ one has the formula

$$\langle \sigma.\epsilon(u), \phi \rangle = - \int_\Omega u.\text{div } \sigma \, \phi \, dx - \int_\Omega \sigma \, u.D\phi \, dx.$$

Moreover, a trace $\nu.\sigma u$ can be defined, and the following Green's formula holds:

$$\int_\Gamma \nu.\sigma \, u\phi \, d\Gamma = \int_\Omega \sigma.\epsilon(u), \phi \, dx + \int_\Omega u.\text{div } \sigma \, \phi \, dx + \int_\Omega \sigma \, u.D\phi \, dx$$

for $\phi \in \mathcal{C}^1(\overline{\Omega})$ with $\phi|_\Gamma$ belonging to $\mathcal{C}^1(\Gamma)$.

10.2.2 Hencky model

We consider a tensor function

$$A_{ij,hk} \in L^\infty \text{ such that } A_{ij,hk} = A_{ji,hk} = A_{ij,kh} = A_{hk,ij},$$

$$\sum_{i,j;h,k=1}^n A_{ij,hk}\tau_{ij}\tau_{hk} \ge \alpha|\tau|^2 \alpha > 0, \quad \forall \tau, \tag{10.1}$$

and functions

$$f_j \in L^2(\Omega), \quad j = 1, \dots, n. \tag{10.2}$$

We consider the set

$$\mathcal{K} = \{\sigma \in \mathcal{L}^2_{\text{sym}} \,|\, \text{div}\,\sigma = f, \nu.\sigma = 0 \text{ on } \Gamma_1\}.$$

Let

$$\mathcal{M} = \{\sigma \in \mathcal{L}^2_{\text{sym}} \,|\, |\sigma_D| \leq \mu, \text{a.e.}\}.$$

We assume that

$$\mathcal{M} \cap \mathcal{K} \neq \emptyset. \tag{10.3}$$

Define on $\mathcal{L}^2_{\text{sym}}$ the functional

$$J(\sigma) = \frac{1}{2} \int_\Omega A_{ij,hk} \sigma_{ij} \sigma_{hk}\, dx.$$

Then the Hencky model corresponds to the variational problem

minimize $J(\sigma)$ over $\mathcal{M} \cap \mathcal{K}$.

It is clear that the functional $J(\sigma)$ has a unique minimum on the nonempty closed convex subset $\mathcal{M} \cap \mathcal{K}$ of $\mathcal{L}^2_{\text{sym}}$. We shall call this minimum σ also. It is the solution of the variational inequality

$$(A\sigma, \tau - \sigma) \geq 0, \quad \forall \tau \in \mathcal{M} \cap \mathcal{K}, \quad \sigma \in \mathcal{M} \cap \mathcal{K}.$$

Remark 10.3. We could consider more general nonlinear elastic energy models and functionals, in the spirit of Chapter 9.

10.2.3 Norton–Hoff Model

Here we assume (a stronger assumption than (10.2))

$$f_j \in L^p(\Omega), \quad j = 1, \ldots, n, \quad \forall 1 < p < \infty. \tag{10.4}$$

We define the functional (for $N \geq 2$)

$$J^N(\sigma) = J(\sigma) + \frac{1}{N\mu^{N-1}} \int_\Omega |\sigma_D|^N\, dx$$

and the set

$$\mathcal{K}^N = \{\sigma \in \mathcal{L}^N_{\text{sym}} \,|\, \text{div}\,\sigma = f, \nu.\sigma = 0 \text{ on } \Gamma_1\}.$$

The Norton–Hoff model (with penalty parameter N) corresponds to the variational problem

minimize $J^N(\sigma)$ over $\mathcal{K}^N$. \tag{10.5}

It corresponds to models studied in Chapter 9, with specific aspects due to the fact that the deviator enters in the nonquadratic term.

Proposition 10.4. *If we assume (10.1), (10.3), (10.4), then the problem (10.5) has one and only one solution, denoted by σ^N.*

Proof. By (10.3), the set $\mathcal{K}^N$ is not empty, since an element of $\mathcal{M} \cap \mathcal{K}$ necessarily belongs to $\mathcal{L}^p_{\text{sym}}$ $\forall\, 1 < p < \infty$, and hence to $\mathcal{K}^N$. It is clearly a convex closed subset of $\mathcal{L}^N_{\text{sym}}$. The functional $J^N(\sigma)$ is strictly convex, continuous on $\mathcal{L}^N_{\text{sym}}$. If one considers a minimizing sequence $\sigma(h)$, then clearly,

$$\|\sigma_D(h)\|_{\mathcal{L}^N_{\text{sym}}} \le C, \|\sigma(h)\|_{\mathcal{L}^2_{\text{sym}}} \le C,$$

where C is a constant not dependent on h. Since $\sigma(h) \subset \mathcal{K}^N$, we can write

$$\operatorname{div} \sigma_D(h) + D \operatorname{tr} \sigma(h) = f,$$

which yields

$$\|D \operatorname{tr} \sigma(h)\|_{(W^{-1,N})^n} \le C.$$

We deduce from the above estimates that the mean $\overline{\operatorname{tr}\sigma(h)}$ of $\operatorname{tr}\sigma(h)$ over Ω is bounded, and that $\operatorname{tr}\sigma(h) - \overline{\operatorname{tr}\sigma(h)}$ is bounded in L^N. Hence

$$\|\sigma(h)\|_{\mathcal{L}^N_{\text{sym}}} \le C.$$

From this estimate it follows that by extracting a weakly convergent subsequence, the minimum of $J^N(\sigma)$ over $\mathcal{K}^N$ is achieved. The minimum is unique, since the functional is strictly convex. $\diamond$

It is easy to derive a necessary and sufficient condition for optimality, for σ^N, namely,

$$\int_\Omega \left(A\sigma^N + \frac{1}{\mu^{N-1}} |\sigma_D^N|^{N-2} \sigma_D^N \right) .\tau \, dx = 0,$$

$$\forall \tau \in \mathcal{L}^N_{\text{sym}} \text{ such that } \operatorname{div} \tau = 0,\ \nu.\tau = 0 \text{ on } \Gamma_1.$$

Since $A\sigma^N + \dfrac{1}{\mu^{N-1}} |\sigma_D^N|^{N-2}\sigma_D^N$ belongs to $\mathcal{L}^{N/(N-1)}_{\text{sym}}$, we deduce from Proposition 9.1 the existence of the displacement. Collecting results, we can state the following result.

Proposition 10.5. *If (10.1), (10.3), and (10.4) hold, then there exists one and only one pair (σ^N, u^N) such that*

$$A\sigma^N + \frac{1}{\mu^{N-1}} |\sigma_D^N|^{N-2}\sigma_D^N = \epsilon(u^N),$$

$$u^N \in \left(W_{\Gamma_0}^{1,\frac{N}{N-1}} \right)^n (\Omega), \quad \operatorname{div} u^N \in L^2(\Omega), \tag{10.6}$$

$$\sigma^N \in \mathcal{L}^N_{\text{sym}}, \quad \operatorname{div} \sigma^N = f\nu.\sigma^N = 0 \text{ on } \Gamma_1.$$

Remark 10.6. Unlike the Hencky model, the Norton–Hoff model leads to the displacement with no difficulty. This is, of course, one of its interesting features.

Remark 10.7. Note the relation

$$\operatorname{div} u^N = \operatorname{tr} A\sigma^N,$$

which in particular yields the property $\operatorname{div} u^N \in L^2$.

10.2.4 Passing to the Limit

Our objective is to prove the following result.

Theorem 10.8. *If (10.1), (10.3), and (10.4) hold, then the sequence σ^N of solutions of the Norton–Hoff model converges in $\mathcal{L}^2_{\text{sym}}$ to σ, a solution of the Hencky model, and σ^N_D converges weakly to σ_D in any space $\mathcal{L}^p_{\text{sym}}, p < \infty$.*

Proof. Take any $\tau \in \mathcal{M} \cap \mathcal{K}$. Note that $\tau \in \mathcal{L}^p_{\text{sym}}, \forall 1 < p < \infty$, and hence in particular belongs to $\mathcal{L}^N_{\text{sym}}$ and to $\mathcal{K}^N$. By definition of σ^N one has

$$J(\sigma^N) + \frac{1}{N\mu^{N-1}} \int |\sigma^N_D|^N \, dx \leq J(\tau) + \frac{\mu}{N}|\Omega|. \qquad (10.7)$$

Take also p fixed, and $N > p$. From the estimate (10.7) it follows that

$$\int |\sigma^N_D|^p \, dx \leq p\mu^{p-1}J(\tau) + \mu^p \text{ meas } \Omega, \qquad (10.8)$$

where we have used the inequality

$$|\xi|^p \leq \frac{p}{N}|\xi|^N + \frac{N-p}{N}.$$

Therefore, as $N \to \infty$, σ^N remains in a bounded subset of $\mathcal{L}^2_{\text{sym}}$, and moreover, for p fixed, σ^N_D remains in a bounded subspace of $\mathcal{L}^p_{\text{sym}}(p < \infty)$. We take a subsequence, still denoted by σ^N, such that

$$\sigma^N \to \sigma \text{ in } \mathcal{L}^2_{\text{sym}} \text{ weakly,}$$
$$\sigma^N_D \to \sigma_D \text{ in } \mathcal{L}^p_{\text{sym}} \text{ weakly } \forall 1 < p < \infty.$$

Note that $\sigma \in \mathcal{K}$. From estimates (10.7), (10.8) it follows that

$$J(\sigma) \leq J(\tau),$$
$$\int |\sigma_D|^p \, dx \leq p\mu^{p-1}J(\tau) + \mu^p \text{ meas } \Omega. \qquad (10.9)$$

From the second estimate in (10.9) we also deduce that

$$\left(\int \left| \frac{\sigma_D}{\mu} \right|^p \, dx \right)^{1/p} \leq \left(\frac{p}{\mu}J(\tau) + \text{ meas } \Omega \right)^{1/p}.$$

Hence

$$\limsup_{p \to \infty} \left(\int \left| \frac{\sigma_D}{\mu} \right|^p \, dx \right)^{1/p} \leq 1,$$

which implies

$$|\sigma_D| \leq \mu \text{ a.e.}$$

Hence $\sigma \in \mathcal{M} \cap \mathcal{K}$. From the first estimate (10.9) it follows that σ is a solution of the Hencky model, and hence by the uniqueness of the solution, the whole sequence converges. Since

$$J(\sigma^N) \to J(\sigma),$$

we also obtain that

$$\sigma^N \to \sigma \text{ in } \mathcal{L}^2_{\text{sym}},$$

and the proof is complete. $\diamondsuit$

10.3 Estimates on the Displacement

Our objective now is to derive estimates for the displacement u^N, in order to derive the existence of the displacement for the Hencky model. We need additional hypotheses for this, similar to those that have been given in the literature, in order to solve the dual variational problem of the Hencky model (which is indeed the problem to which the displacement is the solution).

10.3.1 The f_j Derive from a Potential

We assume here that

$$f = DF, \quad F \in W^{1,p}, \quad \forall 1 < p < \infty, \quad F = 0 \text{ on } \Gamma_1. \tag{10.10}$$

Proposition 10.9. *If (10.1), (10.3), (10.10) hold, then*

$$\|\epsilon(u^N)\|_{\mathcal{L}^1_{\text{sym}}} \le C,$$
$$\|u^N\|_{(L^{n/(n-1)})^n} \le C.$$

Proof. Note that

$$FI \in \mathcal{K}^N,$$

and by the necessary and sufficient condition for optimality of σ^N, we deduce

$$\int \left(A\sigma^N + \frac{1}{\mu^{N-1}} |\sigma^N_D|^{N-2} \sigma^N_D \right) \cdot (\sigma^N - FI)\, dx = 0.$$

Hence

$$\int A\sigma^N . \sigma^N \, dx + \frac{1}{\mu^{N-1}} \int |\sigma^N_D|^N \, dx = \int F \text{ tr } A\sigma^N \, dx.$$

From this estimate and the expression for $\epsilon(u^N)$ (see (10.6)) we deduce the first assertion. The second is an immediate consequence of Section 9.2.2. $\diamondsuit$

10.3.2 Strict Interior Condition

Here we make the following assumption:

$$\exists \tau \in \mathcal{M} \cap \mathcal{K}, \delta > 0 \text{ such that } |\tau| - \mu \leq -\delta. \tag{10.11}$$

Proposition 10.10. *If (10.1), (10.3), (10.4), and (10.11) hold, then*

$$\|\epsilon(u^N)\|_{\mathcal{L}^1_{\text{sym}}} \leq C,$$

$$\|u^N\|_{(L^{n/(n-1)})^n} \leq C.$$

Remark 10.11. The assumption (10.11) corresponds to a condition given in [15] for proving the existence of a solution of the dual variational problem.

Proof. Consider again the optimality condition for σ^N, written as

$$\int \left(A\sigma^N + \frac{1}{\mu^{N-1}} |\sigma_D^N|^{N-2} \sigma_D^N \right) \cdot (\sigma^N - \tau) \, dx = 0$$

for any $\tau \in \mathcal{K}^N$. In particular, we can take any $\tau \in \mathcal{K} \cap \mathcal{M}$. Noting that for such τ one has

$$\left| \int \frac{1}{\mu^{N-1}} |\tau_D|^{N-2} \tau_D \cdot (\sigma^N - \tau) \, dx \right| \leq C,$$

we derive from the optimality condition that

$$\frac{1}{\mu^{N-1}} \int_\Omega \left(|\sigma_D^N|^{N-2} \sigma_D^N - |\tau_D|^{N-2} \tau_D \right) \cdot (\sigma_D^N - \tau_D) \, dx \leq C.$$

From the positivity of the integrand, it also follows that

$$\frac{1}{\mu^{N-1}} \int_E \left(|\sigma_D^N|^{N-2} \sigma_D^N - |\tau_D|^{N-2} \tau_D \right) \cdot (\sigma_D^N - \tau_D) \, dx \leq C$$

for any measurable subset E, the constant C being independent of E. We deduce from this estimate that

$$\frac{1}{\mu^{N-1}} \int_E \left(|\sigma_D^N|^{N-1} - |\tau_D|^{N-1} \right) \left(|\sigma_D^N| - |\tau_D| \right) dx \leq C.$$

Thus

$$\frac{1}{\mu^{N-1}} \int_E |\sigma_D^N|^{N-1} (|\sigma_D^N| - |\tau_D|) \, dx \leq C$$

and also

$$\frac{1}{\mu^{N-1}} \int_E |\sigma_D^N|^{N-1}(|\sigma_D^N| - |\tau_D|) \mathbb{1}_{\{|\sigma_D^N| \geq \mu\}} \, dx \leq C. \qquad (10.12)$$

Applying (10.11) yields

$$\frac{\delta}{\mu^{N-1}} \int_\Omega |\sigma_D^N|^{N-1} \mathbb{1}_{\{|\sigma_D^N| \geq \mu\}} \, dx \leq C,$$

and using (10.12) again, we obtain

$$\frac{1}{\mu^{N-1}} \int_\Omega |\sigma_D^N|^N \mathbb{1}_{\{|\sigma_D^N| \geq \mu\}} \, dx \leq C$$

and

$$\frac{1}{\mu^{N-1}} \int_\Omega |\sigma_D^N|^N \, dx \leq C.$$

So we have obtained the same basic estimate as in Proposition 10.9, and thus the same conclusion follows. $\diamond$

10.3.3 Constituent Law for the Hencky model

Our objective now is to obtain the Hencky model equations with the displacement. We want to prove the following theorem.

Theorem 10.12. *If we make the assumptions of either Proposition 10.9 or Proposition 10.10, then there exists a pair $\sigma \in \Sigma(\Omega), u \in U(\Omega)$ such that for any $\tau \in \Sigma(\Omega)$ belonging to $\mathcal{M}$, one has*

$$(\tau - \sigma).(\epsilon(u) - A\sigma) \leq 0 \text{ in the sense of bounded measures,} \qquad (10.13)$$

$$\sigma \in \mathcal{M} \cap \mathcal{K}, \qquad (10.14)$$

$$u = 0 \text{ on } \Gamma_0 \text{ in the sense } \langle \epsilon(u).\tau, 1 \rangle = 0, \\ \forall \tau \in \Sigma(\Omega) \text{ with } \operatorname{div} \tau = 0, \ \nu.\tau = 0 \text{ on } \Gamma_1. \qquad (10.15)$$

This σ is unique.

Proof. To prove uniqueness, it is enough to notice that

$$\langle \epsilon(u).\tau, 1 \rangle = 0 \ \forall \tau \text{ such that } \operatorname{div} \tau = 0, \nu.\tau = 0 \text{ on } \Gamma_1,$$

and thus if we take $\tau \in \mathcal{M} \cap \mathcal{K}$ in (10.13), and test with 1, we obtain that σ is a solution of the Hencky model and thus unique. Now consider our pair (u^N, σ^N). We know that for a subsequence,

$$\sigma^N \to \sigma \text{ in } \mathcal{L}_{\text{sym}}^2,$$

$$\sigma_D^N \to \sigma_D \text{ weakly in } \mathcal{L}_{\text{sym}}^p, \quad 1 \leq p < \infty.$$

From Proposition 10.9 or Proposition 10.10 we can assert that u_N remains in a bounded subset of $U(\Omega)$, and thus, extracting a new subsequence, we get that

$$u_N \to u \text{ in } (L^p(\Omega))^n, \quad \forall 1 \le p < \tfrac{n}{n-1},$$
$$\operatorname{div} u_N \to \operatorname{div} u \text{ in } L^2(\Omega) \text{ weakly},$$
$$\epsilon(u_N) \to \epsilon(u) \text{ in } \mathcal{M}^1_{\text{sym}} \text{ weakly}.$$

We take $\phi \in C_0^\infty(\Omega)$ and $\tau \in \Sigma(\Omega) \cap \mathcal{M}$, and check that

$$\int \phi(\tau - \sigma^N).(\epsilon(u^N) - A\sigma^N) \, dx \to \int \phi(\tau - \sigma).\epsilon(u) - \int \phi(\tau - \sigma).A\sigma \, dx.$$
(10.16)

But

$$\int \phi(\tau - \sigma^N).\epsilon(u^N) \, dx = -\int [(\tau - \sigma^N)u^N.D\phi + \phi(\operatorname{div} \tau - f).u^N] \, dx,$$

and we can pass to the limit. In the second term on the left-hand side of (10.16) it is also easy to pass to the limit, and so the property (10.16) has been proven. Now we have

$$(\tau - \sigma^N).(\epsilon(u^N) - A\sigma^N) = (\tau_D - \sigma_D^N).\sigma_D^N \frac{|\sigma_D^N|^{N-2}}{\mu^{N-1}}.$$

Since $|\tau_D| \le \mu$, we deduce that

$$(\tau - \sigma^N).(\epsilon(u^N) - A\sigma^N) \le (\mu - |\sigma_D^N|) \left(\frac{|\sigma_D^N|}{\mu}\right)^{N-1} \mathbb{1}_{|\sigma_D^N| < \mu},$$

and it is easy to check that the right-hand side of the above inequality converges to 0 in $L^\infty(\Omega)$ as $N \to \infty$.

Therefore, we get, for $\phi \in C^0(\Omega) \ge 0$,

$$\limsup \int \phi(\tau - \sigma^N).(\epsilon(u^N) - A\sigma^N) \, dx \le 0,$$

and using (10.16) we deduce the result (10.13).

The property (10.15) is clear. $\diamond$

10.4 H^1_{loc} Regularity

10.4.1 Preliminaries

In fact, the regularity theory developed in Chapter 9 does not adapt immediately even to the Norton–Hoff model, because the deviator occurs in the nonquadratic part instead of the full tensor. However, the difference is minimal. We assume that

$$D_j f_k \in L^n(\Omega), \quad \Delta f_k \in L^n(\Omega). \tag{10.17}$$

We shall not redo the theory for the Norton–Hoff model (see also [6]) but assert that

$$\|\sigma^N\|_{H^1_{loc}(\Omega)} \leq C_N, \tag{10.18}$$

$$\||\sigma_D^N|^{N-2}|D\sigma_D^N|^2\|_{L^1_{loc}(\Omega)} \leq C_N. \tag{10.19}$$

We then proceed with a calculation leading to important estimates. We write (10.6) as follows:

$$A\sigma^N + \beta^N(|\sigma_D^N|)\sigma_D^N = \epsilon(u^N),$$

where naturally

$$\beta^N(x) = \frac{x^{N-2}}{\mu^{N-1}}.$$

To simplify the notation, we suppress the explicit N dependence, and write

$$A\sigma + \beta(|\sigma_D|)\sigma_D = \epsilon(u) \tag{10.20}$$

and

$$\text{div } \sigma = f. \tag{10.21}$$

Let θ be a smooth function with compact support in Ω. We are going to check the following inequality.

Proposition 10.13.

$$\int \theta^2 A D_k \sigma . D_k \sigma + \int \theta^2 \beta(|\sigma_D|) D_k \sigma_D . D_k \sigma_D$$

$$\leq \int u_k [D_j \theta^2 D_j f_k + \theta^2 \Delta f_k + f_i D_i D_k \theta^2 + \sigma_{ij} D_i D_j D_k \theta^2]$$

$$+ \int \text{tr } A\sigma [f_i D_i \theta^2 + \sigma_{ij} D_i D_j \theta^2] - 2 \int (A\sigma)_{jk} D_k \sigma_{ij} D_i \theta^2$$

$$- 2 \int \beta(|\sigma_D|)\sigma_{D,jk} D_k \sigma_{D,ij} D_i \theta^2 - \frac{2}{n} \int \beta(|\sigma_D|)\sigma_{D,jk} D_k \text{tr } \sigma D_j \theta^2. \tag{10.22}$$

Proof. We test equation (10.20) with $-D_k(\theta^2 D_k \sigma)$ and integrate by parts. We note that thanks to (10.21),

$$\int D_k \epsilon(u) . \theta^2 D_k \sigma = - \int D_k u_j [D_k \sigma_{ij} D_i \theta^2 + \theta^2 D_k f_j],$$

and we write

$$- \int D_k u_j D_k \sigma_{ij} D_i \theta^2 = -2 \int \epsilon_{jk} D_k \sigma_{ij} D_i \theta^2 + \int D_j u_k D_k \sigma_{ij} D_i \theta^2$$

$$= -2 \int ((A\sigma)_{jk} + \beta(|\sigma_D|)\sigma_{D,jk}) D_k \sigma_{ij} D_i \theta^2$$

$$+ \int D_j u_k D_k \sigma_{ij} D_i \theta^2.$$

We then write, using (10.21) again,

$$\int D_j u_k D_k \sigma_{ij} D_i \theta^2 = - \int u_k [D_i \theta^2 D_k f_i + D_k \sigma_{ij} D_i D_j \theta^2]$$

$$= \int [\operatorname{div} u \, f_i D_i \theta^2 + u_k f_i D_i D_k \theta^2]$$

$$+ \int [\operatorname{div} u \, \sigma_{ij} D_i D_j \theta^2 + u_k \sigma_{ij} D_i D_j D_k \theta^2].$$

We recall that div $u = \operatorname{tr} A\sigma$, and write

$$- \int D_k u_j \theta^2 D_k f_j = \int u_j (D_k \theta^2 D_k f_j + \theta^2 \Delta f_j).$$

We also omit a term involving β' that is positive, thus getting an inequality instead of an equality. Collecting results, we obtain the desired result.   ◇

We also shall use the following useful estimate.

Proposition 10.14. *We have the inequality*

$$\int_\Omega \theta^2 \beta(|\sigma_D|)|D\operatorname{tr} \sigma|^2 \, dx \le 2n^2 \int_\Omega \theta^2 \beta(|\sigma_D|) D_k \sigma_D . D_k \sigma_D \, dx$$
$$+ 2n \int_\Omega \theta^2 \beta(|\sigma_D|)|f|^2 \, dx. \tag{10.23}$$

Proof. We can write

$$\int \theta^2 \beta(|\sigma_D|) D_k \sigma_D . D_k \sigma_D \ge \int \theta^2 \beta(|\sigma_D|) D_k \sigma_{D,kj} . D_k \sigma_{D,kj}$$

$$\ge \frac{1}{n} \sum_j \int \theta^2 \beta(|\sigma_D|) \left(\sum_k D_k \sigma_{D,kj} \right)^2$$

$$\ge \frac{1}{n} \sum_j \int \theta^2 \beta(|\sigma_D|) \left(f_j - \frac{1}{n} D_j \operatorname{tr} \sigma \right)^2,$$

from which (10.23) follows easily.

◇

10.4.2 Uniform Estimates and Main Regularity Result

We can state the following theorem.

Theorem 10.15. *If we make the assumptions of either Proposition 10.9 or Proposition 10.10, and (10.17), then the stress solution of the Hencky model belongs to $H^1_{\mathrm{loc}}(\Omega)$.*

Proof. We have already mentioned (see (10.18), (10.19)) that the stress solution of the Norton–Hoff model belongs to $H^1_{loc}(\Omega)$. However, the norm could depend on N. We want to check that we have an estimate independent of N, so it will apply also to the Hencky model obtained when we let $N \to \infty$. From Propositions 10.13 and 10.14 we know that (10.22), (10.23) hold for σ^N. Using Proposition 10.9 or Proposition 10.10, as well as the estimate (10.8) with $p = n$, we can easily derive from (10.22) a uniform estimate for the H^1_{loc} norm of σ^N. Thus the result is obtained.   ◇

Remark 10.16. The condition (10.17) is more stringent than in [93].

Remark 10.17. If $n = 3$, we conclude, for fixed N, via Sobolev's theorem and the estimate of the penalty term, that $\sigma \in L^{3N}$; hence

$$Du \in L^{3+3\gamma}, \quad \gamma = 1/(N-1), \quad u \in C^\gamma.$$

Further,

$$D^2u \in L^q, \quad q = \frac{3}{2} + \frac{3}{4N-2}.$$

Remark 10.18. The explicit form of $J(\sigma)$ as a quadratic functional can be extended to general strictly convex coercive functionals on convenient spaces.

References

1. R.A. Adams, *Sobolev spaces*, Academic Press, New York (1975).
2. R.A. Adams, A note on Riesz Potentials, *Duke Math. J.* 42 (1975), 765–778.
3. J.P. Aubin, *Mathematical Methods of Game and Economic Theory*, Studies in Mathematics and its Applications, North-Holland, Amsterdam (1976).
4. G. Baccarani, M. Rudan, R. Guerrieri, P. Ciampolini, Physical Models for Numerical Device Simulations, In: *Process and Device Modelling*, ed. W.L. Engl, Elsevier Science Publ. B.V. (North Holland) (1986), 107–158.
5. A. Bensoussan, J. Frehse, On Bellman equations of ergodic control in R^n, *J. Reine Angew. Math.* 429 (1992), 125–160.
6. A. Bensoussan, J. Frehse, Asymptotic Behaviour of Norton–Hoff's Law in plasticity Theory and H^1 Regularity, in Boundary value problems for PDE and Appli. (Volume in honor of Professor E. Magenes), RMA Res. Notes in Appl. Math. 29, eds. P.G. Ciarlet, J.L. Lions, Masson, Paris (1993), 3–25.
7. A. Bensoussan, J. Frehse, Nonlinear elliptic systems in stochastic game theory, *J. Reine Angew. Math.* 350 (1984), 23–67.
8. S. Campanato, P. Cannarsa, Differentiability and partial Hölder continuity of the solution of nonlinear elliptic systems of order $2m$ with quadratic growth, *Ann. Sc. Norm. Super. Pisa*, Cl. Sci., IV. Ser. 8 (1981), 285–309.
9. L. Caffarelli, R. Kohn, L. Nirenberg, Partial Regularity of suitable weak solutions of the Navier–Stokes equations, *Commun. Pure Appl. Math.* 35 (1985), 771–831.
10. F. Chiarenza, M. Frasca, Morrey Spaces and Hardy–Littlewood maximal functions, *Rend. Mat. Appl.*, VII Ser. 7 (1987), 273–279.
11. R. Coifman, P.L. Lions, Y. Meyer, S. Semmes, Compensated Compactness and Hardy Spaces, *J. Math. Pures Appl.* IX. Sér. 72 (1993), 247–286.
12. H.O. Cordes, Über die erste Randwertaufgabe bei quasilinearen Differential-gleichungen zweiter Ordnung in mehr als zwei Variablen. *Math. Ann.* 131 (1956), 278–312.
13. M. Dauge, *Elliptic Boundary Value Problems on Corner Domains*, Lecture Notes in Mathematics 1341, Springer–Verlag, Berlin (1988).
14. E. De Giorgi, Un esempio di estremali discontinue per un problema variazionale di tipo ellittico, *Boll. UMI* 4 (1968), 135–137.
15. G. Duvaut, J.L. Lions, *Inequalities in Mechanics and Physics*, Springer-Verlag, Berlin (1976).
16. C. Ebmeyer, Mixed Boundary Value Problems for Nonlinear Elliptic Systems in n-Dimensional Lipschitztian Domains, *Zeit. Anal. Anwend.* 18 (1999), 539–555.
17. C. Ebmeyer, Nonlinear elliptic problems under mixed boundary value conditions in non-smooth domains, *SIAM J. Math. Anal.* 32, No. 1 (2000), 103–118.

18. C. Ebmeyer, Steady flow of fluids with shear dependent viscosity under mixed boundary value conditions in polyhedral domains, *Math. Models Methods Appl. Sci.* 10 (2000).

19. C. Ebmeyer, Mixed boundary value problems for nonlinear elliptic systems with p-structure in polyhedral domains, (to appear).

20. C. Ebmeyer, J. Frehse, Mixed boundary value problems for nonlinear elliptic equations in multidimensional non-smooth domains, *Math. Nachr.* 203 (1999), 47–74.

21. C. Ebmeyer, J. Frehse, Steady Navier–Stokes equations with mixed boundary value conditions in three–dimensional Lipschitzian Domains, *Math. Ann.* (to appear).

22. J. Eells, J.H. Sampson, Harmonic mappings of Riemann manifolds. *Am. Math. J.* 86 (1964), 109–160.

23. I. Ekeland, R. Temam, *Analyse convexe et problèmes variationnels*, Dunod, Paris (1974).

24. J. Frehse, On two-dimensional quasilinear elliptic systems, *Manusc. Math.* 28 (1979), 21–49.

25. J. Frehse, On the regularity of solutions to elliptic differential inequalities, in: *Mathematical Techniques of Optimization, control and Decision*, ed. J.P. Aubin, A. Bensoussan, I. Ekeland, Birkhäuser-Boston.

26. J. Frehse, A discontinuous solution of a mildly nonlinear elliptic system, *Math. Z.* 134 (1973), 87–107.

27. J. Frehse, J. Naumann, On the Existence of Weak Solutions to a System of Stationary Semiconductor Equations with Avalanche Generation, *Math. Models and Methods in Applied Sciences* 4, 2 (1994), 273–289.

28. J. Frehse, J. Naumann, Stationary Semiconductor Equations Modelling Avalanche Generation, *J. Math. Anal. Appl.* (1996), 685–702.

29. J. Frehse, M. Růžička, On the Regularity of the Stationary Navier–Stokes Equations, *Ann. Sc. Norm. Super. Pisa, Cl. Sci.*, IV. Ser. 21 (1994), 63–95.

30. J. Frehse, M. Růžička, Existence of Regular Solutions to the Stationary Navier–Stokes Equations, *Math. Ann.* 302 (1995), 699–717.

31. J. Frehse, M. Růžička, Weighted Estimates for the Stationary Navier–Stokes Equations, *Acta Appl. Math.* 37 (1994), 53–66.

32. J. Frehse, M. Růžička, Regularity for the Stationary Navier–Stokes Equations in Bounded Domains, *Ach. Rat. Mech. Anal.* 128 (1994), 361–381.

33. J. Frehse, M. Růžička, Existence of Regular Solutions to the Steady Navier–Stokes Equations in Bounded Six–Dimensional domains, *Ann. Sc. Norm. Super. Pisa, Cl. Sci.* IV. Ser., XXIII, Fasc. 4 (1996).

34. K. Friedrichs, Ein Verfahren, das Minimum eines Integrals als das Maximum eines anderen Ausdrucks darzustellen, Göttinger Nachrichten (1929).

35. H. Gajewski, K. Gröger, Initial Boundary Value Problems Modelling Heterogeneous Semiconductor Devices. In: *Surveys on Analysis, Geometry and Mathematical Physics*, Teubner Texte Math., Bd 117; ed. B.W. Schulze, H. Triebel; Teubner Verlagsges, Leipzig (1990), 4–53.

36. G.P. Galdi, *An Introduction to the Mathematical Theory of Navier–Stokes Equations*, Vol. 2, *Nonlinear Stationary Problems*, Springer-Verlag, New York (1993).

37. F.W. Gehring, The L^P-integrability of the partial derivatives of a quasi coformal mapping, *Acta Math.* 130 (1973), 266–277.

38. C. Gerhardt, Stationary solutions to the Navier–Stokes equations in dimension 4, *Math. Zeit.* 165 (1979), 193–197.

39. M. Giaquinta, *Introduction to Regularity Theory for Nonlinear Elliptic Systems*, Lectures in Mathematics, Birkhäuser, Basel (1993).

40. M. Giaquinta, *Multiple Integrals in the Calculus of Variations and Nonlinear Elliptic Systems*, Annals of Mathematical Studies, Princeton University Press, Princeton, New Jersey (1983).
41. M. Giaquinta, E. Giusti, On the regularity of the minima of variational integrals, *Acta Math.* 148 (1982), 31–46.
42. M. Giaquinta, S. Hildebrandt, *Calculus of Variations*, I, II, Springer–Verlag, Berlin (1994), (1996).
43. M. Giaquinta, G. Modica, Almost everywhere regularity results for solutions of nonlinear elliptic systems, *Manuscr. Math.* 28 (1979), 109–158.
44. M. Giaquinta, G. Modica, Regularity Results for some Classes of Higher Order Nonlinear Systems, *J. Reine Angew. Math.* 311/312 (1979), 145–169.
45. M. Giaquinta, G. Modica, Nonlinear systems of the type of the stationary Navier–Stokes system, *J. Reine Angew. Math.* 330 (1982), 173–214.
46. D. Gilbarg, N.S. Trudinger, *Elliptic Partial Differential Equations of Second Order*, Springer-Verlag, Berlin (1977).
47. E. Giusti, M. Miranda, Un esempio di soluzione discontinua per un problema di minimo relativo ad un integrale regolare del calcolo delle variazioni, *Boll. UMI2* (1968), 1–8.
48. E. Giusti, M. Miranda, Sulla Regolarità delle Soluzioni deboli di una classa di sistemi ellittici quasilineari, *Arch. Ration. Mech. Anal.* 31 (1968), 173–184.
49. P. Grisvard, *Elliptic Problems in Nonsmooth Domains*, Pitman Advanced Publishing Program, Boston, London, Melbourne (1985).
50. K. Gröger, On Steady–State Carrier Distributions in Semiconductor Devices. *Aplikace Mat.* 32 (1987), 49–56.
51. K. Gröger, A $W^{1,p}$-Estimate for Solutions to Mixed Boundary Value Problems for Second Order Elliptic Differential Equations. *Math. Ann.* 283 (1989), 679–687.
52. M. Grüter, K.O. Widman, The Green function for uniformly elliptic equations, *Manuscr. Math.* 37 (1982), 303–342.
53. M.E. Gurtin, *Encyclopedia of Physics II, Mechanics of Solids*, ed. Truesdell, Springer-Verlag, Berlin (1972).
54. W. Hao, S. Leonardi, J. Nečas, An example of irregular solution to a nonlinear Euler–Lagrange elliptic system with real analytic coefficients, preprint.
55. G. Hardy, J. Littlewood, G. Pólya, *Inequalities*, University Press, Cambridge (1967).
56. F. Helein, Régularité des applications harmoniques entre une surface et une sphère, *CRAS, Math.* 311 (1990), 519–525.
57. S. Hildebrandt, *Harmonic Mappings of Riemannian manifolds*, Lecture Notes in Mathematics 1161, Springer-Verlag, Berlin (1985).
58. S. Hildebrandt, K.O. Widman, On the Hölder continuity of weak solutions of quasi-linear elliptic systems of second order *Ann. Sc. Norm. Super. Pisa, Cl. Sci.* IV. Ser. 4 (1977), 145–178.
59. J. Jerome, *Analysis of Charge transport. A mathematical study of semiconductor devices*, Springer-Verlag, Berlin (1996).
60. J. Jost, *Harmonic Mappings between Riemann Manifolds*, Proceedings of the Centre for Mathematical Analysis, Vol. 4, Australian National University (1983).
61. J. Kadlec, The regularity of the solution of the Poisson problem in a domain whose boundary is similar to that of a convex domain, *Czech. Math. J.* 89 (1964), 386–393.
62. V.A. Kondrat'ev, Boundary value problems for elliptic equations in domains with conical and angular points, *Trans. Mosc. Math. Soc.* 16 (1967), 227–313.

63. A. Koshelev, *Regularity Problem for Quasilinear Elliptic and Parabolic Systems*, Lecture Notes in Mathematics 1614, Springer-Verlag, Berlin (1995).

64. V.A. Kozlov, V.G. Maz'ya, Singularities in solutions to mathematical physics problems in non-smooth domains, in *Partial Differential Equations and Functional Analysis*, In memory of P. Grisvard, Proceedings of a conference, Birkhäuser (1996), 174–206.

65. A. Kufner, O. John, S. Fučik, *Function Spaces*, Academia, Prague (1977).

66. O.A. Ladyzhenskaya, N.N. Ural'tseva, *Linear and Quasilinear Elliptic Equations*, Academic Press, N.Y. (1968).

67. O.A. Ladyzhenskaya, V.A. Solonnikov, N.N. Ural'tseva, *Linear and Quasilinear Equations of Parabolic Type*, American Mathematical Society, Providence, R.I. (1968).

68. J. Leray, Etude de diverses équations intégrales non linéaires et de quelques problèmes que pose l'hydrodynamique, *J. Maths Pures Appl.* IX. Sér. 12 (1933), 1–82.

69. P. Le Tallec, *Numerical Analysis of Viscoelastic problems*, Masson Springer-Verlag (1990).

70. J.L. Lions, *Quelques méthodes de résolution des problèmes aux limites non linéaires*, Dunod, Paris (1969).

71. W. Littman, G. Stampacchia, H.F. Weinberger, Regular points for elliptic equations with discontinuous coefficients, *Ann. Sc. Norm. Super. Pisa, Cl. Sci.* IV. Ser. 3, 17 (1963), 43–77.

72. P.L. Lions, *Mathematical Topics in Fluid Mechanics, vol 1, Incompressible Models*, Oxford Science Publications, Oxford (1996).

73. P. Markowich, *The Stationary Semiconductor Device Equations*, Springer-Verlag (1986).

74. V.G. Maz'ya, S.A. Nazarov, B.A. Plamenevskii, On the singularities of solutions of the Dirichlet problem in the exterior of a slender cone, *Math. USSR Sb.* 50 (1985), 415–437.

75. V.G. Maz'ya, S.A. Nazarov, B.A. Plamenevskii, *Asymptotische Theorie elliptischer Randwertaufgaben in singulär gestörten Gebieten*, vol. 1, 2, Akademie-Verlag, Berlin (1991).

76. V.G. Maz'ya, J. Rossmann, On the behaviour of solutions to the Dirichlet problem for second order elliptic equations near edges and polyhedral vertices with critical angles, *Z. Anal. Anwend.* 13 (1994), 19-47.

77. N.G. Meyers, A. Elcrat, Some Results on Regularity for Solutions of Nonlinear Elliptic Systems and Quasi-regular Functions (1974), 121–136.

78. C. Miranda, Su alcuni teoremi di inclusione, *Annales Polonaises de Mathématiques* 42 (1965), 305–315.

79. C.B. Morrey Jr., *Multiple Integrals in the Calculus of variations*, Springer-Verlag, Berlin (1966).

80. C.B. Morrey Jr., Partial Regularity Results for Nonlinear Elliptic Systems, *J. Math. Mech.* 17 (1968), 649–670.

81. J. Moser, A new proof of De Giorgi 's theorem concerning the regularity problem for elliptic differential equations. *Comm. Pure and Applied Mathematics* 13 (1960), 457–468.

82. J. Nash, Equilibrium Points in n-Person Games, *Proc. Natl. Acad. Sci.* USA 36 (1950), 48–49.

83. J. Naumann, On the Existence, Uniqueness and Interior Regularity of Weak Solutions to the Stationary Drift-Diffusion Equations in Semiconductor Theory, Vorlesungsreihe 27, Bonn (Jan. 1993).

84. J. Naumann, An existence theorem for weak solutions of the basic stationary semi-conductor equations, *Appl. Anal.* 48 (1993), 157–172.

85. S.A. Nazarov, B.A. Plamenevskii, *Elliptic Problems in Domains with Piecewise Smooth Boundaries*, Walter de Gruyter, Berlin (1994).

86. J. Nečas, I. Hlaváček, *Mathematical Theory of Elastic and Elastico-Plastic Bodies: An Introduction*, Studies in Applied Mechanics, Elsevier, Amsterdam (1981).

87. L. Nirenberg, An extended interpolation inequality, *Ann. Sc. Norm. Super. Pisa, Cl. Sci.* IV. Ser. 20 (1966), 733–737.

88. H.G. Othmer, The Interaction of Structure and Dynamics in Chemical Reaction Networks. In: *Modelling of Chemical Reaction Systems*, Proc. Intern. Workshop, Heidelberg, Sept. 1–5, 1980; ed. K.H. Ebert, P. Deuflhard, W. Jäger, Springer-Verlag, Berlin (1981), 2–19.

89. J. Saranen, On an inequality of Friedrichs, *Math. Scand.* 51 (1982), 310–322.

90. S. Selberherr, *Analysis and Simulation of Semiconductor Devices*, Springer-Verlag, Wien (1984).

91. C. Sbordone, Higher integrability for reverse integral equations. Collection: Methods of Functional Analysis and Theory of Elliptic Equations, Naples (1982), 347–367.

92. S. Semmes, A Primer on Hardy Spaces, and Some Remarks on a Theorem of Evans and Müller, *Commun. Partial Differ. Equations,* 19(1–2), (1994), 277–319.

93. G.A. Seregin, Differentiabilty properties of the stress tensor in perfect elastic-plastic theory, *Differentsial'nye Uravneniya* 23 (1987).

94. H. Sohr, Zur Regularitätstheorie der instationären Gleichungen von Navier–Stokes, *Math. Z.* 184 (1983), 359–376.

95. E.M. Stein, *Singular Integrals and Differentiability Properties of Functions*, Princeton University Press (1970).

96. E.M. Stein, *Harmonic Analysis: Real-Variable Methods, Orthogonality, and Oscillatory Integrals*, Princeton Mathematical Series, 43, Princeton, N.J. (1993).

97. E.M. Stein, G. Weiss, *Introduction to Fourier Analysis in Euclidean Spaces*, Princeton Mathematical Series, 32, Princeton, N.J. (1971).

98. M. Struwe, On partial regularity results for the Navier–Stokes equations, *Commun. Pure and Appl. Math.* 41 (1988), 437–458.

99. M. Struwe, Regular solutions of the stationary Navier–Stokes equations on R^5, *Math. Ann.* 302 (1995), 719–741.

100. S.M. Sze, *Physics of Semiconductor Devices*, 2nd ed. J. Wiley & Sons, New York, Chiehester (1981).

101. R. Temam, *Mathematical Problems in plasticity*, Gauthier-Villars, Paris (1985).

102. F. Tomi, Variationsprobleme vom Dirichlet-Typ mit einer Ungleichung als Nebenbedingung, *Math. Z.* 128 (1972), 43–74.

103. E. Trefftz, Ein Gegenstück zum Ritzschen Verfahren, Zweiter Kongress für Technische Mechanik, Zürich (1927).

104. G.M. Troianiello, *Elliptic Differential Equations and Obstacle Problems*, Plenum Press, New York (1987).

105. K.O. Widman, Hölder Continuity of Solutions of Elliptic Equations, *Manuscr. Math.* 5 (1971), 299–308.

106. M. Wiegner, Ein optimaler Regularitätssatz für schwache Lösungen gewisser elliptischer Systeme, *Math. Z.* 147 (1976), 21–28.

107. V.I. Youdovich, An Example of Loss of Stability and Generation of Secondary Flow in a Closed Vessel, *Math. Sb.* 74 (1967); English Trans.: *Math. USSR Sb.* 3 (1967), 519–533.

108. E. Zeidler, *Nonlinear Functional Analysis and Its Applications IV*, Springer-Verlag, New York (1988).

Index

Applied Mathematical Sciences

(continued from page ii)

(continued on next page)

Applied Mathematical Sciences

(continued from previous page)

Druck: Strauss Offsetdruck, Mörlenbach
Verarbeitung: Schäffer, Grünstadt